GLOBAL INFORMATION WARFARE

How Businesses, Governments, and Others Achieve Objectives and Attain Competitive Advantages

OTHER AUERBACH PUBLICATIONS

ABCs of IP Addressing
Gilbert Held
ISBN: 0-8493-1144-6

Application Servers for E-Business
Lisa M. Lindgren
ISBN: 0-8493-0827-5

Architectures for E-Business Systems
Sanjiv Purba, Editor
ISBN: 0-8493-1161-6

A Technical Guide to IPSec Virtual Private Networks
James S. Tiller
ISBN: 0-8493-0876-3

Building an Information Security Awareness Program
Mark B. Desman
ISBN: 0-8493-0116-5

Computer Telephony Integration
William Yarberry, Jr.
ISBN: 0-8493-9995-5

Cyber Crime Investigator's Field Guide
Bruce Middleton
ISBN: 0-8493-1192-6

Cyber Forensics: A Field Manual for Collecting, Examining, and Preserving Evidence of Computer Crimes
Albert J. Marcella and Robert S. Greenfield, Editors
ISBN: 0-8493-0955-7

Information Security Architecture
Jan Killmeyer Tudor
ISBN: 0-8493-9988-2

Information Security Management Handbook, 4th Edition, Volume 1
Harold F. Tipton and Micki Krause, Editors
ISBN: 0-8493-9829-0

Information Security Management Handbook, 4th Edition, Volume 2
Harold F. Tipton and Micki Krause, Editors
ISBN: 0-8493-0800-3

Information Security Management Handbook, 4th Edition, Volume 3
Harold F. Tipton and Micki Krause, Editors
ISBN: 0-8493-1127-6

Information Security Policies, Procedures, and Standards: Guidelines for Effective Information Security Management
Thomas R. Peltier
ISBN: 0-8493-1137-3

Information Security Risk Analysis
Thomas R. Peltier
ISBN: 0-8493-0880-1

Information Technology Control and Audit
Frederick Gallegos, Sandra Allen-Senft, and Daniel P. Manson
ISBN: 0-8493-9994-7

New Directions in Internet Management
Sanjiv Purba, Editor
ISBN: 0-8493-1160-8

New Directions in Project Management
Paul C. Tinnirello, Editor
ISBN: 0-8493-1190-X

A Practical Guide to Security Engineering and Information Assurance
Debra Herrmann
ISBN: 0-8493-1163-2

The Privacy Papers: Managing Technology and Consumers, Employee, and Legislative Action
Rebecca Herold
ISBN: 0-8493-1248-5

Secure Internet Practices: Best Practices for Securing Systems in the Internet and e-Business Age
Patrick McBride, Joday Patilla, Craig Robinson, Peter Thermos, and Edward P. Moser
ISBN: 0-8493-1239-6

Securing and Controlling Cisco Routers
Peter T. Davis
ISBN: 0-8493-1290-6

Securing E-Business Applications and Communications
Jonathan S. Held and John R. Bowers
ISBN: 0-8493-0963-8

Securing Windows NT/2000: From Policies to Firewalls
Michael A. Simonyi
ISBN: 0-8493-1261-2

TCP/IP Professional Reference Guide
Gilbert Held
ISBN: 0-8493-0824-0

The Complete Book of Middleware
Judith Myerson
ISBN: 0-8493-1272-8

AUERBACH PUBLICATIONS

www.auerbach-publications.com
To Order Call: 1-800-272-7737 • Fax: 1-800-374-3401
E-mail: orders@crcpress.com

GLOBAL INFORMATION WARFARE

How Businesses, Governments, and Others Achieve Objectives and Attain Competitive Advantages

ANDY JONES
GERALD L. KOVACICH
PERRY G. LUZWICK

AUERBACH PUB

A CRC Press Company

Boca Raton   London   New York   Washington, D.C.

Library of Congress Cataloging-in-Publication Data

Jones, Andy, 1952-
  Global information warfare : how businesses, governments, and others achieve
  objectives and attain competitive advantages / Andy Jones, Gerald
  L. Kovacich, Perry G. Luzwick.
      p. cm.
  Includes bibliographical references and index.
  ISBN 0-8493-1114-4 (alk. paper)
    1. Information warfare. 2. Business--Data processing. I. Kovacich,
  Gerald L. II. Luzwick, Perry G. III. Title.

U163 .J66 2002
303.48'33--dc21

2002018570

Visit the Auerbach Web site at www.auerbach-publications.com

© 2002 by CRC Press LLC
Auerbach is an imprint of CRC Press LLC

No claim to original U.S. Government works
International Standard Book Number 0-8493-1114-4
Library of Congress Card Number 2002018570
Printed in the United States of America 1 2 3 4 5 6 7 8 9 0
Printed on acid-free paper

Dedication

Dedicated to all the brave soldiers of all the wars

In wars between nations, the brave soldiers on both sides of the conflict march off to war. They do so because they are patriotic and believe in their nation's cause; they do so to avoid being jailed; they do so to protect their family; and they do so because when all is said and done, they have very little choice.

When the war is over, the winners decide who was right and who was wrong for that is the way of victors. The vanquished were no less brave and no less human.

The politicians vote to send their youth into battle but would be less eager if they had to lead the charge. Yet, when it is over, they speak the loudest of "their" great victory and such a wonderful time for their nation to be proud. Proud of killing, maiming fellow human beings, destroying families; bringing death and destruction to all — victors and vanquished alike.

The victorious generals get promoted, collect more unearned medals, and tell their tall tales of wars that they never fought in. They also brag loudly of the great victories and eye more power in the battlefield of politics. After all, are they not heroes, worshipped and adored by the nation?

And through it all — through the death, destruction, misery, and pestilence — the brave soldiers on both sides lie in the ground. Yes, a few kind words and prayers will be said over their graves and they will be remembered by their nation — once a year, maybe — while their families will grieve a lifetime for them and what could have been.

Those brave soldiers who survived will never be the same. They go home, quietly, wanting to return to a "normal" life. A life as it was before — before the call. They will always remember, although not all will remember them.

This book is dedicated to those brave soldiers who fought the wars. As we enter the information warfare era of the Twenty-first Century, maybe we can fight computer wars where innocence is not lost, only data. Maybe.

Andy Jones
Gerald L. Kovacich
Perry G. Luzwick

Strategy and Force: *According to the laws of warfare, a thousand men can succeed by strategy, ten thousand men can succeed by force. When you apply strategy to others first, enemies will not confront you. Therefore, in warfare, it is important to be first: if you win at this, you will win over others; if you do not win at this, you will not win over others.*

The Art of Attack: *When you attack a country, you must take advantage of its changes. Observe the economy to see its deficits; observe its corruption to see its problems. When those above are perverse and those below are alienated, this is a basis for attacking it.*

Morality of Warfare: *Weapons are instruments of ill omen; contention is a vice. Things must have a basis, so kings strike the violent and unruly, based on humanness and justice. Warring states try to establish prestige, resist adversaries, and plot against each other, so they cannot do without armaments.*

—Master Wei Liao
Excerpts from *Ways of Warriors Codes of Kings:*
Lessons in Leadership from the Chinese Classics
Translated by Thomas Cleary
Published by Shambala Publications, Boston, 1999

Contents

Foreword

As we transition from the information age to the knowledge age, successful organizations are the ones that actively manage their information environments. Because knowledge and the data and information upon which it is built are so important, having the capabilities to conduct offensive and defensive information warfare (IW) is absolutely essential. Many countries, cartels, and terrorist organizations possess sophisticated information warfare know-how and will do what they can — at times and places of their choosing — to disrupt a nation's freedom, safety, economy, and social fabric. The information environment is a battleground of the future, affecting not just the military, but governments, businesses, and citizens as well. The cascading effects of IW attacks could prevent a country from effectively projecting political, economic, military, and social power.

The concepts behind coherent knowledge-based operations detailed in this book operate at the corporate level as well. Blending control of the information environment and knowledge management with an organization's network-centric business processes is absolutely essential in order for organizations to be competitive. The increasing reliance on the Internet indicates that not only will countermeasures and security measures become increasingly necessary and valuable, but so will concepts such as coherent knowledge-based operations that embrace extended enterprises and infrastructure interdependencies.

At Northrop Grumman Corporation, we envision a future in which IW powered by knowledge-based systems will be prevalent on the battlefield. One of our primary organizing strategies has been to create systems central to business and military transformation. We are developing technologies that help protect critical infrastructures by detecting, deflecting, and diverting cyber attacks. We are also working on the new concept of "information resiliency." We want to recover critical data — automatically and in real-time. Extensive cooperation is needed between businesses and governments to make the best use of ideas and resources to solve information warfare issues.

To operate effectively in this new era, governments must be as adaptive and agile as their private-sector counterparts. It is vitally important that the global government, military, and business communities get it right and move forward together. The concepts and recommendations in *Global Information Warfare: How Businesses, Governments, and Others Achieve Objectives and Attain Competitive Advantages* provide a good place to start and are ones that I heartily endorse.

Kent Kresa
Chairman and Chief Executive Officer
Northrop Grumman Corporation
January 2002

Acknowledgments

To successfully take on and complete such a project as writing this book, one must rely on many people who freely give of their professional advice and assistance. The authors are grateful to the following friends and colleagues for their never-ending support and consultations: Motomu Akashi, Debi Ashenden, Anita D'Amico, William C. Boni, Don Evans, Steve Lutz, Colin Nash, Winn Schwartau, Darren Watts, and Paul Zavidniak.

Thanks to Herb Anderson, President, Northrop Information Technology; and Michael Grady, Sector Vice President, Technology, Engineering, and Quality for Headquarters, Northrop Grumman Information Technology, for their support.

Thanks to Diane Davies and Dana Miller, graphics arts specialists at Northrop Grumman Information Technology, who not only are masterful artists, but also do exceptionally well interpreting vague verbal guidance and squiggly lines; and to Jeff Gaynor, Special Assistant, Homeland Security, Deputy Assistant Secretary of Defense (Security and Information Operations), Assistant Secretary of Defense (Command, Control, Communications, and Intelligence) for reviewing what was written and providing outstanding advice.

A special thanks to John Meyer, Open Systems and formerly of Reed Elsevier's Advanced Technology Group, Oxford, England; and to Carola E. Roeder, Permissions, Butterworth-Heinemann, a member of the Reed Elsevier Group, Woburn, Massachusetts, for their years of support and for granting permission to include relevant material from Dr. Kovacich's and Mr. Luzwick's other books and articles published by them.

Special thanks to Rich O'Hanley of Auerbach Publishing for his support and confidence in our knowledge. Regardless of how much good advice we received from our colleagues, this book could never have been successfully published if it were not for Rich, who was willing to risk signing three "crazy info-warriors" to a book contract. Thanks, Rich!

Thanks to Chloe Palmer, Editor, *Computer Fraud and Security,* Elsevier Advanced Technology (Elsevier Science) for her support. And to Greg Swain,

Director, Information Assurance, Northrop Grumman Information Technology, who is an enlightened leader and a great sounding board.

Writing and publishing this book was truly a team effort. Other members of the Auerbach Publishing team who made this all possible deserve a note of appreciation. They include Andrea Demby, Gerry Jaffe, Claire Miller, and Michelle Reyes.

Introduction

This book articulates how warfare in the information age can devastatingly strike in an instant from across the globe, and that traditional military forces may be no match for superior technologists bent on stealing proprietary information, destroying the information infrastructure and thus the economy, or changing the military balance of power between adversarial nations. By scenario, we will demonstrate that those who have the most to gain may be those with the least invested. Scenarios to be illustrated are from regions of the world that include the Americas, Europe, Asia, and the Middle East.

Other books have addressed niche areas on some information warfare (IW) aspects. However, they gloss over many important points, due in part to the authors' lack of real-world experience in actually working in information warfare environments and defending against such attacks, or developing and using offensive tactics.

A systems approach is used, in which the reader is led through Section I (Chapters 1–5) which introduces IW, describes the new environment, and explains how we arrived there. Section II (Chapters 6–14) discusses the past, present, and future use of IW tactics by businesses, government agencies, and others. Section III (Chapters 15–19) discusses how to survive in the new, dynamic, global, and very competitive IW environment. You will note some redundancies in various chapters. This was purposely done so that each could be viewed in as much of a "stand alone" chapter format as possible; it was also done to further reinforce the information presented therein.

Competition, either in the marketplace or on the battlefield, has been part of human nature for more than 10,000 years. Successfully executed IW can be adapted to the business realm. Information is fundamental to both business and the military; therefore, enhancing and protecting the information environment (IE) are vital to achieving global objectives and attaining a competitive advantage.

IW, information assurance (IA), information operations (IO), information superiority, and other constructs popular in the U.S. military are part of the revolution in military affairs (RMA) and revolution in security affairs (RSA).

Although the names are initially obtuse to those who do not work in those areas, they have been a normal evolution in communications and computers and also the dark side move/counter-move/counter-counter-move "cool war."

Physical attack is now complemented by virtual attack. The Love Bug virus came from the Philippines, and the Bubble Boy and Chernobyl viruses came from Argentina. Distributed denial-of-service (DDoS) attacks have come from Canada. India and Pakistan have clashed in the cyber realm, as have the Palestinians and the Israelis. Attacks on NATO servers came from Serbia during the Kosovo Air Campaign. The business world is just as riddled with attacks. Many go unnoticed for months, and some are never detected.

Allegedly, Oracle had private investigators spy on Microsoft. Microsoft servers were broken into for at least three months before the attack was recognized. Indications are that the source of the attack was St. Petersburg, Russia. Steve Balmer, Microsoft's CEO, decried this as a horrific example of industrial espionage. This type of espionage was made famous by the crackers from St. Petersburg and Moscow who in 1994 virtually broke into Citibank and stole more than $400,000. That should have been the trigger for a significant ramp-up in cyber defenses, but businesses have been eating the costs of losses — totaling in the hundreds of billions of dollars — for more than seven years.

Of course, the source of the Microsoft attack should not be surprising — there is a world-class hacker school in St. Petersburg. From the competitive intelligence school in France to a business intelligence company in Florida; from MI5 in England teaching companies how to counter spies to "how-to-hack" courses taught by the person credited with creating the first computer virus; and from cyber gangs such as Cult of the Dead Cow and L0pht Heavy Industries to national data network monitoring entities such as SORM in Russia, it is a minefield out there! Why is outsourcing of information security and cyber insurance on the rise? Because of the lack of enough trained people to keep pace with the threats who are finding and rapidly exploiting physical and virtual vulnerabilities.

For governments, virtual attack can be viewed as a natural progression of electronic warfare (EW); command, control, communications, and counter-measures (C^3CM); and command and control warfare (C2W). Over the past several years, the concepts and implementation of IW and IO have emerged. IW spans the spectrum of conflict, from peace through operations other than war (OOTW) to war and then back to OOTW and peace. IW is reserved for conflict.

What is IW? IW is more than computer network attack (CNA) and defense (CND). That much everyone agrees on. But what else is encompassed by it? Heated debates go on today about what IW should embrace and accomplish. There is the minimalist school that wants to include as few functions as possible. At best, IO would be a small function added on to C2W. The maximalist school wants many functions until now unassociated to be included. Civil affairs, public affairs, physical destruction, legal, law enforcement, intelligence, and other functions would be synchronized. Which is right? Neither, because they are polarized on winning their argument and gaining power, rather than focusing on what needs to be done and planning to make it happen.

We must also advise you that the information provided relative to people, groups, corporations, or governments by no means implies that they have legally committed crimes (been found guilty) unless the courts so state, nor done anything wrong. It is often in the eyes of the beholder, and you must judge for yourself. And though we mention various products, we do not imply any endorsement for or against them.

You may find some views presented among the three authors that, at times, differ. There was no effort to conceal this point, e.g., write in such a manner that it appeared that one individual or three individuals of like mind wrote the book. We believe that by using this approach, it has resulted in a broader view of the IW issues that adds additional value to the information presented. After reading this book, we hope you agree with our approach.

Section I:
How We Came to Where We Are — And by the Way, Where Are We?

This section provides an introduction to global information warfare by first addressing what it is, what it is not, and how we got to where we are in this stage of global information warfare.

Chapter 1: Everything You Wanted to Know about Information Warfare but Were Afraid to Ask

This chapter provides an overall introduction and discussion of global information warfare (IW) and sets the stage for the chapters that follow. The roots of IW, generally accepted definitions, and working definitions are presented. With this foundation in place, the reader is presented with topical issues of the day: What is the global nature of this type of warfare? Is it new or hype? Is it information systems security (INFOSEC) by another name? Is it information superiority with IW as the means? Is it information operations (IO)? Is it an excuse by leftover Cold War spies and security people to keep a job? Is it real and can it happen to me? If so, what is the loss in money, time, and people that I can expect?

Chapter 2: Fuel for the InfoWar Fire: The High-Technology Revolution

Key technological advancements (e.g., microprocessors, GPS, cellular phones, Internet, high-speed modems, etc.) contributing to the revolution in information systems are discussed. Also included are their positive and negative effects on our culture and approach to business.

Chapter 3: The Global and Nation-State Information Infrastructures: Backbone or Achilles' Heel?

This chapter reviews the global information infrastructure (GII), its history, and its impact on government agencies and businesses vis-à-vis industrial and economic espionage, techno-terrorism, and military IW. Individual elements comprising the critical portion of the international information infrastructure (III) (i.e., telecommunications; electrical power systems; gas and oil production, storage, and transportation; banking and finance; transportation; water supply systems; emergency systems; and continuity of government services) are addressed. The discussion includes an overview of what governments, businesses, and others are doing to protect or take advantage of our vulnerabilities. Spain developed a national anti-virus program. The United States has the Critical Infrastructure Protection Office. The United Kingdom implemented the Regulatory and Investigative Powers (RIP) Act, permitting law enforcement very broad powers. Singapore tasked its military to help shore up that nation's cyber defenses.

Chapter 4: Our Changing World, or Is It Really?

This chapter provides a broad background on trends in the world around us. Economics, business, environment, global competition, the "withering" superpowers, economic and industrial espionage, and the blurred distinction between peace and war are all discussed. All of these affect our information environment (IE) and ultimately raise the cost of doing business or running a nation-state.

Chapter 5: Business, Government, and Activist Warfare: The More Things Change, The More They Don't

This chapter discusses business, government, and activist warfare history. Think of a time and a capability that have never been countered. Thick castle walls? Gun powder. Tanks? Bazookas. Better or reactive armor? Thirty-millimeter depleted uranium rounds fired from an A-10 Warthog at a rate of 2000 rounds per minute. Communications? Capturing the transmission. Encrypting the communications? Fast computers and crytpomathematicians. The Germans thought their Enigma encryption machine was unbreakable, but Polish and British cryptographers proved them wrong. Human creativity and technology provide capabilities, and capabilities can be turned into advantages.

Section II: The Past, Present, and Future Use of Information Warfare Tactics by Businesses, Government Agencies, and Assorted Miscreants

Included throughout this section, the computer battlefield objectives and targets will unveil military, civil, and commercial target sets; will introduce the global information infrastructure (GII) and the national information infrastructure

(NII); and will demonstrate the GII and NII integration into the Internet as the primary vehicle for providing the capability and the opportunity for the execution of operations by techno-terrorists, economic espionage agents, activists, revolutionaries, freedom-fighters, military info-warriors, and the like. This section also identifies the probable information warfare combatants. Techno-terrorists, economic espionage agents, and military scenarios are described and discussed. Specific information relative to groups, governments, and individuals are described as well as the rationale for their behavior.

Chapter 6: Information Warfare Tactics: How Can They Do That? It's the COTS, Stupid!

This chapter discusses some of the vulnerabilities of today's hardware, software, and firmware that allow the use of IW tactics, often with little expertise required. For example, commercial-grade products, also known as COTS, are stuffed full of vulnerabilities that crackers and phreakers can take advantage of. The rationale for their use is twofold. First, COTS provide good capabilities at reasonable cost. These capabilities are what enable businesses to make a profit. Second, COTS upgrades and new products are frequently delivered. The timelines are rapidly shortening, with hardware cycles now around nine months and software cycles around eighteen months. Imagine taking a software product, portions of which are developed overseas in countries that either are or may be U.S. competitors, that is several million lines of code (this size is not unusual) and proving it contains no malicious code. This requires a line-by-line code check as well as an understanding of how the lines of code interact. There is no artificial intelligence program that does this. It requires skilled people and time. Now pass this cost on to the consumer and keep pace with the competition as they try to be first to market.

Chapter 7: It's All about Influence: Information Warfare Tactics by Nation-States

This chapter discusses the who, where, when, why, and what of IW tactics by governments, and how government agencies can use IW to defeat an adversary. Tactics include the use of weapons such as logic bombs, viruses, and Trojan horses to manipulate, compromise, and deny use of information and systems to an adversary; monitoring communications (e.g., ECHELON, the collection and sharing between governments of voice and data communications, and Frenchelon); France's alleged actions on Air France aircraft (e.g., planting listening devices and cameras, and using crew members and other people to collect information).

Chapter 8: Information Warfare: Asian Nation-States

This chapter discusses information warfare as it relates to Asian nation-states and includes case studies and commentary.

Chapter 9: Information Warfare: Middle East Nation-States

This chapter discusses information warfare as it relates to Middle Eastern nation-states and includes case studies and commentary.

Chapter 10: Information Warfare: European Nation-States

This chapter discusses information warfare as it relates to European nation-states and includes case studies and commentary.

Chapter 11: Information Warfare: Nation-States of the Americas

This chapter discusses information warfare as it relates to the Americas and includes case studies and commentary.

Chapter 12: Information Warfare: African and Other Nation-States

This chapter discusses information warfare as it relates to African and other nation-states and includes case studies and commentary.

Chapter 13: It's All about Profits: Information Warfare Tactics in Business

This chapter discusses the who, where, when, why, and what of business IW tactics, and how businesses can use IW to their advantage. Tactics include such things as "netspionage," social engineering, misinformation, covert attacks, logic bombs, malicious code, and the like.

Chapter 14: It's All about Power: Information Warfare Tactics by Terrorists, Activists, and Miscreants

This chapter discusses publicly known terrorist nations, drug cartels, and hacktivist (cyber disobedience) capabilities such as those of animal rights groups, freedom fighters, and the like. Examples include terrorists such as Osama Bin Laden using the Internet and encrypted communications to thwart law enforcement; the drug cartels' use of computers to support their money laundering operations; the Zapatista movement in Mexico, outnumbered and outfinanced by the Mexican government, took to the Internet to support its cause. They conducted denial-of-service attacks against the Mexican and U.S. governments.

Section III: Information Warfare Defenses: Countermeasures, Counterattack, or Both

This section discusses how to survive in the dynamic global and very competitive information environment when your organization is being attacked from within and from without.

Chapter 15: Surviving the Onslaughts: Defenses and Countermeasures

This chapter introduces defensive strategies such as defense-in-depth, countermeasures, and counterstrike as critical elements of the any design process intent on negating, thwarting, or delaying the impact of IW strikes from economic espionage agents, techno-terrorists, general miscreants, and military info-warriors.

Chapter 16: Defending against Information Warfare Attacks Begins with Knowledge Management

This chapter discusses a basic approach for defending the business or government agency by establishing a bottom line based on knowledge management (KM).

Chapter 17: What's a CEO to Do? Ya Gotta Have a Plan

This chapter discusses the need for the business or government agency CEO to take an aggressive leadership role if that CEO wants to defend the turf. To do so, a bottom-line, no-nonsense plan is required. This chapter discusses the philosophy, strategy, and planning that are necessary. Throwing technology at a problem is insufficient. A plan to include training, organizational change, projected returns on investment (ROI), competitive advantage, and increase in profits is needed, yet that still is not enough. An approach is needed to bring to bear all the capabilities of the corporation in a coherent and synchronized manner. Coherent Knowledge-based Operations™ (CKO™) does just that. CKO fuses KM, IW, and the business processes of an organization, which are now inextricably linked to networks, to attain and maintain a competitive advantage.

Chapter 18: Those Who Do Not Accept Change and Adapt Will Be Consumed By It; The Enlightened Will Survive

This chapter discusses survival. Traditional security focuses on the physical world, but today that focus may not always be valid. Controlling the IE may be more effective than physical attack and may be able to prevent it. This leads to the conclusion that controlling the IE to successfully attain corporate objectives is the approach we need to pursue. Using CKO, corporate action would be synchronized and coherent, following consistent themes conducted simultaneously.

Chapter 19: The Twenty-First Century Challenge: Surviving in the New Century

This chapter reveals the writers' impressions of the courses that technology, businesses, national objectives, and military strategies have charted for us based on current trends. Will our worlds collide? How will we fare when and if they do? What are our future threats, vulnerabilities, and risks brought on by high technology as a weapon of choice in IW?

HOW WE CAME TO WHERE WE ARE — AND BY THE WAY, WHERE ARE WE?

Section I of this introductory book on global information warfare (IW) sets the stage for a detailed discussion on information warfare around the world. As an introduction to the topic of global information warfare, it is important for the reader to gain an understanding of the world in which we all must live, work, play, and war against each other, groups, and nation-states. This basic understanding of our Twenty-first Century environment explains how we got to the stage of information warfare in which we currently find ourselves. There are five chapters in this section.

Chapter 1: Everything You Wanted to Know about Information Warfare but Were Afraid to Ask

This chapter provides an overall introduction and discussion of information warfare (IW) and sets the stage for the chapters that follow. The roots of IW, generally accepted definitions, and working definitions are all presented.

Chapter 2: Fuel for the InfoWar Fire: The High-Technology Revolution

Key technological advancements (e.g., microprocessors, GPS, cellular phones, Internet, and high-speed modems) contributing to the revolution in information systems, and thus warfare, are discussed.

Chapter 3: The Global and Nation-State Information Infrastructures: Backbone or Achilles' Heel?

This chapter reviews the global information infrastructure (GII), its history, and its impact on government agencies and businesses vis-à-vis industrial and economic espionage, techno-terrorism, and military IW.

Chapter 4: Our Changing World, or Is It Really?

This chapter provides a broad background on trends in the world around us. Economics, business, environment, global competition, the "withering" superpowers, economic and industrial espionage, and the blurred distinction between peace and war are discussed.

Chapter 5: Business, Government, and Activist Warfare: The More Things Change, The More They Don't

This chapter discusses business, government, and activist warfare history.

Chapter 1

Everything You Wanted to Know about Information Warfare but Were Afraid to Ask

---

War does not determine who is right— only who is left.

—Bertrand Russell

This chapter provides an overall introduction and discussion of information warfare (IW) and sets the stage for the chapters that follow. The roots of IW, generally accepted definitions, and working definitions are all presented.

Introduction to Warfare

Wars have been fought ever since there were human beings around who did not agree with one another. These conflicts continue to this day, with no end in sight. The use of information in warfare is nothing new. Those who had the best information the fastest and were able to act on it the soonest were usually the victors in battle.

War is defined[1] as:

- *Armed fighting between groups:* an armed conflict between countries or groups that involves killing and destruction; *The two countries are at war.*
- *Period during war:* a period of armed conflict; *during the Vietnam War*

- *Methods of warfare:* the techniques or the study of the techniques of armed conflict
- *Conflict:* any serious struggle, argument, or conflict between people; *The candidates are at war.*
- *Serious effort to end something:* an effort to eradicate something harmful; *a war against drugs*

Is it any wonder that since we are now in the Information Age that we should also have information warfare? No, that is certainly not surprising — or at least it should not be. Because we now look at almost everything on a global scale, it should also not be surprising then that information warfare is viewed on a global scale. Information warfare is today's much-talked-about type of warfare. A search of the Internet on the topic using google.com disclosed 472,000 hits. Information warfare is becoming an integral part of warfare of all types in the modern era.

Four Generations of Warfare

Military historians and professionals over the years have discussed the various generation of warfare. Some believe there are four generations of warfare to date:[2]

- First-generation warfare started with the rise of the nation-state and included a top-down military structure, limited weapons, and armies made up of serfs that ended in the early Nineteenth Century about the time of the Napoleonic Wars.
- Second-generation warfare began about 1860 in the United States in its Civil War. This generation of warfare included artillery, large armies, frontal assault tactics, machine guns, mass weapons development, and logistics supported by trains. This generation of warfare ended sometime after World War I.
- The third-generation of warfare attributed its beginning to the Germans in World War II, where "shock-maneuver" tactics were used.

In 1989, the U.S. Marine Corps Gazette[3] contained an article by several military personnel. The article, entitled "Changing the Face of War: Into the Fourth Generation," discussed the *fourth-generation battlefield* where it is likely that it will include the "whole of the enemy's society.... The distinction between civilian and military may disappear.... Television news may become a more powerful operational weapon than armored divisions." If one were to have any doubts about the accuracy of that statement, one just has to remember the U.S. television news showing a dead American military man's body being dragged through the streets of Mogadishu. The loss of national will can be closely correlated with how quickly the United States departed that country. This, too, is part of the information warfare campaigns being waged on a worldwide scale.

One can argue that information warfare has existed in all generations of warfare and included spying, observation balloons, breaking enemy codes, and many other functions and activities. True, information warfare is as old as man, but many aspects as to how it is being applied in our information-dependent, information-based world are new.

Introduction to Global Information Warfare

In the early 1990s, several people in the U.S. Department of Defense (DoD) articulated a unique form of warfare termed "Information Warfare," or IW. The Chinese say they were developing IW concepts in the late 1980s. Who is correct? Does it matter? As the areas embraced by IW have been developed over the centuries and millennia, these have been a normal part of human activities from mankind's beginning. What is unique about IW is that it is the first instantiation of trying to tie together all the areas that make up the information environment (IE). The IE runs through every part of your country, organization, and personal life. At the present time, there is no cookbook recipe to do the extremely complex task of bringing together all the areas. This book articulates the full spectrum of IW and offers recommendations and support so that you can tailor this information to your specific situation. IW is both art and science.

What is IW? The general working definition of information warfare employed in this book is as follows: *IW is a coherent and synchronized blending of physical and virtual actions to have countries, organizations, and individuals perform, or not perform, actions so that your goals and objectives are attained and maintained, while simultaneously preventing competitors from doing the same to you.* Clearly, this embraces much more than attacking computers with malicious code. The litmus test is this: if information is used to perpetrate an act that was done to influence another to take or not take actions beneficial to the attacker, then it can be considered IW.

The definition is intentionally broad, embracing organizational levels, people, and capabilities. It allows room for governments, cartels, corporate, hacktivists, terrorists, other groups, and individuals to have a part. It is up to each enlightened enterprise to tailor the definition to fit its needs. This should not be a definition of convenience, to "check the box."

Exhibit 1 shows many information warfare areas.

You are asked, and many times forced by government and businesses, to depend on the Internet; the Internet that is home to hackers, crackers, phreakers, hacktivists, script kiddies, Net espionage (Network-enabled espionage), and information warriors; the Internet that is home to worms, Trojan horses, software bugs, hardware glitches, distributed denial-of-service (DDoS) attacks, and viruses. All this, and the Internet is only a portion of the areas that IW addresses. Although the Internet touches many critical infrastructures, and these in turn affect the many IEs (information environments) with which you interface, most of the IW areas were around before the Internet. We will

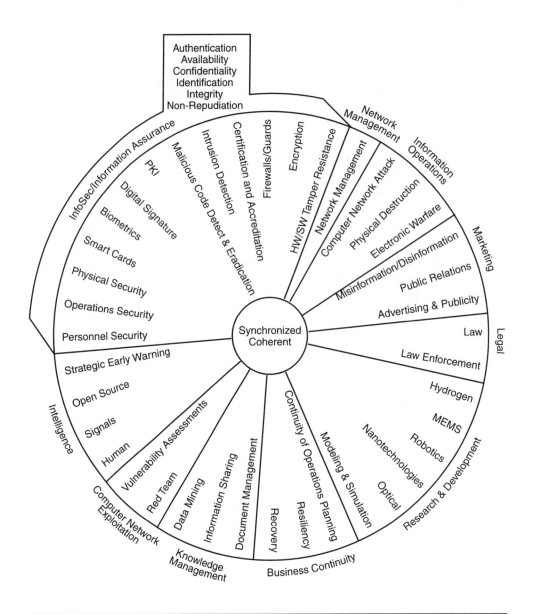

Exhibit 1.   Information Warfare Areas

articulate a global view of IW that is applicable to nation-state governments, their military, groups, individuals, and businesses.

As "competition" is analogous to "enemy" or "adversary," other business–military analogies can be made with profit, shareholder value, competitive edge, and industry rank to achieve brand recognition, customer loyalty, exerting power, influence, and market share. A business leader or military leader must train and equip forces; gather intelligence; assemble, deploy, and employ forces at decisive places and times; sustain them; form coalitions with other businesses and nation-states; and be successful. There are many physical and virtual world parallels, as can be seen in the following headline: "Cisco to use SNA as *weapon* against *competition*…. Cisco believes its experience in melding SNA and IP internetworks can be used as a *weapon* in the company's *battle* with Lucent and Nortel for *leadership* in converging voice, video, and data over IP networks."[3]

Purists will focus on warfare as a state of affairs that must be declared by a government and can only be conducted by a government. Microsoft attacking Netscape, guerrilla warfare, economic warfare (either one country "forcing" another country to spend itself into bankruptcy, as the United States did to the Soviet Union), or a company adjusting prices to damage its competition (e.g., taking a long time horizon to use volume and time to adjust prices downward). "Conflict" or "that's business" does not carry the same sound of ultimate struggle as referring to business as "war." Clausewitz stated, "War is an extension of politics." By analogy, because business is the implementation of a country's laws, economic policy, and values, business is an extension of politics.

In a free market economy, competition is central to business strategy to win customers and market share. Competition, like war, is a struggle for a winning position. The marketplace can then be referred analogously to a battlefield with winners and losers. It follows that business is analogous to war. Therefore, using military phraseology in a business context is appropriate. In fact, one just has to remember September 11, 2001, and New York's World Trade Centers to see that in today's world, warfare is waged on many levels by various adversaries against various targets. These targets can be nation-states, their governments, groups, businesses, or individuals. The tools will be any that can be applied for attackers to successfully attain their goals.

The counter argument is that some insurance companies' contracts state that if a loss is due to an act of terrorism or war, they will not pay for damages. In the United States, attacks on computers by default are criminal acts and are thus in the purview of law enforcement. Only after an investigation determines that the criminal act is a national security issue will the intelligence agencies and other government organizations take the lead.

There are adversaries, winners, and losers. All the writing on IW focuses on weaknesses, defenses, and losses. Despite the gloomy forecasts by government officials and the media, IW is also about strengths, offenses, and gains. These positive features are within the grasp of any government or business organization with a desire to seize and maintain a competitive advantage — to be a winner on the IW battlefield.

What is the motivation for a book on a subject as sweeping, complex, and misunderstood as global IW? What possible application can it have outside specialized military circles? From a practical viewpoint, how does IW shorten decision cycle times, raise revenue, lower or avoid costs, and improve performance? If IW cannot improve effectiveness or efficiency, or bring about innovation, why do it? IW does do these things and ought to be the approach used rather than the pop management fads that come and go like the cute little critters in *Whack a Mole,* leaving businesses worse off for trying them. The purpose of IW is to gain power and influence over others. Power and influence are at the heart of all relationships. Because IW requires effort, the effort needs to resolve into some aspect of power, such as profit, economic or military domination on the battlefield or in the marketplace.

Information Warfare Will Hit You in Your Pocketbook

There are some events that were not expected. Hannibal crossed the Alps. Clay defeated Liston for the heavyweight boxing title. CD Universe did not think crackers would break into its systems. Buy.com did not expect a DDoS attack. On average, new Web sites are discovered and hacked within 15 minutes of being on the Internet. One honey pot project was attacked within five minutes. It will happen: one day your IE defenses are going to be beaten. When they go down, your revenues and profits will go down. The Internet Age has again proven the adage that "time is money." Suppose a company has U.S.$1 billion in electronic and mobile commerce revenue. That equates to $2,739,726 per day, $114,155 per hour, $1903 per minute, and $32 per second.[4] According to *Information Week*, hacker attacks cost the world economy about $1.6 trillion in 2000.[5] That is $4,383,561,644 per day, or $182,648,402 per hour. How long can your business afford to be adversely affected due to an attack? In other words, what are the risks and consequences you are willing to accept?

In a portent of crippling events to come, since early 2000 there have been computer-based automated distributed ("zombie") attacks such as trin00 and Tribal Flood Network, extortion attempts for tens and hundreds of millions of dollars, and posting on the Internet of tens of thousands of supposedly protected credit cards and other private information. The Love Bug virus caused an estimated $10 billion in damage[6] (a Lloyds of London estimate was $15.3 billion[7]). The average major virus during mid-year 2001, like SirCam and Code Red, resulted in about $2 billion in damage. Apparently, the laws and court sentences for computer crimes lack deterrent value. Of course, if hardware and software products, communications systems, E-commerce sites, and other information technology (IT) components were designed with security in mind, we would not have this predicament — something that even Bill Gates of Microsoft finally realized.

In many cases, the dollar loss is secondary to the loss of trust. Banks and insurance companies especially feel customers' wrath. When customers believe their trust has been compromised, they vote with their pocketbooks and take their business elsewhere. That is when revenues and profits decline, which leads to a decline in the stock price, which in the not too distant future will lead to shareholder lawsuits for negligence and other claims.

IW conjures up many images: computer, network, and telecommunications savvy experts in the military and intelligence communities, corporate espionage, and pale fourteen-year-olds looking like they could be the next door neighbor's kids — or yours. Dire prognostications about how an electronic Pearl Harbor threatens national security and the daily media coverage of viruses and denial-of-service attacks interchangeably using phrases like information warfare, cyber warfare, and cyber terrorism may make IW seem distant and surreal.

Some of the attacks, premeditated or unintentional, resulted in billions of dollars in damages. Computer emergency response teams (CERTs) and law enforcement agencies stress protection and defense of information, information infrastructure, and information-based processes to ward off malicious attacks. What do these and many other aspects of operating in the IE have to do with managing a government organization or running a business? For businesses, this may mean new business generation, cost avoidance, profit, customer retention, market leadership, positive, power public perception. For nation-states, this may be economic, political, or military power and influence.

High-profile events such as the Morris worm and Citibank's $400,000 loss ($10 million was stolen, and all but $400,000 allegedly recovered) should have been sufficient warning shots across the bow that a different approach was needed. The much-needed security fixes are years away. Demand is low because the general public appears to be uninterested in cracker exploits, made indifferent by the almost daily news stories. Said differently, the public has come to *expect* cracker activity. Perhaps this is one reason why the public has not warmly embraced electronic commerce and mobile commerce.[8]

Business Is War

An advertising campaign can be considered a subset of an IW campaign. Here is a perhaps not so hypothetical example. Taking grocery store shelf space, due to product or packaging redesign, from a competitor is notionally no different than denying use of a radar or a seaport to the enemy. Instead of cereal boxes that stood and poured vertically, what if they stood horizontally and had spouts for pouring (besides, vertical boxes are prone to tipping)? This would result in more shelf space needed for the same amount of cereal boxes. The packaging will carry a message that conveys "new" and "improved." The boxes will be at eye level — easy for the consumer to spot. In-store advertising will attempt to vector shoppers to the cereal aisle. Newspaper and magazine advertising will attempt to convince customers to try the "new" and "improved" product, and coupons will be used as further enticement. There may even be an in-store demonstration. Because there is limited shelf space and if the cereal company has bargaining power, other cereals have to lose space. Lost space translates to lost product sales, which in turn leads to reduced revenue and profits as well as a lower stock price.

The IW target can be the customer, the competition, or another entity. The purpose of the IW campaign is to have them take action that will result in increased profits for the company. In the best of all outcomes, your revenues go up and the competitors' revenues decline. Even if your sales were constant, just having less space to sell should make competitors' sales decline, so your industry ranking will improve. What will the competition do? Redesign packaging? Alter ingredients? Lower the product's price? Counter with coupons? Have a television campaign employing a doctor to extol the health benefits of their cereal? Play hardball with the supermarket chain? A combination? Nothing, taking a wait-and-see approach? This is physical and virtual IW at the corporate level. It embraces media, perception management, physical operations, intelligence collection, and more.

This is no different than one country observing another and bringing to bear economic, diplomatic, and military means. These means may include very advanced open source searches and analyses, and covert means involving manipulation of the radio frequency (RF) spectrum. From a business perspective, operations, marketing, public relations, manufacturing, finance, transportation, and other parts of the company must operate in a synchronized and coherent fashion. The competition must be monitored so the company can be in position to agilely and effectively respond to any countermoves.

IW Broadly Encompasses Many Levels and Functions

IW is not the sole purview of a government; otherwise, only the wealthy countries could practice it. A narrow interpretation of IW flies in the face of reality. Other than for very unique capabilities that are based on unlimited deep pockets and specialized espionage capabilities, more brainpower, and perhaps more capabilities, reside external to a government. Any organization, and even individuals, can conduct offensive and defensive information warfare. IW is about seizing control of perceptions, physical structures, and virtual assets. Seizing control can be done from both offensive and defensive positions. That squarely puts any organization in control of its destiny. Those that are unenlightened will never perform at or near the top of the pack, and may well go out of existence. Those that embrace IW have a much better chance of surviving and reaping the rewards.

The military, intelligence community, and law enforcement generally do not embrace this perspective. Why? They have capabilities that are highly classified. If used by industry, then "all hell would break loose." Certainly, there are unique offensive and defensive capabilities that only can be developed by the government due to their high risk of success and the necessary funding. However, there has been an explosion of brainpower with regard to physical and virtual capabilities. The majority of brainpower in genetics, robotics, nanotechnology, microelectromagnetic systems (MEMS), and hydrogen technologies resides *outside* the military, intelligence community, and law enforcement. What is to prevent these capabilities from falling into the hands of nation-states, individuals, businesses, and organizations that wish to perpetrate some form of hostile behavior? Absolutely nothing.

Historical and Recent Examples of IW in Action

IW can be inferred in the writings of Sun Tzu 2500 years ago. The ancient Greeks, Genghis Khan, the Medicis, Jomini, von Clausewitz, Mao Tse Tung, Ho Chi Minh, Che Guereva, Fidel Castro, and Slobodan Milosovec all practiced IW. The only difference between historical examples and now is technology, and information technology (IT) in particular, permits communications, computing, and decision making in seconds rather than months. IT is not a requirement for IW, but it is an important factor. An important point to keep in mind is that IW can be physical or virtual; high, low, or no tech; and asymmetric.

In addition to reading the author's work in the original language, it is best to understand the political, military, social, and economic history that serves as the context for the work. Because most of us have neither the language skills nor the time to do this for the many important works, respected translations and insightful thought will have to suffice. While possible to be effective in certain situations, to raise one's self to a higher level and to develop philosophical understanding and intellectual insights requires study. Effective time management includes self-improvement as well as work, exercise, and rest. The following readings put forth concepts that form the basis of many business, military, and political strategies and tactics. It would be best to start with them before going down a focused path.

- The granddaddy is Sun Tzu's "The Art of War" written some 2500 years ago. This timeless classic with hundreds of recommendations that Sun Tzu offered to the king is applicable today. The basis for almost every military doctrine can be traced back to Sun Tzu.
- "A Book of Five Rings" written by Miyamoto Musashi around 1645 is the basis of kendo, the way of the sword. Absorbed by ninjitsu and other martial arts, the principles have been interpreted over the ages and are the basis of many Japanese business strategies. When applied at the higher levels, the principles have an almost mystical basis as a philosophy of life.
- Atilla the Hun controlled most of the known world around 400 A.D. He articulated over several hundred practical elements of guidance embracing areas such as seeking and receiving advice, decision making, delegation, diplomacy, mentoring, and perception management.
- "On War" written by Carl von Clausewitz around 1832 is the German General Staff's answer to Napoleon. Although the work was never completed due to von Clausewitz's untimely death, his insights are applicable today. Some of these are that war is an extension of politics and that total war should be waged judiciously (limited war is a misapplication of combat arms).
- "The Art of War" was written by Baron Antoine Henri de Jomini around 1838 written in response to von Clausewitz. The concepts form a classic treatise that ought to be applied appropriately in every boardroom. Jomini divided war into six areas: strategy, grand tactics, logistics, engineering, tactics, and political and military diplomacy in relation to war.

IW is not unique to governments and the military. Competition, either in the marketplace or on the battlefield, has been part of human nature for over 10,000 years, and IW has been a normal part of these activities. Today, of course, technology provides more capabilities to manipulate information in a shorter period of time. IW is not a "fad-of-the-month" approach. Information is fundamental to both business and government. Protecting and enhancing capabilities in the information environment (IE) is vital to achieving objectives and to attain and maintain a competitive advantage.

Physical attack is now complemented by virtual attack. The Love Bug virus came from the Philippines, and the Bubble Boy and Chernobyl viruses came from Argentina. Distributed denial-of-service (DDoS) attacks have come from Canada. India and Pakistan have clashed in the cyber realm, as have the Palestinians and the Israelis. Attacks on NATO servers came from Serbia during the Kosovo Air Campaign. The business world is just as riddled with attacks. Many go unnoticed for months, and some are never detected. A risk (i.e., the intersection of a vulnerability with a threat) can last for years until all the patches are installed. Some of the threats resurface after several years, and life cycles for some threats have been identified. An information warrior can find a back door anywhere in the world because of the interconnectivity and interdependencies of many infrastructures.

Microsoft servers were broken into for at least three months before the attack was recognized. Indications are that the source of the attack was St. Petersburg, Russia. Steve Balmer, Microsoft's CEO, decried this as a horrific example of industrial espionage. Why does this have to be industrial espionage? Could a government entity not be behind it? After all, if Microsoft software is the *de facto* standard and a country new how to disable it, that country would be capable of disruption and destruction from nation-state or corporation-specific surgical strikes to regional and global attacks. This type of espionage was made famous by the crackers from St. Petersburg and Moscow who in 1994 virtually broke into Citibank and stole more than $400,000. That should have been the trigger for a significant ramp-up in cyber defenses, but businesses have been eating the costs of losses — totaling in the trillions of dollars — for more than seven years.

Of course, the source of the Microsoft attack should not be surprising — there is a world-class hacker school in St. Petersburg. From the competitive intelligence school in France to a business intelligence company in Florida, from MI5 in England teaching companies how to counter spies to "how to hack" courses taught by the person credited with creating the first computer virus, and from cyber gangs such as Cult of the Dead Cow and L0pht Heavy Industries to national data network monitoring entities such as SORM in Russia, it is a minefield out there!

For governments, virtual attack can be viewed as a natural progression of electronic warfare (EW); command, control, communications, and countermeasures (C^2CM); and command and control warfare (C2W). During the past decade, the concepts and implementation of IW, information operations (IO), information superiority, and decision superiority have emerged. Many modern nation-states agree with the U.S military perspective, IO spans the spectrum of conflict, from peace through operations other than war (OOTW) to war and then back to OOTW and peace. The military believes IW is reserved for conflict. This is done for a number of reasons, such as only Congress can declare war and adversaries undertake hostile actions in a peacetime environment. From a business perspective, competitors always attack — computer networks, customers, research and development, contracts, and other areas. Legal and military distinctions between information warfare and information operations do not seem to deter hostile actions. For the purpose of this book, we refer to conflict in the information environment as *information warfare*.

What IW Is...and Is Not

Information warfare is not about a one-time silver bullet for a quick fix and looking good on a quarterly financial report. IW is not restricted to using computers to attack other computers. IW is not confined to the cyber realm. "Virtual" means electronic, radio frequency (RF), and photonic manipulation. Organizations need to use the capabilities within the virtual and physical domains in a manner that optimizes what they wish to do. The best approach for IW, as it should be with a business or government organization, is to conduct physical and virtual operations in a synchronized and coherent fashion. Easier said than done. Goddard's experiments contributed to manned space flight — four decades later. As virtual capabilities become more practical for the government, military, and business, the greater their importance becomes in operations. Fifteen years ago, laptops, mobile phones, and personal digital assistants (PDAs) were bulky, seldom more capable than their traditional counterparts, and much more expensive. For some people, the time-saving and cost-reducing capabilities of the gadgets borders on technological cocaine, and these people almost cannot function without their gadgets. Some business and government organizations have bought into technology, so much so that their operations can truly be termed "network-centric business." What better way to counter this than with IW? Twenty years from now, IW will be mainstream, and those who do not participate will fail.

Much hype surrounds hacker exploits and computer-based viruses. Most hacker, cracker, and phreaker exploits and viruses qualify as falling within IW, albeit at the low end of the spectrum, because there is an attempt to influence, either directly or indirectly, others to take an action. Approaches range from altruistic ("I found a hole in the software. Develop a patch for it.") to anger ("I will make them miserable for firing me.") to social awareness ("Stop drug research on animals.") to criminal ("Here is how to defeat the fraud control and computer security systems at Banco de Agro.") Almost all of the events and attacks fall into the realm of theft, extortion, fraud, and related criminal behavior. Measures must be employed to protect and defend corporate and government systems because individual losses have already been in the tens and hundreds of millions of dollars. Omega Corporation and Emulex Corporation are examples.

Three weeks after Tim Lloyd, Omega Engineering's computer system manager, was fired in mid-1996, all of the programs running the company's manufacturing machines were erased. A jury accused Lloyd of sabotaging Omega's network. This is the first U.S. federal criminal prosecution of computer sabotage. The attack resulted in 80 workers being laid off, $2 million in reprogramming, and a loss of $10 million in other expenses and missed sales.[9]

Emulex's stock plunged $45 in less than an hour. A fake press release purported that wire services were running a story that Emulex CEO Paul Folino resigned because of an accounting scandal. The fiber-optic company's share price plummeted more than 60 percent, a market loss of $2.5 billion, before NASDAQ halted trading.[10] A 23-year-old community college student, a short-seller of the stock who had worked for an online wire service, was found guilty of this sham. Fortunately, major news coverage later that day gave investors confidence, and the stock's price ended within a few dollars of where it had started. Individual shareholders lost about $50 million. The psychology of today's stock market, demonstrated by the rapid selling on unsubstantiated bad news, shows how manipulation of credibility through misinformation can tank a stock. A sophisticated IW attack can ruin a multi-billion-dollar company. In 1999, a PairGain Technologies engineer sent the company's stock soaring 30 percent after he linked a Web posting to a fake Bloomberg News page announcing a takeover bid. Investors were burned when the company issued a denial.[10]

Even if you have taken all the appropriate measures to protect and secure your physical and virtual assets, much falls outside your span of control: protected and secured power, financial, communications, transportation, water, and continuity of government infrastructures; security-rich and bug-free commercial off-the-shelf (COTS) software; and the creativity of crackers and phreakers to find new vulnerabilities in technology to exploit. Also, you probably cannot control your business partners', customers', financial stake-holders', and suppliers' IEs that are connected to yours. If you are an Internet-based company, then electronic and mobile commerce account for the majority of your revenue. Any disruption and your customers will go to your compet-itors. If you are a traditional brick-and-mortar company expanding into the Internet to enhance your customers' ability to do business with you, business interruptions and disclosure of customer data will taint your reputation and credibility. Business interruption can be costly on many levels.[11]

When properly employed, IW is an agile capability that can be tailored to any situation. It can bring a multitude of functions to bear. IW can be implemented in the physical and virtual worlds. Central to IW is how it is used to influence decision makers. Magazines, radio, television, newspapers, leaflets, e-mail, Web pages, and other forms of media can all be used as a vehicle to deliver IW.

IW should not be restricted to a small cadre. Certainly only a few people should know about the sensitive details that will make or break the execution of the IW plan. All parts of an enterprise, not just an organization, need to be linked for the most effective implementation of IW. Any organization has a finite portion of resources. Partnerships, alliances, consortiums, and other relationships can serve to expand an organization's capabilities.

Proper use of information is central to profitable business and successful military operations. IW is used to provide your organization a competitive advantage while at the same time limiting the competition's capability to reduce your advantage and increase their own. Effective IW is not possible without control of your information environment (IE).

An IE is the interrelated set of information, information infrastructure, and information-based processes. Data includes the measurements used as a basis for reasoning, discussion, or calculation. Data is a raw input. Information applies to facts told, read, or communicated that may be unorganized and even unrelated. Information is the meaning assigned to data. Knowledge is an organized body of information. It is the comprehension and understanding consequent to having acquired and organized a body of facts. Information as used here means data, information, and knowledge. No doubt horrific to purists, there is no one good word in the English language that embraces all three concepts together. All three processes exist within any organization. At

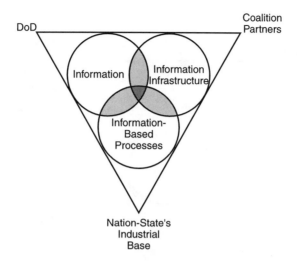

Exhibit 2.   Major Components of an Information Environment (IE)

any given time, one of the processes will be of greater value than the others. Your competition wants your information, so do not believe that "gentlemen don't read other gentlemen's mail."

Information moves across information infrastructures in support of information-based processes. Information infrastructure is the media within which we display, store, process, and transmit information. Examples are people, computers, fiber-optic cable, lasers, telephones, and satellites. Examples of information-based processes are the established ways to obtain and exchange information. This includes people to people (e.g., telephone conversations and office meetings), electronic commerce/electronic data interchange (EC/ EDI), data mining, batch processing, and surfing the Web. Attacking (i.e., denying, altering, or destroying) one or more IE components can result in the loss of tens of millions of dollars in profit, degraded national security, and be more effective than physical destruction. Degrade or destroy any one of the components and, like a three-legged stool, the IE will eventually collapse.[12]

Exhibit 2 shows the major components of an information environment (IE).

Bad things happen, such as floods, hurricanes, and earthquakes; power surges and sags; and fires. Disgruntled employees can steal, manipulate, or destroy information. Crackers work their way through the electronic sieve of protection mechanisms (e.g., firewalls and intrusion detection devices) into information assets.

PriceawaterhouseCoopers, the Computer Security Institute, the Federal Bureau of Investigation (FBI), and others have conducted surveys asking participants if they had experienced IT breaches. Forty-five percent replied that they were attacked and suffered monetary losses; some losses were quite large. The number saying they were attacked is no doubt higher. Many attacks are sophisticated and not readily detected. Companies are reluctant to report computer crimes because of potential shareholder lawsuits and customer loss of confidence and leaving for the competition.

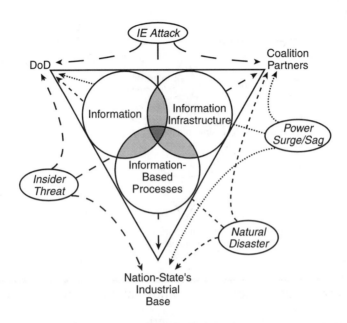

Exhibit 3.   Possible Breakdowns in an IE

Sound disaster recovery, business continuity, and contingency operating plans are essential. For every minute information systems are not up and fully running, revenues, profits, and shareholder value are being lost. The last thing a general counsel needs is a lawsuit from unhappy shareholders who are suing for millions because the corporation did not follow best practices to protect information. One problem is that commercial off-the-shelf (COTS) hardware and software are very difficult to protect. Another concern is that firewalls, intrusion detection devices, and passwords are not enough. The state-of-the-art in information assurance (IA) is against script kiddies and moderately skilled hackers. What about the competition, drug cartels, and hostile nation-states that are significantly better funded? There is no firewall or intrusion detection device on the market that cannot be penetrated or bypassed. Password dictionaries can cover almost any entire language, and there are very specific dictionaries (e.g., sports, Star Trek, or historic dates and events).

Exhibit 3 shows the possible breakdowns in an IE.

IEs exist internal and external to an organization. An IE is tailorable so it can support many actors. The example that follows involves a corporation, its customers, and the government. Another IE can be a military, its allies and coalition partners, and the government. Whatever comprises a specific IE, the important fact remains if its elements are not protected and secured, the consequences can range from irritants to catastrophes.

Exhibit 4 depicts an enterprise IE.

An organization has employees. These employees deliver products, services, and processes to customers. To keep the organization running, suppliers deliver products, services, and processes. Financial stakeholders — venture capitalists, banks, stockholders, and others — provide capital. The public has

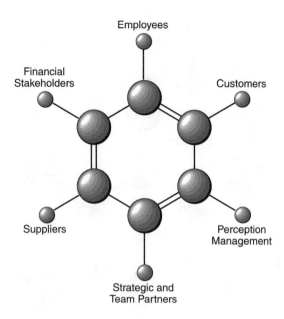

Exhibit 4.   Enterprise IE

a positive, neutral, or negative view of the organization. Strategic teaming partners provide physical, financial, cerebral, and other capabilities. Every entity with which the organization is linked has its own IE. IEs are connected to, and interdependent on, other IEs.

Exhibit 5 depicts an extended IE.

Going Beyond Three Blind Men Describing an Elephant: IW Terms of Reference

IW cuts across national borders, educational background, and cultural views. To ensure a consistent understanding during this discussion, working definitions of IW and many supporting terms are offered. This does not preclude national interpretations and certainly does not attempt to rationalize, harmonize, and normalize definitions. Common terms of reference (TOR) permit a shared understanding, as well as a point of departure for applying the TOR within specific organizations.

George Santayana said, "Those who ignore the lessons of history are condemned to repeat them." Here is an example of how parochialism caused a disaster.

In August and October 1943, the Allies launched air raids against Schweinfurt with disastrous consequences — for the Allies. In the August raid, of 600 planes 60 were lost along with 600 crewmen. Why? There was no long-range fighter escort. Why? In the 1920s and 1930s, resources were allocated for strategic bombardment over pursuit. Why? General Emilio Douhet and others postulated that air power alone could win wars by striking the enemy's strategic

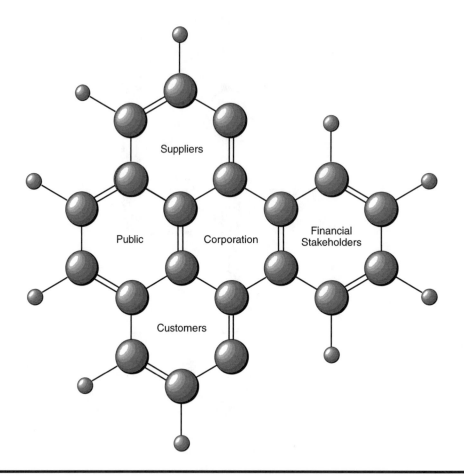

Exhibit 5.   Extended IE

centers. Lesson learned: The decisions made in the 1920s and 1930s led to the wrong tactical employment a decade later. We must not make the same mistake with IW. If we do, national security, economic viability, and corporate capabilities will be lost.

There are as many definitions of IW and related topics as there are people. It is reminiscent of three blind men describing an elephant by touching the animal's various parts. One blind man said, "An elephant is a reptile and is thin and long," as he was touching the tail. Touching the tusks, another blind man said, "An elephant is like a big fish with its smooth and pointed body." The third blind man said, "An elephant resembles a large leaf with a hole in the middle" because he was touching the ears. None of them could extrapolate their interpretations to a real elephant. Similarly, what one sees is not necessarily what one gets. "Ques-que c'est?" will be mispronounced if one does not have a basic understanding of French diction. So, too, is it with terms used to describe various practices in the information realm.

Although the names are initially obtuse to those who do not work in those areas, these information practices have been a normal evolution in communications and computers and also the dark side move/counter-move/counter-counter-move "cool war." These terms include:

- Information security
- Information systems security (INFOSEC)
- Information assurance (IA)
- Information superiority
- Information warfare (IW)
- Information operations (IO)
- Information dominance
- Critical infrastructure protection (CIP)
- Operations Security (OPSEC)
- Communications Security (COMSEC)
- Computer Security (COMPSEC)
- Decision superiority
- TEMPEST

There are many other variations. Little wonder the terms are understood by few people and erroneously used interchangeably. Few understand the difference between a hacker, cracker, and phreaker, much less a white-hat hacker.

- Cyber warfare
- Cyber terrorism
- Cyber vandalism
- Cyber gang warfare
- Hacktivism
- Cyber graffiti

In some cases, more terminology only detracts. "Cyber" is too limiting. It is as if, rather than pushing through difficult points to achieve philosophical insights and technical understanding, people create terms to differentiate themselves without knowing what they are doing.

Information and knowledge are now in vogue. We are in the Information Age, and rapidly transitioning into the Knowledge Age. Acquiring the right data, deriving good information, and applying it to make sound decisions to positively affect the bottom line are essential. Search engines have made finding information on the Internet very simple. Witness during the past 15 years the explosion of terminology related to the protection of information and using information for national security purposes. The most important point is to understand the meaning of these terms and what the different functions can — and cannot — do to make an informed decision whether or not to commit resources (i.e., people, money, and time).

Many countries have developed definitions. IW, information assurance, information operations, information superiority, and other constructs popular in the U.S. military are part of the revolution in military affairs (RMA) and revolution in security affairs (RSA). Government organizations and businesses have developed additional terms, and some do not agree with the national version. So there can be a point of departure for this discussion, definitions accepted by many are put forth. In some cases, working definitions will be used. The following definitions are from the *U.S. Department of Defense Dictionary of Military and Associated Terms:*[13]

- *Command and control warfare (C2W):* The integrated use of operations security, military deception, psychological operations, electronic warfare, and physical destruction, mutually supported by intelligence, to deny information to, influence, degrade, or destroy adversary command and control capabilities, while protecting friendly command and control capabilities against such actions. Command and control warfare is an application of information warfare in military operations and is a subset of information warfare. Command and control warfare applies across the range of military operations and all levels of conflict. Also called C2W. C2W is both offensive and defensive.

- *Defense in depth:* The siting of mutually supporting defense positions designed to absorb and progressively weaken attack, prevent initial observations of the whole position by the enemy, and to allow the commander to maneuver the reserve.

- *Information:* Facts, data, or instructions in any medium or form. The meaning that a human assigns to data by means of the known conventions used in their representation.

- *Information assurance (IA):* Information operations that protect and defend information and information systems by ensuring their availability, integrity, authentication, confidentiality, and non-repudiation. This includes providing for restoration of information systems by incorporating protection, detection, and reaction capabilities.

- *Information-based processes:* Processes that collect, analyze, and disseminate information using any medium or form. These processes may be stand-alone processes or sub-processes that, taken together, comprise a larger system or systems of processes.

- *Information environment:* The aggregate of individuals, organizations, or systems that collect, process, or disseminate information; also included is the information itself.

- *Information security:* The protection of information and information systems against unauthorized access or modification of information, whether in storage, processing, or transit, and against denial-of-service to authorized users. Information security includes those measures necessary to detect, document, and counter such threats. Information security is composed of computer security and communications security. Also called INFOSEC.
  - *An older definition focused on only physical protections:* locks, alarms, safes, marking of the documents, and similar physical world capabilities.

- *Information system:* The entire infrastructure, organization, personnel, and components that collect, process, store, transmit, process, display, disseminate, and act on information.

- *Information warfare (IW):* Information operations conducted during time of crisis or conflict to achieve or promote specific objectives over a specific adversary or adversaries.

We can expand on this because of the definition of IW. What is IW? It is more than computer network attack (CNA) and defense (CND). That much everyone agrees on. But what else is encompassed by IW? Heated debates go on today about what IW should embrace and accomplish. IW is an umbrella concept embracing many disciplines. IW is most effective when performed in a synchronized and coherent fashion. That is why knowledge management (KM) complements it so well. All components of an organization, as well as across the enterprise, need to be included in an IW action plan.

The good news is that IW embraces the marketing, public relations, counterintelligence, and other functions you now perform. IW is not these functions renamed. They continue to be run by the subject matter experts. IW is the coherent application and synchronized approach of these functions. What is needed are experts who, by analogy, are conductors of the orchestra. They know where the expertise resides within the organization, understand what the functions can and cannot do, and bring them to bear for optimum performance. At present, only the military in a few countries comes close to understanding the relationships and functions of linking the physical domain with the virtual realm, and has begun policy development and allocation of resources. The equivalent does not exist in industry — yet.

The purpose of IW is to control or influence a decision maker's actions. An area of control can be directly manipulated, whereas an area of influence can only be indirectly manipulated. Control and influence are the essence of power. From a business perspective, sector and industry-leading market share and profit are the results of proper IW execution.

What would make a decision maker act or not act? Perhaps false or misleading information, an analysis of open source information, documents mysteriously acquired, or intelligence from an employee hired away from the competition. IW at the corporate level manifests itself in marketing, public relations, legal, research and development, manufacturing, and other functions. With the introduction of commercial high-resolution satellite photography, some companies have altered their delivery and shipment schedules, to include using empty rail cars and semi-tractor trailers to mask inventory, production capability, and customer quantities. IW is a full spectrum of capabilities. Ingredients are carefully selected and tailored to each case.

IW can be conducted without using physical destruction. Both military psychological operations (PSYOPS) and commercial advertising heavily depend on psychology and sociology, the study of individual and group behavior. The implications of this insight are enormous. Businesses engage in IW all the time, or is it that only the effective ones do?

IW enables direct and indirect attacks from anywhere in the world in a matter of seconds. Physical proximity to a target is not necessary. How is this possible? Because we have made conscious and unconscious decisions to have speed and connectivity without complementary security. In Sun Tzu's and Genghis Khan's eras, physical, personnel, and operational security were all that was needed for protection. Today we have fiber optics, satellites, personal digital assistants (PDAs), infrared and laser communications, interactive cable television, mobile phones, and a host of other technology marvels that allow us in a few seconds to reach anywhere. Now, in seconds, our information can be intercepted, modified, manipulated, and stolen.

No simple sentence or paragraph effectively describes IW. There are broad and narrow interpretations within national and international government, business, and academic communities, and some even totally reject the notion of IW. The overall view of IW must be expansive. Information is everywhere. We find information, for example, in mass media such as radio, television, and newspapers, at World Wide Web (WWW, or Web) sites, in communications systems, and in computer networks and systems. Any and all may be subjected to attack via offensive IW (OIW). It follows that all these areas must be defended with defensive IW (DIW).

- *Offensive IW* can make a government, society or nation, or business bend to the will of the attacker. Attacks can be very large, devastating, and noticed, such as economic or social disruption or breakdown, and denial of critical infrastructure (e.g., power, transportation, communications, and finance) capabilities. They can also be small, low key, and unassuming, such as a request for publications and telephone calls (as the basis for social engineering). Businesses do not have the deep pockets of a government, but that does not restrict them from engaging in IW. A business wants to deny the competition orders, customers, and information about its research and development (R&D). Industrial espionage has its share of illegal activities: theft, monitoring communications, and denying use of servers to conduct electronic commerce. Governments engage psychological operations (with the subsets of mis-/dis-information, propaganda using leaflets, television and radio broadcasts). Businesses must identify when disinformation is being used to lure customers away and have the means to counter it. Of course, that is starting from a position of weakness. What is a proactive, defensive IW approach to counter the attack? Inoculate the customers, suppliers, business partners, and others in the IE.
- *Defensive IW* is the ability to protect and defend the IE. Defense does not imply reactive. Measures can be taken to forewarn of attacks and to preposition physical and virtual forces. Examples of virtual forces are software and brainpower. The acme of skill is to present a posture to prevent a competitor from attacking and to achieve victory without having to attack. Perception management is as important as demonstrable physical and virtual capabilities.
- *Information operations (IO):* As stated above, for the purposes of this book, IW is not restricted to war, so IO as described below is included in IW. Actions taken to affect adversary information and information systems while defending one's own information and information systems. Also called IO.
  - *Defensive IO:* The integration and coordination of policies and procedures, operations, personnel, and technology to protect and defend

information and information systems. Defensive information operations are conducted through information assurance, physical security, operations security, counter-deception, counter-psychological operations, counterintelligence, electronic warfare, and special information operations. Defensive information operations ensure timely, accurate, and relevant information access while denying adversaries the opportunity to exploit friendly information and information systems for their own purposes.

– *Offensive IO:* The integrated use of assigned and supporting capabilities and activities, mutually supported by intelligence, to affect adversary decision makers to achieve or promote specific objectives. These capabilities and activities include, but are not limited to, operations security, military deception, psychological operations, electronic warfare, physical attack or destruction, and special information operations, and could also include computer network attack.

- *Information superiority:* The degree of dominance in the information domain that permits the conduct of operations without effective opposition. Information superiority is the relative state of influence and control of the IE between two or more actors. Some argue the opposite of "superiority" is "inferiority." This is not the case. All actors have equal access to open source information. Restricted, sensitive, and classified information can be acquired through overt or covert operations. Having the data, information, and knowledge is not the key to attaining and maintaining information superiority. What is done with the information and the speed at which it is done is the gold nugget. Information sharing, automation, cross-platform information sharing, automating processes (like air traffic control; sales-manufacturing/production-inventory-transportation; and military intelligence-platform maneuver-weapons selection and release-battle damage assessment) are essential to have execution cycles faster than those of the competition.

- *Operations security:* A process of identifying critical information and subsequently analyzing friendly actions attendant to military operations and other activities to: (1) identify those actions that can be observed by adversary intelligence systems; (2) determine indicators that hostile intelligence systems might obtain what could be interpreted or pieced together to derive critical information in time to be useful to adversaries; and (3) select and execute measures that eliminate or reduce to an acceptable level the vulnerabilities of friendly actions to adversary exploitation. Also called OPSEC.

- *Vulnerability:* In information operations, a weakness in information system security design, procedures, implementation, or internal controls that could be exploited to gain unauthorized access to information or information system.

In addition to the above definitions, the U.S. National Security Telecommunications and Information Systems Security Committee (NSTISSC) 4009, National Information Systems Security (INFOSEC) Glossary[14] offers the following:

- *Attack:* Type of incident involving the intentional act of attempting to bypass one or more security controls.
- *Confidentiality:* Assurance that information is not disclosed to unauthorized persons, processes, or devices.
- *Critical infrastructure:* Those physical and cyber-based systems essential to the minimum operations of the economy and government.
- *Integrity:* Quality of an IS reflecting the logical correctness and reliability of the operating system; the logical completeness of the hardware and software implementing the protection mechanisms; and the consistency of data structures and occurrence of the stored data. Note that, in a formal security mode, integrity is interpreted more narrowly to mean protection of unauthorized modification or destruction of information.
- *Non-repudiation:* Assurance that the sender of the data is provided with proof of delivery and the recipient is provided with proof of the sender's identity so that neither can later deny having processed the data.
- *OPSEC:* Process denying information to potential adversaries about capabilities or intentions by identifying, controlling, and protecting unclassified generic activities.
- *Probe:* Type of incident involving an attempt to gather information about an IS for the apparent purpose of circumventing its security controls.
- *Risk:* Possibility that a particular threat will adversely impact an IS by exploiting a particular vulnerability.
- *Risk management:* Process of identifying and applying countermeasures commensurate with the value of the assets protected based on a risk assessment.

Neither NSTISSC 4009 nor the U.S. DoD Dictionary of Military and Associated Terms define consequence and consequence management. Risks are the intersection of threats and vulnerabilities. Residual risks are those that remain after mitigating actions. To plan effectively, decision makers need to know the consequences of various courses of actions. The residual risks influence the outcomes. The outcomes are best represented via consequence management cascading effects. Third- and fourth-order effects, or further, need to be well estimated for the best course of action to be chosen.

Information Warfare Is a Powerful Approach for Attaining and Maintaining a Competitive Advantage

The purpose of a business is to create value for its shareholders, and the purpose of a government is to provide for the common good. From a business viewpoint, being effective and efficient in current markets and opening new lines of business are key to sustained revenue generation and profits. From a national security perspective, we should expect the military, intelligence community, and law enforcement to develop and use capabilities to maintain sovereignty, create and sustain peace and economic prosperity, and ensure public safety from criminals and monopolies. These entities cannot survive

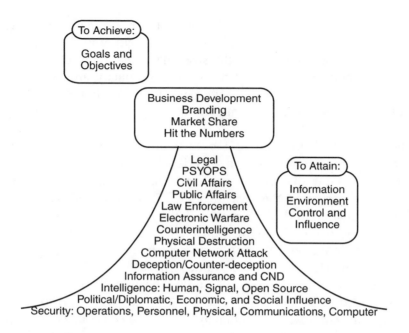

To Achieve:

Goals and
Objectives

Business Development
Branding
Market Share
Hit the Numbers

Legal
PSYOPS
Civil Affairs
Public Affairs
Law Enforcement
Electronic Warfare
Counterintelligence
Physical Destruction
Computer Network Attack
Deception/Counter-deception
Information Assurance and CND
Intelligence: Human, Signal, Open Source
Political/Diplomatic, Economic, and Social Influence
Security: Operations, Personnel, Physical, Communications, Computer

To Attain:

Information
Environment
Control and
Influence

Exhibit 6.   How to Use IW to Achieve Goals and Objectives

by insulating themselves. They must embrace, within their value system, whatever it takes to go beyond surviving to "thrive."

Exhibit 6 shows how to use IW to achieve goals and objectives.

Complexity interwoven across government, industry, and society presents a daunting challenge for IW. It is in the best interest of any government, business, and other organization to take prudent action to defend against information warfare attacks and to be able to launch them.

The advanced cracker breaks into online shopping exchanges, manipulates orders, steals merchandise, plunders credit card numbers — the modern-day pirate, highway robber, and Wild West outlaw. Those who would be part of the online shopping population come to expect this malicious behavior and are dissuaded from shopping online. The result: "dot.coms" go out of business; venture capitalists withdraw support and are reluctant to back new start-ups; and, in the very near future, insurance either will not be written or the premiums will be too high and there will be lawsuits. It appears that IW is tailor made to handle such a situation.

Espionage, disinformation, physical destruction (normally permitted by law only for the military and law enforcement), and other actions are a means to an end. IW is a higher level, cerebral activity. The target can be a population (the national will or a specific political, religious, or ethnic group), a despot, a general, or anyone in an organization. How, then, should IW be applied to industry? After all, is war not a declaration of Congress, Parliament, or other government entity? If a business is destroyed by an act of war or terrorism, it will not be remunerated by insurance. Is this a misnomer? By no means!

Because business is war, the principles of war normally associated with the military ought to be applied. These are not rigid, and their application is tailored to each use. Objective, offensive, mass, economy of force, unity of leadership, maneuver, security, surprise, and simplicity are generally recognized principles that will benefit any organization. Applying the principles to coherent and synchronized IW will produce a positive return on investment (ROI).

In the IT world, determining ROI is considered the Holy Grail. The problem for quantitative metrics for IW is that orders of magnitude are more difficult because of the many disciplines, organizational levels, and the sheer scope involved. Some prefer it that way because it allows them to hide behind classified information and black magic. If IW is to be successful, metrics are necessary. Existing traditional measures are a good start (e.g., How many probes did our intrusion detection system pick up?), but are not sufficiently expansive and precise. What is the value of a database? What it the value of that database after it has been successfully data mined? Because quantitative metrics still need to be developed, qualitative ones will need to be used.

Information warfare is an embracing approach, customizable to produce positive results in any organization, and tailorable to met the demands of the marketplace. By balancing tried and true capabilities with leading-edge technologies and concepts, IW remains a fresh and useful approach for achieving goals and objectives on the way to attaining and maintaining a competitive advantage.

Coherent Knowledge-Based Operations (CKO)

Information warfare for information warfare's sake is senseless. IW must help countries achieve their national security objectives and help businesses attain their goals. When IW is combined with knowledge management (KM) and how business is done, the combination provides a powerful capability. Applying IW with KM results in information superiority. When KM is applied to how business is done, situational awareness will result. Combining IW with how business is done delivers tactics, techniques, and procedures to attain a competitive advantage. The intersection of IW with KM and how business is done is Coherent Knowledge-based Operations. CKO enables a country or a business to attain and maintain a competitive advantage through the synchronization and coherent application of all of its capabilities in the extended IE.

Organizations dabble in many pop management fads. Well-intentioned or not, these often are stovepipe solutions that divert finite resources — people, money, and time — from the organization's central interests and objectives. CKO brings together what appear to be several disparate components. Coherent means an orderly or logical relation of parts that affords comprehension or recognition. The parts are network-centric business (NCB) (how business is done), KM, and IW. When used in concert, their sum is far more powerful than the individual components, creating a powerful means of attaining and maintaining a competitive advantage. CKO can be used to execute and to survive IW attacks.[16] CKO is discussed in detail in Chapter 17.

Network-Centric Business (NCB)

We are told that we are in the Information Age, ride the information highway, and are part of the knowledge-based economy. We conduct electronic commerce (EC), have electronic data interchange (EDI) between computers, allow employees to telecommute and have remote access, and spend millions of dollars on Web sites to attract customers to sell products and services. Computers and robots are in the manufacturing plants, personnel and medical records are automated, and many of us participate in automated deposits and bill payments. If the computers stopped, not enough trained and skilled people could take over the functions, and many businesses and government functions would quickly come to a halt. Computers, databases, and networks are as vital to a business as the circulatory and nervous systems are to your body. Computers and networks have become as ubiquitous as toasters, and network-centric appliances are in the works. Handheld PDAs are the forerunners of tools with tremendous capability, limited only by human creativity. If you do not quickly gain control of your IE, doing so in the future will be exponentially more difficult — and expensive. The main advantage of controlling your IE is that your bottom line will improve.

There is no faster, more effective, or more efficient means to beat the competition than to use NCB. NCB allows an organization to take maximum advantage of its business processes: taking and placing orders, using the supply chain, conducting just-in-time production, and using distribution channels to field products and services. NCB leverages not only all the resources within an organization, but also its customers and business partners. They are all part of the solution set that drives the bottom line. The resources within the organization — people, money, and time — are finite, but can be effectively and efficiently allocated to provide optimal support to customers and to maximize the bottom line.[12]

Knowledge Management (KM)

KM integrates technologies, processes, and cultural changes to provide a means for well-informed, rapid decision making via collaborative information and knowledge sharing by varied and dispersed organizations and individuals. KM tenets include support for organizational processes, tailored content delivery, information sharing and reuse, capturing tacit knowledge as part of the work process, situational awareness of information and knowledge assets, and valuation. KM enables an organization to be more agile, flexible, and proactive. The approach is ideal for integrating, for example, intelligence (e.g., economic and open source) and security (e.g., physical, personnel, and operations); sales and production; and research and development with business development.[12]

Summary

Information warfare (IW) is an embracing concept that brings to bear all the resources of a nation-state or business organization in a coherent and synchronized manner to control the information environment and to attain and maintain a competitive advantage, gain power and influence. Judicious use of IW, when coupled with knowledge management and network-centric business, lead to reduced or avoided costs, increased revenues, more satisfied customers, and larger profits and national security. Governments and businesses can use IW offensively and defensively in the physical and virtual domains. Counters to IW do not have to be in-kind; they can be no, low, or high technology and they can be asymmetric. Not conducting IW will result in a reduced market presence and lower national security. Although the name may change over the years, IW will evolve from its nascent stage and become mainstream in 20 years.

IW occurs when, in the physical and virtual domains, you attack your competition or they attack you. IW is about synchronized and coherent relationships and capabilities. As previously discussed, central to IW are those physical and virtual capabilities to control the IE.

CKO couples IW in a useful approach with KM and how the organization does business. Not only is the corporation's IE engaged, the resources of its enterprise are brought to bear to use all its capabilities in a coherent and synchronized manner to seize as great a competitive advantage as possible. In this fashion, a country can call on its allies and coalition partners, and a business can call on its suppliers and business partners so as much knowledge and as many capabilities as possible can be brought to bear.

Notes

1. Definition from *Microsoft's Encarta World English Dictionary.*
2. Taken from a Gannett News Service article, September 27, 2001.
3. *Network World,* August 10, 1998.
4. If Most of Your Revenue is From E-Commerce, then Cyber Insurance Makes Sense, Perry Luzwick, "Surviving Information Warfare" column, *Computer Fraud and Security,* a Reed-Elsevier publication, March 2001.
5. Year's Hack Attacks to Cost $1.6 Trillion, Tim McDonald, www.Ecommerce-Times.com, part of the NewsFactor Network, July 11, 2000.
6. Hackers say "love bug" was ammo in cyberwar: Virus was meant to disable rival cybergang, Uli Schmetzer, *Chicago Tribune,* February 8, 2001; http://sns.chicago-tribune.com/technology/sns-lovebug.story?coll=sns%2Dtechnology%2Dheadlines.
7. Virus claims threaten insurers, Reuters, May 8, 2000.
8. Crackers and Phreakers Conduct Better Coherent Knowledge-based Operations than Most Companies, Perry Luzwick, "Surviving Information Warfare" column, *Computer Fraud and Security,* a Reed-Elsevier publication, March 2000.
9. Network Manager Found Guilty of Sabotage, Sharon Gaudin, *Network World Fusion,* online, May 9, 2000.
10. The Latest Investor Threat: Infobombs, Brian O'Keefe, *Fortune,* October 17, 2000.

11. If Most of Your Revenue is from E-Commerce, then Cyber Insurance Makes Sense, Perry Luzwick, "Surviving Information Warfare" column, *Computer Fraud and Security,* a Reed-Elsevier publication, March 2001.
12. What's a Pound of Your Information Worth? Constructs for Collaboration and Consistency, Perry Luzwick, American Bar Association, Standing Committee on Law and National Security, National Security Law Report, August 1999.
13. *Department of Defense Dictionary of Military and Associated Terms,* April 12, 2001.
14. National Security Telecommunications and Information Systems Security Committee Publication 4009, September 2000.
15. You Can't Attain and Maintain a Competitive Advantage if You Don't Control Your Information Environment, Perry Luzwick, "Surviving Information Warfare" column, *Computer Fraud and Security,* a Reed-Elsevier publication, January 2000.

Chapter 2

Fuel for the InfoWar Fire: The High-Technology Revolution

What hath God wrought?

—Samuel F. B. Morse
The first telegraph message ever sent, 1844

The revolution in technology, especially in high technology, has caused nation-states, corporations, and individuals to become more technology driven, technology supported, and technology dependent. This chapter provides an overview of the history and revolution in technology that has changed the world and how high technology has driven the changes in the ways we live, work, and prosecute wars.[1]

What Is Technology?

When we speak of technology, what exactly does that mean? After all, if it was not for technology — high technology — information warfare would still be relegated to the world of propaganda, written and verbal communication attacks. So, technology is obviously a significant piece of information warfare. In fact, one cannot discuss modern information warfare without discussing high technology. It is important to first define what is meant by technology before proceeding any further in the discussion of how the revolution in technology/high technology has impacted how we conduct warfare, especially information warfare. (Note: Throughout this book the terms "technology" and "high technology" will be used interchangeably.)

One would expect that by now there would be a common definition for both technology and high technology. However, as with most things related to technology today, there is no easy answer to what appears to be a simple question: what is technology?

According to one dictionary,[2] technology is defined as follows:

> tech·nol·o·gy [tek näl′ ə jē] (plural tech·nol·o·gies) noun
>
> 1. Application of tools and methods: the study, development, and application of devices, machines, and techniques for manufacturing and productive processes • recent developments in seismographic technology
> 2. Method of applying technical knowledge: a method or methodology that applies technical knowledge or tools • a new technology for accelerating incubation "...Maryland-based firm uses database and Internet technology to track a company's consumption of printed goods...." *Forbes Global Business and Finance,* November 1998.
>
> [Early 17th century. From Greek *tekhnologia,* literally "systematic treatment," literally "science of craft," from *tekhne* "art, craft."]

Well, that is fairly general, as should be expected from a general dictionary, but how do those who are involved in technology define it? What better place to look for the answer about technology than by using technology as a tool to find a definition of itself — on the Internet, of course. One company's Web site[3] defined it this way:

> *Technology includes all intellectual property and know-how.*
>
> —http://www.pax.co.uk/ttdefine.htm

Also noted on the Web site were others' definitions of technology:

> *Technology is information embodied in a device, the production, use or sale of which may be restricted by law.*
>
> —Jonathan Putnam

> *Technology is the practical application of knowledge gained through theoretical research to a specific production problem.*
>
> —Michael J. Patrick
> Filament, Renton, Washington

> *Technology is born of a re-search process; institutionalized, glamorized and isolated in its nuclear R&D sheath.*
>
> —Desmond R. Jimenez
> Senior Staff Researcher
> AgraQuest Inc.

Technology = knowledge to do something

—Alan S. Paau, Ph.D.
Director, Technology Transfer
University of California, San Diego

What it seems to come down to is that the definition of technology is based on one's own perspective that is biased toward one's own profession. Thus, there are multiple definitions and all can be applied in discussions of information warfare.

From Cave Warrior to Information Warrior

The world is rapidly changing. We humans are in the midst of, or have gone through, a hunter-gatherer period, an agricultural period, an industrial period, and now the modern nation-state and its society is in an information-based and information-dependent period. Our global society can no longer function without the aid of automated information and high technology — computers and networks. With computers and global networks such as the Internet come opportunities to make life better for all of us. However, it also makes each of us more vulnerable and increases the risk to the high technology we depend on, as well as increased risks to information warfare, our personal freedoms, and our privacy.

Throughout human history, technology has played a role in the development of our species, and it has played a major role in how we prosecute wars. Even the making of fire was probably seen as a technological wonder in the early history of the human race — and also used as a weapon of war such as by setting fire to the enemy's fortifications, houses, and crops. It was also used to help forge tools as weapons of war.

A short look back at that history is appropriate for as some historian once said: "If you don't know where you've been, you don't know where you are going" — and one might add, "you don't even know where you are." And if you do not know where you are, your survivability in an information warfare environment is not good. Appendix C provides a short history of technology and its relationship to warfare over the years. The remainder of this chapter is devoted to high technology as a basis for information warfare tactics and weapons.

Technology drives change.

—Andrew Grove
CEO, Intel Corporation

Twentieth Century Technology and the Advent of High Technology

The use of technology during the agricultural and industrial time periods saw great numbers of new inventions and improvements in old technologies. This was also the time of the building of the great cities of the world, as well as their total destruction in global wars. Thus, technology for warfare had truly come of age. With the advent of the atomic and subsequent bombs, the entire world could now literally be destroyed. The period also saw great improvements in technology inventions and new inventions such as the telegraph, telephone, air transportation, and computers. This period saw increases in education, mass transportation, and exponential growth in communications — the sharing of information.

During this period, the sharing of information became easier due to the improvement of communications systems, new communications systems, and the increased consolidation of people into large cities. This also made it easier to educate the people, a needed skill for working in the more modern factories and offices of the period, and needed skills to develop, improve, and implement technologies.

The transition period from the Industrial Age to the Information Age in world history varies with each nation-state. In the United States, the well-known authors, the Tofflers, estimated the transition to take place about 1955 when the number of white-collar workers began to outnumber the blue-collar workers. Some nation-states are still in various phases of transition from the agricultural period to the industrial period to the information period.

However, no matter when a nation experiences this technology-driven transition, it will see, as the United States and other modern nation-states have seen, the most rapid changes in all aspects of human existence since humans first walked on this Earth — to include how wars are prosecuted.

The Twentieth Century (1900–1999) saw the rapid expansion and use of technology and high technology more so than all past centuries combined. It was also the beginning of the concentrated development of technology to specifically develop new and improved weapons of war on a massive scale. This ushered in the era of modern warfare, an era that was sponsored primarily by governments that had the will and the means for such development, and these governments were able to use these new technologically developed weapons to cause death and destruction on a massive scale.

Thus, the Twentieth Century was the true beginning of technology-based warfare. Due to the technological improvement of older inventions (e.g., submarine, machine gun) and new inventions such as the nuclear weapons, never before could so many be killed by so few. There were also the tanks, hand grenades, poison gases, land mines that gave way to the chemical/biological/nuclear weapons, carpet bombings, smart bombs, and the beginning of true information warfare.

> *In 1962…the CIA quietly contracted the Xerox Company to design a miniature camera, to be planted inside the photocopier at the Soviet Union's embassy in Washington. A team of four Xerox engineers… modified a home movie camera equipped with a special photocell that triggered the device whenever a copy was made. In 1963, the tiny Cold War weapon was installed by a Xerox technician during a regular maintenance visit to the Soviet embassy.[4]*

This period included many significant technology-driven inventions too numerous to mention here in their entirety. In the medical field alone, we have seen the rapid invention of literally thousands of new drugs, procedures, and devices, many of which saved possibly millions of lives of the warfighters over the years. Some other significant technologically driven inventions during this century that contributed to changes in warfighting include:

- Zeppelin
- Radio receiver
- Polygraph machine
- Airplane
- Gyrocompass
- Jet engine
- Synthetic rubber
- Solar cell
- Short-wave radio
- Wirephoto

The Twentieth Century also saw the development and improvement of more modern warfighting offensive and defensive devices, equipment, and weapons such as those noted below:

- Kevlar
- Hovercraft
- Atomic, hydrogen, and other sophisticated bombs
- Helicopter
- Liquid-fueled rockets
- Nuclear submarine
- Jeep
- Tommy gun
- Tanks
- Gas mask
- Vast array of weapons too numerous to list

This period also began the development of our modern era's amazing electronic inventions leading to the computer and its peripherals. Although not all-inclusive, these included the ones listed in Exhibit 1.

Other Significant Twentieth-Century Technological Developments and Events

Some of the other significant technological events and inventions that took place in the Twentieth Century, and have led to our rapidly changing information-based societies, information dependencies, and assisted in the development of new methods of prosecuting warfare, include:[5]

Exhibit 1.   The Twentieth Century's Technologies and Consequences of Technologies

Electronic amplifying tube (triode)	Photocopier
Radio tuner	Computer
Robot	Integrated circuit
Digital computer	BASIC language
UNIVAC I	FORTRAN
Sputnik	Compact disk
Explorer I satellite	Computer mouse
Laser	Computer with integrated circuits
OS/360 IBM operating system	RAM, ROM, EEPROM
Minicomputer	ARPANET
Optical fiber	Daisy wheel printer
Cray supercomputer	Floppy disk
Space shuttle	Dot-matrix printer
IBM PC	Liquid-crystal display (LCD)
Videotape recorder	Computer hard disk
Graphic user interface (GUI)	Modem
Cathode ray tube	Mobile phones
Television	Transistor
FM radio	World Wide Web
Voice recognition machine	Browsers

- *1930:* Shannon's doctorate thesis explains the use of electrical switching circuits in modern Boolean logic.
- *1934:* Computing-Tabulating-Recording becomes IBM.
- *1936:* Burack builds the first electric logic machine.
- *1940:* Atanasoff and Berry design a computer with vacuum tubes as switching units.
- *1943–1946:* Mauchley, Eckert, and Von Neumann build the ENIAC, the first all-electronic digital computer.
- *1947:* The transistor is perfected.
- *1955:* Shockley Semi-Conductor founded in Palo Alto, California; Bardeen, Shockley, and Brattain share the Nobel Prize for the transistor.
- *1957:* Fairchild Semi-Conductor founded.
- *1962:* Tandy Corporation buys chain of Radio Shack electronic stores.
- *1964:* Kemeny and Kurtz, Dartmouth College, develop the BASIC computer language.
- *1968:* Intel founded.
- *1969:* Intel produces integrated circuits for Japanese calculators; Data General releases Nova.

What Is High Technology?

We could not come up with a nice, clean definition of technology, but what about "high technology"? Someone once said it was "technology before it became high." That is not bad, but unfortunately it does not help in this instance. One of the better discussions on the topic came from a government Web site[6] and is quoted in Exhibit 2.

As we have found, trying to get a handle on this thing called technology, any kind of technology, is like grabbing air. Even low technology was once considered high technology in its day. For example, when the first plow was invented, it was probably considered a technological wonder. Then, after hooking it up to a horse or water buffalo, it increased the productivity of the farmers and it certainly drastically changed farming methods. When the wooden plow was integrated with a steel blade, certainly that was considered high technology in its day. One must remember that high technology today will undoubtedly be considered low technology 25 to 50 years from now. So, high technology is also based on a reference point and that reference point is time — perception and time are also key factors in information warfare.

As we see, it is not easy to come to grips with this phenomenon called high technology. For our purposes, a narrowly focused definition is better. In today's world, the microprocessor drives the technological products that drive the Information Age and information warfare. So, we will define high technology based on the microprocessor.

> *High technology is defined as that technology which includes a microprocessor.*

The Microprocessor

In 1971, Intel introduced the Intel 4004 microprocessor. This was the first microprocessor on a single chip and included the central processing unit (CPU), input and output controls, and memory. This made it possible to program "intelligence into inanimate objects," and was the true beginning of the technology revolution that has caused so many changes in the world and ushered in the beginnings of the age of information warfare.

The microprocessor was developed through a long line of amazing inventions and improvements on inventions. Without these dramatic and often what appear to be new, miraculous breakthroughs in microprocessor technology, today's information warfare phenomenon would still be only in the dreams of science fiction writers, the likes of Jules Verne and George Orwell. However, because of the amazing developments in the microprocessor, information warfare is beginning to come to the forefront in modern-day warfare.

Today, because of the microprocessor, its availability, miniaturization, power, and low cost, the world is rapidly developing new high-technology devices, procedures, processes, networks, and of course information warfare and conventional warfare weapons. The global information infrastructure (GII) is just one example of what microprocessors are making possible. GII is the massive international connections of world computers that carry business and personal communication as well as that of the social and government sectors of nations. Some say that GII will connect entire cultures, erase international

Exhibit 2.   High-Tech: A Product, a Process, or Both?

There is no universally accepted definition of "high-tech," nor is there a standard list of industries considered to be high-tech. Today nearly every industry contains some element of technology, and even the most technologically intensive industry will include low-tech elements.

Nevertheless, several groups have developed lists of industries they consider high-tech using U.S. Standard Industrial Classifications (SIC).

The breadth of these lists depends on two factors: (1) the goals of the organization and its customers and (2) whether the organization ascribes to the argument that only industries that produce technology can be considered high-tech or to the argument that industries that use advanced technology processes can also be categorized as high-tech.

Any industry-based definitions of high-tech will be imperfect, but none of the definitions discussed here should be considered incorrect. The important factor to consider is the perspective from which any list is derived.

Most high-tech industry classifications have common elements, yet may vary significantly in scope. Let's consider four classifications of high-tech industries developed by the following respected and often quoted organizations: the American Electronics Association (AEA), RFA (formerly Regional Financial Associates), One Source Information Services Inc. (formerly Corp Tech), and the U.S. Bureau of Labor Statistics (BLS).

The different missions of these four organizations influence how they define high-tech. AEA is a trade association made up of mostly electronics and information technology companies. Its members generally produce technology and ascribe to the limited definition of high-tech based only on the nature of an industry's product rather than its process. RFA is a national consulting firm. Its clients include builders and contractors, banks, insurance companies, financial services firms, and government. The industries with the greatest growth potential and those reflective of their clients' interests are included in RFA's list of high-tech industries. While both AEA and RFA have narrowly defined high-tech, One Source and BLS use broader definitions that include industries with both high-tech products and processes.

One Source gathers and sells corporate information on technology firms for use in sales and marketing. As it has built its database of firms, One Source has expanded its list of what should be considered a high-tech industry. BLS is a federal agency responsible for collecting and analyzing data on the national labor force. It has defined those industries with the highest concentration of technology-based occupations, such as scientists and engineers, as high-tech industries.

The Trade Association: AEA

AEA recently released Cyberstates 4.0, its annual report on technology employment, based on AEA's limited definition of high-tech industries, which fall into three categories: (1) computer, communication, and electrical equipment, (2) communication services, and (3) computer related services. AEA's list is the most restrictive of the four classifications. Absent from the list are areas such as drug manufacturing, robotics, and research and testing operations.

The Consulting Group: RFA

RFA's high-tech sectors are similar to those selected by AEA. However, RFA does not include household audio and video equipment or telephone communications, but adds drugs and research and testing services.

Exhibit 2.   High-Tech: A Product, a Process, or Both? (Continued)

Information Provider: One Source

Unlike the short lists compiled by AEA and RFA, the One Source list classifies 48 sectors as high-tech. Major additions include a number of manufacturing industries, such as metal products and transportation equipment, and several service industries.

The Research Group: BLS

BLS has further refined its high-tech industry definition by separating sectors into two groups. Those industries with a high concentration of research-oriented occupations are labeled intensive, while those with a lower concentration are considered non-intensive. The differences shown here illustrate why knowing how data are defined is essential to understanding what the data mean. Once again, those wishing for a simple answer will be frustrated. It is not the data that has failed them, but the reality of a complex system (the economy) and the human factor that must determine how to best reflect that system using data.

From "INCONTEXT: The Indiana Economy," http://www.incontext.indiana.edu/june 2000/.

---

borders, support "cyber economies," establish new markets and change our entire concept of international relations.

The GII is based on the Internet and much of the growth of the Internet is in developing nations such as Argentina, Iran, Peru, Egypt, Philippines, Russia, Malaysia, and Indonesia. The GII is not a formal project but it is the result of thousands of individuals', corporations', and governments' need to communicate and conduct business by the most efficient and effective means possible. The GII is also a battlefield in the information warfare arena. More on the GII in Chapter 3.

Moore's Law

No discussion of high technology and information warfare weapons would ever be complete without a short discussion of Moore's Law. In 1965, Gordon E. Moore, Director, Research and Development Laboratories, Fairchild Semiconductor, was asked by *Electronics* magazine to predict the future of semiconductors and its industry during the next ten years. In what became known as Moore's Law, he stated that the capacity or circuit density of semiconductors doubles every 18 months or quadruples every three years.[7]

The interesting thing about Moore's comments is that they became sort of a high-technology driver for the semiconductor industry and, even after all these years, it has been pretty much on track as to how semiconductors have improved over the years. Its power, of course, depends on how many transistors can be placed in how small a space. The mathematical version of Moore's Law is:

$$\text{Bits per square inch} = 2 \ (\text{time-1962})^{7a}$$

Some of the-high technology "inventions" of the Twentieth Century that depended on the microprocessor include:

- Ethernet (1973)
- Laser printer (1975)
- Ink-jet printer (1976)
- Magnetic resonance image (1977)
- VisiCalc (1978)
- Cellular phones (1979)
- Cray supercomputer (1979)
- MS-DOS (1981)
- IBM-PC (1981)
- Scanning tunneling microscope (1981)
- Apple Lisa (1983)
- CD-ROM (1984)
- Apple Macintosh (1984)
- Windows operating systems (1985)
- High-temperature superconductor (1986)
- Digital cellular phones (1988)
- Doppler radar (1988)
- World Wide Web/Internet protocol (HTTP); HTML (1990)
- Pentium processor (1993)
- Java computer language (1995)
- Digital versatile disc or digital video disc (DVD)(1995)
- Web TV (1996)

The Pioneer 10 spacecraft used the 4004 microprocessor. It was launched on March 2, 1972, and was the first space flight and microprocessor to enter the Asteroid Belt.[1]

Other Significant Twentieth Century High-Technology Developments and Events

Some of the significant high-technology computer events and inventions that took place in the Twentieth Century and led to our rapidly changing methods of prosecuting a war include:[8]

- *1971:* Intel develops the 8008; Wozniak and Fernandez build the "Cream Soda Computer."
- *1972:* Kildall writes PL/1, the first programming language for the Intel 4004 microprocessor; Gates and Allen form "Traf-O-Data"; Wozniak and Jobs begin selling Blue Boxes.
- *1973:* Wozniak joins HP; Kildall and Cooper build "astrology forecasting machine."
- *1974:* Intel invents the 8080; Xerox releases the Alto; Torode and Kildall begin selling microcomputers and disk operating systems.
- *1975:* Microsoft (previously known as "Traf-O-Data") writes BASIC for the Altair; Heiser opens the first computer store in Los Angeles.
- *1976:* Kildall funds Digital Research; work on the first Radio Shack micro-computer started by Leininger and French; first sale of the CPM operating system takes place.

- *1977:* Apple introduces the Apple II. TRS-80 developed.
- *1978:* Apple ships disk drives for the Apple II and begins development of the Lisa computer.
- *1980:* HP releases the HP-85; Apple III is announced; Microsoft and IBM sign an agreement for IBM's PC operating system.
- *1981:* Osborne I developed; Xerox comes out with the 8010 Star and the 820 computers; IBM presents the PC.
- *1982:* Apple Lisa is introduced; DEC develops a lines of personal computers (e.g., DEC Rainbow 100).
- *1983:* IBM develops the IBM PC Jr.; Osborne files for Chapter 11 as the microcomputer market heats up.
- *1984:* Apple announces the Macintosh microcomputer.
- *1986:* Intel develops the 8086 chip.
- *1987:* Intel develops the 8088 chip.
- *1990s:* Intel, already the leader in microprocessors, announces the 286, 386, and 486 chips, followed rapidly by the Pentium chips now reaching speeds of 1.7 GHz as we enter the Twenty-first Century.

Moore's Law is still holding true although some believe we will soon hit the silicon wall, based on the laws of physics. Some of these doomsayers have been saying such things for years. Others are more optimistic and believe that other material will be found to replace silicon or that silicon will be somehow enhanced to "defy" the laws of physics. If the past is any clue to the future, the future of high technology will not be impaired by such minor impediments such as the laws of physics.

According to the U.S. General Accounting Office in a report to Congress,[9] the rapid developments of the telecommunications infrastructure in the United States have resulted in the creation of three separate and frequently incompatible communications networks, including:

- Wire-based voice and data telephone networks
- Cable-based video networks
- Wireless voice, data, and video networks

In the future, this problem will become a non-issue as integration and commonality, forced by business and government needs for total information compatibility, will take place. It is already happening in many areas (e.g., cellular phones and notebooks computers, television, and Internet access).

The Internet

The real issue is control. The Internet is too widespread to be easily dominated by any single government. By creating a seamless global-economic zone, anti-sovereign and unregulatable, the Internet calls into question the very idea of the nation-state.[10]

—John Perry Barlow

It is in the context of this phenomenal growth of high technology and human knowledge that the Internet arises as one of the mechanisms to facilitate sharing of information and as a medium that encourages global communications. The Internet has already become one of the Twenty-first Century's information warfare battlefields.

The global collection of networks that evolved in the late Twentieth Century to become the Internet represent what could be described as a "global nervous system" transmitting from anywhere to anywhere facts, opinions, and opportunity. However, when most people think of the Internet, it seems to be something either vaguely sinister or of such complexity that it makes it difficult to understand. Popular culture, as manifested by Hollywood and network television programs, does little to dispel this impression of danger and out-of-control complexity.

The Internet arose out of projects sponsored by the Advanced Research Project Agency (ARPA) in the United States in the 1960s. It is perhaps one of the most exciting legacy developments of that era. Originally an effort to facilitate sharing of expensive computer resources and enhance military communications, it has over the last 14 years, from about 1988 until 2002, rapidly evolved from its scientific and military roots into one of the premier commercial communications media. The Internet, which is described as a global meta-network, or network of networks,[11] provides the foundation on which the global information superhighway is being built.

However, it was not until the early 1990s that Internet communication technologies became easily accessible to the average person. Prior to that time, Internet accesses required mastery of many arcane and difficult-to-remember programming language codes. However, the combination of declining microcomputer prices and enhanced microcomputer performance, and the advent of easy-to-use browser[12] software were key enabling technologies that created the foundation for mass Internet activity. When these variables aligned with the developing global telecommunications infrastructure, they allowed a rare convergence of capability.

- *E-mail.* Although e-mail was invented in 1972, it was not until the advent of the "modern Internet system" that it really began to be used on a global scale. In 1987, there were approximately 10,000 Internet computer hosts and 1000 news messages a day in 300 newsgroups. In 1992, there were more than 1,000,000 hosts and 10,000 news messages a day in 1000 newsgroups. By 1995, the number of Internet hosts had risen to more than 10 million, with 250,000 news messages a day in over 10,000 newsgroups.[13]
- *Internet protocols.* In the 1970s, the Internet protocols were developed to be used to transfer information.
- *Usenet newsgroup and electronic mail.* Newsgroups and electronic mail were developed in the 1980s.
- *Gopher.* In 1991, personnel at the University of Minnesota created the Gopher as a user-friendly interface that was a menu system for accessing files.

- *World Wide Web.* In 1991, Tim Berners-Lee and others at the Conseil Européene pour la Recherche Nucleaire (CERN) developed the Web. In 1993, the Web had approximately 130 sites; in 1994, about 3000 sites; and in April 1998, more than 2.2 million.[14] But that was just the beginning!

The most commonly accessed application on the Internet is the World Wide Web (Web, WWW). Originally developed in Switzerland, the Web was envisioned by its inventor as a way to help share information. The ability to find information concerning virtually any topic via search engines, such as Alta Vista, HotBot, Lycos, InfoSeek, and others, from among the rapidly growing array of Web servers is an amazing example of how the Internet increases the information available to nearly everyone. One gains some sense of how fast and pervasive the Internet has become as more TV, radio, and print advertisements direct prospective customers to visit their business or government agency Web sites. Such sites are typically named www.companyname.com where the business is named "company" or www.government-agency.gov for government agencies.

From the past century until now, the Internet has rapidly grown from an experimental research project and tool of the U.S. government and universities to the tool of everyone in the world with a computer. It is the premier global communications medium. With the subsequent development of search engines and, of course, the Web, the sharing of information has never been easier. Sites such as Google.com state that they search 1,346,966,000 Web pages!

It has now become a simple matter for average people — even those who had trouble programming their VCRs — to obtain access to the global Internet, and with the access search the huge volume of information it contains. Millions of people around the world are logging in, creating a vast environment often referred to as cyberspace and the global information infrastructure (GII), which has been described as the virtual, online, computer-enabled environment, and distinct from the physical reality of "real life."

By the end of the Twentieth Century, worldwide revenues via Internet commerce had reached perhaps hundreds of billions of dollars, an unparalleled growth rate for a technology that has only been really effective since the early 1990s! The "electronic commerce" of the early Twenty-first Century is expected to include everything from online information concerning products, purchases, services, to the development of entirely new business activities (e.g., Internet-enabled banking).

An important fact for everyone to understand, and which is of supreme importance to those interested in information warfare, is that the Web is truly global in scope. Physical borders as well as geographical distance are almost meaningless in cyberspace; the distant target is as easily attacked as the local one.

The annihilation of time and space makes the Internet an almost perfect environment for information warfare. When finding an adversary's[15] desired server located on the other side of the planet is as easy and convenient as calling directory assistance to find a local telephone number, information warriors have the potential to act in ways that one can only begin to imagine.

Undeterred by distance, borders, time, or season, the potential bonanza awaiting the information warrior is a chilling prospect for those who are responsible for safeguarding and defending the assets of a business or government agency.

Because of religious beliefs, Internet access to material considered pornographic is not acceptable. One of society's struggles will be how to provide access to the world's information without causing some moral decay of society? This will be a struggle for many countries and it is believed that the information warriors will have a major impact on the society of such developing countries.

The Internet is the latest in a series of technological advances that is being used not only by honest people to further their communication, but also by miscreants, juvenile delinquents, and others for illegal purposes. As with any technological invention, they can be used for good or for illegal purposes. They are really no different than other inventions such as the handgun. The handgun can be used to defend and protect lives or to destroy them. It all depends on the human being that is using the technology.

The High-Technology-Driven Phenomenon: Internet Service Providers

Using an Internet search engine and searching for "Internet service providers (ISPs)," 1,330,000 "hits" were identified in a search that took, according to Google, 0.20 seconds! Through a process of elimination, clicking on Google's Web Directory afforded 16 ISP categories. Clicking on the Business category afforded 49 categories, one of which was Internet, which had 11,930 hits. Clicking on Access Providers afforded 721 hits. That led to AOL (46); By Region (343); Cable (30); CompuServe (6); Cooperatives (6); Directories (17); DSL (121); Free Internet Access (18); Resources for ISPs (56); Reviews (41); UNIX Shell Providers (72); and Wireless (41). Clicking on By Region (343) afforded Africa (8); Asia (9); Caribbean (1); Central America (5); Europe (41); Middle East (3); North America (263); Oceania (2); and South America (2).

In other words, there are thousands of ISPs operating and connected all across the globe. We all know by now (hopefully) that our e-mails do not go point-to-point, but hop around the Internet. They are susceptible to be gleaned by all those with the resources to read other people's mail or steal information to commit crimes (e.g., identity theft, competitive intelligence information collections, and, of course, useful information for information warriors).

So, what is the point? The point is that there are ISPs all over the world with few regulations and absolutely no protection and defensive standards. Some ISPs may do an admirable job of protecting our information passing through their systems, while others may do nothing. Furthermore, as we learn more and more about netspionage (computer-enabled business and government spying), we learn more and more about how our privacy and our information is open to others to read, capture, change, and otherwise misuse.

In addition, with such programs as SORM in Russia, Internet monitoring in China and elsewhere, global Echelon, and the U.S. FBI's Carnivore (still Carnivore no matter how often they change the name to make it more politically correct), we might as well take our most personal information, tatoo it on our bodies and run naked in the streets for all to see. Well, that may be a slight exaggeration; the point is that we have no concept of how well ISPs are protecting government, business, individual, or information of associations. Through your ISP, how susceptible are you to the threats of information warfare? Do you know if your ISP is protecting or monitoring you? If it is monitoring you, for whom?

Faster and More Massive High-Technology-Driven Communications

We are quickly expanding into the world of instant messages (IMs) through ISPs. After all, the more rapidly our world changes, the more rapidly we want to react and we want everything — now! IMICI.com's Web site stated that it expects there to be more than 200 million users sending two trillion messages per year by 2004, and that IM is the fastest growing Internet technology. Furthermore, it can be used to transfer files, send graphics, and unlike the telephone and normal e-mails, with IM one knows whether or not the person being contacted is there. Interesting ramifications — check to see if a person is online, if not (after already setting up a masquerade or spoof), take over that person's identity and contact someone posing as the other — instantly. Of course there are perhaps hundreds, if not thousands, of examples of ISPs being penetrated or misused. As far back as approximately November 1995, for example, the *Wall Street Journal* ran a story entitled "America Online To Warn Users About Bad E-Mail." We all know about the basic issues of viruses and other malicious codes being sent via ISPs. So, the problem has existed for quite some time.

> ### Solar Storms Could Affect Telecommunications
> *Intense storms raging on the sun...could briefly disrupt telecommunications.... The eruptions triggered a powerful, but brief, blackout Friday on some high-frequency radio channels and low-frequency navigational signals...forecast at least a 30 percent chance of continuing disruptions.... In addition to radio disruptions, the charged particles can bombard satellites and orbiting spacecraft and, in rare cases, damage industrial equipment on the ground, including power generators and pipelines.*[16]

High technology is vulnerable to nature and the universe in general. What a great time to launch an information warfare attack on an adversary. Is it sunspots or an adversary causing these outages? By the time the adversary finds out it is you and not three days of sunspots, the war could be over.

The Beneficial Effect of Hacker Tools and Other Malicious Software on Network Security with Dual Roles as Information Warfare Tools

The following malicious software was selected as a representative sample of those that are available and have been selected for their range of functionality and additionally for their range through time from 1991 to present, and which can be and are being adapted and adopted for use in information warfare.[17]

- *Hacker tools*. Of the hacker tools that were reviewed, while the intentions of the originators of the tools were mixed, with some being malicious and some well intentioned, they can all be used to strengthen the security of a network or to monitor the system for illicit activity. This can be achieved if the system owner uses hacker tools to identify the weaknesses that exist in the security of the system, to identify appropriate remedial action, before a person with malicious intent attempts to exploit the weaknesses. A number of the tools can also be used to monitor the system for illicit activity, even before software patches are available, so that the system owner can make informed decisions on appropriate action to prevent or minimize damage to their system. As an information warrior, how will you use such tools to attack an adversary?

- *Viruses*. Viruses have no direct beneficial effect on the security of a system except to provide a visible indication that there has been a breakdown in procedures for the transfer of software or data between systems. The negative effect of viruses is the cost in terms of time and the anti-virus software to check data and software being imported or exported to and from the system, as well as the cost of rectifying a problem when an infection has occurred, which can be considerable. In an abstract way, the advent of the virus has actually been beneficial to the computer security manager because the impact of a virus on the user is a visible and constant reminder of the need to observe good computer security practices. In the majority of cases, the virus is detected before it can activate its payload, so the damage is normally limited to the inconvenience and cost of the cleaning up the system to remove the virus. As an information warfare weapon, it is a valuable and cheap weapon that can cause devastating results against the unprepared information systems of an adversary.

- *Worms*. The release onto the Internet on November 2, 1988, of the Internet Worm, written by Robert T. Morris, Jr., quickly caused widespread disruption and the failure of a large proportion of the network that existed at that time. The problem was compounded by the fact that some of the servers that had not been affected were taken offline to prevent them from becoming infected, thus placing a higher load on already-affected sections of the system and denying those elements of the network that had gone offline access to the patches that would protect them, as the normal distribution method for patches was over the Internet itself. To date, there have been no security benefits derived from worms, other than, in the case of the Robert T. Morris worm, to highlight the urgent need for effective and early communication of information on incidents. The potential for the use of this type of program in a way that would aid the security of systems has been postulated, in the form of autonomous intelligent agents that would travel through the system and report back predefined information, such as the system assets, the condition and identity of system elements, and the presence or absence of specific types of activity. As a weapon for prosecuting information warfare, worms have excellent potential and may even be considered a "weapon of mass destruction" because of the damage they can cause a high-technology, information systems-dependent adversary. Of course, we now have many "colored" worms being written and traveling around the GII, NIIs, and other networks.

- *Easter eggs.* Easter eggs have no beneficial effect other than to highlight that even proprietary software can have large sections of code included in them which are redundant to the functionality for which they were intended and also that the quality control procedures for the production of software by well-known organizations is poor if the Easter eggs were not detected during production. Can you think of any way to use the these "eggs" in an information warfare battle?

- *Trojan horses.* The Trojan horse, by definition, carries out actions that are normally hidden from the user while disguising their presence as a benign item of software. They are difficult to detect because they appear to be a legitimate element of the operating system or application that would normally be found on the system. Given that the purpose of a Trojan horse is to hide itself and its functionality from legitimate users, there have been no beneficial effects derived from them — unless you are an information warrior.

- *Logic bombs.* Logic bombs, as with Trojan horses, carry out actions that are unexpected and undesirable. Some may cause relatively minor damage, such as writing a message to a screen, while others are considerably more destructive. They are normally inserted by disaffected staff or by people with a grudge against the organization. Again, they are difficult to detect before they have been activated and, as a result, can be expensive to rectify. Logic bombs are correctly named as they can have the same effect against the system of an adversary as a physical bomb might have against a building — Boom! It is gone!

The clear implication from the issues discussed above is that some hacker tools can have a beneficial effect on the security of computer systems if they are used by the system staff before they are used by personnel either within the organization or outside it to identify shortcomings or flaws in the operating system or applications software, the configuration of the system, or the procedures used to secure it. Viruses, while providing no direct benefit, do provide a detectable indication that there has been a breach in the security of the system, either by an exploitation of a flaw in the security procedures or by a shortcoming in the system software (it allowed a virus through any barriers that had been created to prevent access to the system). Worms currently have no beneficial effect on the system security management. However, the concept that was used to disseminate the Robert T. Morris worm may have an application in the mapping of large networks if applied to autonomous agents. The Trojan horse and the logic bomb, which, by their very nature, are covertly inserted into the system without the owner's knowledge, have no beneficial effect and have only malicious applications. However, as noted above, for purposes of prosecuting information warfare, these have some excellent potential and are, in fact, being refined and improved on for the information warfare arsenals of individuals, corporations, and nation-states.

Other High-Technology Tools in Information Warfare

Information warfare through high technology is being fought on many fronts — on the personal privacy, corporate netspionage,[18] and nation-state battlefields of the world. Even such innocent-sounding words as "cookies" take on new meaning in the information warfare arena.

These cookies — the computer kind, not the ones you eat — are beneficial, except when they are used to profile customer habits and gather an individual's private information, which is then sold. High-technology cookies are files that a Web site can load onto a user's system. They are used to send back to the Web site a user's activity on that Web site, as well as what Web sites the user has previously visited. They are also a potential tool of the information warrior.

Intel's Pentium III included a unique processor serial number (PSN) in every one of its new Pentium III chips. Intel claimed that the PSN could identify an individual's surfing through electronic commerce and other Net-based applications. It was noted that by providing a unique PSN that can be read by Web sites and other application programs, it could make an excellent information warfare tool. Although this number is designed to be used to link user activities on the Internet for marketing and other purposes, one can easily imagine other uses, from an information warfare perspective, that can be made of this high-technology application. And as for Microsoft's new operating system, XP, imagine the IW possibilities.

Steganography is another use of high technology that can be used in information warfare:[19]

> *Hiding information by embedding a file inside another, seemingly inno-*
> *cent file is a technique known as "steganography." It is most often used*
> *with graphics, sound, text, HTML, and PDF files. Steganography with*
> *digital files works by replacing the unused bytes of data in a computer*
> *file with bytes that contain concealed information.*

Steganography (which translated from Greek means covered writing) has been in use since about 580 B.C. One technique was to carve secret messages into wooden objects and then cover the etched words with colored wax to make them undetectable to an uninitiated observer. Another method was to tattoo a message onto the shaved messenger's head. Once the hair grew back, the messenger was sent on his mission. Upon arrival, the head was shaved, thus revealing the message — obviously not time-dependent. The microdot, which reduced a page of text to the size of a typewriter's period so that it could be glued onto a postcard or letter and sent through the mail, is another example.[20]

Two types of files are typically used when embedding data into an image. The innocent image that holds the hidden information is a "container." A "message" is the information to be hidden. A message may be plaintext, ciphertext, other images, or anything that can be embedded in the least significant bits (LSB) of an image.[21]

Steganographic software has some unique advantages as a tool for netspionage agents. First, if the agents use regular cryptographic software on their computer systems, the files may not be accessible to investigators but will be visible, and it will be obvious that the agents are hiding something. Steganographic software allows agents to "hide in plain sight" any valuable digital assets they may have obtained until they can transmit or transfer the files to a safe location or to their customer. As a second advantage, steganography can be used to conceal and transfer an encrypted document containing the acquired information to a digital dead drop. The agents could then provide the handler or customer with the password to unload the dead drop but not divulge the steganographic extraction phrase until payment is received or the agents are safely outside the target corporation. As a final note, even when a file is known or suspected to contain information protected with steganographic software, it has been almost impossible to extract the information unless the passphrase has been obtained.

Summary

If you are involved in any activity in which technology is used as a tool to help you accomplish your work, you are aware of the tremendous and very rapid advances that are being made in that arena. It is something to behold.

We are in the middle of the most rapid technological advances in human history, but this is just the beginning. We are not even close to reaching the potential that technology has to offer, nor its impact on all of us — both good and bad.

It is said that there have been more discoveries in the past 50 years than in the entire history of mankind before that time. We have just to read the newspapers and the trade journals to look at every profession and see what technology is bringing to our world. There are new discoveries in medicine, online and worldwide information systems, the ability to hold teleconferences across the country and around the globe, and hundreds of other examples that we can all think of.

High technology is the mainstay of both our businesses and government agencies. We can no longer function in business or government without them. Pagers, cellular phones, e-mail, credit cards, teleconferences, smart cards, notebook computers, networks, and private branch exchanges (PBXs) are all computer based and all are now common tools for individuals, businesses, and public and government agencies. Information warriors are also relying more and more on computers. As computers become more sophisticated, so do the information warriors. As international networks increase, so does the number of international information warriors.

Networking and embedded systems, those integrated into other devices (e.g., automobiles, microwave ovens, medical equipment), are increasing and drastically changing how we live, work, and play. According to a study financed by the United States Advanced Research Projects Agency (ARPA) and published in the book *Computers at Risk*:

- Computers have become so integrated into the business environments that computer-related risks cannot be separated from normal business risks, or those of government and other public agencies.
- Increased trust in computers for safety-critical applications (e.g., medical) leads to the increased likelihood that attacks or accidents can cause deaths (*Note:* It has already happened).
- Use and abuse of computers is widespread with increased threats of viruses, credit card, PBX, cellular phones, and other frauds.
- Unstable international political environment raises concerns about government or terrorist attacks on information and high technology-dependent nations' computer and telecommunications systems.
- Individual privacy is at risk due to large, vulnerable databases containing personal information, thus facilitating increases in identity theft and other frauds.

If I want to wreak havoc on a society that, in some cases, has become complacent, I am going to attack your quality of life.

—Curt Weldon, R-PA.
U.S. House, Armed Services Committee[22]

Personal computers have changed our lives dramatically and there is no end in sight. High technology in general has improved the quality of life for societies, made life a little easier, and yet makes an information-dependent way of life more at risk than ever before. The use of modems is now commonplace with all newly purchased microcomputer systems[23] coming with an internal modem already installed and ready for global access through the Internet or other networks. Therefore, home computers and long-distance telephone networks potentially represent some of the most serious and complex crime scenes of the Information Age. This will surely increase as we begin the Twenty-first Century.

...it is computerized information, not manpower or mass production that...will win wars in a world wired for 500 TV channels. The computerized information exists in cyberspace — the new dimension created by endless reproduction of computer networks, satellites, modems, databases, and the public Internet.[24]

—Neil Munro

High technology development continues to play a dual role in information-based nation-states. The high-technology devices have been turned into tools that have been used to determine the adequacy of information systems' security and defenses, have been adopted and adapted by global hackers and other miscreants. They now have been using those tools for probing and attacking systems, especially through the Internet interfaces of corporations, nation-states, as well as the GII and NIIs of nation-states. These same hacker techniques have been readily adopted and enhanced by the information warriors of nation-states and others.

Notes

1. It is not the intent of the authors to provide a detailed history of technology, especially high technology. The intent is to provide a brief, not all-inclusive historic overview. Readers interested in a detailed account of technology history can find numerous Web sites and books that can provide that detail. However, this overview is provided because it is important for those interested in information warfare to get some idea as to how we got to our current state of information warfare through technology, and learning from the past helps us look at where information warfare will be heading in the future.
2. *Encarta World English Dictionary,* 1999, Microsoft Corporation. All rights reserved. Developed for Microsoft by Bloomsbury Publishing Plc.
3. http://www.pax.co.uk/ttdefine.htm.
4. From an article by Dawn Stover in the January 1996 issue of *Popular Science,* entitled The CIA's Xerox Spy-cam.
5. See P. Freiberger and M. Swaine's book, *Fire in the Valley, The Making of the Personal Computer,* Osborne/McGraw, Berkeley, 1984.
6. http://www.ibrc.indiana.edu/incontext/june2000/.
7. The Origin, Nature and Implications of "Moore's Law," a paper written by Bob Schaller, September 26, 1996.
7a. http://www.tuxedo.org/~esa/jargon.html/moore's-law.html.
8. See P. Freiberger and M. Swaine's book, *Fire in the Valley, The Making of the Personal Computer,* Osborne/McGraw, Berkeley, CA, 1984, and http://www.island-net.com for additional details of computer history.
9. Information Superhighway: An Overview of Technology Challenges, GAO-AIMD 95-23, p. 12.
10. John Perry Barlow, Thinking Locally, Acting Globally, *Time,* January, 1996, p.57; as quoted on p. 197, *The Sovereign Individual,* by James Dale Davidson and Lord William Rees-Mogg, published by Touchstone, New York, 1999.
11. Ibid., p. 11.
12. Software that simplifies the search and display of World Wide Web-supplied information.
13. *Internet Guide* by Microsoft Personal Computing at http://www.microsoftt.com/magazine/guides/internet/history.htm.
14. *Internet Guide* by Microsoft Personal Computing at http://www.microsoftt.com/magazine/guides/internet/history.htm.
15. The term "adversary" is used more often these days to describe an enemy than the word "enemy" because it seems it is not as harsh a term, although the intent is still to disable or kill them.

16. CNN.com/Space, March 31, 2001 (http://www.cnn.com/2001/TECH/space/03/31/solar.flare.ap/index.html).

17. A number of other tools were reviewed but contain no obvious property or functionality that was considered to be both beneficial and a potential information warfare weapons; that is, they modified the system to exploit vulnerabilities or they were purely malicious and caused a denial-of-service. These are tools that are "pure" information warfare tools.

18. See the book, *Netspionage: The Global Threat to Information,* published by Butterworth-Heinemann in September 2000.

19. Excerpt taken from the book, *Netspionage: The Global Threat to Information,* published by Butterworth-Heinemann in September 2000, and reprinted with permission.

20. http://webopedia.internet.com/TERM/s/steganography.html.

21. http://www.jjtc.com/steganography/.

22. Speaking at an InfoWar Conference in Washington, D.C., in September 1999.

23. Microcomputers had been a term used to differentiate them from minicomputers and mainframe computers. The computers' power and what the manufacturer decided to call them differentiated these systems. However, with the power of today's microcomputer equaling that of larger systems, the issue is unclear, and basically no longer very relevant. What these systems are called, coupled with notebooks, PDAs, workstations, desktops, etc., are not that important because they all basically operate the same way.

24. Neil Munro, The Pentagon's New Nightmare: An Electronic Pearl Harbor, *Washington Post,* July 16, 1995, p. C3.

Chapter 3

The Global and Nation-State Information Infrastructures: Backbone or Achilles' Heel?

Power always has to be kept in check; power exercised in secret, especially under the cloak of national security, is doubly dangerous.

—William Proxmire

This chapter reviews the global information infrastructure (GII), its history, and its impact on government agencies and businesses vis-à-vis industrial and economic espionage, techno-terrorism, and military information warfare (IW). Individual elements comprising the critical portion of the integrated information infrastructure (III) (i.e., telecommunications; electrical power systems; gas and oil production, storage, and transportation; banking and finance; transportation; water supply systems; emergency systems; and continuity of government services) are addressed. Discussions include an overview of what governments, businesses, and others are doing to protect, or take advantage of, our vulnerabilities. Also discussed are the related information infrastructures, (i.e., NII, DII, MEII, III, MEDII, and RII).

What Is the Global Information Infrastructure?

The term global information infrastructure (GII) conjures up an image of a grand edifice that has been considered and constructed to meet the needs of the world and humanity. If we look at the individual words, "global" means all encompassing or worldwide, "information" is the all-encompassing word for data that has meaning, and "infrastructure" means the underlying foundation or basic framework of a system or organization — all of which would tend to support the image. However, if we look at what is actually meant when people use the term, what they are really trying to highlight are the systems and networks that have become increasingly essential to the well-being and development of all aspects of life. We have now all come to accept this connected world in which we live where we can instantly communicate with anyone, anywhere in the world, at will and at an affordable cost. We have access to more information than we could ever have imagined in the past, and our lives are enriched and improved by the technologies being developed and used.

The realization that there might be such an entity is relatively new and has only been spoken of for about a decade, but there had been an understanding prior to this period that there was an increasing dependence by society on things that we no longer fully understood. To put this in perspective, 30 or 40 years ago, the electricity you used was provided by the local or regional power company. The water you used was also provided from local resources, and the telephone operator who connected your long-distance calls was probably someone that you knew if you lived in a rural environment. If the electricity supply failed, it was a local issue; and despite the fact that it might be inconvenient and cause local problems, the source of the failure was certainly within a well-defined and understood scope (the boundary of the system). As these systems became more sophisticated, we started to replace the people who controlled these systems with computers, which were cheaper and more efficient. Then we started to get really clever and discovered a way to be able to trade our surplus capacity with our neighbors and we began to connect our systems with their systems to allow the two-way flow of information and power. All of this was essential to keep pace, economically, with the growing demands of the populace and to remain competitive by being able to meet the fluctuating demands without the need to increase production capacity to meet the peaks of demand. The more aware people put security measures in place to protect their systems and had to trust that the people they conducted business with had done the same. What has in fact happened as a result of this is that we now have an infrastructure that underpins businesses, governments, and global trade that we do not fully understand.

The paper entitled "Living in the Information Society," states that:

> *The world economy of the 21st century will not be possible without integration in the GII. For developing countries, projects funded by MDBs (multilateral development banks) can make that integration a reality. For example, farmers can connect in real-time with agricultural experts at the U.S. Department of Agriculture, or anywhere around the world.*[1]

This paper provides insight into one of the potential benefits of the GII that span national boundaries and the government/industry divide. It shows the possibility for the improvement of life in the developing world being delivered in a cost-effective way from an organization that has the required knowledge in one country, to the farmers who need it in another country, while being funded by the development banks in possibly multiple countries.

Some Definitions of Common Terms

As might be expected with something as vast and all encompassing as the global information infrastructure, there are numerous terms that relate to its elements and constituent parts that are in common use and, unfortunately, misuse. The terms in most common usage include NII, III, DII, MEII, MEDII, and RII; and to understand the perspective in which they are commonly used, they are defined below.

GII — Global Information Infrastructure

For every country, august body, and organization, there is a definition of what constitutes the GII. While they vary to cater to the cultural and environmental issues, they all identify or allude to a number of common factors. The elements of the GII most commonly referred to are:

- The convergence of technologies
- Communications
- Information technology
- Information
- People

Several different definitions of the GII, providing insight into the differing perspectives of the infrastructure of government, industry, and international bodies, some of which date back to 1994, are given below.

The former U.S. Vice President, Al Gore, at the First World Telecommunications Development Conference in Buenos Aires in 1994, gave the United States' vision for the GII with the words:

> *Let us build a global community in which the people of neighboring countries view each other not as potential enemies, but as potential partners, as members of the same family in the vast, increasingly interconnected human family.*[2]

This very early and generalist description captures, in part, the essence of the potential of what the nascent GII should be. It identifies the need for both connectivity and international collaboration.

The U.S. National Institute of Science and Technology (NIST) describe the GII as:

> *Governments around the globe have come to recognize that the telecommunications, information services, and information technology sectors are not only dynamic growth sectors themselves, but are also engines for the development and economic growth through the economy.... The United States is but one of many countries currently pursuing national initiatives to capture the promise of the Information Revolution. Our initiative shares with others an important, common objective: to ensure*

> *that the full potential benefits of advances in information and telecom-*
> *munications technologies are realized for all citizens.*
>
> *The GII is an outgrowth of that perspective, a vehicle for expanding*
> *the scope of these benefits on a global scale. By interconnecting local,*
> *national, regional, and global networks, the GII can increase economic*
> *growth, create jobs, and improve infrastructures. Taken as a whole, this*
> *worldwide "network of networks" will create a global information mar-*
> *ketplace, encouraging broad-based social discourse within and amongst*
> *all countries.[3]*

This definition by the NIST is probably one of the most comprehensive and all-encompassing and captures not only the technological aspects, but also the environmental changes that will be required and the potential benefits that humanity should gain.

The GII is described by the International Standards Organization (ISO) as an infrastructure that is capable of supporting the development or the implementation and interoperability of information services and applications that currently exist, or may do so in the future, through information technology, telecommunications, consumer electronics, and also the content provision industries.

It identifies that "The infrastructure consists of interactive, broadcast, and other multimedia delivery mechanisms coupled with capabilities for individuals to securely share, use and manage information, anytime and anywhere, at acceptable cost, quality, level of security, and privacy protection"[4] This ISO definition deals primarily with the technology convergence aspects of the GII but, as it deals with standards, does not address the socio-political aspects.

The Computer Systems Policy Project (CSPP), which represents a significant part of the IT industry, describes the GII as *"more than a network of networks,"* and explains that the GII is a worldwide collection of connected systems that integrate five components described as essential and identified as:[5]

- Communications networks, which would include fixed telephone networks and cellular, cable, and satellite networks
- Equipments used for the provision of information, which will include computers, televisions, radios, and telephones
- Information resources, which will include educational, medical, entertainment, and commercial programs, and databases
- Applications, such as those used for telemedicine, electronic commerce, and digital libraries
- People

Strangely for an organization such as the CSPP, while the main thrust of the definition refers to the individual technologies and their convergence, there is also recognition of the importance of people in the GII.

It is noteworthy that while the technologies have developed tremendously during the period from which these definitions have been taken, the overarching vision that they portray has remained remarkably consistent.

The GII gives the potential for huge benefit and future development and in doing so also gives an attacker, potentially, all of the facilities, applications, and access he or she will need to conduct an attack at the very root of the infrastructure on which we now rely so heavily.

NII — National Information Infrastructure

The national information infrastructure (NII) is the set of information systems and networks on which a nation depends to function. The NII consists of more than the physical structures and elements used to transmit, store, process, or display data, images, or voice. As technologies develop, the NII encompasses an ever-increasing range of devices such as cameras, flatbed scanners, bar-code readers, magnetic strip readers, compact disks, DVDs, computers, video tapes, wireless local area networks (LANs), infrared communications, cable networks, satellite systems, television and set-top Internet devices, digital video recorders, microwave links, mobile telephones and Wireless Application Protocol (WAP) devices, short message system (SMS), and an endlessly increasing set of developing technologies.

To date, at least 53 countries, including the United States, Canada, most European countries, a number of African and South American countries, and a large proportion of the countries around the Pacific Rim, have identified that they have NIIs. All have started to study how they can maximize the benefits of their infrastructures, while at the same time making them safe, secure, and reliable.

A 1996 U.S. Department of Defense (DoD) report[6] identified that the United States' NII was vulnerable, in that it had not been designed for resiliency and to facilitate repair and that it was unable to fully depend on the public switched network.

It is notable that the DoD has identified that one of the problems with the U.S. NII is that it was not "designed" to achieve one of the major requirements of a system — availability — in that it cannot guarantee that there will be accessibility when required.

III — Integrated Information Infrastructure

In a paper by the International Telecommunications Union, which is based in Geneva, Switzerland, it was identified that there was no single existing definition for the nascent III.[7] The paper postulated three alternative definitions for an III; but unsurprisingly, given the source of the definitions, they are all concerned with networks and the media. The first of the offered definitions is for a high-performance computer network. This infrastructure would facilitate high-speed data access and retrieval. The definition postulates that the Internet might be a precursor to the III.

The paper continues along this line of argument with the proposal that if the Internet could successfully be extended from the academic and scientific research communities and then supported to satisfy a broader commercial marketplace while maintaining the openness and innovation that had been fundamental to the success of the Internet, then the Internet could form the basis for future network development.

The second model put forward was that of a multimedia network for which the primary use would be conveying data streams that might consist of data, text, video, or voice. In this definition, many of the potential applications were considered to likely arise from the entertainment, education, health care, and business markets.

The third potential definition put forward is as a medium for interactive television, in which a smart television set or equipment, rather than a PC or a videophone, is the main communications channel. The different technological solutions envisaged were considered to be of help to consumers in coping with the plethora of new television channels, video-on-demand, home shopping, and other services. Teenagers playing video games could be using the network alongside multinational corporations holding videoconferences.[7]

Because this definition is from the telecommunications sector and deals with the concept of an integrated infrastructure, it is not surprising that this definition deals with the technologies and their convergence.

DII — Defense Information Infrastructure

It is perhaps most appropriate to use the U.S. Department of Defense's (U.S. DoD's) own definition of the DII to explain the concept that has been accepted in most countries around the world:[8]

> *The DII is a seamless Web of communications networks, computers, software, databases, applications, data, and other capabilities that meets the information processing and transport needs of DoD users in peace and in all crises, conflict, humanitarian support, and wartime roles. It includes:*

- The *physical facilities* used to collect, distribute, store, process, and display voice, data, and imagery;
- The *applications and data engineering practices* (tools, methods, and processes) to build and maintain the software that allow C2 (command and control), Intelligence, and Mission Support users to access and manipulate, organize, and digest proliferating quantities of information;
- The *standards and protocols* that facilitate interconnection and interoperation among networks and systems and that provide security for the information carried; and
- The *people and assets* which provide the integrating design, management and operation of the DII, develop the applications and services, construct the facilities, and train others in DII capabilities and use.

The DII is required to provide information services and products for the DoD Services and Agencies. The current DII consists of many elements, with each of these elements being connected, much like the pieces of a puzzle. However, similar to a puzzle, the DII is not complete without every piece in place. No single element is meaningful when it is separated from the rest, and a single missing element could impact the entire picture.

The DII definition above acknowledges the importance of a full range of aspects of an infrastructure and, interestingly, also recognizes the importance of standards and protocols.

As the DII will form a subset of the NII and GII, it will inherit the benefits that are to be gained from the development of the technologies and the bandwidth and redundancy that the scale of the overall potential structure generates. In doing so, the very infrastructure that will be relied on to protect the nation will, at the same time, be vulnerable to attack through the same medium.

MEII — Minimum Essential Information Infrastructure

The MEII is the minimum subset of the NII and describes the smallest subset of the NII that would be required to maintain the effective working of a nation. The description given in the Information Systems Trustworthiness Interim Report by the Committee on Information Systems Trustworthiness of the National Academy of Sciences, Washington, D.C., in 1997 provides a good discussion of the complexity of the problem of deciding what elements of the infrastructure will be essential and in what context.[9]

The report proposed the notion of a "minimum essential information infrastructure" (MEII) as an alternative to hardening the entire NII to provide a communications infrastructure for implementing trustworthy systems. The concept deals, in a large part, with the availability dimension of trustworthiness. As a subset of the NII, the MEII would have to retain sufficient capability to permit essential services to continue despite failures and attacks. But "minimum" and "essential" are not consistent across the wide range of areas that might rely on the infrastructure. What constitutes "essential" will depend on the context. Losing one of the utilities (e.g., water or power) for a day in one city is inconvenient but tolerable, but the loss of such a utility for a week would be unacceptable. The loss of the same utility to a state or entire country for even a short period of time, perhaps minutes or hours, would also be unacceptable. What constitutes "minimum" also depends on context: a hospital has a different minimum infrastructure that it needs for work during a normal period (e.g., patient health records, billing and insurance records) than it does during a civil disaster. Finally, the facilities that should be provided by the MEII are dependent on the role of the organization: the police and the military may not require confidentiality in their communications when dealing with the effects of a civil disaster but would in day-to-day operations.

It is clear from the report that a single monolithic MEII would probably not be desirable or achievable because the needs of the different elements of the infrastructure are too varied. As an example of this, for a national government, the "elements of the infrastructure," might consist of communications and information needs of the national government to interact with the military and local governments, to coordinate civil responses to natural disasters, and to direct law enforcement against internal threats, terrorists, and organized crime. This may include the infrastructure required to rapidly communicate national issues to the public. Current examples include national radio and television networks (both broadcast and cable) and the national emergency broadcast program and national newspapers.

For a military example where, in the simplest of terms, the military has the task of mounting and conducting operations, whether they are of aid to the civil community or warfighting, the MEII might consist of the short-term strategic and tactical communications and information management needs of the armed forces, as required, to operate national defense systems, gather intelligence, and move logistics and troops.

Other areas that will have an identified MEII include the power and telecommunications services, which will require communications to operate electric power distribution grids and the public switched telephone network, albeit possibly at a reduced level to allow power for the civil community and non-military communication. In addition, the economy will require power and communications to operate public and private banking systems, stock exchanges, and other economic institutions, as well as shops, markets, and other essential mercantile infrastructure.

Non-government agencies and voluntary organizations such as the Red Cross, hospitals, and ambulance services, will require power, water, and communications and will have information management needs. Finally, transportation and logistics will require power and a communications infrastructure to manage air traffic, control rail traffic, and the road infrastructure traffic control for vehicular traffic to maintain the movement of people and goods. [9]

MEDII — Minimum Essential Defense Information Infrastructure

This term has, to date, only been used by the U.K. Ministry of Defence (MoD) to define the subset of the MEII that is essential to defense being able to carry out its role. [10]

RII — Regional Information Infrastructure

A regional information infrastructure (RII) will consist of a subset or a superset of an NII. The RII will be identified as a result of common cultural, commercial, or geographic needs. It may come into being in a part of a country or in several countries in the same area. The RII is, perhaps, most comprehensively described by an Australian Group, Deep Thought Informatics, as:

> *A community ideal, or vision, in terms of a technological strategy. Technical architectures stemming from the requirements represent a repository of solution patterns which can be re-used and evolved.*

In the RII concept postulated by Deep Thought Informatics, it is argued that the concept of "community" is one potential response to global and local issues. The paper goes on to explore the concept that with such an information infrastructure in place, physical communities would have the potential to develop. From this, communities would potentially evolve from individuals and other small groups. The RIIs would eventually support the development of agriculture, trade, local government, environmental monitoring, the delivery of health and education, and social and cultural activities.

The group identifies the most significant features of an RII as:

- Being knowledge based rather than technology based
- Supporting a geographic region
- Being based on interactive rather than mono-directional communications
- Ubiquitous and closely integrated into the community rather than elitist IT centers
- Encouraging the sharing of information between different disciplines within a geographical area

The paper goes on to identify that, if correctly implemented, an RII could provide a means to facilitate, on a local basis, the interaction between those concerns that affect the environment, the economy, and society and encourage potentially more effective, sustainable, and relevant social structures.[11]

There is a noticeable difference in the definition of the RII from other information infrastructures (IIs). The RII definition is generated from a different starting point to all of the other definitions given above. All of the other IIs are based on the founding pillar of the nation-state, whereas the RII is based on environmental, cultural, commercial, or social foundations.

The Relationships between the Infrastructures

With the recognition that there is a GII has come a rash of terms to describe elements comprising it. If we accept that there is a GII that spans the world and covers all aspects of our lives, then it is not too difficult to understand that there will be national information infrastructures (NIIs). While no single country, or group of countries, can own the GII, an NII is under the control, protection, and authority of one government. If we accept that there are NIIs, then it does not stretch the imagination too far for us to understand that some elements of this NII will be essential to the well-being of the nation-state and this, therefore, can be termed the critical national infrastructure (CNI). In addition to this, there will be other purpose-specific infrastructures, such as the one for defense (the DII) and other cultural, social, geographic, or environmental RIIs. While we can look at these in isolation, they are all part of the GII. Are you confused yet? Imagine having the job of identifying those elements that comprise an information infrastructure on which your organization or nation depends. One thing is for certain: the information infrastructure on which a nation-state depends will be a prime target of the adversary in

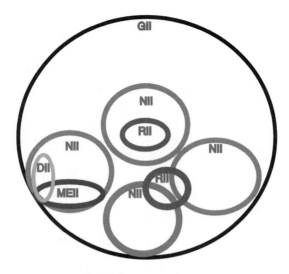

Exhibit 1.   Infrastructure Relationships

the event of information warfare. In fact, it already is. Exhibit 1 illustrates the relationship of the different infrastructures to each other.

Connectivity: An Example

Take a look at an example at the lowest level. You are in business and run a warehouse. For the stock in the warehouse, you have stock control computers that tell you what you have in stock and where it is stored. If you are very lucky (and can avoid pilfering, spoilage, and mislabeling), it will be reasonably accurate.

However, the goods are of no use if they are in the warehouse, so you must have a distribution system for the goods that leave the warehouse. It is sensible to connect the two systems so that as goods are loaded for dispatch, the warehouse stock holding is modified at the same time (think of the benefit of only entering the data once). However, your business is quite large and you need to move some of the stores by air and some by sea. To do this, air waybills and bills of lading must be produced; thus, it is sensible and efficient to connect your distribution system to the air carrier's system and the sea shipping company. This makes the whole movement of the goods more efficient and allows you to track them through their journey.

The air carrier, in providing a cost-effective service, does not run its own system for the production of air waybills, but in fact shares a system with all the major airlines. The sea shipping company has adopted a similar approach and employs a shipping agent in the local port who deals with a number of carriers and its system is connected to a number of shipping line systems.

This is the way business is actually conducted and is a very simplified example. The outcome of this is that an almost infinite number of people, of whom you as the owner of a warehouse system have no knowledge, are connected to your warehouse stock control system. When, as the owner of the warehouse, you started to connect your stock control system to your distribution system, you did not have any concept of the eventual outcome.

Now imagine that you are a government and are trying to identify all of the systems that are critical to the national well-being (the easiest part of the job) and the systems to which they, in turn, are connected or are dependent on.

What Are the Origins of the GII?

The GII is not a single entity and has no single root in history. In reality, it is still in the early stages of development and, as described in the previous paragraphs, is actually the sum of many things, with the terms used to describe it meaning different things to the different groups involved. In addition, it was not planned and is not structured. As mentioned, the GII owes its existence, in part, to the small community, regional, or organizational systems that were initially developed to meet local needs. If we look for some of the elements that have combined to make up the GII, then we must look at the introduction of radio, telecommunications, satellite communications, television, computer networks, and a host of other technologies. If we look at more recent history, then we have to include the development of the Internet from:

- The 1967 development of the first packet-switched network at the National Physical Laboratory in Middlesex in the United Kingdom,
- To the 1969 introduction of the ARPANET,
- To the 1970 development of ALOHAnet in Hawaii of the first packet radio network,
- To the 1973 introduction of the first international connection to ARPANET,
- Through to the 1991 introduction of the World Wide Web

as significant points in the evolution of the GII.

The last five years, in particular, have seen a shift in the use of the Internet. With maturity and an increasing accessibility by the public has come the commercialization of its use. Today we have E-commerce, the electronic shopping mall, and E-business, enabling the use of the infrastructure for businesses to communicate and conduct business 24 hours a day, and M-commerce, providing the ability to carry out commerce on the move such as buying an airline ticket from your mobile phone. While it is true to say that commerce has grasped the new opportunity, it has not yet fully embraced it and the next few years will see significant further development. The late 1990s and early 2000s saw the advent of the dot.com (or, in many cases, the dot.bomb), where entrepreneurs have attempted to change the buying habits of the public. To date, none of these have achieved a significant profit and there have been many high-profile failures. The main reasons for this have been the lack of confidence by the buying public in the safety of online monetary transactions, the credibility of the companies that are offering the products, and a lack of awareness by the buying public, the element of which

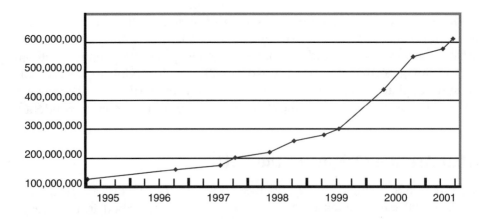

Exhibit 2.   Internet Usage Numbers from 1993 to 2000

has the greatest disposable income, being computer illiterate. As more of the established and trusted "bricks-and-mortar" companies increase their Web presence, as the public becomes more confident in online trading, and as the spending power of the "computer generation" increases, this situation will improve.

If we view the GII through the Internet, the element that is most visible and to which most of society has access, then we can get an impression of the rate of change that is taking place.

From its historically very recent origins and the initial limited vision of its potential, the Internet has developed and migrated in a number of notable ways. The growth in the number of users has been described as exponential by a number of authors and this may be true. It is certainly true that the growth, by any description, has been phenomenal. When describing the time taken to develop new technologies or businesses, it is now common parlance to refer to "Internet time" to describe the speed that must be achieved from concept to delivery to remain competitive in today's electronic marketplace. Exhibit 2 gives some indication of how rapidly the Internet has grown.

That part of the global population that has English as its first language tends to think that the language of computing and the Internet is exclusively English. However, Exhibit 3 reminds us that more than half of the users, and the information that is available, on the Internet may be in another language. It is of interest to note that the number and quality of language translation applications has dramatically increased in the past few years. Examples of some of the more commonly used applications are Babelfish (http://babelfish. altavista.com), CyberTrans (http://lexica.epiuse.co.za/CyberTrans.html), Free-translate (http://www.freetranslation.com/default.htm), and ajeeb ((Arabic–English) http://tarjim.ajeeb.com/ajeeb).

In its early days, the Internet was the preserve of academia and the military, but this has now changed to the point where the military has less than a one percent stake in the infrastructure, and academia has a minority stake with less than ten percent (see Exhibit 4). The major single user group is now the commercial sector, with a stake of nearly 50 percent.

Any view of Internet usage would be incomplete without a breakdown of the demographics (see Exhibit 5). It is significant, but not surprising, that the

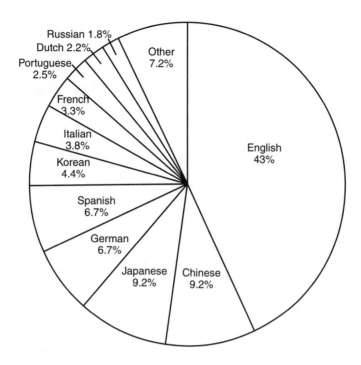

Exhibit 3.   Breakdown of Internet Usage by Language Used
(Statistics courtesy of Global Reach, http://www.glreach.com/globstats/)

highest percentage of Internet use is in the younger age groups, but it is worth noting that it will be the parents, one of the groups that have a lower rate of usage, who have purchased the computers that are being used. This is probably the first generation where the elders will be taught and guided by their juniors.

How the GII Has Developed: Technology Convergence

The convergence of technology can be seen in a large number of developments over the years. In the 1970s, we heard integrated services digital networks (ISDNs) mentioned; then in the 1980s, we were introduced to the concept of fiber-optic cables to give a huge increase in the bandwidth that was available. In the 1990s, we saw the introduction of the advanced digital subscriber line (ADSL), the Internet TV, voice-over-IP, digital TV channels, streaming video, the mobile telephone, the Wireless Application Protocol (WAP), and a host of other developments.

Over the past three decades, we have seen the convergence of industries and technologies that have delivered telecommunications, the media in all of its forms and information. This has led to the integration of networks that were previously used to separately deliver voice, TV, video, and data to the point where today, all of the transmission mediums are used to deliver a range of services. The current decade will see the rate at which information

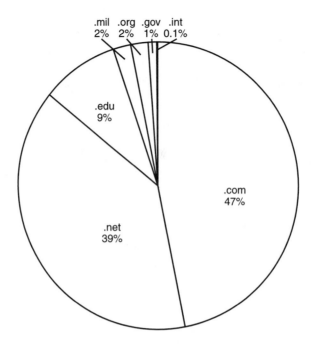

Exhibit 4.   Breakdown of Internet Usage by Organization Type
(Based on data from the Internet Software Consortium,
http://www.isc.org/ds/www-200101/dist-bynum.html)

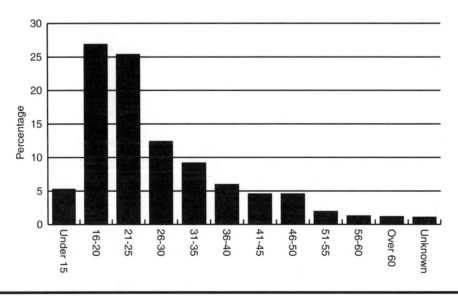

Exhibit 5.   Demographics of Internet Usage

is delivered to the user — whether a government, a commercial organization, an academic institute, a service, or an individual — increase dramatically. It will see the available types of information expand to meet the new demands being generated and the quality, speed of gathering, processing, fusing, presentation, and suitability of the information will be improved.

While this period has seen technologies develop and converge, the past few years have witnessed the growing importance to the global economy of the information itself. The increasing volumes of information available, together with its accessibility, have started to change the balance of control toward a truly democratic environment. Unfortunately, as with all developments, there is the chance of misuse and abuse of the potential, as well as benefit. The potential to use the global reach of the information infrastructure to communicate with like-minded individuals and groups, in conjunction with the availability of the means to do so, is already causing significant social change. This will continue at an increasing rate as the technologies and capabilities mature.

Criticality of the GII

The United States was the first country to publicly identify that it has a "critical national infrastructure" (CNI); this is not surprising because it is also the country that is most dependent on it. What is difficult to determine is whether the CNI is different to the MEII, or whether, in the minds of those responsible, these terms are interchangeable. There is evidence that other countries, including the United Kingdom and Sweden, have now also agreed that they have CNIs, but the overwhelming evidence is that none of them can identify what these infrastructures really consist of or what the dependencies and interdependencies are. In this international world with multinational companies, it is common for the companies that are running critical elements of a nation's infrastructure to be owned and controlled by foreigners. There is also further confusion caused by the difficulty in identifying what is critical to the well-being of a nation and what is essential to the commercial sector.

Work is currently underway in a number of countries to identify not only those elements that comprise the CNIs, but also to determine ways and means of measuring the criticality of systems. Once this work has been further developed, work will begin on identifying measures that can be taken to either protect these systems or make them less critical, for example, by building alternative systems that provide diversity and redundancy.

The response in the United States to the recognition of the CNI was the establishment, in February 1998, of the National Infrastructure Protection Office, known as the National Infrastructure Protection Center (NIPC), which describes its role as an organization that *serves as a national critical infrastructure* (note that the critical national infrastructure is referred to here as the national critical infrastructure — so much for consistency) *threat assessment, warning, vulnerability, and law enforcement investigation and responding entity. The NIPC provides timely warning of international threats, comprehensive analysis and law enforcement investigation and response.*[12]

The response in the United Kingdom was the establishment of the National Infrastructure Security Coordination Centre (NISCC). The role of the NISCC is more conservative than that of the US NIPC but with the task of achieving the protection of the CNI against electronic attack. As with the NIPC, the NISCC is a multi-agency and multi-organizational effort. However, in the case of the NISCC, there is no grand establishment; it remains a virtual organization. The establishment of the NISCC in December 1999 was announced by the Right Honourable Tom King, MP, in the U.K. Intelligence and Security Committee Annual Report for 1999-2000[13] as:

> *A single point access to the Government's Critical National Infrastructure Protection (CNIP) arrangements. The NISCC, which is largely resourced by the existing Security Service and CESG baselines, acts as an umbrella organization coordinating relevant activities in several departments and Agencies, including the Security Service, CESG, Cabinet Office, Home Office, MoD, DERA, DTI, and the Police.*

The report goes on to describe one of the main activities of the NISCC to be to issue alerts and warnings about security incidents and vulnerabilities. These were to be initially for use within the United Kingdom, and after this, for use by a wider community and to other international organizations, by means of a scheme that is known as the Unified Incident Reporting and Alert Scheme (UNIRAS).

The report went on to describe that while the U.K. Intelligence and Security Committee had not taken evidence that related to IW during this year (1999–2000), it did investigate the way in which the Critical National Infrastructure and Government had responded to the Love Letter virus. It is interesting that the report notes that the committee expressed disappointment that a warning was not sent out until 12:45 p.m. on May 4. It noted that the government, in the form of the House of Commons, received a warning on the virus at 10:33 a.m. It also noted that by that time the virus had infected a large number of the government's computer systems, the effect of this delay in issuing warnings was that the United States did not receive a warning of the virus before the U.S. departments had started work. From this it can be seen that the British government, as early as 1999, was conscious that the response to incidents had to be improved and also that the global nature of the problem would allow nations to assist each other by exploiting the different time zones.

The approach adopted in the United Kingdom is similar to that taken in the United States.

The aim of the NISCC is to co-ordinate activities in support of the government's objective to achieve an effective level of protection from electronic attacks against critical systems within the United Kingdom. As with the United States, this task requires effective partnerships with the owners of the infrastructure (non-government owned) and an exchange of information, particularly of material relating to threats. This in itself poses a problem for government because it will potentially involve the dissemination of sensitive information on threats to organizations that may not be owned by the governments, nationals.

It is worthy to note that, in the second paragraph of the quote above, the concept of timely passage of warnings on potential problems, from government to government, has been identified as important. This type of cooperation will be essential if the GII is to be used to its full potential.

In addition to the top-down approach of the central authorities, populated by members of law enforcement and security organizations, a bottom-up approach of raising awareness of the importance of information security in general is expected to encourage the development of good commercial security products and sensible security procedures within government, academia, and the civil community. This two-level approach will be essential if the CNI is to achieve a more survivable and secure status.

Within the United Kingdom, other initiatives have been taken to improve the general standards that have been applied to the security of information systems. This has been spearheaded in the civil sector by the development of the British Standard BS 7799, Information Security Management, which has been adopted, in modified form, as an international security standard by the International Standards Organization Standard (ISO) 17799.

In the defense arena, one initiative that has been adopted is the *domain approach* to security that was pioneered at the Defense Evaluation and Research Agency (DERA) now known as QinetiQ. This approach provides a methodology for determining, in a system, where security measures can most effectively be deployed.

In addition, within the United Kingdom, many of the sectors of society have combined for the common good with the creation of the Information Assurance Advisory Council (IAAC). This is a "think tank" that includes representatives from government agencies, the telecommunications and computer industries, and the research community. IAAC defines its role as:

> *The provision of objective research and analysis into the strategic policy implications of protecting critical information infrastructures. The IAAC is a global centre of excellence and provides a forum for development of international as well as national policies.*[14]

These initiatives will, in due course, ensure that the U.K. networks and systems will be better protected against electronic attack. This will include an early indication and warning of potential and actual attacks, and together with the dissemination of actions that can be taken to mitigate or minimize the effects of an attack and the response arrangements, should help counter any attacks that do occur.

The U.K. government is keen to create the electronic infrastructure of the island nation as a safe and desirable place to conduct electronic business. As businesses move toward the E-economy, with this ambition in mind, the business case for security is easy to make because it becomes clear that security is vital to successful transactions in the new environment. Furthermore, the U.K. concepts and processes can be adapted and adopted by other nation-states.

Why Has the GII Developed in the Way That It Has?

The GII has evolved to where it is today, partially as a result of the drive by all the interested parties to make a better environment in which to operate and partially as a result of the growing availability of technologies that will enable it. The global economy, the digital revolution, and the developments in telecommunications, together with the developments in the Internet, are all having a profound effect on society, especially in the developing nations. As a result, communities, regional economies, and social groupings are all becoming increasingly interdependent. The availability of these technologies is providing the means to increase the rate of economic development and education and to significantly reduce the poverty and lack of access to education that currently affects a huge proportion of the global population. The potential of the GII in the developing world was expressed at a conference in South Africa[15] where the country's president, President Mbeki stated:

> *The information and communications revolution offers even more powerful and enhanced capabilities, affecting and transforming patterns of work, education and health delivery, entertainment, public opinion and so on.*

Real and Potential Benefits of the GII

The development of the GII is, effectively, the result of the potential released by the advancement and convergence of technologies that has been taken up in a number of areas. Probably, and understandably, the first area to embrace and make use of the new potential was the academic arena. The opportunity for the sharing of what were, until then, rare resources, and the chance to collaborate and to provide educational resources to that portion of the population that had previously been excluded, were quickly taken up and developed.

An interesting downside of this "progress" in the academic use of the Internet has been the phenomenon that new developments in research are now aired and debated over the Internet, rather than as the result of papers published in academic circles. The outcome of this change in the way in which developments are debated is that there is no permanent record of the comments made as a result of the peer review.

International and non-government agencies had identified the potential benefits that could be derived from the use of these resources. A 1996 example of this was seen in a Survey of Information Communication Technology (ICT) in Sub-Saharan Africa.[16] This had the objective of identifying initiatives to utilize ICTs that added value to development programs in the area of natural resources (in particular, water, forestry, environmental management, and agriculture/food/fisheries). The survey discovered 17 relevant projects, of which 11 were sponsored by non-government organizations (NGOs), of which two were international, one was a research project, three were government-sponsored projects, and two were of undetermined origin.

Another early example was the use of the radio networks in Australia to provide distance learning to school-age children who lived too far from the nearest school to be able to receive lessons in the school. With the development of computer networks and the increasing availability of access to the Internet, the radio school concept has been updated to the distance learning programs, such as the Open University in the United Kingdom, that are available from many countries and accessible worldwide today.

Governments were also quick to understand the potential that the GII had released, and many of them have initiated programs to allow their citizens easier and greater access to the government itself. An example of this is the Modernizing Government program, also known as "joined up government" in the United Kingdom, which has mandated that all dealings with government departments in the United Kingdom will be electronically achievable by 2008 and that public services will be available 24 hours a day, 7 days a week.[17]

With the huge rise in the number of Internet users, commerce is now taking an increasing interest in the potential of the GII — the financial markets around the world operate 24 hours a day (the money moves around the globe with the sun). The world markets are now available to buyers who previously had no option but to buy from their local area. A classic example of this is Amazon.com, the online bookstore. It is now as easy for a buyer in Europe to log on to this U.S.-based store and benefit from the prices in the United States as it is for them to order from their High Street store, and the delivery time is likely to be very similar. It is of note, however, that as well-known as Amazon.com is, at the time of this writing, it has still not achieved profitability.

The speed and pervasiveness of communications that the GII has provided means that it is now possible to remain in contact, either by voice or data, from almost anywhere in the world. In addition, with the diversity of communications systems that exist, the redundancy that has been achieved means that, even after a major disaster, it is almost certain that some of the communications routes will have survived or it will be possible to establish them quickly on demand (e.g., satellite communication systems).

This diversity of communications also has the effect of providing surplus bandwidth with the result that the cost of communications is reduced and the types of media that can effectively be used has increased and will now cater to videoconferencing, streaming video, groupware, and similar bandwidth-intensive applications.

What Are the Problems That It Has Caused?

Governments have grasped the benefits that the GII can provide, but for them, it caused a dichotomy — with all of the benefits that could be gained, there were also penalties. Who is responsible for this new, undefined infrastructure? Who has control of it? Who is to protect it? Furthermore, there is the problem at the national level that if all nations have equal access, then in times of conflict one or more nations may attempt to use the infrastructure to attack another nation.

With the GII have come new risks and threats. The globalization of infrastructures has increased the exposure of individual elements to potential harm. The vulnerabilities of infrastructures are enhanced by easy accessibility via the Internet. The interdependency of systems potentially makes the consequences of an attack more severe and more difficult to predict. Harmful tools that can be used to attack computer and telecommunications networks are freely available, easy to use, and their use is unlikely to be detected.

Types of Computer Crime That Affect the GII

The birth of computing was followed, shortly after, by the birth of computer crime. With the development of the GII has come the use of it by criminals to gain power, influence, and financial advantage. It is a historical fact that when you introduce a new technology, criminals will quickly adapt to take advantage of it. An example of this was the introduction of the railroads in the United States and their resultant use to transport mail and money, which was quickly followed by criminals who took advantage of the new opportunity to intercept the mail and money in transit. The following discussion provides several types of crime that have been identified in the computer environment.

One might wonder why a book on global information warfare would discuss computer crime. There are several reasons for this, to include those noted below:

- One may be investigating a computer crime; however, it may be one that supports an adversary in its information warfare strategy. The investigator, info-warrior, security professional must always keep this in mind. For example:
    - It is fairly well-known that some terrorist groups commit computer crimes to fund their terrorist activities.
    - During a U.S. local police investigation of the theft of computer chips from Intel Corporation, they found that they were not dealing with ordinary criminals but with a sophisticated covert operations network controlled by the former U.S.S.R. Their goal was the theft of chips to reverse-engineer and also to use them in their high-technology war with the United States during the Cold War.
- Computer crimes may actually be more than violations of the laws of a nation-state but rather an indication of an information warfare attack. For example, unauthorized access to a computer may be a violation of law; however, the reason for it may be a prelude to an act of information warfare.
- When a computer crime or some security violation occurs, one never knows whether it is just a hacker having a "joy ride" or an actual nation-state-sponsored or terrorist attack on a system by an information warrior.

In a bulletin titled "How High Tech Criminals Operate," Carter[18] put forward the following definitions of computer crime types:

- Computer as a target. This includes:
    - Hacking
    - Denial-of-service attacks
    - Intellectual property theft
    - Spying
    - Industrial espionage

While the computer as a target of crime is a variant of crime that could not exist until the advent of the computer, crimes such as intellectual property theft, spying, and industrial espionage predate the computer and have been adapted to make use of the new environment.

An example of this type of crime was the distributed denial-of-service (DDoS) attacks targeted, in February 2000, at Yahoo.com, Buy.com, Amazon.com, e-Bay, CNN, and possibly UUNet. At that time, these organizations were able to thwart the attacks within about an hour of them starting. DDoS attacks are not particularly new, but because they can affect public perceptions of the targeted organization, they may not always be reported. DDoS attacks use multiple Internet sites to send requests for data very quickly. The attacks that manage to utilize the highest number of sites to launch the attack are likely to be the most disruptive and also the most difficult to terminate. These high volumes of traffic can easily overwhelm a site, thereby making it unavailable to other users. The incidence of this type of attack has continued to increase over the past few years and, as more and more sites are connected to the Internet, the likelihood is that they will continue to be a problem.

- Computer as an instrument of crime. This includes:
    - Terrorism/communications
    - Pornography
    - False credit card number generation
    - Capturing mobile phone details
    - Fraud
    - Cyber stalking

All of the crimes in this group existed in forms that predate the computer; but again, the perpetrators of the crimes have utilized the new environment to take advantage of the changed methods of business.

An example of this type of crime was the first case filed under a new California state cyber-stalking law, when, in 1999, a North Hollywood man was charged with using the Internet to attempt to set up the rape of a woman who had spurned his advances. In an article in *The Los Angeles Times,* it was reported that prosecutors had said that Gary S. Dellapenta, 50, stole the identity and privacy of his victim and paraded her on the Internet as a woman with fantasies of being raped. While the report explained that the woman had not been physically harmed, a total of six men had arrived at her apartment between April and July 1998, claiming that they were responding to online personal adverts that she had placed. At that time, the woman did not even own a computer. The article went on to report that the e-mails claimed that the woman was into "rape fantasy and gang-bang fantasy." It was alleged in the article that Dellapenta not only described the victim to a number of men and provided them with her address and telephone number, but also explained to them how to bypass her home security system.

> *Dellapenta, who was employed as a security guard, was arraigned in February 1999 on one count of stalking; another count of using a computer to commit fraud, deceive, or extort; and three further counts of solicitation to commit sexual assault, district attorney's spokeswoman*

Sandi Gibbons told The Los Angeles Times. *He remains jailed in lieu of $300,000 bail.*[19]

A second example of this is the "Wonderland" pedophile case that involved more than 100 suspects around the world. They are reported[20] to have gathered together and traded in excess of 120,000 pedophile images.

You might question what these types of crime have to do with IW and why it has been included here, but the reality is that the only thing that separates this fairly sad type of crime from IW is the motive.

In the examples above, the identity theft was carried out to conduct a cyber stalking, but what if the woman had been the chief executive of a bank and the motive was related to her job and not to her as an individual. In the "Wonderland" case, it was about pedophilia, but there is well-documented evidence that Osama Bin Laden has used steganography to hide his communications. What if the "sex ring" had actually been created as a communications facility?

- Computer is incidental to other crimes. This includes:
  - The storage of criminal information on computers. (e.g., pornography, details of planned and actual crimes, fraudulent accounts, and trade secrets).

Crimes in this group do not, as a matter of course, directly involve computers. In the case of these crimes, the computer or computer media has replaced paper as the storage medium for the record of the crime.

An example of this type of crime would be the case in Georgia (United States), in which Caroll Lee Campbell, Jr., and Susan Campbell were indicted for violations of the Economic Espionage Act (EEA) and wire fraud statutes for conspiring with a third party to convert trade secrets relating to the *Gwinnett Daily Post.* Mr. Campbell pled guilty to conspiring to steal trade secrets and was sentenced in 1998 to three months' imprisonment, four months' home confinement with electronic monitoring, three years' supervised release, and a fine of $2800.[21]

- Crimes associated with the prevalence of computers. This includes:
  - Software piracy
  - Copyright violations
  - Counterfeit equipment
  - Counterfeit software
  - Computer theft

This group of crimes has become prevalent as a result of the huge increase in the number of home computers and their components.

An example of this class of computer crime would be the rash of laptop computer thefts that were reported in airports around Europe. According to reports, laptop computers had become a premium target for theft throughout Europe. International travelers are warned to remain on constant alert. Airport security authorities have identified two primary methods of theft involving two or more thieves working together. One method took place at Brussels Airport at the point where travelers undergo the preboarding security inspection. The first thief would walk through the metal detector at the security checkpoint ahead of the victim. The second thief ensured that he was directly in front of the victim when he placed his laptop computer on the x-ray conveyor belt and then purposely set off the metal detector. While the victim was delayed by the second thief having to pass through the x-ray machine again and then possibly be hand-checked, the first thief retrieved the laptop from the other end of the conveyor and quickly disappeared. At the International Airport at Frankfurt, Germany, a different type of theft occurred while the traveler was in a busy public area of the airport terminal. The first thief through a crowd of people preceded the traveler, who was carrying his computer in a roll-type bag. As the traveler made his way through the crowd, the first thief stopped abruptly and caused the victim to stop. During this delay, the second thief quickly removed the traveler's computer from the roll bag and disappeared into the crowd.

The development of the computer and related technologies, together with the dependence on them of organizations in the developed world, has, in the case of some crimes, changed either the target of the crime, the perpetrator, or the vehicle used to carry out the crime.

In 1998, Rapalus[22] reported that the annual cost of computer crime had risen to $136 million, an increase of 36 percent over the 1997 reported figure. The 2001 CSI/FBI Survey[23] gives the cost of cybercrime in the United States in the year 2000 as $378 million, a rise of nearly 300 percent in only two years.

The Current Problem

One of the fundamental problems with the policing of the Internet is that of development speed. The Internet was "born" in 1969 from the research and military ARPA and DARPA networks and has seen unprecedented growth and diversification of use. The number of hosts, according to Netsizer,[24] has grown from an estimated 10 million in November 1994 to 164.426 million in February 2002, and is expected to continue to grow at an increasing rate for the foreseeable future. The number of users is estimated as being in the region of 425 million.

Because the public has gained easy and affordable access to the Internet, and crime began to occur as a result, the legal systems and law enforcement bodies have been expected to develop and enforce laws that are appropriate to the new environment. Given that the environment is changing at a pace that even the preeminent research establishments cannot predict, it is not surprising that the legal system has not yet provided the policy, laws, tools, and capability required to effectively police the environment.

As mentioned previously in this chapter, one of the major factors that has not yet been addressed is that the Internet is a global facility and that it is not "owned" by any country or organization. One impact of this is that a person in one country can carry out a crime in another country without ever physically entering that country and, in most cases, without breaking any laws in the country from which they carried out the crime. One of the few exceptions to this situation is the United Kingdom, where the act of "hacking" is a crime even if the target machine is in another country.

This globalized system creates significant law enforcement problems because there are often problems of jurisdiction. The nearest comparable situation is the law of the sea, where international agreement took more than 40 years to achieve and which all nations have still not yet ratified. While there is considerable attention being given to the subject of international coordination for the investigation and prosecution of computer crimes, few new laws have, to date, been passed.

Looking at the situation within the United Kingdom as an example, there are 57 separate police forces, all of which (with the exception of a small minority such as the Metropolitan Police, Transport Police, and Ministry of Defense Police) are accountable to local authorities for the maintenance of law and order and are budget restrained. There is little incentive for them to become involved in policing the Internet for a crime that does not directly concern their geographical area. The pressure on the police is from the local authority to address local issues that affect the people who fund the police service, such as reducing street crime, burglary, and car theft. This situation is also true in most democratic countries and, as a result, even if the laws and international agreements should exist, the motivation for a locally financed police force to undertake significant effort on behalf of another country will always be lacking.

This is an abstract view of the subject. Crimes on the Internet will vary in the degree of severity. At one extreme, the use of the Internet to pursue terrorist activity will be vigorously investigated within the country that is affected by it and other countries that have an interest in the activity. Emotive crimes such as paedophilia will also normally be pursued with vigor, as public demand will put pressure on the relevant law enforcement agencies to pursue, arrest, and prosecute those involved. At the other end of the scale, crimes against companies, such as copyright infringement, are difficult to investigate and are not perceived as serious, except by the industries that are losing a significant portion of their revenue in lost sales. As a result, such crimes do not attract high levels of interest.

One difficulty experienced by law enforcement agencies around the world is the capture and presentation of forensically sound evidence that can be presented to a court. In non-computer-based crimes, there has normally been a detectable, visible connection between the person perpetrating the crime and the scene of the crime. In computer crimes, it is possible for the criminal to never enter the country in which the crime is committed, but also to easily adopt a number of aliases. In addition, evidence of their activity may be held on computer systems located outside the jurisdiction of the investigating agency and may be extremely transient and difficult to preserve.

Another problem being encountered in the prosecution of computer crimes is that, when the evidence is presented to a court, judges and juries have difficulty in understanding the complex details of the evidence. This is further complicated because the laws in the United Kingdom, for example, have not yet been rigorously tested in the courts. What this means to info-warriors is that they are practically immune from prosecution, an incentive to be more aggressive.

The Effects of High Technology

Policing on the Internet is reactive to changes in the technology available to the community. Three of the most significant technical advances that have caused problems in policing the Internet in the recent past include:

- *The ability of users to anonymize their activity on the Internet so that they cannot easily be traced.* This can be achieved in a number of ways, such as using one of the many anonymous re-mailers that are publicly available or through the use of free ISPs where there is no requirement to verify identities. This, in nontechnical terms, would equate to a person having an endless supply of false identities.
- *High-grade cryptography.* This publicly available facility (e.g., Pretty Good Privacy (PGP)) allows people not only to encrypt all of their electronic communications to a level that is unlikely to be broken by law enforcement, but also to encrypt all their stored information and voice communications to the same level.
- *Prepaid mobile telephones.* These provide an individual with the capability to communicate, either by voice or over the Internet, without the possibility of being traced through the fixed telephone system. For those who can afford this option, it is one of the up-to-date equivalents of phreaking, in that it makes the tracing back to the source of an action more difficult.

Changes in the Types of Crime Resulting from the New Technologies

There are three main types of crime that have resulted from the new technologies:

- Technology-supported terrorism
- Online fraud
- Online pornography

Techno-Terrorism

It is now strongly suspected that terrorists have been using the Internet to pass encrypted messages among themselves. This provides them with secure communications that not only allow for the anonymous transmission of messages from one individual to another, but also allows for the replacement of the traditional "dead letter box" where messages could be deposited until the recipient could collect them. Establishing a chain of continuity for a message from the originator to the recipient is thus, now, more difficult. Terrorists have also used the Internet to publicize their cause to the global population. An example of this can be seen in the Internet Web presence of the Irish Republican Army (IRA), AN PHOBLACT.[25] Terrorists have also used the Internet to apply global public pressure to governments to prevent or minimize the reaction to incidents in which they are involved. This was demonstrated by the Mexican Zapatista terrorist group that used the medium extremely effectively against the Mexican government.[26]

According to a 1998 paper by Jones,[27] there is now significant concern that terrorists have access to the required technology and communications and that they will use the Internet to launch attacks on the infrastructure of a nation to cause disruption and to further their cause. One further factor that has become apparent in the past four years is that technology and communications have reduced the entry level for crimes that have the potential to have a significant impact on an industry or a country, to a level where it is possible for individuals of school age to have a significant impact.

Online Fraud

The use of the Internet to carry out fraudulent attacks is well-documented. The future potential for fraud that will come about as electronic commerce increases will pose a major problem for law enforcement and the commercial institutes that are dependent on this method of trading. The passage of credit card numbers over the Internet, and the storage of personal details on computers that are accessible from the Internet, are now commonplace. However, the lack of adequate security and encryption, coupled with the failure to thoroughly check credit card purchases, will continue to provide major opportunities for criminals. The prevention and detection of this type of crime will largely be in the hands of the commerce and banking communities, which will have to improve their verification and checking systems to reduce the potential for abuse. Such activities can also financially support techno-terrorists' activities.

Online Pornography

The Internet has been cited in a number of highly publicized pornography incidents. These have shown that the ability for large numbers of people to access information and communicate with others of similar interests, without having to attend in person, have increased the level of this type of crime, as they perceive that the risk of being caught is reduced. The Internet removes the stigma and guilt from individuals who would, in the past, have had to visit a shop in person or have the ubiquitous brown paper parcel delivered to their residences. Now they can visit a huge number of Internet Web sites, with apparent anonymity, and view their preferred type of pornography. The global reach of the Internet also means that material that would be illegal in one country can be stored and accessed in other countries where the material is not illegal, or the law is not strictly enforced, thus allowing individuals access to material that they would previously have been prevented from seeing, at a minimal risk.

Action that Is Being Taken to Address the Problems

In the European Union (EU), the Council of Europe resolution of January 17, 1995, on the lawful interception of telecommunications looked at ways of harmonizing law within Europe to facilitate the detection of computer-related crime. Later resolutions have dealt with problems of criminal procedure with information and computer-related issues.

The United Nations, at the Cairo meeting in 1996, highlighted the subject of Internet crime in relation to international terrorism, organized crime, and drug trafficking.

An EU conference, entitled Policing the Internet — First European Conference on Combating Violence and Pornography on the Internet, took place in London in February 1997. The conference dealt with "the technical and moral issues around policing, legality and censorship to tackle the growing amount of and ever easier access to violence and pornography on the Internet."

Within the developed nations, in December 1997, the justice ministers from the Group of Eight (G8) nations met in Washington to devise a strategy on how to defeat Internet crimes. The subject was again discussed at the G8 meeting in Birmingham, United Kingdom, and a report[28] from that conference stated that it had produced solid results in fighting global financial crime.

In June 1998, a press statement from INTERPOL addressed the need for international cooperation in the fight against organized crime. This has been followed by a series of initiatives that, while still developing, have already resulted in the production of a computer crime law enforcement manual. Statements made by INTERPOL representatives during 2000 and 2001 make it clear that they have understood that the policing of the GII is not an issue that law enforcement can achieve without the cooperation of the financial and commercial sectors or the assistance of assets from other governments. A June 2000 announcement by INTERPOL Secretary General Raymond Kendall indicated that there are plans to create a global intelligence network to tackle cybercrime. A press release in February 2001[29] announced that INTERPOL and a company called AtomicTangerine were exploring plans to make information gathered by INTERPOL available to private firms in return for information gained by those private firms on Internet threats.

In reality, this type of initiative will almost certainly encounter problems with the data protection and human rights legislation of a number of countries.

In the United States, the White House has published a paper on "Defending America's Cyberspace." This paper, Program 3: Developing Robust Intelligence and Law Enforcement Capabilities to Protect Critical Information Systems, Consistent with the Law,[30] outlines research that must be conducted to establish a legal infrastructure capable of dealing with cybercrime into the next century.

In the United Kingdom, the Telecommunications (Fraud) Act 1979, which as amended in 1984, is still the primary law invoked to prosecute the theft of telecommunications "phreaking." The United Kingdom's Interception of Communications Act 1985 (IOCA) and the controversial Regulation of Investigative Powers (RIP) Act (2000) provide the legal framework for authorizing interception of public telecommunications networks; however, any information gained can only be used for intelligence purposes and cannot be presented as evidence in a court. There is currently no statutory basis for interception of non-public networks. The IOCA was initiated with telecommunications providers in mind; however, the situation has changed considerably from the time this act was passed. In 1985, there were two telecommunications providers within the United Kingdom; but by 2001, after deregulation, there were 167 licenses to supply telecommunication services. In addition, in 1985, public access to the Internet was minimal and there were no mobile telephone networks or private international corporate data networks. This law is now ineffective in the developing environment and will have to be rethought if it is to deal with the future.

Also in the United Kingdom, the initial response to the problem of hacking was the passing of the Computer Misuse Act (CMA) of 1990. The CMA is "an act to make provision for securing computer material against unauthorized access or modification; and for connected purposes." It is the primary vehicle for the prosecution of computer crimes and, to date, is probably unique in that it also contains additional provisions with regard to jurisdiction and extradition. In effect, this means that an offense has taken place if either the offender or the computer was located in the United Kingdom when the offense occurred.

Within the United Kingdom's London Metropolitan Police, a specialized unit (the Computer Crime Unit) was created as part of the Fraud Squad to start to tackle computer crime. However, the control area of the unit was that of crimes that originated, or had a victim, within its geographic area of jurisdiction. Since then, a number of police forces within the United Kingdom have also created teams to tackle the problem. Unfortunately, this does not generate a coherent capability because the levels of commitment and capability vary from force to force, there is no formalized policy, and there are no cooperation agreements in existence. The creation of the U.K. National Criminal Intelligence Service and the U.K. National Crime Squad have provided the foundation for the National High Tech Crime Unit which, since its creation in April 2000, has national authority and responsibility.

Unfortunately, this will still not provide a single source of effort to tackle computer crime, because, in common with the United States, other agencies will become involved, depending on the type of computer crime taking place. If, for example, the crime were that of terrorism or spying, then in the United Kingdom, it would fall under the responsibility of the Security Service. Given the scope of the GII, the obvious conflicts that will occur in determining who has authority to investigate or prosecute a crime that spans national borders is guaranteed to reduce the likelihood of a concerted approach to the enforcement of law.

Spain, where there has been a significant problem with computer viruses, launched a national anti-virus campaign[31] in July 2000, through a collaboration between the Ministry of Science, a software house (Panda Software), and the Association of Internet Users. The response to the initiative was extremely positive, with more than 400,000 users downloading the anti-virus software, of which 320,000 were private users and 80,000 were commercial users.

In Singapore, according to an Associated Press report,[32] the Defense Minister, Mr. Tony Tan, stated that the military would be used to step up the country's response to "cyber-attacks." He stated that it would "enhance Singapore's capabilities to deal with a range of non-conventional threats, such as terrorism, piracy and cyber-attacks."

At the time of writing this book, 37 countries[33] have either already enacted computer crime laws or are in the process of creating legislation. While this is only a relatively small fraction of the countries connected to the GII, the situation is not as disappointing as it might seem. Other countries, such as the Philippines, have successfully dealt with computer crimes such as the release of the Love Bug virus, allegedly by Onel de Guzman under legislation covering the illegal possession of passwords, which is normally used to tackle credit card fraud.

Other Initiatives

If the Internet is to be a safe place for recreation and business in the Twenty-first Century, it will not, in the foreseeable future, be as a direct result of law enforcement and government-led policing. The problem belongs to the global community and will, in part, be solved by the equivalent of the United Kingdom's and United States' neighborhood watch type of system, where responsible citizens will be the eyes and ears who will report unacceptable or suspicious behavior to the authorities. There are already groups of people who are active on the Internet, who monitor for the activities of pedophiles and report them, publicly post the offenders' personal details, or close down offending sites. The latter two options are a form of Internet vigilantism, and are themselves, in most countries, probably illegal; however, in some groups, this is perceived as an acceptable response.

Within the international community, a number of organizations have been created to monitor and report unacceptable behavior on the Internet. These are normally targeted against specific activities. Examples of these include organizations such as the Cyberangels, who monitor for pedophilia; the Internet Watch Foundation, which monitors for pornography; and the International Chamber of Commerce — Alliance against Commercial Cybercrime, Cybercrime Unit, which collects and disseminates information on fraud and money laundering.

Corporate entities that have suffered or are likely to suffer the financial losses of fraud have already started to take action. Not only have they taken action to improve the security of their systems and thus make fraud more difficult, but they have also formed associations and groups that will share knowledge of computer crime and thus prevent others from suffering from a crime that has already been observed elsewhere. Hoey[34] in 1996 concluded that:

> *It is imperative that law should give industry and commerce clear guidance on how to make their records acceptable to the courts.... As technology develops evidential practice will need to be evolved to accommodate it.*

Police forces around the world are increasingly developing their own Web presence to provide online advice about computer crime and to give the "citizens" of cyberspace a point of contact to report computer crimes. Examples of this can be seen at the Columbus, Ohio, Department of Police,[35] the Easton, Massachusetts, Police Department,[36] the Beaumont, Texas, Police Department,[37] the Tempe, Arizona, Police Department,[38] Royal Canadian Mounted Police Web page,[39] and the U.K. Hertfordshire Police Web page.[40]

Limiting the use of the Internet by terrorists is the area that will be most problematical because terrorism affecting one country is perceived in another country as freedom-fighting. Most terrorist activity is sponsored by other states, so while it may be agreed in international arenas that the terrorism is unacceptable, there will always be pariah states that will support it and provide safe havens for the perpetrators. In the United Kingdom, the law has recently been widened to incorporate "terrorist" other than that which is state sponsored.

It is perhaps promising that the recent terrorist trial of the Lockerbie bombers took place under Scottish law, with the trial taking place in the Netherlands. We are truly becoming a global community.

Computer Crime Summary

The policing of the Internet will remain a challenge in the Twenty-first Century as the speed of change in high technology and diversification of use continues to accelerate. The creation of new laws and international agreements, which historically have been developed over long periods, will have to occur within a realistic time period. The law enforcement agencies throughout the world currently have neither the levels of skill, the resources, nor the motivation to deal with the range of computer crimes that do, and will, exist.

The effective policing of the Internet in the Twenty-first Century will rely on a variety of resources. These will range from skilled computer enthusiasts who have the ability to understand the complexity of the systems and the time to investigate misuse of the Internet, to commercial and academic organizations that have a material interest in reducing the level of crime, to police agencies that have a responsibility to enforce the law. The endpoint will be government agencies that have a responsibility to ensure the well-being of the nation-state and to enable international trade.

If the policing of the Internet is to be successful, there will need to be a major shift in a number of areas. The "citizens of the Internet," commercial organizations, law enforcement agencies, governments, and international organizations will all have to develop new relationships that utilize the strengths and capabilities of the individual groups. New methods of capturing and preserving evidence that are acceptable to the courts of all countries will have to be devised and tested.

In summary, much as the Internet has developed and continues to develop rapidly, the policing of the Internet will have to change and develop at a pace to meet the new environment. There will also have to be a concerted program of education for new users of these resources in the moral and ethical issues of their use.

Summary

The developing GII is the sum of many elements, some of which are controlled by governments, some by commerce and industry, and some by academia. The main issues that currently, and for the foreseeable future, affect it are that while the drive to interconnect systems continues to increase, there is no method for accurately predicting what the impact on the confidentiality, availability, and integrity of the interconnected systems will be. Systems that may be critical to the GII are owned and managed by groups that have differing pressures for their management and use. The terminology in use is still developing and, while common terms are used, it is clear that the meaning of the terms is not the same to the different interest groups. There has been a recognition that some of the elements are critical, but there is presently no methodology to determine what is critical to the infrastructure and when. The use of the GII by criminal interests and the lack of international laws or law enforcement bodies means that the prosecution of network criminals and infowar criminals is difficult.

It would be fair to say that if you were planning an infrastructure on which you knew that you would, in the future, be critically dependent, you would not develop what is currently considered to be the GII.

Notes

1. Living in the Information Society, Understanding the Global Information Infrastructure.
2. Paper entitled The Global Information Infrastructure: Agenda for Cooperation, http://www.iitf.doc.gov/documents/docs/gii/giiagend.html.
3. What is the GII? National Institute of Science and Technology, 1996, http://nii.nist.gov/gii/whatgii.html.
4. ISO/IEC JTC 1/SWG-GII N 190. Special working Group on Global Information Infrastructure, December 1997 (updated January 1998).
5. CSPP Report entitled Perspectives on the Global Information Infrastructure, http://206.183.2.91/reports/perspectives.html.
6. Report of the Defense Science Board Task Force on Information Warfare — Defense (IW — D), November 1996.
7. Technology, Standards, and Regulatory Principles for the Developing and Developed Nations, an IIIC Common Views Paper adopted in October 1996.
8. The Defense Information Infrastructure (DII), Strategic Enterprise Architecture, Coordination Draft, dated May 31, 1995.
9. Information Systems Trustworthiness Interim Report, Computer Science and Telecommuications Board, Commission on Physical Sciences, Mathematics and Applications, National Research Council, United States, 1997.
10. Presentation entitled "Military Operation and Their Reliance on the National Information Infrastructure and Minimum Essential Defense Information Infrastructure in an Information Warfare Scenario, given at the 1998 Command and Control Research and Technology Symposium in Monterey, California, by M. Corcoran and J. MacIntosh.
11. A Regional Information Infrastructure (RII) for Sustainable Communities, T. Beale, March 2000.
12. National Infrastructure Protection Center home page: http://www.nipc.gov/.

13. Intelligence and Security Committee Annual Report, 1999–2000, HM Stationary Office.
14. The Information Assurance Advisory Council home page: http://www.iaac.ac.uk/l2.htm.
15. The Global Information Infrastructure — What Is at Stake for the Developing World? Address to the Infodev Symposium, November 9, 1999.
16. A Survey of Information Communication Technology in Sub-Saharan Africa, Peter Van Heusden, August 1996.
17. U.K. Government White Paper — Modernising Government, March 1999.
18. Carter, D.L. *How Techno-Criminals Operate,* School of Criminal Justice, Michigan State University.
19. *Los Angeles Times,* posted 5:45 a.m., January 22, 1999.
20. Article entitled 100 Arrested in Global Sweep of Internet Pedophiles, by Jill Serjeant, Reuters, September 1998.
21. Reported Criminal Arrests under the Economic Espionage Act of 1996, R. Mark Halligan, September 2000.
22. Rapalus, P. *Issues and Trends: 1998 CSI/FBI Computer Crime and Security Survey,* Computer Security Institute.
23. CSI/FBI 2001 Survey, www.gocsi.com/press/20020407.html.
24. Statistics from http://www.netsizer.com/, March 2001.
25. http://www.utexas.edu/students/iig/archive/aprn/95/July06/index.html.
26. http://flag.blackened.net/revolt/zapatista.html.
27. Jones, A. Warfare and Extortion, *Journal of Financial Crime,* 6(2), October 1998. Henry Stewart Publications.
28. 1998 G8, The Birmingham Summit, release on drugs and international crime, http://www.g8summit.gov.uk/docs/crime.shtml.
29. INTERPOL press release CPN°02/00/COM&PR, INTERPOL and AtomicTangerine, dated February 21, 2001.
30. Hearing before the House Government Reform Committee, Subcommittee on Government Management, Information and Technology, March 9, 2000, statement of John S. Tritak, Director Critical Infrastructure Assurance Office.
31. Finaliza con Éxito la Primera Campaña Nacional Antivirus Informáticos, http://www.sgc.mfom.es/sgcinfor/articulos/antivirus3107.htm.
32. Singapore Military to Tackle New Threats, "Cyber-Attacks," Associated Press, March 14, 2000.
33. The Legal Framework — Unauthorized Access to Computer Systems. Penal Legislation in 37 Countries, Stein Schjolberg, January 2001.
34. Hoey, A. Analysis of the Police and Criminal Evidence Act s.69 — Computer Generated Evidence, *Web Journal on Current Legal Issues,* Blackstone Press Ltd.
35. http://www.columbuspolice.org/Home/.
36. http://eastonpd.com/htcu/intcybersnitch.htm.
37. http://www.ih2000.net/bpd/.
38. http://www.tempe.gov/police/.
39. http://www.rcmp-grc.gc.ca/index_e.htm.
40. http://www.herts.police.uk/.

Chapter 4

Our Changing World, or Is It Really?

Today, after more than a century of electric technology, we have extended our central nervous system itself in a global embrace, abolishing space and time as far as our planet is concerned.

—Marshall McLuhan[1]

This chapter provides a background on trends in the world around us because the world environment at all levels has an impact on why, when, where, and how information warfare is prosecuted. Economics, business, global competition, the "withering" superpowers, terrorism, regional conflicts, economic and industrial espionage, and the blurred distinction between peace and war are all discussed. All of these affect our information environment (IE) and may ultimately lead to various types of information warfare as a subset of economic, business, political, or military warfare.

Is Our World Really Changing?

The first three chapters of this book discussed this new phenomenon called information warfare, technology and high technology, what they are and what changes they are causing in our world, to include those brought on by the GII, NIIs, and the Internet. So, yes, there are changes occurring — always have been and always will be. However, when one looks at the world today and the world a hundred or more years ago, one sees that many things have not changed.

We have today, as in years past and probably in years into the future, what is called "systemic problems." That is, once one gets past what appears to be the major problems, we find that they are not really problems but rather the symptoms and results of systemic problems. These systemic problems should be obvious to us, but often they are not. They are the human-related, human-caused problems that have not been solved. In fact, in some ways we are making them worse. Systemic problems have consequences, one of which is conflict — a conflict in which information warfare plays a vital role. The two basic human traits that are or cause systemic problems are greed and the need to survive. Along with these, add the ability that humans have to rationalize their actions, even when others consider them irrational. Couple all that with one's perception of how things are and one has the makings for what we have today — conflicts between regions, nation-states, businesses, groups, associations, and individuals.

The human race seems by nature to be a violent race driven by wants and desires. We are a species of wants and desires. We want to survive. Thus, the need to satisfy our inner hunger often causes conflicts. If one is hungry, even if that hunger is caused by one's own doing, and sees that the neighbors have food but will not share it, we rationalize warring with them to get food, to survive. Once we are past the basic needs of food, clothing, and shelter, we expand our needs to wanting more — bigger shelter, more materialistic things, more food, and more than that of our neighbors. Even that cannot satisfy us humans. We then want power. Ever wonder why a person who is a multi-millionaire wants to run for public office at such a comparatively low salary? There are those who want to give back to society that which had allowed them to take so much. Yes, there are probably a few of them — or at least some start out that way. However, when they have everything else, they want power. If they have power, they want more power.

Fueled by egotism, jealousy, and revenge, the quest for power at any level of the human race causes conflicts. These conflicts can be applied across all religions, races, and nationalities. And because the world (nation-states, businesses, associations, groups) is made up of people, these human traits are transferred to the various levels of our societies and these entities. Thus, these various entities also become greedy, have the need to survive "instincts," and the quest for power. Yes, the world, because of high technology, is changing; but our basic, human-driven world does not appear to be changing. So, one can assume that although environmental changes have occurred (e.g., from one of hunter-gatherers, to agricultural, to industrial, to information-based); however, we humans really have not changed all that much — at least not as much as our environments have changed. Therefore, although information warfare, based on high technology, has changed how wars are being prosecuted and will be prosecuted in the future, it has not changed the reasons for wars.

Information warfare is being used by nation-states to further their national objectives, whether they are economic, military, or political. Corporations are using information warfare tactics to assist in gaining a competitive edge in the new battlefield — the global marketplace. Groups and individuals are using high technology to conduct information warfare to assert their independence, their power, and their influence on others.

The changes taking place in this our Twenty-first Century can be categorized and viewed from three basic revolutions,[2] driven by high technology, that are taking place today on a global scale. They are the:

- Economic revolution
- Information technology revolution
- Revolution in military affairs

One can no longer look at the world from a local or nation-state perspective, or even think in those terms. One must view their place not from that of living in an isolated nation-state, but living in the entire world. One must not think nationally. One must now think globally because what happens in another part of the world will more than likely have an effect on one's own part of the world. This impact felt half-way around the world often impacts us adversely or favorably because we are members not only of a nation-state, but also employees of corporations, members of a group, or individuals. The following are just a few examples:

- Asian economic crisis having economic impacts (e.g., imports and exports around the world)
- Price of oil or oil output from the Middle East having political and economic impacts on nation-states, corporations, and individuals
- Nation-states support for another nation-state against a third (e.g., the United States' unbalanced support of Israel in the Arab-dominated Middle East; United States troops were sent to Lebanon and subsequently hundreds died in a terrorist attack on their barracks; United States corporate offices and citizens attacked around the world)
- Any major crisis in the world affects the financial markets and thus stockholders, whether they are corporations, associations, or individuals

Why are groups as disparate as national governments, law enforcement and security officers, terrorists, organized criminals, and businesses all investing heavily in the cyber domain? The reason is readily apparent: civilization is experiencing a fundamental transition, from a world managed by and devoted to physical "atoms" to the rapidly unfolding world focused on bytes and electronics. Information warfare must inevitably become more common in this new, highly competitive global environment.

We have an unprecedented pace of change in communication driven by high technology and, more specifically, the global information infrastructure based on the Internet. One study, conducted by the U.S. Department of Commerce, looked at how many years it took various technologies to achieve 90 million users. The results were:

- Broadcast radio — 38 years
- Broadcast TV — 14 years
- Personal computers — 16 years
- Commercial Internet — 4 years

The bottom line is that the world wants to talk to the world. Although there have been some setbacks with such pioneer systems as GlobalStar and Iridium,[3] the goals are the same: to talk to anyone, anytime, and anywhere. Even now, one cannot travel to any nation in the world without seeing people communicating through their cellular phones, notebook computers, and PDAs. There is a "need," a "drive" for people, businesses, and governments to communicate information and profit from rapid receipt and dissemination of information — or the withholding of information. Furthermore, they want to be able to communicate literally at the speed of light. This need and drive cause the users of these communications systems to be vulnerable to information warfare tactics because the high technology is vulnerable.

Demise of the Superpowers

One of the changes that has recently taken place in our world is the gradual demise of the superpowers — or at least much of their power to act independently. The days of the true superpowers are quickly fading. The world is rapidly changing and it seems to be changing at an increasingly rapid rate. One of the most tremendous and most visible changes to recently occur was the breakup of a major world superpower — the Union of Soviet Socialist Republics (U.S.S.R.). It was as though a small crack opened up in the Communist wall and then the entire system collapsed. The thought of this even happening, and happening with such speed, remains almost impossible to believe.

What caused such a breakup? Basically, it appeared that the Communist system imploded. It imploded due to changes in other nation-states caused by their (other nation-states) use of high technology and the free flow of information. The Communist system could no longer compete in a rapidly changing world where economic power, backed not by guns and military might but by high technology, was the key to success. The historians will probably study and argue for decades as to what caused the breakup of the U.S.S.R.

One certain reason was that the U.S.S.R. could not compete economically with the other modern nation-states of the world, especially its archenemy, the United States. In addition, the United States, and to a lesser extent its allies, had upped the military and economic ante and thus the U.S.S.R. could not afford to compete, as much as it tried to do so. Was this a type of economic warfare, with early forms of information warfare as a subset, which was the major cause of the U.S.S.R.'s demise? We may never know for sure. However, we did learn that economic warfare supported by high technology in this global economy is now a form of warfare that has more serious consequences for victors and victims alike than ever before. However, economic power alone, or coupled with military strength, cannot make one a true superpower today. We may never again see a true superpower in the world.

Even the United States with both its military power and more important economic power has not been immune to these global changes. No longer can the United States dominate the free world or act on its own. The key that is needed today is not the old military power, which is fading in its importance, not political power, and not economic power, but all three combined with high technology power. One can see examples of that in the past major wars, from World War I, World War II, and Korean War, through Desert Storm, the Yugoslavia–NATO War, and the War on Terrorism. No nation-state could have succeeded if it had acted alone. It needed to build up a coalition of forces at least to give the appearance of a united front. Could the United States military acting alone have won the war with Iraq? Probably; but politically, without the support of some of Iraq's Arab neighbors to give at least the appearance of a united effort, the United States could not have been successful. For example, they would not have had a staging area in the region. The United States learned the difficulty of not having a regional staging area when it bombed Libya during Operation Eldorado Canyon, and France allegedly forbid the United States' military aircraft on that mission to fly over France en route to Libya.

Another example is the United States, a world leader in pushing for human rights throughout the world, being voted out of the United Nation Human Rights Commission in May 2001, and replaced by those nations that, shall we say, have a less than glowing human rights record.

> *UNITED NATIONS*— *In what amounts to a stinging rebuke, the United States has been voted off the UN Human Rights Commission in Geneva. This marks the first time the United States will not be represented on the commission since its inception in 1947. The commission investigates human-rights abuses around the world. "It's a stunning development," one council ambassador said.*[4]

We also see that the United Nations is being called on more and more to provide the leadership and forum in the global community. The United Nations today has peacekeeping forces throughout the world.[5] Since 1948, the United Nations has undertaken a total of 54 peacekeeping operations. However, 36 of them have been established since 1991. The number of active operations hovers between 14 and 17, with 15 active peacekeeping operations at the end of 2000.[6] Of interest is the fact that many of the smaller, developing nation-states have begun to play more of a global role, filling the void left by the decreasing political clout of the superpowers.

We see more and more conflicts in which nation-states once held together by strong authoritarian leadership are breaking apart and various ethnic groups are seeking their own independence (e.g., post-World War I breakup of the Austro-Hungarian empire). Two more recent examples:

- The breakup of the U.S.S.R. and the regional conflicts developing through their former regions, (e.g., Chechnya).
- The breakup of the former Yugoslavia once held together by strongman Josef Tito.[7]

Thus, when strong, centralized governments of modern nation-states run by despots allow their citizens some freedoms, it opens up the floodgates and causes the downfall of these nation-states. Is it any wonder then why today's similar types of nation-states are trying to control the flow of information into and out of their borders (e.g., The People's Republic of China (PRC))?

> *War to the hilt between communism and capitalism is inevitable. Today, of course, we are not strong enough to attack. Our time will come in fifty to sixty years. To win, we shall need the element of surprise. The Western world will have to be put to sleep. So we shall begin by launching the most spectacular peace movement on record. There shall be electrifying overtures and unheard of concessions. The capitalist countries, stupid and decadent, will rejoice to cooperate to their own destruction. They will leap at another chance to be friends. As soon as their guard is down, we shall smash them with our clenched fist.*
>
> —Declaration of Dmitry Manuilski
> Professor of the Lenin School of Political Warfare in Moscow, 1930[8]

While these breakups are occurring, there is another revolution taking place in the world and that is a religious revolution. Those of the Islamic faith are leading this revolution. It is not only growing in the Middle East, but also in Europe and Asia. The call to arms in the name of a religion is not new and has been around for as long as there have been competing religions. The Crusades of old are a prime example of warfare in the name of religion. Thus, this is nothing new in our changing world. What is new is that high technology is being used to support these wars and information warfare is a key subset.[9]

These groups are also in conflict with each other; and although they have much in common (e.g., language, culture, needs, and resources), they cannot agree. If nations such as Iran, Iraq, Saudi Arabia, Syria, Kuwait, and others ever come together to form a sort of Middle East Union, similar to that of the European Union, we may see the emergence of not a superpower, but a "super-region." The world may then be ripe for conflict with the "non-believers" of the both the West and the East. The Middle East Union would have the resources to mount attacks against other nations and by withholding its oil to the West, while also striking those nation-states where they are most vulnerable — in their information-based systems (e.g., their NII).

> *The first United Nations peacekeeping operation was "an attempt to confront and defeat the worst in man with the best in man" to counter violence with tolerance, might with moderation, and war with peace. Since then, day after day, year after year, UN peacekeepers have been meeting the threat and reality of conflict, without losing faith, without giving in, without giving up.*
>
> —Kofi A. Annan, United Nations Secretary General[10]

The Nation-States: Trying to Cling to Power, Control, and Survive in a Changing World

Throughout history, groups, clans, empires, and colonies have given way to this entity we today call the nation-state. One thing we have learned by studying history is that this idea of a nation-state is not fixed in time. Nation-states are often born out of warfare and die out in warfare (e.g., Yugoslavia). There are no guarantees that they, even as a concept, will exist forever. However, we do know that they will not dissolve peacefully, and there seems to be more and more people employed by governments and who rely on governments for their very survival. Conflicts will arise between these, between those who, even with good intentions, want the nation-state to survive in order for them to survive, and those who want more freedoms and liberty than the nation-states are willing to give.

The nation-state, just as we individual humans, will fight to survive. They will fight to survive although that will mean more human rights violations within its own borders, and more conflicts with its neighbors and others. It also means that they will use all means at their disposal to assist in that conflict for survival, much of which is high technology and information warfare based. Those citizens of the nation-states who want less government control and more individual freedoms are even today fighting daily for that freedom and independence.

In the past, citizens of the various nation-states were able to live in a world that was somewhat isolated. Communications were slow and, thus, news traveled slowly. One example of that as it related to warfare is the United States in the War of 1812. The battle, known as the Battle of New Orleans, was fought after the war was over, simply because the combatants did not receive the news that the war had ended. In the early Twentieth Century and before, one was more involved in the internal affairs of one's own nation-state and also what was happening in the region around them. When one thought of warfare, one first thought of defending the home front, and that meant defending the physical borders. The majority of political, economic, and warfare decisions were made looking at their nation-state as more or less the center of the world. Yes, for some nations there were the colonies. The colonies expanded the outlook of nations such as Britain, France, Spain, Portugal, the Netherlands, and other predominantly European nations. However, the use of the colonies was not for the greater good of the colonies, but for the greater good of the homeland. The main thrust for most nations was the homeland, and to a certain extent that is still true today, although there are more global influences of major proportions on a nation-state than ever before.

With the demise and reduced influence of the world's superpowers, it appears that there are more conflicts than ever before. These, however, are at least on more of a "controlled and limited" regional scale and therefore do not, at this time, appear to be about to literally obliterate life on Earth. However, there is one thing that we have all learned in information warfare: always expect the unexpected. That is, hope for the best but plan for the worst.

Just as our world is changing, so are the nation-states. Smaller nations are gaining a more level world playing field in dealing with global, regional, and national issues. Many of the world's smaller nations, especially those in similar regions, with similar goals, or with similar beliefs, are joining together and are having a greater influence on the world today. The cases alluded to above (e.g., the United States losing its ability to influence human rights) are indications of more power transitioning into the hands of smaller nations.

> *The last decade also showed a substantial increase in troop contributions of developing countries in United Nations peacekeeping operations. For example, in the beginning of 1991, out of the top ten troop contributors, only two were developing countries, Ghana and Nepal, while by the end of 2000, eight out of the top ten contributors are developing countries: Bangladesh, Ghana, India, Jordan, Kenya, Nepal, Nigeria, and Pakistan.[10]*

At the same time, there are other trends taking place that greatly affect nation-states — in fact, their very existence. When we look back at the history of nation-states, we find that they have a relatively short history. Even today, new nations are forming while others are fading. This trend will have a significant effect on global peace and stability.

In their book *The Sovereign Individual,*[11] Davidson and Rees-Mogg explain why the nation-states will eventually be replaced or at least drastically changed from how we know them today. Furthermore, the transitioning of the power from the nation-states to the "sovereign individuals" will not be done peacefully as the nation-states use their power to survive. It is believed that information warfare is and will continue to play a key role in the nation-states' action to survive. One can see examples of this today. Many nation-states continue to try to find ways to control the information that their citizens can receive and also control what those outside the nation-state can receive. After all, in today's information environment, information is power and nation-states do not want to relinquish that power — even to their own citizens, except as the nation-states deem appropriate.

In the days prior to the advances in high technology, GII, and the Internet, this task was somewhat easier, especially in a closed society such as the former U.S.S.R. and The People's Republic of China. However, because of today's massive communications systems and devices such as cellular phones, faxes, computers, and PDAs, this is an impossible task.

Even in the United States, the industrial period's processes and philosophies related to power are being applied to information period issues — and without much success. For example, how can the United States control Internet gambling? It thinks it can do so by passing laws outlawing it. However, if one were to set up a gambling Web site in another nation and also use a foreign bank where Internet gambling is legal, how does the United States prevent its citizens from gambling on that Web site? To catch such people, federal agents would have to monitor e-mails and Web site traffic, possibly breaking into and monitoring the ISP, although it is in a foreign nation-state and it is probably illegal to do so without the permission of the ISP and its government.

A good example of this was seen in the United Kingdom, where the tax on gambling made it sensible for the major gambling organizations to move offshore to places such as Gibraltar and Malta. In 2001, this caused the British government to change the law and abolish the tax.

Someone once said that every form of government has one characteristic peculiar to it and if that characteristic is lost, the government will fall. In a monarchy, it is affection and respect for the royal family. If that is lost, the monarch is lost. In a dictatorship, it is fear. If the people stop fearing the dictator, he'll lose power. In a representative government, it is virtue. If virtue is lost, the government falls.

—U.S. President Ronald Reagan[12]

It is not possible to effectively control access to information (e.g., the Internet). All one needs are a computer device and a modem. Such things as today's cell phones and PDAs also interface with the Internet, and so can faxes for receiving information. Thus, the freedom-fearing governments of the world must effectively stop, monitor, or control all information-carrying devices. Is this possible? No. However, the world's dictators will continue to try. That only shows their lack of understanding of the world we all live in today. This is a type of information warfare between nation-states and individuals.

The Nazis tried to control information (e.g., radios) during World War II in their occupied lands. They were not successful, as the members of the French Resistance units can tell you. So, can any government leaders anywhere in our information-rich world, regardless of their form of government, effectively control the flow of information? No. The Communist Chinese despots tried it at Tien An Men Square. Were they successful? No — and they had much better equipment than any of the other Asian or world despots who are today still trying to control their citizens' access to information, such as the use of the Internet.

In order to gain control of information, those currently in control, whether they be nation-states, corporations, groups, or individuals, must agree to relinquish control or have it taken from them by force.

—Author Unknown

One must remember that the Internet is a communications tool but there are others. In today's global information-driven environment, one cannot stop people from getting information. If one stops the Internet, one must also stop cellular telephones, faxes, magazines, and newspapers — all things that provide information and many of which can be connected to the Internet. Furthermore, if such access was limited to only the "trusted" few in government, who does one trust? Despots are by their very nature paranoid and trust no one. It is like squeezing the air in a balloon; squeeze it at one end and it expands in another. Squeeze it too hard and it will eventually be destroyed. So will governments that do not have the support of the people. Those that try are only fooling themselves. In the end, they will lose. It is just a matter of time.

If governments truly want to maintain a leadership role in their nation and stay in power, then they should lead the people to their new-found freedoms. They should embrace the Internet and expand its use throughout their nations. As the people gain in knowledge, they can help bring their nation into the

Twenty-first Century while still maintaining their religious beliefs and culture. For example, the nations of Central Asia have a long, great cultural history and strong religious beliefs. The power of the nation, its government, and its people is based on that cultural heritage and religion. These can stand the test of time and are strong enough to withstand any negative external influences. Leaders must have confidence in their people, their culture, and their religion. They have nothing to fear; but then again, because they use fear as a weapon, maybe that is all that they know. However, more sophisticated information warfare tactics driven by faster, cheaper, more powerful, and more available high technology is beginning to change the power balance from almost exclusively in the hands of the nation-states to those of groups, associations, and individuals.

With their great cultural and religious strengths, these nation-states can also use the Internet to let the world know of their nations, promote free trade, and raise the educational and economic levels for all their people. They have a unique opportunity in history to bring their culture, their beliefs, and their knowledge to the world through the Internet and other high technology. This subset of information warfare (i.e., propaganda) is not new, but it can also be used for the greater good. By doing so, they enrich all of mankind and we can learn more about them and they about us. In that learning, communications and understanding take the place of misunderstanding and warfare. A worthy goal, and another reason for these despots to support the use of the Internet by their people.

They should embrace it, nourishing its growth throughout their nations. If they do so, they will be considered great leaders. They will be welcomed into the community of freedom-loving nations of the world. One cannot think of a better legacy than to be known as the nation's leader who brought prosperity and freedom to the people.

Yes, even if one has omnipotent powers, one can try to stop the flow of information and the Internet. Those that try are spitting into the wind. The strong wind of freedom is blowing across this world and nothing can stop it. Many of the nations in this category were part of the former Soviet Union. Can they not see? Can they not learn from the mistakes of the Soviets? Where are the statutes of Marx and Lenin today? These dictators cannot live forever, but the will of the human spirit and right to basic human freedoms will never die. The despots of the world's nation-states have a choice: change and use the Internet to bring peace and prosperity to their nation-states or have their statues wind up on the trash heaps of history. Is that what they want for their legacy and their families? For after they are long gone, their peoples' human spirit will eventually overcome all obstacles. One cannot stop the will of the people forever — especially in our ever-growing, information-rich environment.

It is not true to say that governments that control the Internet really have the power, and not the people. Can the governments really maintain total control of those systems? As everyone knows, even 13-year-old kids are breaking into computer systems. What would stop someone from spoofing a system and masquerading as an authorized user? No systems are that secure on the Internet. Based on the vulnerabilities of the Internet and networks in general, one cannot stop the flow of information from those who want to use it to spread the spirit of freedom.

If governments refuse to let their citizens access information on the Internet, for example, how can it help the nation-state grow and compete in a world where access and the free flow of information is a key ingredient? Governments can try to stop their people from getting Internet access but, as stated previously, there are ways around that. And there are other ways of getting information to the people. Some governments believe that by refusing to let their people access the Internet they can control the flow of information. As stated earlier, they cannot. They just do not understand this information-based and information-driven world we live in. It seems that by trying to keep their people in ignorance, they have also done the same to themselves. How else can they be so blind as to their unique opportunity to help their nation?

And what about the possibility down the road that the Internet could include these people in international electronic and mobile commerce, thereby increasing the opportunities for democracy and liberty? If these government leaders were visionaries, cared about their people, their heritage, their culture, and their religious beliefs, they would embrace the Internet. Are these despots so blind? Are they so insecure that they cannot see how great they can be and how great they can make their nations? If only they would lead their people into the Twenty-first Century and use the Internet as a tool instead of pushing them back into the poverty and ignorance of the early Twentieth Century.

U.S. President Ronald Reagan once called the Nicaraguan rebels "freedom fighters" while their government called them criminals and terrorists. The same holds true throughout the world. Those in power want to keep it and make the rules. Those who are left out of the governmental process want their voices to be heard — thus, the conflict. As the nation-states begin to have less importance, its employees — those who work for the government or have power and influence because of their relationship with the government — will fight to keep their power and conflicts will rise at the expense of personal liberties. Information warfare tactics will be used extensively in these conflicts.

To counter such "revolutions" by individuals, the nation-state is increasingly using high technology to fight its information warfare battles. George Orwell's vision of the future in his book, *1984*,[13] may have been a little premature, but it is by no means invalid. Exhibit 1 shows some examples of what some nation-states are trying to do to maintain control and not relinquish power to the "sovereign individual," sometimes under the guise of protecting their citizens (see Exhibit 2).

> *When governments try to tightly control the flow of information — including information on the Internet — they only show a fundamental misunderstanding of the Information Age. What governments and others who want to control information don't understand is this — they are no longer in control of the information. Power is moving into the hands of individuals and away from the exclusive control of governments or corporations.*

> —Dr. Gerald L. Kovacich
> From a Radio Free Europe/Radio Liberty Article[14]

Exhibit 1.   Nation-States Trying to Maintain Control and Not Relinquish Power to the Sovereign Individual

Iranian Police Close Hundreds of Internet Cafés
Tehran (Reuters) May 13, 2000: http://www.infowar.com, May 14, 2001
Iranian police have closed down more than 400 Internet cafés in the capital Tehran, demanding that the owners obtain licenses to stay in business...telecommunications authorities had banned the use of Internet sites offering cheap telephone connections to relatives abroad, citing a state monopoly on long distance calls...

Comment
According to the article, Iran has approximately 1500 Internet cafés. As with most new technologies, the Internet is being used by the younger generation in Iran as in other nation-states. Shutting down these cafés is being done in fear by government agencies. They fear the free flow of information to the youths of Iran. Such free flow of information, they fear will jeopardize their control of information and thus the power and influence in that nation-state. This is not limited to Iran and Iranian youth. It is happening in a number of nation-states controlled by a select few at the expense of the majority. This is another example in the information warfare between individuals and nation-states.

Russian SORM
Jen Tracy, Feb. 4, 2000, Moscow and posted on infowar.com:
http://www.motherjones.com/news_wire/sorm.html
The Russian government has just authorized itself to spy on everything its citizens do on the Net — and to punish ISPs that won't help.... Russia's KGB successor agency, the FSB, launched a grand project — code-named SORM — to spy on its citizens' Internet transmissions.

Comment
This new law is alleged to allow numerous government agencies to spy on Internet users, read their e-mails, and do so without a search warrant or any type of legal probable cause. It should come as no surprise that even democratic nation-states are doing the same. For a nation-state to survive it must control the flow of information and that which it cannot control it must monitor. This is perceived by the leadership of the nation-states as a requirement for survival. They fear the power of individuals at their expense.

United States Lead High Technology Global Spying Network
Echelon
ECHELON attempts to capture staggering volumes of satellite, microwave, cellular, and fiber-optic traffic,.... This vast quantity of voice and data communications are then processed through sophisticated filtering technologies...

Note: *The above information, taken from the cited Web site, is designed to encourage public discussion of this potential threat to civil liberties, and to urge the governments of the world to protect our rights.*
http://www.aclu.org/echelonwatch/index.html.

Comment
Is this new? Should this shock us? Not very likely. Even as early as 1947, the United States' National Security Agency (NSA) and others were obtaining communications traffic over the latest technology-driven communications devices such as cable traffic. This was done under the code name "SHAMROCK." This was kept covert until

Exhibit 1. Nation-States Trying to Maintain Control and Not Relinquish Power to the Sovereign Individual (Continued)

approximately 1975, when the NSA Director, Lt. General Lew Allen, told the Pike Commission of the United States House of Representatives: "NSA systematically intercepts international communications, both voice and cable." So, times have not changed, just the technology — the methods of interceptions and what is being intercepted. Was this a 1947 version of information warfare? If so, why is the use of information warfare today considered a new phenomenon by some? *See also Blind Man's Bluff, The Untold Story of American Submarine Espionage, published by Harper Paperbacks, New York, 1998.*

United States' CARNIVORE

http://www.aclu.org/echelonwatch/highlights.html

The "Carnivore" Internet surveillance program, which is currently being used by the United States government, is somewhat similar to ECHELON....all TCP communications on the network segment being sniffed were captured by Carnivore...default configuration.... When turning on TCP full mode collection and not selecting any port, the default is to collect traffic from all TCP ports...

Comment

Again this is not surprising. The FBI says that they must obtain a court order to collect the information from a network. However, this is not that difficult a task and the FBI allegedly has very, very few requests rejected. It should be noted that under the United States' Federal Intelligence Surveillance Act of 1978, a secret wiretap court was created. The court approved over 13,600 wiretap requests in 22 years and rejected one.[a]

Once the FBI's request for an Internet wiretap is approved, they can simply connect a computer to the network in question and begin to vacuum all the information the FBI wants — allegedly only saving that information needed and provided for in the court order. However, in the interest of national security, even that court approval may not be required. Is this a change for the FBI? No, not really, just using high technology to do the same job as before but in a high technology-based information environment. The difference is that it offers a government agency the ability to harvest more information, faster than ever before. As we saw with the NSA, nothing has really changed, only the environment and tools used by nation-states. Again, this is a good example of information warfare brought into the Twenty-first Century's information environment.

China: Beijing Developing Internet Control Device

Hong Kong iMail Web site/BBC Monitoring Media, 3/21/2001

Beijing is working on developing a system similar to "black box" flight data recorders capable of monitoring Internet traffic as it seeks to tighten its surveillance of cyber world activities. In March 1998, Shanghai entrepreneur Lin Hai was jailed for furnishing 30,000 mainlanders' e-mail addresses to an overseas electronic dissident newsletter.... Members of the banned Falun Gong group have also been arrested for using the Internet to spread information about their activities and government efforts to crush the movement.

Comment

While the Chinese government continues to try to control the information accessed by its citizens and also what information its citizens send to the world, there are more and more Chinese citizens gaining access to free information from around the world. In fact, according to a *Newbytes* article,[b] *China continues to experience an overall surge in household connectivity, with users buying goods and services online and*

Exhibit 1.   Nation-States Trying to Maintain Control and Not Relinquish Power to the Sovereign Individual (Continued)

e-mail dominating Internet usage. This is according to a survey by netvalue.com. However, it is not just the United States, its allies, and the larger nations of the world who are attempting to monitor, filter, and/or block information that its citizens access. All of the world's nation-states appear to be using information warfare tactics in their vain attempts to control information. Even Myanmar's despots are trying to keep their citizens in "information ignorance" as have others, such as Indonesia, in its lost battle to keep East Timor.

Myanmar's Rulers Block Internet

Matthew Pennington, Associated Press Writer, YANGON, Myanmar (AP) Sunday April 23 12:17 PM ET: http://dailynews.yahoo.com/h/ap/20000423/tc/internet_dreams_1.html

Dozens of key-tapping students stare intently at computer screens in a cramped classroom three floors up a crumbling colonial terrace in downtown Yangon...these eager teens learn everything from operating systems like Microsoft Windows to programming.... But there's one glaring gap in the curriculum: the Internet.

Comment

As noted earlier, most nation-states appear to be attempting to control the flow of information to its citizens. Unfortunately, in today's highly competitive global economic marketplace, high technology and the flow of information are required for success. The despots in Burma would rather keep power and control than to bring the nation-state into the Twenty-first Century.

East Timor Cyberwar on New Global Battlefield: The Internet

Jakarta Post newspaper article, 8-7/99

Indonesia may be on the brink of a cyber war, but no blood will be shed in this conflict nor lives lost due to bullets. The battlefield of such a war is the Internet, with bullets the words, and soldiers the hackers.... The global communication system we have rapidly adopted...is becoming a conceptual space or site of political struggle, and a global high-tech battleground which transcends national geographic boundaries.

Comment

The battle for the freedom of East Timor is pretty much over; however, it was an interesting study in the use of IW tactics from attacking Indonesian government systems and Web sites to letting the world know the truth as to what was happening in that new nation-state. These attacks were staged from Portugal, Germany, the United States, Australia, and other nation-states based on a global "call to arms" of hackers to help the people in East Timor in their fight for freedom. The use of IW tactics seems to have played a major role in this success story. At the same time, it undoubtedly frightened other despots around the world causing some to further increase their strangulation of the free flow of information in and out of the nation-states.

---

[a] *Wiretaps sought in record numbers*, Richard Willin, U.S.A. *Today*,
     http://www.usatoday.com/life/cyber/tech/cti018.htm
[b] *Newsbytes*, Beijing, China, 16 May 2001, 6:41 A.M. CST.

Exhibit 2. European Actions Raise U.S. Free Speech Concerns

A landmark international case pitting the French courts against a huge American online company has resulted in a trans-Atlantic conflict over freedom of expression on the Internet and who, if anyone, controls it.... Correspondent Julie Moffett reports. Washington, Jan. 2001, (NCA/Julie Moffett) — U.S. free speech monitors fear French efforts to control what may be sold in France over the worldwide computer network known as the Internet may set a precedent that jeopardizes freedom of speech. They are carefully watching developments arising from November's decision by a French judge ordering an American Internet firm to block access in France to Internet sites that offer Nazi memorabilia for sale.

Americans, and others, say the action violates the right to freedom of expression. The French contend they are seeking to uphold a French law that makes it illegal for anyone to sell or display anything that incites racism. The sale or display of Nazi memorabilia is strictly forbidden.

Adam Clayton Powell, Vice President for Technology and Programs at the Freedom Forum in Washington, told RFE/RL that the case sharply highlights the differences in how nations view the issue of free speech, especially on the Internet. For Americans, he says the matter is simple. "One of the founding fathers of the U.S. who helped write the Bill of Rights said, 'We are not protecting respectable speech, we are not protecting speech that is polite.' Because respectable and polite speech is usually accepted, you don't have to protect it. What we are protecting is people who are rude, who say things that others may not like, because — and this is a crucial point here — the solution to bad speech, the solution to speech which is not correct, the solution to speech which is insulting, is viewed [in the U.S.] as more speech. More speech is better speech."

The case began last April when the American Internet company, Yahoo! Inc., became the target of a lawsuit in France because Nazi memorabilia items were available online in locations that could be accessed by French Internet users.

In November, the judge ruled that Yahoo had three months to find ways to block access to the Web sites. If Yahoo did not meet the deadline, the judge ordered for the company to be fined 13,000 U.S. dollars for each day it did not comply.

In the United States, freedom of speech is fiercely guarded. It is protected by the First Amendment of the Constitution, and has long been a cornerstone of American law, culture and politics.

Free speech advocates decried the ruling, saying it created a potentially dangerous legal precedent by giving one country the right to impose laws on Internet locations, called Web sites, that are based in other nations.

Yahoo lawyers claimed that France had no jurisdiction in the case and refused to abide by the ruling. The *Times of London* called the ruling an attempt by France to "impose international censorship" on the Internet.

On the other side, people supporting the decision said France has a legal right as a sovereign country to determine what materials can or cannot be accessed from its own soil, including those in the realm of the Internet, referred to as cyberspace. Internet experts also say the case clearly shows the difficulties of developing an international legal code for the Internet. The European Union is currently attempting to draft such a code, but many experts in the U.S. are skeptical. Part of the reason for that skepticism, says Powell, is that the Internet was technically designed to circumvent

Exhibit 2.   European Actions Raise U.S. Free Speech Concerns (Continued)

any form of blockage or censorship. Problems arise, he says, when nations start trying to impose restrictions on the Internet anyway. In a worst-case scenario, he says, successful implementation of any such regulations could severely limit freedom of speech — a dangerous precedent for the survival of democracy. "When you consider that for true democratic debate and dialogue, you have to have people who are willing to step forward and say things that are not only dissenting but sometimes very vigorously dissenting."

Powell also said that trying to hold millions of organizations on the Internet legally responsible for a variety of laws from different countries could have a chilling, not to mention, chaotic effect on freedom of speech worldwide.

> *I think that most people in the world are not sympathetic to Nazism or Nazi causes. But there are other variations of this. As soon as France figures out how to do this, or Yahoo figures out how to block certain things for France, you know that other countries that are not necessarily democracies are going to come right behind them and say, "All right, we don't want to have any criticisms of our government anywhere in our country."*

In a surprise move, Yahoo announced on Jan. 2 that it was issuing a blanket ban on auction items containing any symbols or materials associated with hate groups. Yahoo lawyers issued a statement insisting that the French court ruling played no part in the new policy.

"We decided we do not necessarily want to profit from items that promote hatred or glorify hatred and violence," the statement said. Regardless of the outcome, Powell says the case has illuminated the need for all nations — especially democracies — to work harder to preserve, not regulate, freedom of speech on the Internet. "In the civil voices that take place in political debate — in that multiplicity of voices — we believe there is truth, more than if there is an attempt to find just one voice that is the truth. It means it is not always as neat or cut-and-dried as certain centrally organized governments and societies. But we think it has worked pretty well for the past couple of centuries." Powell said that preserving free speech on the Internet will help all free societies, not just the U.S., thrive in years to come.

> *While the State exists, there can be no freedom. When there is freedom, there will be no State.*
>
>   —Lenin, *State and Revolution*, 1919[15]

Information Warfare Through Economic Espionage and Netspionage

When we combine our human instincts and desires with high technology, we find that we keep doing the same old things but in a high technology-driven, information-based environment. When it comes to spying by nation-states (e.g., stealing proprietary information, trade secrets, and government secrets), this certainly holds true.

As we know, the global Internet has rapidly been transformed from an academic playground into a global communications medium, the mainstream of government and business. One of many areas that have already changed dramatically is the area of gathering and using information and intelligence. Reports have already begun to circulate of how sophisticated criminals, terrorists, and especially the intelligence agencies of all major nations, are exploiting the new "cyber world" to their various ends. These organizational actors and an increasing number of individual "information brokers" are using the Internet to commit the old crimes of economic and industrial espionage in a revolutionary new way: netspionage (network enabled espionage).[16]

As we have seen over the past several years, nations are globally entwined in the marketplace. Such things as electronic commerce, balance of trade deficits, the Asian economic crisis, trade wars, dumping of goods and products, violation of copyright statutes, competitive intelligence collection, thefts of trade secrets, patent lawsuits, and global mergers are examples of this new business environment in which various forms of espionage are flourishing.

Broadly defined, classic espionage is the essential act of spying and has four major purposes, depending essentially on the sponsorship of the activity:

- Defense of a nation
- Assist in defeating an adversary
- Expand a businesses market share
- Increase economic power of a nation or business

However, as many analysts have noted, military power is now largely, if not wholly, dependent on economic strength. This strength is based on the strength of corporations that are increasingly rooted in the world of bits and bytes and less in the industrial age of mass (physical) production. During the past Asian economic crisis, many nation-states in that region cancelled or delayed military expenditures, as well as suffered setbacks in the drive to modernize their nation-states into more information-based and high technology-supported nations (e.g., Malaysia).

In this new information-driven age, the line between the espionage motivated purely by military advantage and the quest for market dominance is blurred, if not completely eliminated. The spy is held in high esteem in many parts of the world, and has been a decisive asset throughout history, as noted by Sun Tzu.[17] The Twenty-first Century, which many prognosticators expect to be the Information Age or the Age of Technology, may instead come to be known as the Age of the Techno-Spy.

This "second-oldest profession" thrives on demand for information. So, in a time when high technology and technology-based information equates to power, and the Internet provides unprecedented access to information, we expect the *sine qua non* of spying to be the skillful collection of information for every purpose via networks and other computer systems.

The popular media, including TV, newspapers, and trade journals, have already noted how hackers, governments, and businesses are already using the Internet as a platform that supports a wide range of computer and other crimes. Many criminal schemes depend on efficient gathering and transmission of information for their success.

Nation-states are focusing their economic espionage operations more and more on collecting information on corporations that are competing with their country's corporations for global market shares, a key to economic power.

Information Warfare and Corporations and Other Forms of Business

The world of business has gradually grown over its long history, from proprietary, family-owned small business, through medium and national business, to multinational conglomerates. Businesses come in all sizes. More and more of them have interests and trade all over the world; but this is hardly new, is it? The merchants of Asia and Europe traveled throughout the world. World trade is not something that was just invented in the Twentieth Century. Merchants go anywhere and everywhere they can to trade their goods. What has gradually changed over time is, of course, the same thing that has changed all our lives, and that is high technology.

From sailing ships, steamships, and diesel-powered ships, and from backpacks, horses, camels, trucks, and rail, goods have always moved around the world. The difficulty was finding merchants selling the lowest-cost raw materials and the markets where the most profits could be made. Yes, there were the few companies that traded tea from China and furs from Canada, but they were small in number. Along came radio and television to expand the market coverage through advertising. However, that still basically limited the market to a regional or national market. Then along came the Internet.

The Internet, as we all know, began as a United States military and academic tool. When it was being opened to others, there were many objections by academia and some government officials who wanted to maintain it exclusively for the use of a select few — academia for research and the military for communications. When businesses were first allowed to use the system, there were many who were afraid that it would be taken over by businesses because they had more resources to devote to using it and also were more aggressive in its use as they became aware of its business potential. And you know what? Those who objected to the Internet's use by businesses were right and businesses were also right when they saw its potential to expand their business. In just three years (1993–1996) the use of the Internet's Web capabilities by businesses (.com) went from 1.5 percent to 68 percent.[18]

To anyone who spends time surfing the Internet, it is quite obvious that the number of business Web pages (.com) by far surpasses the .org, .gov, .edu, and others. The Internet has changed the way businesses communicate, buy raw materials and other products, advertise, market, and sell products — in fact, every phase of business. Furthermore, the Web sites can be accessed by literally million of potential customers, and at a fraction of the cost of advertising in magazines, on television, radio, and newspapers, which have limited audiences. What has also changed is that a one-person business can look and act as large as a multinational conglomerate through its Web pages. It is also important to remember that this opportunity is being taken by businesses around the world, competing with others around the world. Thus, market competition among businesses selling competing products — or even the same products — is now global. Electronic commerce (E-commerce), business-to-business commerce (B2B), mobile commerce (M-commerce), or whatever you call it, is a rapidly changing and growing phenomenon, but then again so is everything associated with high technology.

More than ever in our history, businesses are now responsible for the economic power of a nation, with or without the help of their nation-states. And as noted earlier, the economic power of a nation has become more important than ever in the world community of nation-states. Tariffs, imports versus exports, and balance of trade matters have also taken on increased importance as all those involved must understand the global marketplace on almost a daily basis to take advantage of its opportunities and protect against falling markets due to falling currency prices, regional conflicts, etc. At the same time, the very competitive global marketplace has a growing number of conflicts, especially when the economy declines and customers are few in number.

A more recent phenomenon is the increase in global corporations whose tentacles are spread throughout the world. These mega-corporations have more power and control over citizens of some nation-states than does their government. In fact, these mega-corporations often employ and "take care" of more of a nation's citizens than does the nation-state's government. Could these mega-corporations replace the nation-states in the future? It is quite possible because they now often provide them with health and welfare benefits, wages, sometimes even housing, transportation, and a place to shop.

As we enter the Twenty-first Century and the "fourth-generation warfare" environment, businesses will have increased threats as they become lucrative targets in the information war and other forms of warfare.

High-Technology Professionals and Companies

As everyone knows, the advent of high technology has brought with it entirely new professions. In fact, high technology has matured to the point where governments have seen fit to develop and implement separate sections or entirely new directives associated with high technology. For example, the British have developed some employment guidelines relating to high technology, a portion of which is quoted below as an example:

> *B.C. Reg. 396/95, the Employment Standards Regulation, is amended by adding the following section:*
> *Exclusions — high technology companies*
> *37.8 (1) In this section "high technology professional" means a person who*
> > *(a) is a computer systems analyst, manufacturing engineer, materials engineer, Internet development professional, computer programmer, computer science professional, multimedia professional,*

> *computer animator, software engineer, scientific technician, scientific technologist, software developer, software tester, applied biosciences professional, quality control professional, technology sales professional (other than a retail sales clerk), electronics engineer, or any similarly skilled worker.*
> *(b) in addition to a regular wage, receives a stock options or other performance based compensation package set out in a written contract of employment, and*
> *(c) has one of the following qualifications:*
> > *(i) a baccalaureate or licenciature degree;*
> > *(ii) a related post-secondary diploma or post-secondary certificate;*
> > *(iii) equivalent relevant work experience.*

These professionals are not only employed to enhance the competitiveness of a business or groups, but are also becoming highly trained information warriors inducted into special units to fight the information warfare battles of the future. They are not only defending information systems, but also developing sophisticated attack tools to be used against an adversary — foreign or domestic.

Information Warfare and the Business World

In the Twenty-first Century, the control, power, and money will be in the information coursing through cyberspace and the Internet, global information infrastructure (GII), and national information infrastructures (NIIs). By some estimates, intangible assets of the enterprise (i.e., the knowledge, trade secrets, and sensitive proprietary information) already constitute 50 percent or more of the value of a modern business organization. Such assets are now most commonly found in digital form somewhere in the organization, and often on the corporate Web site. This makes the "crown jewels" ever more accessible to those with motive, opportunity, and the capability to work with network-enabled tools to capture information. In short, they are perfect targets for various information warfare tactics.

Because businesses have grown so dependent on the Internet, GII, and NIIs as an integral part of their businesses, they too are vulnerable on a global scale due to the vulnerabilities of the high technology that supports their businesses. One might ask, "What does this have to do with information warfare?" History is replete with examples of wars between nation-states where the businesses were also the victims. One just has to look at the Allies bombing of German industries in World War II to see that businesses of a nation-state at war are considered military targets. When information and high technology-dependent nation-states become embroiled in conflict, the Web sites of not only government agencies, but also businesses become prime targets. Furthermore, the massive telecommunications, microwave, and satellite systems used by and often owned by businesses are and will undoubtedly be attacked. More than 90 percent of the United States' critical infrastructure is owned by businesses.

> *"Rapid growth coupled with a huge pool of potential users will work to ensure Asia's place at the forefront of the Internet revolution," said Douglas Jaffe, analyst with IDC Asia/Pacific's Internet Market research. "Internet users and E-commerce revenues are increasing rapidly, and the wireless Internet is poised to transform Asia into a key global growth center for M-commerce."[19]*

The shift to Asia for business is changing the global economic power base. This may cause increased friction among nation-states. After all, today's Western nation-states do not want to lose their world leadership and power bases, yet ironically at the same time they are driving the shift in economic power to Asia. With economic power comes the ability to increase high technology as well as military spending. It follows then that the political power of those Asian nations affected will also grow. Where this potential power shift and potentially increased friction may subsequently conflict is in the The People's Republic of China (based on its current form of government and attitude). China may once again become the world's center as the "Middle Kingdom." The following is an example of how the Western nation-states are helping make all this possible.[20]

China Unicom, China's second-largest telecom carrier, awarded $1.46 billion in wireless-infrastructure contracts:

- Lucent Technologies Inc. reportedly received ten contracts worth an estimated $420 million.
- Motorola Inc. received 11 contracts totaling $407 million.
- Canada's Nortel Networks Corp. was awarded $275 million worth of new business.
- Sweden's Ericsson AB received more than $200 million in commitments.
- The remainder was divided among a handful of Chinese equipment makers.

At a time when some nation-states have a slowing economy and new business contracts are also subsequently slow, such awards are great news to corporations such as Motorola and Lucent who have seen their stock values and business decline. However, these contracts come with a price because China often expects high-technology transfers to take place with the award of these contracts. Thus, businesses are stuck between accepting such contracts, even if they must release trade secrets and proprietary information, or have foreign businesses, often backed by their home governments, go elsewhere. Thus, corporations, many times with the support of their governments, may be providing the high technology to a potential adversary, but it sure is difficult to resist over one billion potential customers, is it not?

On the bright side, the more a sophisticated, high technology-based nation-state provides high technology to a potential adversary — whether that adversarial relationship be one based on economics, military power, or a different form of government — the more that nation-state makes its adversary vulnerable to information warfare attacks. In addition, the modern nation-state, acting as the seller, is already quite familiar with these vulnerabilities and how to attack a nation-state relying on them. That would include anything from insecure Web sites of government agencies and businesses to changing economic data, scrabbling database information, denial-of-service attacks, etc. If one were to do this to a competitive business, the impact could be devastating, with losses of revenue and time. That old adage is truer now than ever before: in the business world, time is money.

Another interesting phenomenon associated with high-technology business is a foreign government's ability to dictate information protection to others. For example, in the European Union (EU), there is a drive for privacy protection. In doing so, if one were to connect to certain European Union member networks, one might be required to install certain privacy protection controls dictated by the European Union or one of its nation-states. Thus, a business physically located in one part of the world might be required to conform to the controls established by a foreign power physically located at the other end of the world.

The issue of privacy is being talked about throughout the world. It is bad enough that no communications sent and received via high technology are safe from government surveillance. They are not even safe from the prying eyes of businesses. In fact, information brokers are busily collecting information on people, associations, or businesses for which they can find a customer. Global businesses are not only collecting information on their customers and competitors, but they are often also selling that information stored in their massive databases to others that are willing to pay for them. Is this conflict between privacy advocates and the collectors and sellers of private information a concern? Yes, of course it is; but is this conflict a form of information warfare? Is it an information warfare concern? It is based on how one defines information warfare as discussed in Chapter 1. What do you think?

> *Amazon continues to publish the purchase circles (albeit less prominently). Systematic tracking of user behavior and purchasing is key to most current business models on the Internet.*
>
> *Firms hope that by learning more about surfers, overtly or covertly, they can precisely target ads and E-commerce opportunities. Putting cookies onto surfers' browsers and hence building deep profiles of consumer behavior is a key objective of almost every high-traffic site.*[21]

Mergers, Acquisitions, and Joint Ventures

Another possible form of information warfare taking place is when businesses are in the process of merging with others, acquiring others, or even just involved in joint ventures. If you were competing with a rival firm to buy another rival firm, would it not be nice to covertly determine its cash flow, customer database information, and trade secrets? Yes; but in most nation-states, such activity may be illegal. However, some believe it is only illegal if you get caught and only then if you are found guilty in a court of law. Even then, maybe the "downside" is slight and the risk worthwhile. Information protection and defenses as part of any businesses warfare or information warfare strategy may help prevent this loss of information from occurring.

Unfortunately, most businesses systems, especially those connected to public networks such as the Internet, are vulnerable due to a lack of adequate protection and defenses, or maybe just because some of the high technology is just inherently vulnerable.

> *Brian Dunham has a hot Internet business idea, but he worries that someone will steal it...blocked potential competitors from finding his brand-new Web site. When the rest of the world clicks on eframes.com, it sees a Web business that frames and ships digital photographs overnight. But likely rivals get only a dummy site sporting this message: "Coming in time for Christmas!" Known to insiders as Web-access blocking...this maneuver is made possible by the growing ability of computer programs to identify Internet users — technically called "domain-name identification" — Web sites can secretly see where visitors are coming from the moment they click on....[22]*

Nation-States and Businesses Information Warfare Cooperation

Nation-states and businesses have been cooperating for years. That is nothing new, and has led us into the high-technology revolution through such things as the United States sponsoring the development of computer systems and the greatest project of them all to date — the Internet. Cooperation has, of course, also led to the development of military weapons. With the high-technology revolution, it only follows that this cooperation has been extended into the era of information warfare weapons.

Yes, even the telegraph and telegram companies and the post offices have also cooperated with government agencies when it came to "reading other people's mail." Prior to the advent of the Internet and more sophisticated monitoring, eavesdropping, listening, and intelligence collection devices — or whatever you want to call them — the task was more labor intensive and also more specific when targeting an individual, government agency, or association. However, as discussed above with Echelon and Carnivore, the collection has grown not only more sophisticated, but also more massive in scope. One reason is that it is so easy to do and difficult to resist the temptation of narrowly focusing this information warfare weapon.

The development of high technology-based, information warfare-driven weapons has bred entirely new high-technology corporations and government contractors. A "high-technology company" implies a company in which more than 50 percent of its employees meet the definition of a high-technology professional, are managers of persons meeting the definition of a high-technology professional, or are employed in an executive capacity.[23] These high-technology businesses are often hidden in high-technology parks and are therefore difficult to differentiate from the "ordinary" high-technology company. Some are even general high-technology companies that have some small, classified areas where the covert development of high-technology weapons are developed for nation-states' use in the information warfare environment. Such weapons may be offensive or defensive. Because of the ability of offensive weapons to almost totally devastate a modern nation-state dependent on high technology, they are considered as powerful as nuclear weapons and protected with the same type of security.

One example was the push by approximately 20 or so nation-states to require that manufacturers and operators build in "interception interfaces" to the Internet and all future digital communications systems.[24] Do the manufacturers and operators have any choice in the matter? Must a court order be obtained first? That all depends on the nation-state involved. However, if one obtains a lucrative contract from the government, will a business pass up the opportunity to build the hardware and software requested by the nation-state? Of course not! The rationale often is that "if we don't do it, someone else will." What is interesting to note is that this type of covert project also has not only information warfare potential as a tool for intelligence collection, inserting false information, etc., but it may also be vulnerable to penetration by an adversary who could also make use of it. It is suspected that the businesses of the world and nation-states are developing many such information warfare weapons that may go unheard of for decades. After all, many military weapons have been in that "black closet" for years. Why would it not be so in the future?

Organizations, Associations, and Other Groups

We humans band together whenever we find commonality of purpose, a cause worth fighting for, or just share information and views on specific topics. Even nation-states form such groups (e.g., United Nations, Association of Southeast Asian Nations, and North Atlantic Treaty Organization). People with common skills, needs, and purposes form trade unions. These groups have been around much longer than high technology; however, they too have transitioned to using high technology as a tool. (See Chapter 14 for additional information on the use of IW by others, e.g., terrorists and hacktivists.)

One of the most powerful tools that all of these groups have in common is, of course, the Internet. Now, people from all around the world can share information and work on common causes. The Communist slogan, "Workers of the World Unite!," was difficult to envision actually taking place in the Twentieth Century; however, today such things are not only possible, but are actually taking place.

There are many groups, organizations, and associations around the globe that are using high technology and even information warfare tactics to reach their goals. Who are these people? Anyone and everyone who knows how to use high technology and who has a cause. These groups range from the world organizations that some nation-states classify as terrorists to animal rights groups who deface some furriers' Web sites. Their use of the high technology to support their objectives has been coined "hacktivism." The groups using the Internet's Web pages, e-mails, and chat rooms can get what they consider to be the truth out to the public, or they can also spread misinformation.

A lie told often enough becomes the truth.

—Lenin

Is It Cyber-Terrorism, Techno-Terrorism, Information Warfare, or None of the Above?

One of the best-known groups in existence today is the terrorist group. There has been a lot of talk over the last couple of years about "cyber-terrorism" — the use of cyberspace by terrorists. All the so-called terrorists experts and consultants are just waiting for any indication of such attacks so they can say, "I told you so. I told you it was coming." Many of them can already hear their cash registers ring as they ply their trade to corporations and government agencies. But alas, the only publicly documented announcement of the use of cyberspace, the Internet, for such an attack really was not — massive e-mails for a denial-of-service attack.

Some, looking for sensationalism, think of every attack against a computer as a terrorist act. When we hear of such talk or read about it in a newspaper column, the first thing we must do is consider the source and what motives they may have for saying such things. Let us begin with reporters. They may not be so much interested in factual reporting as getting their column accepted by their boss (who is only interested in selling newspapers) and a little publicity for themselves. After all, they have career goals just like the rest of us. The only thing they know about terrorism is what someone tells them. So, who is doing the talking? Consultants, so-called experts in terrorism and information systems security, and law enforcement and criminal investigators. Are they giving us advanced warning for our own safety, making a pitch for more budget, or trying to get a little publicity by hyping the threats, and of course, maybe a lucrative consulting contract?

Take a look at one of these articles, which has banner headlines that begin with the words, "Web of fear." We will not further identify the reporter or the newspaper because there is no sense in embarrassing them (if that is at all possible); they have plenty of company out there in selling sensationalism, which sells newspapers.

The article listed three examples of "cyber terrorism":

- *IRA sympathizers at the University of Texas disclosing sensitive details of British army bases in Northern Ireland on the Internet.* So, because they transmit the information via the Internet, is it a dire warning of cyber-terrorism? Is an IRA sympathizer a terrorist?
- *A "denial-of-service" attack by Tamil Tigers, who bombarded Sri Lankan embassies with e-mail, causing their Internet systems to crash.* Tamil Tigers are considered terrorists by those in power and "freedom fighters" by those who support them — as with any so-called terrorist group. Now, is this really a cyber-terrorist attack? If so, then we hope that all terrorists will come online and use this technique. Imagine the number of lives it would save!
- *During the recent bombing campaign in Yugoslavia, NATO found itself under attack by a flood of 2000 virus-laden e-mails a day.* Was NATO, during its bombing campaign in Yugoslavia, attacked by a cyber-terrorist? Why stop there? Was the latest global "attack" by the "Love Bug virus" a terrorist act? Some may want you to think so as they sit waiting for some

real terrorist attack to take place in cyberspace. The bombing campaign
in Yugoslavia was war. Is warfare terrorism? Remember, anyone can call
anyone a terrorist, or any incident a terrorist act.

The article went on to talk about viruses and hackers, *not* terrorism as
defined by most knowledgeable people.

An American television show, *America's Most Wanted,* aired an episode
about cyber-terrorism. *TV Guide* magazine said it was about, "Examining U.S.
government efforts to stop criminal computer hackers, who can disrupt vital
city services." It included an overview of cyber-terrorism and "easy targets;
wide-ranging effects; and a profile of a dangerous hacker." Now hackers are
terrorists. We seem to see all kinds of dangers in "hackers."

Well, before the sensationalists, "terrorist experts," and others begin to
increase their campaign of fear and hype, take a look at this entire matter of
terrorism as an example of an information warfare tactic and what is called
techno-terrorism.

To do that, start with some basic definitions. After all, the understanding
of what is meant by the terms "terrorists" and "terrorism" will help keep us
focused on the issue. We are using the United States definitions, quite frankly
because they were the easiest to find.

The U.S. FBI defines terrorism as follows: Terrorism is the *unlawful* use
of force or violence against persons or property to intimidate or coerce a
government, the civilian population, or any segment thereof, in furtherance
of political or social objectives. So, when those in power do it, it is lawful. It
is defending the nation-state and its citizens. When someone or group does
not like the government in power, it would of course be called unlawful by
the government in power. This is an important point because it depends on
what side of the government you are on. Often, that is all that separates
someone from being a terrorist and a government employee!

The U.S. Central Intelligence Agency says that international terrorism is
terrorism conducted with the support of *foreign* governments or organizations
and/or directed against foreign nations, institutions, or governments. A little
different twist because they look at it from the standpoint of terrorism against
the United States by outsiders.

The U.S. Departments of State and Defense define it this way: Terrorism
is premeditated, politically motivated *violence* perpetrated against a noncom-
batant target by sub-national groups or clandestine state agents, usually
intended to influence an audience. International terrorism is terrorism involving
the citizens or territory of more than one country. Interesting point here.
Therefore, by this definition, the Tamil Tigers e-mail flooding was not terrorism
because it was not violent.

What Are Terrorists Anyway?

They are whoever the people in power say they are. Do not confuse terrorists
with the "normal" criminals. Criminals are those that violate the laws of society,

usually for personal gain. A terrorist is one who causes intense fear; one who controls, dominates, or coerces through the use of terror. The same can be said for a child abuser. Shall we begin to call them terrorists? Actually, although not politically motivated, they are more of a true terrorist than those folks hacking the Internet.

Why use terrorist methods? Why do people resort to fighting against the massive powers of a nation-state? Most do so for one or more of the following reasons:

- When those in power do not listen
- When there is no redress of grievances
- When individuals or groups oppose current policy
- When no other recourse is available
- When a government wants to expand its territory (yes, a nation-state can be considered a terrorist)
- When a government wants to influence another country's government

What Are Terrorist Acts?

As stated earlier, they are what those in power say they are. Anything and anyone can be classified as a terrorist because those in power make the rules. Some questions to ponder: What is the difference between a terrorist and a freedom-fighter? Does "moral rightness" excuse violent acts? Does the cause justify the means?

Results of Terrorist Actions

When there are so-called terrorist acts, it often plays right into the hands of those in power. In fact, those in power may commit terrorist acts and blame it on their opposition. Why? Because it will call for increases in security — usually demanded by the people. Usually death, damage, and destruction cause governments to decrease freedoms in the interest of security. Note what happened after 9/11. On the other hand, it may cause awareness of grievances by the population; it may cause governments to listen; and it may lead to social and political changes. Some believe the bombings in Russia attributed to the Chechen rebels were in fact done by the Russia security personnel to get the Russian people to support the intervention of Russia into Chechnya, maybe a lesson they learned after their debacle in Afghanistan.

Terrorists' Technology Threat Environment

So are there cyber-terrorist threats or potential threats? Yes, because of the following reasons:

- More reliance on information to run businesses and governments
- Larger concentration on information that can be accessed

- Security is an add-on to technology — more weaknesses
- Destruction of automated information can cripple a government, a business, and the economy
- Information can be stolen and the theft is not known
- Computer electronic circuitry is vulnerable to interference
- A weapon that transmits a high-energy radio frequency beam can disable computer systems

Techno-Terrorists

It is easy to see threats everywhere, from anyone and at any time. However, potential threats do not imply that there are cyber-terrorists at work today. However, there are those who use technology for terrorist activities — what are called techno-terrorists. These people use technology to support or commit terrorist acts. The use of encrypted e-mails by terrorist groups; the use of Web sites to get their messages across to others; and the use of hacking techniques, theft, and fraud to raise money for their cause are all examples of the use of technology to further their cause.

Why Techno-Terrorism?

Techno-terrorism is easier, causing more disruptions. It promotes their cause with less negative public image and can be used by more terrorists with less funding. In addition, the punishment, if caught, for this type of activity is orders of magnitude less serious than for acts considered to be conventional terrorism.

Techno-Terrorist Possibilities

Some examples of techno-terrorism activities include:

- Use a computer to penetrate a control tower computer system and send false signals to aircraft, causing them to collide in mid-air or crash to the ground.
- Use fraudulent credit cards to finance their operations.
- Penetrate a financial computer system and divert millions of dollars to finance their activities.
- Bleach U.S.$1 bills, use a color copier, reproduce them as $100 bills, and flood the market with them to destabilize the dollar.
- Use cloned cellular phones and computers over the Internet to communicate using encryption to protect their transmissions.
- Use virus and worm programs to shut down vital government computer systems.
- Change hospital records resulting in patient deaths because of an overdose or the wrong medicine.
- Penetrate a government treasury department computer to issue checks to all citizens.
- Destroy critical government computer systems processing tax returns.
- Penetrate computerized train routing systems, causing passenger trains to collide or hazardous materials to be released.

- Take over telecommunications links or shut them down.
- Take over satellite links to broadcast their messages to televisions and radios.

Yes, all these things are possible and some may have already occurred. However, most are examples of using technology for terrorist support or acts, and not cyber-terrorism — the use of the Internet or other "cyberspace" for terrorism.

So, why are we not we seeing actual cases of pure cyber-terrorism? No one really knows for sure, but one can certainly speculate. Physical acts of violence have more impact than knocking out a government or corporate computer system. Such attacks will not currently help the terrorist groups reach their goals. Those of us living in an information-based society and surrounded by high technology tend to lose sight of the fact that most of the world is not systems dependent — nor are their governments. Until that day comes, brutal violence will continue to be the vehicle of choice for a terrorist group — however they are defined and by whom.

Individuals

High technology has played a major role in the global revolution. Communications are reaching more parts of the world than ever before. We see what is happening around the world as it is happening. High technology is being developed and integrated into the infrastructure of even some of the world's poorest nations. More and more people around the world are gaining access to computers and, through computers, to the Internet. As they become more computer literate and access more information on a global basis, they are beginning to learn and to find others who share their goals.

The Global Hacker and Hacktivism

The global hacker community, although often misguided as to worthwhile objectives in their attacks, are at least beginning to establish better lines of communication among themselves. The hackers of the world are using the Internet to communicate and attack systems on a global scale. Many of the attacks are aimed at totalitarian governments, government agencies, political parties, and against the slaughter of animals for their fur, all of which can be considered politically motivated attacks. These are the worst kind and most feared by nation-states. The mounted attacks by global hackers, based on a "call to arms" by other hackers, against the government of Indonesia's Web sites due to the East Timor issue is an example of what they can do and, more importantly, what is yet to come.

These hacker "freedom-fighters" are the ones who will be needed in the Twenty-first Century to protect our freedoms on a global scale against current trends of governments to control the Web and thus the information flow to the people; the governments that want more and more laws to control us; the governments that want to prosecute citizens if those citizens criticize their governments; and those governments that want to deprive us of the pure freedom that we all inherited as a God-given right when we were born human beings. We will not be able to rely on our governments or the military's info-warriors because they will be protecting the nation-state and the status quo — and it will not be the first time that has happened! *Hactivism* is the new term for old human traits of attacking what they do not like, such as violations of human rights. As one writer put it:

> *These are the self-proclaimed freedom fighters of cyberspace. They've even got a name for it: hactivism. And political parties and human rights groups are circling around to recruit hactivists into their many causes.... The government tries to put electronic activism into the peg of cyberterrorism and crime with its Infowar eulogies.*[25]

The fight between nation-states and individuals has many heroes and many villains. This battle has entered the information warfare arena. For some, the price has been very high; but if you ask any of them, they will say the cost is worth it. Chinese dissenter Guo Qinghai, who posted Internet writings with titles such as "Be an eternal dissenter," and "First democratize your own thinking," is a good example of such an individual, of an info-warrior. Although in the past they were fighting their battles in the information warfare arena, often alone and without defenses; however, because of today's GII, NIIs, and Internet, they are no longer fighting those battles alone.

> *Americans, Burmese, and others are waging an elaborate Internet assault against Burma's (Myanmar's) military regime, which has retaliated by e-mail and a Web site to deflect charges of mass executions, torture, and other abuses.*[26]

Of course, those in power in the nation-states do not agree with the hacktivists. For example, over East Timor's fight for independence, a call went out to hackers around the world to attack the Indonesian government's Web sites. Over 47 were attacked within a few days. "This is terrorism against democracy," responded a senior official of Indonesia's foreign ministry, Dino Patti Djalal.[27] Other articles are listed in Exhibit 3.

Summary

When we look back to the times of the caveman and up to today, we often fail to realize that the entity we call the nation-state has a very short history of only several hundred years. Some, such as Davidson and Rees-Mogg, believe that:

> *...Something new is coming. Just as farming societies differed in kind from hunting-and-gathering bands, and industrial societies differed radically from feudal or yeoman agricultural systems, so the New World to come will mark a radical departure from anything seen before...*[11]

Exhibit 3. Hacktivism in Print

Love Bug Variant Targets Echelon Spy System
Kevin Featherly, Newsbytes, 17 May 2001
A variant on the so-called Love Bug virus has been detected, and its creators apparently hope that by disseminating it on the Internet, the controversial Echelon eavesdropping system will become overloaded and rendered inoperable.

Comment
This again is a further example of global attacks against nation-states; however, this time it may be a "global citizen" and not one residing in one of the nation-states contributing to Echelon.

How the Government Will Track You Down
Paul Somerson, Small Business Advisor (date unknown)
…The government wants to use technology to track you down without a warrant. Do you trust these clowns?

Comment
The truth is that what was once the purview of moviemakers is becoming real life. Information warfare between individuals and corporations and between individuals and nation-states are a reality today. Privacy is the battleground. For more information, see http://dailynews.yahoo.com/h/zd/20010518/tc/how_the_government_will_track_you_down_1.html

---

Davidson and Rees-Mogg, among others, believe that the "sovereign individual" will rise out of the nation-states. If what they say is true, then we can all probably agree that the nation-state will not go quietly. Just as individuals fight for survival, so will nation-states. There are many rather recent examples, such as the former Yugoslavia, the former Union of Soviet Socialist Republics, and Indonesia.

There will be those citizens whose patriotism and "love of country" will support the nation-state over its citizens, to include many of our fellow managers, auditors, security, and law enforcement professionals. They will join with those in the government who do not want to lose power, as well as those corporations who find the power of the nation-state to be a powerful ally in the competitive world of global business.

> *The real issue is control. The Internet is too widespread to be easily dominated by any single government. By creating a seamless global-economic zone, anti-sovereign, and unregulatable, the Internet calls into question the very idea of a nation-state.*[28]
>
> —John Perry Barlow

From the looks of the legislative attempts by nation-states to control the Internet, GII, and NII, they are not getting the message. In any case, the collection of information via netspionage is a crucial ingredient, whether you are "Big Brother" or a "Freedom-Fighter." Like many of our significant changes throughout history, they are often violent as those in power fight to keep it. Therefore, the use of spies and spying on individuals through netspionage will most assuredly take place.

The old world is rapidly changing due to high technology and yet the human instincts and spirit remain the same, as does the drive for the businesses' competitive edge. The Twenty-first Century is ushering in a vastly new environment than was the case in the beginning of the Twentieth Century. Global competition, aided and driven by information and high-technology-based nations and corporations, is rapidly increasing.

To maintain control, gain power and market share, and beat competitors, nations and businesses are using information and technology as the new instruments of business and economic warfare — and information warfare is their main process.

More on IW-related individuals, groups, terrorists, and other miscreants can be found in Chapter 14.

Notes

1. Marshall McLuhan, *Understanding Media*. Signet, New York, 1964, p.19.
2. These three revolutions will be discussed or alluded to throughout this book. Some of the information presented was taken from the book, *Netspionage: The Global Threat to Information*, published by Butterworth-Heinemann Publishers and reprinted with permission.
3. Corporations that have launched satellites in order for anyone to talk to anyone anywhere in the world through wireless communications.
4. U.S. Ousted from U.N. Human Rights Commission; May 3, 2001; Web posted at 4:34 p.m. EDT (2034 GMT), CNN.com.
5. http://www.un.org/Depts/dpko/dpko/pub/pko.htm.
6. See http://.un.org/Depts/dpko/dpko/pub/pko.htm. 5/14/01.
7. Josip Broz Tito (Josef) b. May 7, 1892. d. May 4, 1980. Leader of Yugoslavia, 1945–1980. Prime minister from 1945–1953 and Communist President of Yugoslavia, 1953–1980, established it as a Communist state independent of the U.S.S.R. Led the Communist guerrilla resistance to German occupation and had the country 80 percent liberated at the time of the Normandy Invasion in France. Since WWII, celebrated his birthday on May 25. Yugoslavia was a country from 1918–1991, and included the republics of Bosnia and Herzegovina, Croatia, and Serbia. NATO bombed Tito's burial site in 1999 and the main residence was destroyed, taking with it a huge collection of items from Marshal Tito's reign. Information came from http://www.findagrave.com/pictures/8062.html.
8. There have been allegations that this statement is a fabrication and, in fact, no such person exists. The authors have tried unsuccessfully to verify the source. However, the philosophy if not the exact words was a valid U.S.S.R. goal — and to some in Russia, it still is. It was also quoted in the *Dictionary of War Quotations*, 1989, lists address of 1931; Brassey's Soviet & Communist Quotations #1512, quote in *Newsweek*, 1955; *Dictionary of Military and Naval Operations*.
9. See Web sites angelfire.lycos.com; www.iap.org; kosovo.net as examples of the use of high technology to fight one aspect of information warfare.
10. See http://www.un.org/Depts/dpko/dpko/home_bottom.htm.

11. *The Sovereign Individual, Mastering the Transition to the Information Age,* by James Dale Davidson and Lord William Rees-Mogg, published by Touchstone, New York, 1999.

12. Taken from page 16, *REAGAN, In His Own Hand,* The Writings of Ronald Reagan That Reveal His Revolutionary Vision for America, published by The Free Press, New York, 2001.

13. Originally published in 1949 by Harcourt Brace Jovanovich, Inc., and renewed in 1977 by Sonia Brownell Orwell.

14. Published by Radio Free Europe/Radio Liberty in April 2001 as part of their broadcast.

15. http://www.quotationspage.com/quotes.php3?author=Lenin.

16. For additional information on netspionage, see *Netspionage, The Global Threat to Information*, published by Butterworth Heinemann, Woburn, 2000.

17. Sun Tzu, a Chinese general, is credited with writing *The Art of War* for Ho Lu, King of Wu, who reigned in China from 514 to 496 B.C.

18. See http://www.mit.edu/people/mkgray/net/; Matthew Gray of the Massachusetts Institute of Technology provided this information.

19. http://cyberatlas.internet.com/big_picture/geographics/article/0,,5911_767371,00.html.

20. Information provided by William C. Boni, Motorola Corporation.

21. Reported by Stephen Ellis, *Australian,* 14/10/1999.

22. From an article by Michael Moss; *Wall Street Journal Europe.*

23. http://www.productmktg.com/product.html.

24. Article by Duncan Campbell, *The Guardian,* 29/04/1999.

25. Inside the World of a 'Hactivist', October 18, 2000, by Deborah Radcliff for *ComputerWorld.*

26. Internet Attacks Burma!, by Richard S. Ehrlich, Asia correspondent.

27. *Jakarta Post* column on East Timor, August 27, 1999.

28. John Perry Barlow, Thinking Locally, Acting Globally, *Time* magazine, January 15, 1996, p. 57.

Chapter 5

Business, Government, and Activist Warfare: The More Things Change, The More They Don't

To change and to change for the better are two different things.

—German proverb

This chapter provides an overview of how businesses, governments, and activists conduct warfare. There is a high positive correlation between business, government, and activist warfare in that the intelligence collection, physical destruction, and even computer network attacks have many similarities. All use information warfare (IW) to achieve their goals.

Introduction

Think of a time and a capability that have never been countered. Knight's armor? Crossbows and then long bows. Thick castle walls? Gun powder. Tanks? Bazookas. Better or reactive armor? Thirty-millimeter depleted uranium rounds fired from an A-10 Warthog at a rate of 2000 rounds per minute. Communications? Intercepting and copying the transmission. Encrypting the communications? Crytpomathematicians using computers with encryption-breaking algorithms. The Germans thought their Enigma encryption machine was unbreakable, but Polish and British cryptographers proved them wrong. Human creativity and technology provide capabilities, and capabilities can be turned into advantages. The nonlinear compression rate of change in technology — computer hardware cycles about nine months and software cycles about 18 months — capabilities and competitive advantages today are short-lived.

Over the centuries as new technologies were introduced, people and businesses applied them to support their causes. Information warfare has always been an essential part of military, business, and activist operations. Spying has been around for thousands of years. Today's business intelligence, for example, gathered via spyware, clickstream analysis, data mining, and automatic software registration, is simply a modern, technology-assisted twist on spying.

For centuries, the horse was the fastest method of transportation and means of communication. Large-scale manipulation of public perception was slow and almost impossible. Today, the media using satellites with semi-global footprints transmitting images and stories almost instantaneously are capable of shaping public perception. How about the magazine cover that depicted an emaciated man behind barbed wire in the former Republic of Yugoslavia? The article led readers to think that he was being intentionally malnourished and tortured, and the intent was to swing public opinion to support U.S. or NATO military action. The real story is he was a captured criminal with tuberculosis. Misinformation travels as fast as real information, but few are willing to verify the first information they read.

Misinformation is a subset of propaganda. Mao Tse Tung believed perception management was essential to his cause and information environment control was so important that he had his loyal supporters disassemble, transport, and reassemble printing presses while government forces pursued him through the mountains.

Governments, businesses, and other actors such as terrorists and activists today use physical and virtual means to manipulate the perceptions of target groups. High technology has made these means faster and more lethal. In the Middle Ages, biological warfare was used. Corpses with bubonic plague were catapulted over castle and city walls. Today, the capability exists to use aircraft, helicopters, and artillery shells to deliver scientifically engineered chemicals and biologicals. In the virtual world, the capability exists to send computer-altered photos and misinformation to a billion people in a matter of seconds. How will you inoculate your people against this modern electronic form of biological warfare?

Attacks Have Been Variations on Themes Throughout History

The marketplace is almost as cruel and dangerous as the battlefield. There is a very high positive correlation between IW and the military, business, and activism. This makes sense because there is no difference between bits and photons in the marketplace and on the battlefield. Here are some examples, all of which can be found in news articles from around the world during any week for the past two years:

- E-mail is intercepted
- Web pages are hijacked
- Spyware is used to capture an individual's information
- High-resolution satellite photography is used to determine manufacturing and sales volumes
- Trade secrets and other intellectual property are physically and virtually stolen
- Radio and mobile phone transmissions are pirated
- Misinformation is used to deceive the competition (and at times consumers)
- Clandestine means are used to defeat physical and virtual security

Reflection on the previous list shows that there is nothing new under the Sun — only in the way they are performed.

Intercepting mail is nothing new. The Athenians, Spartans, Egyptians, and other ancients routinely waylaid each other's couriers to obtain scrolls. Couriers were (and are) prime targets in any war. To let the owner know that the contents of his message were unseen, wax seals were affixed. Alas, that did nothing to protect against viewing the contents by those who may have been so brazen as to break the seal. Cryptography was invented to scramble the contents of the message so that either it appeared as gibberish or gave a completely different story unless you had the key to decipher the real message. Cryptography does not prevent interception, but may add significant delays to understanding the contents.

The hijacking of a Web page means the target site's Web address was changed to something else. Gambling and pornography sites have used this tactic.

In 1905, the Japanese Navy intercepted Russian naval high-frequency communications, and then used the information to soundly defeat the Russian Navy in a well-executed "crossing the T" maneuver. In the United States in the mid-1800s, train robbers and Native American Indians listened to the rails to determine when a train was approaching. Today, a 2.4-gigahertz receiver system and an Oronoco card hooked up to a laptop can help intercept mobile telephone calls.

Before electronics, the way to capture an individual's information was to send people to spy on them. Human intelligence, or spying, is a profession with thousands of years of history. "The walls have ears" means it was likely one could be overheard and, as a result, secrets could be compromised; but also, in some circumstances, if what was heard was not "appropriate," it would be used against you as blackmail, get you sent to jail, or executed. In World War II, Hitler Youth turned in their parents for disagreeing with the Fuehrer.

Before satellites, salesmen and others in the field derived manufacturing and sales volumes by checking rail cars, watching the numbers of trucks and cars going into and out of a factory, surreptitiously checking bills of lading and manifests, and listening to workers in bars.

Sun Tzu wrote over 2500 years ago that spying was a high art and necessary for the state to survive. Ninjas during the Japanese Middle Ages were hired out because of their exceptional reconnaissance capabilities. In more modern times, is it not an amazing coincidence that Russian military aircraft, such as their large transports, supersonic bombers, and fighters resemble U.S. aircraft? Of course, the Russians say the same thing about U.S. aircraft!

Propaganda and misinformation are frequently used in attempts to alter perceptions. The Allies had a masterful misinformation campaign to let the Germans believe the invasion of Europe would take place at Pas de Calais. The misinformation so strongly affected German perceptions that Panzer units that could have helped repel the attack were never ordered to attack. The Russians had an equally masterful misinformation campaign at Kursk, creating with radio traffic a fictitious army to trick the Germans.

Perception management is as old as man. In business, advertising campaigns are designed to influence entire cultures and subcultures. Clothing, alcohol, and cigarette ads are designed to let you know what is cool and tough. Other campaigns are designed to have guilt or fear be a motivating factor. In the former East Berlin in 1985, posters and ads showed a skeleton draped in an American flag, and the text stated the United States would start nuclear war.

There is no limit to what can be done to get around restrictions. When swords were outlawed in Japan, blades were hidden in bamboo and other wooden shafts. How does a terrorist get a gun through an x-ray device? By making the gun of components transparent to x-rays. Ninjas disguised themselves as beggars, magicians, musicians, and others so as to have mobility and penetrate into the enemy's rear areas. The Trojan horse is perhaps the best example of clandestine means to defeat physical security.

Information warfare is as old as man. Human creativity is boundless, and ancient methods are given new life via high technology. Businesses, governments, and activists take full advantage of high technology in order to apply IW to achieve their goals.

An Estimated Cost Due to Malicious Activity Is Large in Absolute and Relative Terms

The Internet has made corporate espionage faster, easier, nearly anonymous, and potentially more damaging. Corporate espionage, successfully attempted or thwarted, is almost never disclosed, so monetary values are difficult to attribute to this area with any degree of precision. Because corporate spies should be skilled and have many capabilities on which to draw, it is reasonable to assume that corporate espionage monetary loss exceeds, for example, malicious activity such as hacking and computer viruses. Yes, we are talking about a very large sum of money.

A study covering 30 countries and nearly 5000 IT professionals calculated that hacker attacks cost the world economy U.S.$1.6 trillion in 2000. "These estimates are based on the broadest sampling ever achieved in the security industry," said Rusty Weston, editor of *InformationWeek Research,* which carried out the study for PricewaterhouseCoopers.[2] A contemporary study by

Reality Research suggested that businesses worldwide would lose $15 trillion in 2000 due to computer viruses. The range of estimates for the Love Bug virus primarily fell in the $8–11 billion range. The Bubble Boy and Chernobyl viruses were in the $2–5 billion range. Taking double accounting, improved research methodology, and other factors into account, suppose the total sum of damage due to malicious computer-based behavior, which means not the full spectrum of IW, was $1 trillion and there are 200 countries (for ease of math) in the world. That amounts to a loss of $5 billion for each country. Some countries obviously suffered far greater losses than others. What do you think corporate spying cost? Would twice that amount (i.e., $2 trillion) be unreasonable? After all, they had more skill and capabilities on which to draw, and could be more targeted in their attacks.

Techniques Are Bold and Creative, but Often a Technology-Assisted Twist on Established Practices

Businesses will do almost anything to find out what competitors are doing and to steal intellectual property. The following examples are not unique.

Drive By Hacking

With the innovation of wireless networks that are rapidly being deployed by corporations and which the military is starting to consider for use, an entirely new range of vulnerabilities has emerged. Silicon Valley and a number of other high-tech areas where they have been deployed have already seen the equivalent of old-style "war dialing," in which the modern-day attacker is driven around the neighborhood in the vicinity of the target organization with a laptop and a wireless receiver card and the attacker looks to see whether he or she can acquire a signal from the wireless network. It is surprising the number of organizations that have installed wireless systems and have not taken even the most rudimentary precautions. In one small "high-tech" town in the United Kingdom, a test was carried out, and in one drive-through, more than 20 networks were identified, of which four (20 percent) could have easily been compromised.

Because almost none of the mobile communications are encrypted, contract, payroll, business strategy, research and development, and other valuable information is in the clear, and free to be taken. In the United States, there is a law that states it is illegal to capture and record wireless communications, an attempt to provide wireless security the same protection as landline communications. For those who engage in drive by war-dialing, the law pertains; but from their perspective, only if they are caught. Chances of this are negligible. What value should be placed on the captured communications by those who are losing them (we assume they are aware of the risk)? Such a value would be useful in establishing security policy, architecture, and products, services, and processes to protect the information.

Cyber-Extortion

Cyber-extortion is another example of those who are technically gifted and resort to crime. Instances of cyber-extortion are increasing, according to FBI agent Dave Marziliano, due primarily to an increase in hackers, particularly in underdeveloped countries. Most incidents involve relatively small amounts of money ($50,000 to $100,000), which many companies would rather pay than take the chance of losing competitive advantage.[2]

Spyware

Corporate IW can be insidious. Computer users were warned that secret codes, known in the trade as spyware, could be used to record their conversations. E-mails, screen savers, and electronic greeting cards can carry secret code that is able to switch on the computer's microphone, make a recording, and forward it without the user's knowledge.[3] Phillip Loranger, currently the U.S. Army's Director of Biometrics, demonstrated this capability more than three years ago to senior Army officials.

A form of spyware used against consumers is automatic hardware and software registrations, allowing companies to read all the information on a hard drive. Another type is clicking stream analysis; could you imagine a sales clerk recording every move you did in a store? This is what is done during online shopping. What is the correlation between this business technique and government espionage? There is a 100 percent correlation.

We Have Met the Enemy and He Is Us: The Lack of Electronic and Mobile Commerce Security Mirrors the Lack of IW Emphasis

Retailers, especially dot.coms, wonder why the public has not embraced electronic and mobile commerce. In a highly mobile and convenience-oriented society, the retailers thought E- and M-commerce was "the next big thing." In general, because security is always low or non-existent in corporate meetings, these retailers did not adequately address consumer concerns. Consumers are more than aware that their security is at risk. The following six studies — a representative sample — reinforce consumer concerns.

1. "Top of the E-Class: Ranking and Best Practices of over 170 Web Sites," a study performed for Ryerson Polytechnic University's Centre for the Study of Commercial Activity, found that a third of online retailers fail to provide E-shoppers with adequate security and privacy protection. Another third of E-tailers fail to meet minimum standards.[4,5]
2. A survey by KPMG showed that 27 percent of organizations never tested their Internet security. This is in spite of research from the United States that shows that the average new Web site receives some kind of attack within its first five hours of existence.[5]

3. According to the survey "Keeping the Faith: Government Information Security in the Internet Age," the Internet would meet resistance from the general public because of security concerns: 72 percent said they would not feel safe using a secure digital signature to sign a legal document.[6]

4. The 1999 Vulnerability Index, a study (note that this was a data network and not an IW study) released by Comdisco Inc. and BellSouth Corp., identified that 30 percent of business organizations have plans in place to protect against breakdowns in their Internet applications; 54 percent have computer network recovery plans; and, of these, only 14 percent have plans that really work.[7]

5. Banks are considered to be a bastion of physical security, but in the cyber realm more than a comfortable amount have been hit. Four Norwegian banks admitted leaving the financial details of one million customers exposed on their Internet sites for two months. The flaw was only discovered by one of the banks when a 17-year-old boy contacted a Norwegian newspaper to explain how it was possible to see the details.[8]

6. According to A.T. Kearney, Inc., in the July/August 2001 edition of *ComputerWorld ROI,* only three percent of mobile phone users in the United States say they have any intention of using the devices (i.e., mobile phones and personal digital assistants) to purchase something via the Internet.

Businesses and consumers read and hear the international news, and find little to give them confidence that it is safe to engage in online business. Said another way, E- and M-commerce are suffering from a death of a thousand cuts. According to the FBI, nine out of ten organizations have reported computer security breaches since March 1999. That study was based on more than 600 companies and government agencies. In a separate study, 33 percent of corporate organizations did not know if they had ever been attacked; and of the ones who knew, 13 percent could not quantify the extent of loss.[9] These negative statistics will only become worse as more businesses and consumers turn to always-on connections, such as digital subscriber line and Internet cable systems.

What prudent consumer or business would risk loss in the face of these statistics? Inadequate systems security and management may lead to companies being blacklisted by "city investors," according to the report "A Risk Too Far," commissioned by Vistorm. Ian McKenzie, managing director, said, "...if E-security isn't given attention at the director level, there will be enough high-profile security breaches to damage the development of E-business in this country."[10]

To survive, much less make a profit, businesses need to be aggressive — aggressive to the point of putting the competition out of business. Businesses will employ many IW capabilities (such as psychological operations, signal intelligence, spying, and computer network attacks) to achieve their goals and objectives on their way to attaining and maintaining a competitive advantage. Governments are also aggressive, but to protect sovereignty and to defend national security. And activists?

To get attention, activists would normally have to do something physical in nature, such as blowing something up or storming a building. Examples include the American colonists throwing British tea overboard into Boston Harbor and French truckers shutting down highways to protest fuel costs.

Activists know not to use their Web sites for direct communications within their groups. The FBI has made several announcements that activists and terrorists are increasingly making use of encryption in their e-mail. Law enforcement and intelligence organizations have the ability to intercept, collect, and decrypt these communications.

Computer hackers with political agendas have become a fast-growing threat to governments and big companies worldwide. The Internet-era activism brings the methods of guerrilla warfare, grass-roots organizing, and graffiti to cyberspace. Although resource constrained, hacktivists do reasonably well addressing the major IW functions of intelligence, psychological operations, computer network attack, and malicious code development and propagation.

Summary

Businesses compete for profits, governments conduct war as an extension of national security, and activists protest. The difference between 3000 years ago and today is that high technology enables all three to perform more efficiently, effectively, and innovatively. Businesses use high technology to discern as much as possible about the competition to gain a competitive edge. This allows them to develop, field, and withdraw highly targeted products, services, and processes. Governments have spied on each other and attacked each other since there were clans; they have used IW for thousands of years to maintain sovereignty and ensure their survival. Activists have expanded their efforts into the virtual domain; they can bring together in virtual space tens of thousands of people from around the world to take a unified action that in physical space would be impossible to do.

High technology now enables sophisticated intelligence, physical and virtual action, and perception management to be available to almost any group and individual. Using these IW capabilities, businesses, governments, and activists can use ancient techniques with modern twists and innovative approaches — previously only a dream — to achieve their goals and objectives.

Notes

1. Report: Year's Hack Attacks to Cost $1.6 Trillion, Tim McDonald, www.EcommerceTimes.com, part of the NewsFactor Network, July 11, 2000.
2. *InformationWeek Online,* John Soat, *InformationWeek,* July 3, 2000, p. 150.
3. Internet spies' new tactics, http://news.bbc.co.uk/hi/english/sci/tech/newsid_537000/537520.stm, BBC News, November 26, 1999.
4. E-tailers failing to protect shoppers; Study warns security, privacy not adequate, Marina Strauss, Retailing Reporter, June 22, 2000.

5. Egg Raid Shows Cracks in E-Security, Nigel Cope, *The Independent,* August 29, 2000.
6. E-Shoppers Rattled by U.S. Security Snafus, http://www.ecommercetimes.com/news/articles2000/001016-6.shtml, James M. Morrow, E-Commerce, October 16, 2000.
7. Companies Aren't Ready for IT Network Breakdowns, Ann Keeton, Dow Jones Newswires, November 2, 1999.
8. Online Bank Security Breach Hits Norway, http://uk.news.yahoo.com/001023/80/amwc2.html, Nick Farrell and Matt Chapman, U.K. Internet.com, October 23, 2000.
9. Threat of Network Break-ins Looms Large, M K Shankar, *The Hong Kong Standard,* September 2, 1999.
10. Lax E-security Hinders Dotcom Funding, http://www.vnunet.com/News/1114718, Ian Lynch, Vnunet.com, November 30, 2000.

THE PAST, PRESENT, AND FUTURE USE OF INFORMATION WARFARE TACTICS BY BUSINESSES, GOVERNMENT AGENCIES, AND ASSORTED MISCREANTS

II

Included throughout this section, the computer battlefield objectives and targets will unveil military, civil, and commercial target sets; will explain the global information infrastructure (GII) and the national information infrastructure (NII); and will demonstrate the GII and NII integration into the Internet as the primary vehicle for providing the capability and the opportunity for the execution of operations by techno-terrorists, economic espionage agents, activists, revolutionaries, freedom-fighters, military info-warriors, and the like. This section also identifies the probable information warfare combatants. Techno-terrorists, economic espionage agents, and military scenarios are described and discussed. Specific information relative to groups, governments, and individuals are described, as well as the rationale for their behavior. This section consists of nine chapters.

Chapter 6: Information Warfare Tactics: How Can They Do That? It's the COTS, Stupid!

This chapter discusses some of the vulnerabilities of today's hardware, software, and firmware that allow the use of IW tactics, often with little expertise required.

Chapter 7: It's All about Influence: Information Warfare Tactics by Nation-States

This chapter discusses the who, where, when, why, and what of IW tactics by governments, and how government agencies can use IW to defeat an adversary.

Chapter 8: Information Warfare: Asian Nation-States

This chapter discusses information warfare as it relates to Asian nation-states and includes case studies and commentary.

Chapter 9: Information Warfare: Middle East Nation-States

This chapter discusses information warfare as it relates to Middle Eastern nation-states and includes case studies and commentary.

Chapter 10: Information Warfare: European Nation-States

This chapter discusses information warfare as it relates to European nation-states and includes case studies and commentary.

Chapter 11: Information Warfare: Nation-States of the Americas

This chapter discusses information warfare as it relates to the Americas and includes case studies and commentary.

Chapter 12: Information Warfare: African and Other Nation-States

This chapter discusses information warfare as it relates to African and other nation-states and includes case studies and commentary.

Chapter 13: It's All about Profits: Information Warfare Tactics in Business

This chapter discusses the who, where, when, why, and what of business IW tactics, and how businesses can use IW to their advantage.

Chapter 14: It's All about Power: Information Warfare Tactics by Terrorists, Activists, and Miscreants

This chapter discusses publicly known terrorist nations, drug cartels, and hacktivist (cyber-disobedience) capabilities such as those of animal rights groups, freedom-fighters, and the like.

Chapter 6

Information Warfare Tactics: How Can They Do That? It's the COTS, Stupid!

There are two ways of constructing a software design: one way is to make it so simple that there are obviously no deficiencies, and the other way is to make it so complicated that there are no obvious deficiencies. The first method is far more difficult.

—C.A.R. Hoare

The issues addressed in this chapter relate to the information warfare element of computer network attack (CNA) from a COTS (commercial off-the-shelf) perspective. The range of issues encompasses the hardware, firmware, operating systems, applications, and network infrastructure, as well as some of the vulnerabilities of today's hardware, software, and firmware that allow the users of information warfare tactics, often with little expertise required, to succeed.

Why Commercial Off-The-Shelf Software (COTS)?

In the past, government agencies and corporations developed their own software, purchased some generic software and modified it for their specific use, or purchased some limited commercial software such as an accounting software package that could be easily applied to their environment. Commonality of software became a necessity as the "world began to talk to the world." However, with this newfound process came problems. For example, commercial-grade products, also known as commercial off-the-shelf (COTS), are stuffed full of vulnerabilities of which information warriors can take advantage.

The rationale for the use of COTS is twofold:

- COTS provides good capabilities at reasonable cost. These capabilities are what enable business to make a profit.
- COTS upgrades and new products are frequently delivered. The timelines are rapidly shortening, with hardware cycles now around nine months and software cycles around 18 months.

Imagine taking a software product, portions of which are developed overseas in countries that either are, or may be, U.S. competitors, that is several million lines of code (this size is not unusual), and proving it contains no malicious code. This requires a line-by-line code check, as well as understanding how the lines of code interact. There is no artificial intelligence program that does this. It requires skilled people and time. Now pass this cost on to the consumer and keep pace with the competition as they try to be first to market.

Yes, much of today's software comes with some security features. There are at least 300 security features in Windows NT that can be turned on or off. Complexity is a hallmark of modern software. Information warriors constantly probe for weaknesses. It takes just one weakness not detected and resolved in one system to make all users connected to it vulnerable to exploitation and attack. Because of the trusted relationship between systems and networks in our highly interconnected infrastructure, achieving and maintaining control over our environment is difficult — no, today, it is impossible!

The Way Software Is Developed: The COTS Way

The four main proprietary operating systems (OSs) are:

- Microsoft Windows family
- UNIX in its many forms
- Macintosh operating system
- Novell

However, another operating system that most people do not consider is IOS, the Internet Operating System used in Cisco routers, without which the Internet would not operate in the manner we know. This OS will not be discussed further, but is mentioned for completeness.

Yes, there is also Linux, and although it is growing, it has not found its way into today's mainstream for businesses and corporations. Many other operating systems exist, but the ones listed above are the primary ones used in business and by government agencies around the world.

While each OS is distinctly different in the way it functions, they all must achieve similar functionality to meet the requirements and expectations placed on them. In the competitive market of operating systems, the commercial drivers to bring the new, feature-rich operating system to market at the right time is paramount. The market for OSs demands that each new software version be more feature-rich than its predecessor. In attempting to deliver this, the manufacturers have incorporated more and more features in each release, with the result that the size of the OS grows each time. In doing this, the manufacturers are satisfying the demands of the wider market, in essence trying to be all things to all people. However, operating systems are now so large that it is impossible to verify in any formal way that they are carrying out the operations that they are supposed to or doing so in a way that is "safe."

There is a direct correlation between the development of hardware in terms of the processing speed of the computer processing unit (CPU), the storage space, and access speeds of the storage media. It is difficult to remember that, for the home user, less than 15 years ago, 512 kilobytes (kB) of random access memory (RAM) and 10 megabytes (MB) of hard disk storage were all that could be either expected or afforded. In addition, if you wanted to install new software on a system, the most likely media that you would be using would be either a 180-kB single-sided 5¼-inch floppy disk or one of the latest 360-kB double-sided floppy disks. These constraints concentrated the minds of the software developers to produce programs that were both efficient and compact. Imagine how many 5¼-inch floppy disks would have been needed to load the Windows 2000 operating system, even if the hard disk was large enough to be able to hold it.

Microsoft Windows currently has by far the largest share of the microcomputer OS market, with approximately 86 percent of the market.[1] Although there are reports that the source code for Windows 2000 was stolen in a hacking attack on Microsoft in October 2000, the benefit of stealing this code is not as obvious as may appear at first sight.

Windows 2000 is said to consist of in excess of 40 million lines of code and was developed by a large team of software engineers over a lengthy time period. In the same way that it is not possible to mathematically verify the code, the benefit to an individual or small group having access to the code to determine any weaknesses is limited. What is clear from previous versions of Microsoft OSs is that there is a considerable amount of redundant code contained within it. This has been demonstrated in one way through the discovery of "Easter eggs." In addition to code that does not have an obvious function, Easter eggs have a number of purposes, most of which relate to satisfaction of the egos of the authors of the software. Examples of Easter eggs can be found at the Easter Egg Archive,[2] the Easter Egg Page,[3] or the PC Win Resource Center.[4] It is amazing that such software should have been allowed into the commercial product and even more surprising that any quality control process did not pick them up. It is incredible they have been allowed to remain once they were discovered. Exhibit 1 shows a few examples of Easter eggs, with one from each of a number of the OSs, that you may want to try.

It might appear excessive to show so many Easter eggs, but it was done to include the range of examples in the hope that it reinforces the point that there is the potential for software to contain code that does not perform the role for which it was developed, even in packages from respected commercial development houses. Imagine what you might be getting from less-responsible developers. An Easter egg can easily be an Easter bomb!

Exhibit 1. Easter Eggs

Microsoft Windows 2000
Name: The Volcano Screen Saver
To access the Easter egg:

1. Right-click on the desktop and choose Properties.
2. Go to the Screensaver tab.
3. Select the 3-D text screen saver and open the Properties.
4. In the text box, type "volcano."

Microsoft Windows NT 4
Name: The Volcano Screen Saver
To access the Easter egg:

1. Right-click on the desktop and select Properties, or open Control Panel from the Start button/Settings menu.
2. Select Display.
3. From the Display Properties, select Screen Saver.
4. Select 3D Text (OpenGL) as your screen saver type.
5. Click on the Settings button to configure the screen saver.
6. Change the text that the screen saver will display to "Not Evil."
7. Click OK to save the changes.
8. Click the Test button or wait until the screen saver kicks in the next time.
9. This will show the names of the members of the Windows NT developer team.
10. You can also try the changing the text to "Volcano," which will display the names of a number of volcanoes from the Cascade Mountains in the United States.

Microsoft Windows 3.5
Name: Not Evil
To access the Easter egg:

1. In the control panel, select Desktop.
2. Select Open GL 3D Text screen saver.
3. Go to Options (setup).
4. Change text to "not evil" (no quotes).
5. Click the Test button to see the egg.

Microsoft Windows 98
Name: The Volcano Screen Saver
To access the Easter egg:

1. Right-click on desktop; choose Screen Saver tab, then choose 3D Text.
2. Type in the word "volcano" in the 3D Text box settings.
3. Click OK, then preview to see the names of volcanic mountains.

Comment: Works on 98 and SE.

Microsoft Windows 95
Name: Developer Credits
To access the Easter egg:

1. Select Help from the Start menu.
2. Click Find.

Exhibit 1.   Easter Eggs (Continued)

3. Click Options.
4. Select the first checkbox.
5. Select "begin with the characters you typed."
6. Click OK.
7. Type this in the top box. "Who knows who built this tool?"
8. Hold down Ctrl, Shift, and click the Clear button.

Comment: Works on 95 and SE.

Microsoft DOS

Name: Joke
To access the Easter Egg:

1. Start the MSD.EXE program located in the DOS directory.
2. Select the menu Help, About.
3. Press F1 to get to the joke.

Comment: Only works on versions 6.2 and below.

BeOS

Name: Hidden Menu Item
To access the Easter egg:

1. Hold down Ctrl, Alt, Left Shift and click the Be button on the desk bar.
2. A secret menu will display called "Window Decor."
3. Play around with the options for some fun, such as changing the style of BeOS!

Comment: Only works on versions 4 to 5.

Linux

Name: The Printer Fire Message
To access the Easter egg:

1. Print from the network and force a real printer jam.
2. Issue a Print command.
3. Watch the output of the console, it says lp0 printer on fire!

Comment: Needs Linux version 2.2.1 with bi-directional printer support.

Mac OS

Name: Macintosh Classic Secret Disk
To access the Easter egg:

1. Hold down Command/Option/X/O (the code name for the Classic's was apparently XO) and turn the computer on to start up System 6.0.3, Finder 6.1x, and AppleShare from a ROM-disk. This secret disk is contained in the read-only memory of the Classic and cannot be altered.
2. Apparently, the ROM-disk was supposed to allow the Classic to be sold as a diskless workstation, but Apple abandoned that marketing angle without removing the capability from the machine.
3. Using a program able to see invisible files (like ResEdit, Norton Utilities, or Mac-Tools), examine the ROM-disk for a "Brought to you by" folder containing more hidden folders bearing the names of the Classic designers.

Exhibit 1.   Easter Eggs (Continued)

IBM OS2
Name: Developer Credits
To access the Easter egg:

>   1.   Click the desktop with your left mouse button.
>   2.   Hold down Ctrl, Alt, Shift, 0 for the credits.

PalmPilot
Name: Developers Credits
To access the Easter egg:

>   1.   Start the Memory App.
>   2.   Place your stylus on the top of the screen, then press the DOWN button. You'll see a long list of the names "Development Team Credits."

U.S. National Security Agency and Outsourcing

Another example is the use of outside sources to develop code for the U.S. National Security Agency (NSA). The NSA discovered that some of the software that was developed for it, and of course used as part of very sensitive national security work, contained trap doors. It was subsequently learned that this was not done for any malicious reason. It allows the programmer who developed the code to make it easier to get into the source code to make changes, as required. However, trap doors can also be found by information warriors whose goal is to destroy or modify the code and to steal national security and other sensitive information.

Outsourcing Telecommunications Systems Maintenance

The use of outsourcing systems and program maintenance of telecommunications systems and programs are not immune to similar problems. For example, representatives from two U.S. telecommunications service providers learned that they both had outsourced their systems' maintenance to the same company. In fact, they no longer even knew how to maintain the systems themselves. Thus, they were at the mercy of the company performing their maintenance. Because such systems are computers run by software, telecommunications could easily be routed through various other locations where the information could be monitored and read by other than the intended people.

If that was not bad enough, the company doing the telecommunications maintenance was from a foreign nation-state with a known history of sophisticated espionage and intelligence collections operations. One important point to remember: if you outsource some of these types of services, look beyond the lowest bidder criterion. Of course, they will bid low to ensure they obtain the contract. The information that can be gleaned from the systems are much more important than any money losses due to operating at a loss on the contract. Always investigate the background of the business before signing the contract.

The Dilemma

Operating Systems

Why is it that an operating system, which has so many security features that could be activated, is delivered with them all switched off? Well, the answer is because the vast majority of users do not have the knowledge that they would need to set the appropriate security features for the way they use their systems and do not want to spend the time learning how to use them. Furthermore, information technology specialists do not want to be bothered by security, which they believe complicates their work, slows processing time and system throughput, and numerous other excuses that they can come up with at a moment's notice.

If an OS was delivered with all of its security functionality enabled, the majority of the users would not be able to use it the way they wish (switch the computer on and have it do what you want it to with a very intuitive interface). They also argue they do not have the skill and knowledge to make it usable. The result would be that they would find it difficult to use and would look for an alternative OS that was easy to use. No manufacturer is going to take that risk.

The Windows NT operating system, which has in excess of 300 security settings that can be selected, is delivered with them all set to OFF. In this way, it is easily loaded and fully functional. If the everyday user of the operating system is to be expected to configure the system in a secure manner, then there is a lot of education to be provided. All of this assumes that the everyday user cares, which, for the most part, they do not. The private user and a large number of commercial users do not consider the security of the system very important. There is the commonly held view that they hold nothing of importance on the systems and that a security event has never happened to them, so what is the problem? Furthermore, this is not a problem confined to the business world. Employees of government agencies also feel the same way. In October 2001, Jupiter Media reported that less than 50 percent of U.S. companies with Web sites were appropriately concerned with the security of their online data.

Even if they do care, there is little chance that they will have the skill to ensure that they configure the system in a secure manner. Additionally, there is a considerable investment in maintaining the security of the system as fresh vulnerabilities are discovered, which is an almost daily occurrence. The prevalent commercial approach is that, as the software manufacturer, you deliver it in a state that will suit the majority and then let the minority of users who would want to enhance its security do so. The fact that the vast majority of the Internet is totally insecure is not just the fault of the producer, but also the fault of the consumer.

Applications

Thus far we have referred only to the operating systems, but the computer is of no use to anyone with just an OS. It is the applications that are loaded on top of the OS with such enthusiasm that make it do what users actually want it to do. There are applications for just about anything, from the obvious and well-known word processing packages and spreadsheets to the totally obscure. There are packages for predicting your horoscope and there is software that you can load onto your system that uses the spare capacity on your computer to process, as part of a cooperative effort, the signals that the SETI (Search for Extra Terrestrial Intelligence) intercept from space. These applications are written, in some cases, by teams of software engineers who work for the same company that produced the OS. The obvious example is the Microsoft Office suite, but also hackers who produce tools to make it easier to do what they want to do by automating some of the repetitive processes.

It would be reasonable to expect that applications produced by Microsoft did not intentionally undermine any security features that had been built into the operating system. However, there have been occasions in the past where this has happened. A good example of this is the remote denial-of-service vulnerability that was discovered in a component of the NetMeeting application.[5] The denial-of-service can occur when a malicious client sends a particular malformed string of code to a port that the NetMeeting service is listening to and when the Remote Desktop Sharing is enabled. The NetMeeting application had to be downloaded separately for Windows NT 4, but now has been included in Windows 2000. On the plus side, the application and affected component were not enabled in the default configuration, and users who have not enabled it were not at risk from this vulnerability.

Now look at the other end of the scale: at applications produced by hackers (yes, hackers do produce software; take a look at any of the good hacker sites to see the software that is available). It is normally the case that they are trying to undermine any existing security features by exploiting flaws in the logic or the preparation of the OS or legitimate applications. Examples of hacker sites that hold a good selection of software include:

- *Hideaway.net* — http://www.hideaway.net/
- *The L0pht site* — http://www.l0pht/com/blkcrwl/hack.html
- *Hackers paradise* — http://www.hackersparadise.org.uk/
- *Hackers Toolz* — http://www.hackers-toolz.com/
- *The Underground News* — http://www.undergroundnews.com/
- *The Cult of the Dead Cow* — http://www.cultdeadcow.com/

This is a tiny sample of the sites on the Internet where this type of software is available. One information systems security professional maintains a database in excess of 6000 such sites, which in itself is only a small fraction of the better sites.

So, when you download the "freebie" from the Internet that looks so promising, how do you actually know what effect it is having on your system? Well, the answer is that you do not.

If we were all good, honest citizens and were prepared to pay for the software that we use, the situation would be less problematical; but in reality, a large portion of the software in use is illegal. It has been copied and redistributed countless times. A few years ago in Hong Kong, if you had a computer built, you would tell the person making it what software you wanted on it and it would be delivered with it, at no additional charge. This was not a back-street operation — it was the norm. Although the police and authorities tried to stop it, they were singularly unsuccessful for a number of years. On one occasion in 1998, customs officials in Hong Kong seized more than 100,000 suspect CDs from just one shopping arcade in Kwai Chung. The problem with this was that, in most cases, not only did you get the software that you had asked for, but you also got a cocktail of viruses and other malicious software.

At the same time in Hong Kong, you could go into one of the large shopping arcades that specialized in computers and buy, for about 50 Hong Kong Dollars (at that time the exchange rate was about 8 Hong Kong Dollars to U.S.$1) any of the commercially available packages. As writeable CDs became popular, the situation changed slightly because it was easier for the people who pirated the software to copy a CD full of software with everything that was available on it — it meant that they could mass produce the pirated disks instead of producing them to order. As a result, for about 100$HK, you could have all of the software that you would ever want, including viruses. Microsoft reports[6] that in Vietnam, 97 percent of the software in use is not legitimately purchased. In Russia and Indonesia, 92 percent of all software has been pirated. Interestingly, this is not just a problem in the East. The United States, Japan, Germany, France, Italy, the United Kingdom, and Canada are among the top-ten countries for dollar losses due to pirating.

If we operate in a culture in which "freebie" and pirated software are OK to use, we should not be surprised when we get what we paid for. In fact, on the chance the customers work for a government agency or a targeted company, what better way to infect the system? Then, in case of conflict, the adversary can send an e-mail or allow access to a Web site that would "awaken" a logic bomb, Trojan horse, or other malicious code. This could be used to send all the information from the targeted machines to the adversary or destroy or corrupt information on the targeted machines.

Even in the applications that we do buy, there is more than we asked for or expected. Once again, the Easter eggs are there, for example:

Microsoft Office 97

Name: Splash screen
To access the Easter egg:

1. Load the Microsoft Office 97 Shortcut Bar.
2. Click the small Office 97 icon in the upper-left corner.
3. Select About Microsoft Office.
4. Hold down Ctrl-Alt-Shift and double-click the puzzle piece.
5. It then loads the Office 97 Banner.

And also remember that software may appear in places that you least expect it. One of the favorite Easter eggs is actually on a digital video disk (DVD).

The Abyss DVD

Name: Hidden Trailers
To access the Easter egg:

1. At the Trailer section on the second DVD, look at the posters behind the three Abyss ones, you will see that there are movie posters for Aliens, True Lies, and Strange Days.
2. When your cursor is on the Reviews Trailer item, press the UP button and you will access an Aliens trailer.
3. Place your cursor on the Main Trailer item, press the DOWN button and you will get to a True Lies trailer.
4. From the True Lies item, press the RIGHT button and you will get to the Strange Days trailer.
5. Go to "The Abyss In-Depth — Mission Components" section. At the next menu, press the UP button until you get to "The Wave" entry and press the UP button one more time. There is a Harrier aircraft in the sky that you can select. It will take you to another True Lies trailer.

Why Can We Not Do It Better?

The basic problem is that the people who write the software are largely anonymous. It is human nature to want to make your mark, especially when you have invested hundreds or thousands of hours in the task of developing that software. The problem is compounded by the fact that software today is very feature-rich and, as a result, complex.

Storage space and memory availability on systems are increasing, with the result that, unlike in the early days of computing, there is now little requirement for software to be efficient either in its memory usage or the amount of disk space that it occupies. When software packages are so large and have to carry out so many functions, it is difficult to predict the interaction of one part of the code on another.

It also becomes impossible to mathematically prove all the outcomes of the operation of different sequences of the code and, as a result, any scientific examination of the code for efficiency is impossible. The software of today and in the future is not produced by individuals or by small teams — the complexity and size of the program suites will mean that they are produced by large teams. These teams are often in foreign nations, nations that may now be or may one day be your adversary. There may be malicious code just waiting to be activated. In this environment, quality control of the code becomes increasingly difficult, as the process of understanding and reviewing the code is extraordinarily difficult. In addition, it is not a strong driver in the commercial market; the imperative is to maintain or increase your market share and to keep delivering the updated and more feature-rich versions of the software to the market at the appropriate time.

The driver is first to market; beat the competitors. If it has bugs in the code, so what? A patch can fix that later. No time now; worry about it later when enough complaints are received — that appears to be the software developer's motto. So, what has all this to do with information warfare? Obviously, it makes attacking systems of the competitors and adversaries so easy. After all, if 12-year-old "script kiddies" can do it, think what a Ph.D. computer scientist in the employment of a competitor, a foreign government, a hacktivist, or terrorist group can do? And they are out there doing it as you read this!

The Way Software Is Developed: The Open Source Way

Open source, by definition, means that the source code of the software is open for inspection by anyone who chooses to look at it. This has the advantage that a wide peer group with the result can review that, at the end of the review process, the software should be as near perfect as it is possible to make it. Among the disadvantages of this approach is that it does not belong to any individual or organization and, as a result, there is no economic benefit in promoting the software. With no commercial benefit and limited or no financial backing, the development time for open source software is likely to be far longer than for a commercial version of software. In addition, the software source code is available to an aggressor, who can spend time analyzing the software to identify any shortcomings or vulnerabilities, or add that "something special" to it.

The GNU Software Project

A project named GNU was launched in 1984 to develop a complete UNIX-like operating system that was available for no cost to whoever wanted it (freeware). The GNU OS (GNU is a recursive acronym for "GNU's Not UNIX," and is pronounced "guh-NEW") and variants that use the Linux kernel are now widely used. However, although these systems are often referred to as Linux, they should, more accurately, be called GNU/Linux.

For the project, the development of a complete operating system was, potentially, a huge undertaking. To bring it into a realistic scope, available free software was used whenever it was possible. Examples of this are the TeX text editor and the X Window graphic interface system.

The result of this is that while the GNU operating system is a cost-free system, it is built in part from other, existing components. While this provides a free and effective operating system, the use of the elements that were not built for the project have unproven origins. Therefore, they must inevitably weaken the whole system.

Linux

Linux is another free UNIX-type operating system that was originally created by Linus Torvalds. In the development of the Linux operating system, Torvalds gained huge support from other software developers from all around the world. The system was developed under the GNU General Public License; and in the spirit of this project, the source code for Linux is freely available to everyone to download or copy. An information warrior can easily modify the code and send it on its way to populate systems all over the world. Using your information warrior imagination, think of all the various types of malicious codes you could insert in the software for later activation.

Shareware

Shareware is a hybrid that sits between freeware and commercial software. The concept behind shareware is that you make it available free on the Internet for people to download and evaluate, and then invite them to pay a nominal sum to receive updates and support. Some very good software has been distributed in this way and for those applications that do not attract the attention of the major software houses, it can be a reasonable way to generate an income (a few thousand contributions of U.S.$15 can very quickly add up). Again, there are no controls for preventing the insertion of malicious codes that can find their way into your systems and networks for later activation.

The Dilemma

Most of the software developed under the open source philosophy is produced by people who are not trying to make any or much commercial gain. This largely means that the parties that are involved are individuals, or academic or research institutes. Their aim is that by releasing the source code for inspection by potentially countless millions of users of the Internet, some of whom are highly skilled, good, free software can be produced that has been subject to widespread testing and improvement. Unfortunately, the downside of this concept is that no one "owns" the problem and there is no commercial drive to "productize" and support the software. Also, by its very nature, when the source code is available to all, it can be modified and have unwanted features added by anyone who so desires.

If the average person downloads the source code to compile it, are they really likely to have the knowledge or desire to check that the source code only does what it says on the label? There are notable exceptions to this generalization and operating systems such as Linux have established a huge following and good reputation; but what of the other obscure tools that we see? The best that you can hope for is that if you obtain your freeware or shareware from a reputable source, the provider will have taken reasonable care. Again, unfortunately, on the Internet, with its huge connectivity, it is unrealistic to expect that everyone else will have taken the same precautions or have the same intentions as you.

Bespoke Software

Bespoke software is software that is developed for a specific purpose for a specific customer. In the past, the military was the primary customer for bespoke systems, which is natural because you would not want an "off-the-shelf" missile defense system or command and control systems. To have such systems that were at the leading edge of capability, vast amounts of money were invested in having systems designed and built to meet specific requirements.

The disadvantage of this approach is that the systems were vastly expensive, took a long time to develop, and rarely met the requirements of the user by the time they were delivered. It is more normal in today's environment to take COTS software and modify it to meet specific needs. This is a much lower risk option and has a shorter development cycle, allowing more modern technologies to be introduced. The downside of this is that by taking COTS and modifying it, you have to take all of the additional functionality that is offered by the package and, as a result, are exposed to any shortcomings inherent in those features. Take a look at the generic UNIX operating system. On most systems, the software is loaded out-of-the-box with all its functionality. If, for example, you have no requirement on your system for the File Transfer Protocol (FTP) or the remote connection capability of Telnet, why should you leave this software on the system? It means that any miscreant or information warrior who gains access to your system can exploit these facilities when they offer you nothing of benefit in terms of the way you run your system.

In a truly bespoke system, the only functionality that should be built into the system is that which is required: what else would you pay for and why would the developer spend time and effort in producing extra functionality? Always assuming that they are building it for you from scratch and not re-using software modules that they have previously produced for another customer. Also, because the software is being developed for one customer, the quality controls that can be imposed are strong.

The Dilemma

The main disadvantage of bespoke software is the cost of obtaining and maintaining it. If one organization takes on the entire cost of developing an application, it will need to have very deep pockets. In commercial software, the development cost is spread across the entire population of users who buy it; thus the cost can be shared across millions of purchases. The second disadvantage is that any software house that takes on the task of developing a bespoke system for a client will do so based on the experience it has gained over a number of similar projects in the past. It is not surprising that, given the opportunity, the software house is going to incorporate code into the system that it had developed in the past for another project. After all, it is pre-tested, so why re-invent the wheel? The problem is that it probably will not be used in exactly the same way as it was designed to operate when it was developed and may, as a result, be less than perfect for the task at hand. The upside for the developer is that it will have saved huge amounts of development and testing time and will have increased its profit margin (assuming it was a fixed-price job).

Hardware

What do we mean by computer hardware? In these days of convergent technologies, this is perhaps not such a naïve question. Not so long ago we would have assumed that it meant the physical computer, the input devices such as a keyboard, mouse, and monitor, and an output device such as a printer. Now, where do we draw the line? Do we include the handheld microcomputer that communicates by infrared with the network? Do we include the mobile phones such as the Nokia 9110 (Communicator) that has its own keyboard and screen and is capable of acting as a network browser in one mode or as the modem for a laptop in another? We will not be too discerning for the purposes of this discussion and will look at the general questions and problems that relate to bits of metal, wire, printed circuits, and integrated circuits.

We all assume that the hardware we are using is benign and only carries out the function for which it was designed. However, there is the possibility that a piece of hardware could be modified to carry out a function for which it was not developed. There were a number of reports — some supporting the possibility[7,8] and some debunking it[9] as a hoax — both during and after the Gulf War, that an attempt was made to have printers modified before they were delivered to Iraq. The story contends that a shipment of printers from a French source were intercepted and fitted with modified chips (supposedly developed by the NSA). The purpose of the modified chips was reported as being to inject a virus into the systems used by the Iraqi Air Defense. It is of note that while this is within the bounds of possibility, no report has ever been circulated on how the virus was to be activated or how the coalition forces would have known that the target systems had been disabled.

This type of attack is not a major consideration for the average user, but what if the hardware were to be modified so that at some predetermined event or time, they would perform a specific function or even cease to function? Do we really have control of the hardware that we use? The answer is that we do not. Look at the labels on the circuit boards that are in your computer and see where they were fabricated. We do not produce our own circuit boards or the majority of the chips that are on them. We buy them, mostly from Pacific Rim countries. In addition to leaving ourselves vulnerable to market shortages, such as the one triggered in the availability of random access memory (RAM) as a result of the earthquake in Kobi, Japan, what would be the result if we found ourselves in dispute with the countries that produce these essential components?

Let us consider the possibility of a dispute with China in which the Chinese decided that it was to their advantage to disable the Western economies. Tom Clancy, in his novel *Debt of Honor*,[10] postulated a similar scenario, based on a conflict with Japan, but the twist here is that a large proportion of the integrated circuits used in the West are made in China. If China had decided on this course of action, what would there be to prevent it from programming the chips to carry out the specific operation that they required at a predetermined time or on receipt of a predetermined sequence (after all, all of our systems are connected at one level or another these days)? The impact would be devastating. Look at the cost and effort that were invested in trying to prevent the disaster that was expected as a result of the change of date to the year 2000 (the Y2K bug). This was for something that was brought about as the result of bad or naïve programming and was a potential problem that had been known about for a number of years. Imagine the impact of a sudden and unexplained failure of a large proportion of the computers available to us. It would take a long time to determine what had actually happened. If it was as a result of such a hostile act, by the time we had figured out what had happened, it would certainly be too late; after all, the opposition would not be affected and would know when the event was due to occur. The counter side of the scenario is that the United States is home to the corporations that control all of the major operating systems.

Firmware

Firmware is included here because, logically, firmware resides on the hardware. Firmware is a special-purpose module of low-level (hexadecimal or machine code) software that serves to coordinate the function of the hardware during normal operation and contains programming constructs used to perform those operations. As an example, in a typical modem, firmware can be used to establish the modem's data rate, command set recognition, and special feature implementation.

Firmware is stored on a type of memory chip that does not lose its storage capabilities when power is removed or lost. This basic type of *nonvolatile* memory is classified as "read-only" memory (ROM) because the user, during normal operation, cannot change the information stored there. The basic type of chip is called a PROM, which is programmable by any technician who has a programming console with the appropriate equipment. A basic PROM receives one version of firmware. That code is "burned in" to the PROM and cannot be changed. To update the firmware, the PROM must be physically removed from the device and replaced with a new chip. A more sophisticated version is the EPROM, which stands for erasable programmable read-only memory. This chip can be updated. What this means is that, embedded within a large number of the electrical/electronic equipment that we use are embedded items of firmware. As mentioned, the modem is one, but the list is almost endless. These chips are programmed in low-level languages, so even if the average user could gain access to the code, he would not be able to understand it. Even worse, when the Y2K problem was looming, people started to realize that there were embedded chips in far more items than they had thought and that, in many cases, people had forgotten not only where they were but, more importantly, where they had been produced and by whom. After all, if people cannot see them and they are reliable, why should they give them any thought?

Now, imagine that a mole (covert agent or an adversary) is involved in programming firmware that is to be used in navigation systems of weapons such as modern aircraft and missiles. The mole embeds malicious code in the firmware that would cause the on-board computer systems to shut down if a certain longitude and latitude is crossed. Thus, any weapon being used against the mole's employer would simply fall out of the sky. Again, with all the lines of code in these programs, is anyone really checking for such things? No; they are only verifying and validating that the programs are operating as required under the contract.

Can this be checked? Yes, but it is costly and time-consuming; thus, the developer and the customers assume that the possibility of such scenarios is low risk and not cost beneficial. However, taking the firmware at random off the production line and running it through multiple wartime scenarios could be done in parallel with production. Expensive, but is that not better than no testing for information warfare vulnerabilities?

How an Attacker Can Take Advantage

The options for an attacker are almost endless and we have repeatedly seen that as soon as one vulnerability in a system is addressed, another is discovered. Exhibit 2 shows the Computer Emergency Response Team (CERT) statistics for the number of vulnerabilities that were discovered over a number of years and the number of reported security breaches in which they resulted. Note that the reference is to "reported" security breaches, as this is significant. The actual number of breaches is unknown. Even in the U.S. Department of Defense (DoD), which is structured, security conscious, and aware of the problems, tests carried out in the mid-1990s to test the security of DoD systems by its own staff revealed that only around four percent of attacks were reported. When you apply these kinds of metrics to industry and private usage, can we truly believe that even a fraction of one percent of security breaches make their way into the statistics of reported attacks?

Their capability, their intentions, and the type of equipment and software that you are using will determine the choice of approach that the attacker takes. The use of the single term "attacker" is inadequate to describe the range of potential adversaries and their capabilities. The potential is that an adversary could be anything from a juvenile using an out-of-date microcomputer in a bedroom, to a sophisticated, nation-state sponsored organization. As a result, there are likely different ways in which these varied adversaries will attempt

**Exhibit 2. Number of Incidents
and Vulnerabilities Reported**

Year	Incidents
Incidents Reported	
1988	6
1989	132
1990	252
1991	406
1992	773
1993	1334
1994	2340
1995	2412
1996	2573
1997	2134
1998	3734
1999	9859
2000	21,756
2001	52,658

Total incidents reported (1988 through 2001: 100,369)

Year	Incidents
Vulnerabilities Reported	
1995	171
1996	345
1997	311
1998	262
1999	417
2000	1090
2001	2437

Total vulnerabilities reported (1995 through 2001: 5033)

CERT/CC Statistics from http://www.cert.org/stats.

to exploit any vulnerability. At the low end will be the "script kiddy" that will use tools and scripts that are available on the Internet to attack your systems. The script kiddy will have little or no skill, but will rely on the work that has been carried out by more experienced information warriors. Be careful not to underrate the impact that such an attack can have.

Common Types of Attack

There can be as many types of attacks as there are imaginations in information warriors. However, there are several, basic common attack methods. These are discussed below.

Denial-of-Service Attacks

A denial-of-service (DoS) type of attack is used to deny the owner of a system the facilities of the network. The scope of the attack and its sustainability will depend on the resources that the attacker can muster and the reason for carrying out the attack. If it is script kiddies with a cause or a grudge, the impact may be minimal and the organization little affected. After all, most large organizations have considerable bandwidth at their disposal; thus, for an individual or a small group to totally deny service for any length of time is difficult.

A recent example of denial-of-service attacks is Trinoo. Trinoo is one of a number of distributed denial-of-service (DDoS) tools that have gained notoriety because of its rumored involvement in the large-scale DDoS attacks that took place in February 2000. Trinoo can be traced back to at least July 1999, when it was observed in use in a number of DDoS attacks that took place in Europe. The Trinoo software has two components: a master and a slave. A Trinoo attack is initiated when the attacker locates and compromises a number of suitable hosts. A Trinoo slave is installed on the system. Additional systems are compromised in order to install the master component. Once sufficient systems have been compromised to establish a network of one or more masters and many slaves, all the attacker needs to launch a DoS against a site, network, or several sites is to instruct the master systems on which target(s) to attack. The masters forward the commands to their list of slaves, which start the attack for real.

Trinoo is suspected of being involved in attacks that started during the week of February 7, 2000, and saw DDoS emerge as a major new type of attack on the Internet. The attacks seriously affected a number of sites, including Yahoo, eBay, Amazon, E*Trade, and CNN. Each of the "victims" was not accessible from the Internet for several hours. When you have a business that is totally dependent on your connectivity to the Internet, this type of disruption could be catastrophic. Most disturbing of all is that, with the way in which Trinoo is implemented, using masters and slaves to mount the attacks from compromised systems, there is little likelihood of catching the attacker or of preventing similar attacks in the future. After all, there are a lot of systems out there that can be compromised with little effort.

Since the first observed DDoS tool in late 1999, at least eight new tools have been discovered, providing a range of new attack capabilities and using new communication techniques. They have also improved on the user interface of the older tools. Although these tools have been seen in use since the February 2000 attack, they have not been used on targets as prominent or as widespread as the initial group.

One of the features that have resulted from the development of the DDoS attack technique is a marked increase in the level of Internet Relay Chat (IRC) wars. IRC has a wide variety of devotees who use the facility in the way that they find most suitable, but IRC channels that are devoted to hacking are the most affected by DDoS attacks. Battles for control of these channels have been seen for a long time with different users and groups struggling to become

Exhibit 3.   Web Defacement

Location	Defaced Web Site
International	UNICEF
	Amnesty International
United States	The White House
	CIA
	Department of Justice
	NASA
	The Pentagon
	New York Times
United Kingdom	Labour Party
	Conservative Party
China	Chinese Agricultural University
Republic of Indonesia	Department of Foreign Affairs
Greece	Ministry of Foreign Affairs

Source: http://onething.com/.

channel operators so that they can evict people they do not like and take control of the channel for their own ends.

To achieve this, one technique is for a group of IRC users to coordinate an attack over one channel against the person in control of another channel. By coordinating a packet flood from several sources, it is relatively easy to cause the operators' system to freeze and deny them service. After all, in most cases, you are talking about a PC and a connection, not a sophisticated network with lots of bandwidth.

For a roundup of known DoS attacks, there are a number of news sources that will provide up-to-date information, including *ZD News.*[11]

Web Defacement

This is the equivalent of a teenager with a can of spray paint who leaves his mark (tag) on the wall of a building or on the side of a bus. The defacement of a Web site is a very visible sign of an attack on the system. This type of attack is now so prevalent that there are sites on the Internet that will provide up-to-date status reports on which Web sites were defaced that day. Perhaps the most telling comment on how widespread this problem is, is that probably the best-known site that used to provide a list of hacked and defaced Web sites has now given up due to the volume of sites — it just could not keep up. For an up-to-the-minute list of defaced Web sites, it is now necessary to look at a German site,[12] which gives details not only of the site that has been defaced, but also contains considerable additional information about the site. Well-known examples of this are too numerous to go into detail, but the list includes those listed in Exhibit 3.

System Modification

With the difficulties that exist in verifying the probity of the software that is in use, due to its complexity, size, and variety of sources, it is not surprising that the problem of system modification is one of the major problems that system owners face in maintaining a secure system. One of the favorite tricks of an attacker is to subvert the system by exploiting vulnerability in one application or function of the operating system to subvert another part of the system. The usual reason for this is to gain additional privilege on the system to allow the attacker the ability to carry out his desired actions. The "Holy Grail" of most attackers is to gain "root" access to the system. Root access in UNIX is the highest level of access to a system that gives the user total control of the system and all of its applications. Once the attacker has achieved this, they "own" the system and can do anything they want on the system, including hiding and erasing all obvious traces of their actions.

Theft

This activity falls into two separate areas: the theft of hardware and the theft of software or intellectual property. The theft of hardware will be either (1) carefully targeted to ensure that the system that is stolen is the correct one and will normally take place when the hardware is of a high value or the information that the system contains is valuable, or (2) it will be opportunistic.

A computer that was stolen from a company in Scotland provides a good example of opportunistic theft. The laptop, belonging to ROV Networks Limited which designs and builds electronic software control systems for mechanical equipment in the oil and other industries, was stolen from the National Hyperbaric Centre in Aberdeen, Scotland, after a thief sneaked through a hole in a fence. The owner described the laptop as, "It is where all the control systems are developed and is used for downloading and upgrading information to computers." The computer had a ten-gigabyte hard disk that he estimated contained five years' worth of information and that the owner calculated was worth hundreds of thousands of dollars.

When you look at the targeted theft of hardware, a classic example was the 1996 theft from Centon Electronics Inc. in Irvine, California, of large quantities of computer chips worth $9 million. At the time of the theft, the chips were worth more, weight for weight, than gold. Another example is the theft from the Intel Corporation in Portland, Oregon, of SIMM (single in-line memory modules) chips worth $1.4 million.

When you look at the theft of a specific system for the information that it contains, then the examples that spring to mind are the 35 laptops that were reported as stolen from government ministers and officials in the United Kingdom[13] during the past three years. Among the information that was lost were files relating to a fighter aircraft.

The theft of software and intellectual property is rife throughout the world. The value of the losses is immense, both in terms of the value of the lost information and the cost of recovery. If a company has invested billions of dollars in developing a new car or new drug and the information is stolen, not only is the cost of development lost, but also any potential revenue. One example of this was the alleged theft by McAfee Associates of code from the Symantec Corporation,[14] who said that it had discovered code in the McAfee PC Medic 97 product that had been copied from the Symantec Norton CrashGuard.

The theft of information is an increasingly common crime that has, potentially, huge consequences. While the number of recorded crimes of this type is still relatively small, it is growing rapidly. Examples from the range of crimes that fall into this category include:

- A Cisco employee exceeded his authorized access to Cisco's systems to obtain information on its products and ongoing development projects, which was valued at more than $5000.[15]
- An Orange County man, Jason Allen Diekman, pleaded guilty to a misdemeanor charge of accessing NASA's Jet Propulsion Laboratory and a number of other computers. He was also charged with the felony of unauthorized use of a credit card to obtain approximately $6000 worth of electronic equipment.[16]
- A New York man, Jesus Oquendo, was convicted of computer hacking and electronic eavesdropping after he hacked into the computer of a venture capital company and stole password files. He also broke into the computer systems of another company and deleted their database, the repair of which cost $60,000.

Radio Frequency Weapons

There have been rumors for a number of years of radio frequency weapons capable of disabling or destroying computer systems. A radio frequency weapon (most commonly referred to as HERF [high-energy radio frequency] weapons) theoretically has the capability of interfering with information systems by affecting the electronic circuits within the computer. In 1996 there were reports[17] that financial institutions in the City of London and New York had been subject to a number of HERF attacks since 1993, and that vast amounts of money had been paid out in blackmail as a result. These reports were never substantiated and were subsequently vigorously denied. The NSA (National Security Agency) is reported to have commented that the criminals may have penetrated computer systems by using what are called "logic bombs." These are coded devices that can be detonated remotely.

The international investigation firm, Kroll Associates, confirmed at that time that they had been called in to assist a number of financial firms as a result of these blackmailing schemes. In New York City, a Kroll Associates spokesman stated, "One of the problems we face is that the potential embarrassment from loss of face is very serious. The problem for law enforcement is that the crime is carried out globally, but law enforcement agencies stop at each frontier."[18]

According to the same report:

> *A Bank of England spokesman confirmed that his bank has come under attack by the cyber-terrorists. Scotland Yard has assigned a senior detective from its computer crime unit to take part in a European-wide operation that has been codenamed Lathe Gambit.* (Authors' note: *Lathe Gambit is not a European-wide operation* — *it is a NATO computer security forum.) In the United States, the Federal Bureau of Investigation (FBI) has three separate squads investigating computer extortion.*

Since that time there have been HERF weapons demonstrations at a DEFCON and INFOWARCON conference, in United Kingdom (pictures of the device are available online), and at the U.S. Army's Aberdeen Proving Ground (using COTS equipment in a van). A *Popular Mechanics* (September 2001) article, "E Bomb," written by Jim Wilson, states that a weapon could be built for only U.S.$400.

TEMPEST

TEMPEST is an acronym that stands for (according to some) either Transient Electromagnetic Pulse Emanation Standard or Transient Electromagnetic Pulse Surveillance Technology — probably different depending on whether you are an engineer or an intelligence collection government agency.

The TEMPEST phenomenon has been well-documented for a number of years and is also known as the van Eck effect. There are two important research papers that together provide some of the first unclassified descriptions of the formerly highly sensitive area known as TEMPEST. The first paper (published in 1985 by Wim van Eck, a Dutch scientist) was titled "Electromagnetic Radiation from Video Display Units: An Eavesdropping Risk?" The second was published in 1990 by Professor Moller of Aachen University and was titled "Protective Measures against Compromising Electromagnetic Radiation Emitted by Video Display Terminals."[19] Together, these papers describe how it is possible to obtain meaningful information by intercepting the electromagnetic signals given off by various parts of the computer, especially the monitor.

There is much that is now published on the subject and "The Complete, Unofficial TEMPEST Information Page"[20] gives a wide range of resources that can provide information on the subject, including declassified American policy and advice on the protection of systems from this type of attack. In practical terms, this means that at the design stage of a system, a solution can be achieved through the design of the buildings in which the systems will be used, the purchase of systems that are protected from this type of emanation, or through the location of the building in which the system is to be used. If this problem is addressed after the system has been designed, then the options are more limited and, as with most retrofitted security solutions, will be extremely expensive.

It is possible to intercept sensitive information from computers by capturing radio-frequency emissions. Furthermore, the equipment is available for less than $60,000. But it is possible that a "software radio," specially designed to let computers tune in to radio signals in any waveband, could make eavesdropping on computer signals easy and inexpensive. Some believe such a system could cost as little as $2000 to $3000. It could be available in the very near future (if it is not already on the market).[21]

When a commercial system is developed, it will allow an attacker to capture the contents of computer monitors. If such a system is created, an attacker could monitor everything from passwords and user IDs for approving wire transfers in banks to the keyboard input for anything of value. This will all be done remotely without any need for physical access to the computer or perhaps the building itself, depending on the range and size of the interception unit. The ability to intercept all types of sensitive information from the monitors and microcomputer peripherals will make this type of device a very powerful and dangerous new tool in the hands of attackers.

Wim van Eck demonstrated the phenomenon and the capability to be intercepted in a documentary program called "Tomorrow's World" for the British Broadcasting Corporation (BBC). The demonstration included a van equipped with an antenna to intercept the emissions from the screens of computers inside buildings in the City of London.

There is no evidence that this effect has ever been exploited by a nation-state and, after the fall of the Berlin Wall and the end of the Cold War, Marcus Wolf, the former head of the East German Intelligence Service stated, "...but while it is true that in the age of satellites and computer hackers, technological intelligence, which does not come cheap, has to be used, human agents cannot be completely replaced. Technology can only establish the situation of the moment...," a reminder that it is still easier to use an insider to gain the information that you want to get from an organization.

As the use of information technology becomes increasingly pervasive, it is possible that corporations are also acquiring the capability to intercept their opponents' information. While the suppliers of the type of equipment that can be used to intercept the emissions all say they sell only defensive (protective) equipment, some of them admit that their products can be easily adjusted to allow them to be used for offensive surveillance.

In July of 1999, the FBI arrested an Israeli citizen, Shalom Shaphyr, who was in the United States on a business visa, for attempting to export a monitoring system that could be used to spy on computers. It was reported that Shaphyr was in the United States acquiring equipment on behalf of the Vietnamese government.

According to the FBI affidavit, Shaphyr agreed to pay $30,000 for "computer-intercept equipment," after a meeting with an undercover FBI agent, who posed as a salesman for surveillance equipment. Shaphyr was sentenced to 15 months in federal prison in January 2000, after pleading guilty to attempting to export defense equipment without a license. According to the FBI affidavit, Shaphyr had stated that the monitoring equipment "would be used in an urban environment to view computer screens in buildings and offices without the knowledge or consent of the computer users."

Bugs

The insertion of an eavesdropping bug into a system is still thought of as being very much like a line from a James Bond movie. However, if you walk into the Spy Shop in any major city or the shops in the electronics area, bugs that will intercept a telephone or a spoken conversation are readily available. You can also purchase small devices that will let you intercept every keystroke that is made on a keyboard. The potential for these to be used by industrial spies is huge and while most of them can be detected with reasonably inexpensive equipment, how often are corporate meeting rooms and offices checked for them? The answer, in most cases, is never.

Summary

While this chapter may seem to be all gloom and doom and lead you to think that disaster is inevitable, this is not true. COTS products provide what we are prepared to pay for and to tolerate; so in reality, you could say that it is our own fault. The manufacturers are capable of providing us, the users, with a far higher level of security in the products that we purchase and can also ensure that they are far more robust than we currently accept as the norm, but they will only do so when there is a financial imperative to make them. If we are prepared to keep on buying the software in the state that it is currently offered, which is largely bug laden and not fully tested, they will continue to sell it to us. Why not? They can catch you twice with the upgrade to fix the things that should not have been wrong in the first place. Again, with the security features that are provided, if the user insists on them being switched off because he does not know how to use them and how to configure his systems, is it surprising that what is delivered to us is set for the lowest common denominator?

It is possible, with public support, to influence the manufacturers, but only if there is a strong and coherent voice.

In the field of open source software, there is one particular endeavor that merits mention and that is the collaboration between Linux and the NSA to produce a secure version of Linux. The concept of a government security agency collaborating with and supporting an open source venture would not have been considered but a few short years ago. It is nice to see that dogma has been cast aside to support a venture that is hopefully in the public good.

Notes

1. *PC World* magazine, July 1998.
2. http://www.eegs.com/.
3. http://www.htsoft.com/.
4. http://pcwin.com/easter.egg.html.
5. Microsoft Security Bulletin (MS00-077), November 14, 2000.
6. http://www.Microsoft.com/piracy/basics/worldwide.

7. *The Next World War,* James Adams, 1998.

8. Center for Strategic Studies report: Cybercrime, Cyberterrorism, Cyberwarfare.

9. *Crypt Newsletter,* A Good Year for the Gulf War Virus Hoax, 1999.

10. Tom Clancy, *Debt of Honor,* Berkley Publishing Group, June 1995.

11. http://zdnet.co.uk/.

12. http://www.alldas.org/ or http://www.alldas.de/.

13. Mystery of the stolen laptops. Article in the U.K. Politics section of the BBC Web site. http://www.bbc.co.uk/hi/english/uk_politics/newsid_765000/765433.stm.

14. Symantec Alleges More Code Theft by McAfee, Martin Veitch, July 24, 1997.

15. U.S. Department of Justice Press Release dated March 21, 2001: http://www.usdoj.gov/criminal/cybercrime/MorchPlea.htm.

16. U.S. Department of Justice Press Release dated April 18, 2001, Orange County Computer Hacker Arrested in Scheme to Use Stolen Credit Cards to Make Wire Money Transfers via Western Union.

17. Excerpted from EmergencyNet News Service, Tuesday, June 4, 1996, Vol. 2–156, The Cyber Terrorists, Steve Macko, ENN Editor.

18. The Cyber Terrorists, excerpted from EmergencyNet News Service, Tuesday, Vol. 2–156, by Steve Macko, ENN Editor, June 4, 1996.

19. http://www.thecodex.com/c_tempest.html.

20. The Complete, Unofficial TEMPEST Information Page, http://www.eskimo.com/~joelm/tempest.html.

21. http://www.newscientist.com/ns/19991106/newsstory6.html New-wave spies.

Chapter 7

It's All about Influence: Information Warfare Tactics by Nation-States

States having the largest coercive means tended to win wars; efficiency (the ratio of output to input) came second to effectiveness (total output).... Only big governments with ever-greater command of resources could compete on the battlefield.[1]

—Charles Tilly, Historian

This chapter identifies and describes information warfare as it relates to nation-states; discusses why nation-state conflicts could occur; and discusses information warfare methods that can, and are being used in nation-state conflicts.

Introduction

Charles Tilly's comments, as stated above, were once valid in the industrial period of human history; but in this the information-knowledge period, "command of resources" does not mean iron, coal, and steel, but hardware, firmware, and software. Therefore, any nation-state, no matter what its size, can be in the possession of these cheap and powerful resources — and information warfare tools.

Any individual, group, corporation, or government can and does use information warfare tactics to promote their causes. Among the most active are, of course, the nation-states. This stands to reason because they have the most power, high technology, funding — and may be the most to gain or lose.

This and subsequent chapters do not discuss every nation-state in the world because many are still caught between the agricultural and industrial periods.

Their information infrastructure is primitive compared to the more modern nation-states of the world. Therefore, their information warfare capabilities are also relatively primitive, or nonexistent. A regional concept is used to discuss information warfare of nation-states and more detailed information provided relative to those specific nation-states with more advanced information warfare capabilities.

Defining Nations, States, and Nation-States

To ensure that we all understand what is meant by nations, states, nation-states, countries, and their associated governments, a few basic definitions[2] are in order.

- *A nation is...a people in land under single government: a community of people or peoples who live in a defined territory and are organized under a single government; people of same ethnicity: a community of people who share a common ethnic origin, culture, historical tradition, and, frequently, language, whether or not they live together in one territory or have their own government...*
- *A state is...a mostly autonomous region of federal country: an area forming part of a federal country such as the United States or Australia with its own government and legislature and control over most of its own internal affairs;...a country or nation with its own sovereign independent government;...a country's government and those government-controlled institutions that are responsible for its internal administration and its relationships with other countries...*
- *A nation-state is an independent state: an independent state recognized by and able to interact with other states, especially one composed of people who are of one, as opposed to several, nationalities...*

As these terms relate to information warfare (IW), we can see that they are often used interchangeably. Throughout this and subsequent chapters, the discussions will relate to IW using all three terms — nation, state, and nation-state — as well as their associated governments. However, we will also be discussing IW as it relates to other related categories that the U.S. Central Intelligence Agency's (CIA) *World Factbook*[3] calls "entities." These entities are:

> *...Some of the independent states, dependencies, areas of special sovereignty, and governments included...are not independent, and others are not officially recognized..."Independent state" refers to a people politically organized into a sovereign state with a definite territory. "Dependencies" and "areas of special sovereignty" refer to a broad category of political entities that are associated in some way with an independent state.*

The World Factbook lists 267 separate geographic entities.[4] They are categorized as follows:

- Independent states (191)
- Other (1)
- Dependencies and areas of special sovereignty (64)
- Miscellaneous (6)
- Other entities (5)

So, as one can see, the world is made up of many "entities," all of which are or have the potential of being involved in conflicts between nation-states. There are landmasses that are claimed by more than one nation-state (e.g., Spratly Islands). There are some dependencies whose people may some day want to form their own independent nation-state. This is important to understand because conflicts can break out at any time, for any reason, and in any part of the world. We are all aware of the conflicts, including the use of IW tactics, being used in such places as the West Bank. However, other places where today's nation-states may have an interest can also erupt into conflict, drawing in other states as allies and adversaries. As one can easily understand, where conflicts break out, a nation-state will use all means at its disposal to successfully obtain its objectives. IW tactics can, are, and will be used to support the national objectives of nation-states. After all, why not? It is just another tool to help a nation-state gain an advantage over its adversary, whether that adversary is an individual, group, another nation-state, or other entity.

History of the Nation-State

Because humans have a relatively short lifetime in the scheme of things, we tend to see nation-states as having always been in existence. Well, those who are alive today, yes, they have. However, in the history of human life, nation-states have been around for a relatively short time — and they may be on their way out. In their chapter, "The Life and Death of the Nation-State," Davidson and Rees-Mogg.[1]

> ...*the fall of the Berlin Wall in 1989 culminates the era of the nation-state, a peculiar two-hundred-year phase in history that began with the French Revolution. States have existed for six thousand years. But before the nineteenth century that accounted for only a small fraction of the world's sovereignties. Their ascendancy began and ended in revolution. The great events of 1789 launched Europe on a course toward truly national governments.... Those two revolutions, exactly two hundred years apart, define the era in which the nation-state predominated the world, spreading or imposing state systems on even the most tribal enclave...*

History has shown us that nation-states have come and gone throughout history. One does not have to look any further than the Twentieth Century to see that although the concept of the nation-state has existed for a few thousand years, individual nation-states are always changing (e.g., the collapse of the Austro-Hungarian Empire, the Soviet Union, and Yugoslavia). These changes have usually occurred as a result of war. As nation-states are pressured to give up some or all of their powers to the "sovereign individual"[1] and global corporations, it is unlikely that most if not all will do so without trying to retain their power and authority. In doing so, they will undoubtedly call for support from other nation-states that also are in a similar position. This support will come in the form of alliances and through nation-state associations. One thing is certain: a nation-state will not die quietly.

Some of the associations that may become embroiled in these conflicts include the United Nations (UN), which has been very active in peacekeeping missions; the Association of Southeast Asian Nations (ASEAN); the Organization of Petroleum Exporting Countries (OPEC); the European Union (EU); and the North Atlantic Treaty Organization (NATO), just to name a few. In total, the United States' Central Intelligence Agency's *World Factbook* lists approximately 280 associations or other groups of entities (e.g., nation-states). When viewing the listing, one can easily see where conflicts between individual nation-states can quickly erupt into regional or even global warfare. Association members call on their fellow members for help against its adversary (e.g., the United States calling on NATO and others in their fight against global terrorism). Conversely, their opposition will likely do the same.

The United Nations and Its Information Warfare Role

Because the United Nations (UN) is playing an ever-increasing role in the resolution of conflicts around the world, both as peacekeepers and as a forum for discussions between nation-states, it is important to understand the UN and its global information warfare role.

> *The United Nations was established on October 24, 1945 by 51 countries committed to preserving peace through international cooperation and collective security. Today, nearly every nation in the world belongs to the UN: membership now totals 189 countries.*
>
> *When States become Members of the United Nations, they agree to accept the obligations of the UN Charter, an international treaty which sets out basic principles of international relations. According to the Charter, the UN has four purposes: to maintain international peace and security, to develop friendly relations among nations, to cooperate in solving international problems and in promoting respect for human rights, and to be a centre for harmonizing the actions of nations.*
>
> *UN Members are sovereign countries. The United Nations is not a world government, and it does not make laws. It does, however, provide the means to help resolve international conflict and formulate policies on matters affecting all of us. At the UN, all the Member States — large and small, rich and poor, with differing political views and social systems — have a voice and vote in this process.*
>
> *The United Nations has six main organs. Five of them — the General Assembly, the Security Council, the Economic and Social Council, the*

> *Trusteeship Council and the Secretariat — are based at UN Headquarters in New York. The sixth, the International Court of Justice, is located at The Hague, the Netherlands.*[5]

As conflicts continue throughout the world, one sees the UN becoming more and more active as peacekeepers. However, that role is based on Industrial Age warfare of bombs, small arms, and infantry, but only if supported by the major nation-states. As nation-states begin using more and more information warfare tactics in conflicts with their adversaries, there is little that UN forces can do to keep the adversaries at arm's length, except to offer a forum for discussions. That, to be sure, is an important role; however, it cannot stop the fighting when the weapons are the information warfare weapons of denial-of-service attacks and the use of logic bombs and other forms of malicious code as information warfare offensive weapons. One cannot envision the UN using information warfare offensive and defensive weapons to keep the peace between warring nation-states or other entities. However, that possibility cannot be ruled out in the very distant future. After all, anything is possible.

Organization of Petroleum Exporting Countries (OPEC)

OPEC is an organization of countries that share the same policies regarding the sale of petroleum — at least overtly. The members are Algeria, Gabon, Indonesia, Iran, Iraq, Kuwait, Libya, Nigeria, Qatar, Saudi Arabia, the United Arab Emirates, and Venezuela.[2] Although OPEC member nation-states have sometimes been, and still often are, at odds with each other, they have also worked together more often than the oil importing nation-states would have liked. Although nation-states have fought among themselves (e.g., Iraq and Iran, Iraq and Kuwait), that has not stopped them from working together when they each see it as being in their own best interests.

As the global economy and the economic power of a nation-state vice military power continue to increase in importance as the world's priority and as a gauge of world power and influence, associations and groups of nation-states working together have formed economic alliances. Such alliances, when confronted with adversarial conflicts (e.g., from oil importing nation-states) pressuring OPEC and individual nation-states to increase production, lower prices per barrel and the like, some of the nation-states within OPEC or others may resort to the use of covert information warfare weapons in their conflicts. Add to that the power of oil and OPEC's role or non-role in fighting terrorism, and one can see conflicts brewing for decades.

North Atlantic Treaty Organization (NATO)

Just as the UN has used military forces as peacekeepers, so has NATO. However, the Serbians may argue that NATO was an attacking, aggressive force and not a peacemaking or peacekeeping force. As in all conflicts, those in power set the rules — and define the terms on which their actions are based.

During the recent conflict between the Serbs and NATO, both sides used information warfare tactics. For example:

> **Cyber Attacks against NATO Traced to China**
> **(IDG, 9/2/99, by Bob Brewin)**
> *Hackers with Chinese Internet addresses launched coordinated cyber-attacks against the United States and allied forces during the air war against Yugoslavia...cyber attacks emanated from the Serbs,...U.S. military traced the attacks back to more than one Chinese IP address.*[6]

Such information warfare attacks offer some interesting dilemmas:

- Are the Chinese true allies of the Serbs, or did they attack because their embassy in Serbia was bombed?
- Were the attacks actually coming from China, or were the defenders spoofed into believing that the attacks actually came from China?
- Were the attacks made to look as if they came from China but in reality came from somewhere like Russia which supported the Serbs?

Because of today's primitive information systems security methods, hardware, and software, coupled with the vulnerabilities of today's high technology, can one ever be sure? There are some specialists who might say that there was no doubt that the attacks came from China. That may or may not be true. If, in fact, the attacks came from China, is it because the Russians had compromised the Chinese systems and used them to launch such attacks? If so, why? One possibility is that they did so to have NATO expand the war to China or at least diminish the attacks against their friends the Serbs.

Whether one is on land, at sea, or in the air in combat, one can be fairly certain as to who their adversary is that they are fighting. There are flags on the weapons systems and other markings to identify them as belonging to one group, nation-state, etc. In addition, one can always identify the adversary because that would be the one shooting at you. However, as with terrorist and other such attacks, one often does not know the identity of the true attacker or entity. Some may falsely claim credit for attacks while others falsely claim "they didn't do it." As nation-states rely more and more on IW weapons and tactics, it will become much more complicated because one may not know who the attacker really is.

Association of Southeast Asian Nations (ASEAN)

The Association of Southeast Asian Nations (ASEAN) was established in August 1967 and formed "to encourage regional economic, social, and cultural cooperation among the non-Communist countries of Southeast Asia."[3] Interestingly, China is a "consultative partner" and, of course, controlled by a Communist government. Russia is also a consultative partner. Papua New Guinea is an "observer" and there are also eight "dialogue partners" (Australia, Canada, the European Union, India, Japan, South Korea, New Zealand, and the United States). ASEAN is composed of the following Asian nations:

- Brunei
- Burma
- Cambodia
- Indonesia
- Laos
- Malaysia
- Philippines
- Singapore
- Thailand
- Vietnam

One can see some obvious adversarial relationships when noting the nation-states involved in ASEAN. However, they become more obvious when one looks at the potential conflicts of these names over territories (e.g., the Spratly Islands are being disputed by the Philippines, China, Malaysia, Taiwan, Vietnam, and possibly Brunei).

Many of these nation-states, such as Singapore, have a relatively advanced national information infrastructure (NII), while others are slowly building up their NIIs. Consequently, IW tactics are not out of the question. Nation-states such as Singapore, with the more advanced NIIs may be more vulnerable and consequently may lose an IW conflict.

Indonesia has for some time been involved in IW with global hackers over East Timor independence — but more on that later.

International Monetary Fund (IMF)

Another interesting group is the International Monetary Fund (IMF). The IMF was established in July 1944 "to promote world monetary stability and economic development; a UN specialized agency member."[3] It has 182 members, some of which are or have been involved in conflicts of various intensities (e.g., Croatia and Serbia, Iraq, and Iran).

How can the IMF become a "player" in warfare? It can be a nation-state's reason for initiating a conflict. In fact, IW would be an excellent method of waging such a war in a very covert fashion based on the vulnerabilities of financial networks.

A reason to initiate a conflict? On December 17, 1997, a U.S. newspaper, *The Orange County Register* (Irvine, California), printed: "ASEAN leaders offer nothing new, ask more help — the Associated Press, Kuala Lumpur, Malaysia.... Throughout the summit, the region's leaders warned the United States and other economic superpowers they should do more or face the consequences..."

What do they mean by "face the consequences"? Does that mean there will be serious economic problems that affect the superpowers also, or is there a more sinister implied threat? If one were to look at that region today, one can see that little has changed since 1997. The nation-states are still trying to recover from their economic woes. Will that lead to more festering of hostility toward the economic superpowers? If so, information warfare may provide a cheap and covert way of getting the assistance these nation-states need — or at least that may be their rationale for using IW tactics against the IW-vulnerable, economic superpowers.

European Union (EU)

The European Union was born out of the European Community (EC),[3] with its headquarters in Brussels, Belgium. It was established in February 1992 with an effective starting date of November 1, 1993. The aim of the EU is:

> *To coordinate policy among the 15 members in three fields: economics, building on the European Economic Community's (EEC) efforts to establish a common market and eventually a common currency to be called the 'Euro', which will supersede the EU's accounting unit, the ECU; defense, within the concept of a Common Foreign and Security Policy (CFSP); and justice and home affairs, including immigration, drugs, terrorism, and improved living and working conditions.*

There are 15 member nation-states: Austria, Belgium, Denmark, Finland, France, Germany, Greece, Ireland, Italy, Luxembourg, the Netherlands, Portugal, Spain, Sweden, and United Kingdom, with another 13 membership applicants: Bulgaria, Cyprus, Czech Republic, Estonia, Hungary, Latvia, Lithuania, Malta, Poland, Romania, Slovakia, Slovenia, and Turkey.

The EU, of course, is somewhat new but conflicts among its members go back centuries. However, the EU conflicts of today center more on financial, trade, and privacy/security matters.

Governments and Information Warfare

As one can see, there are various types of associations, alliances, conflicts, and adversarial relationships around the world that might easily erupt into "hotter" conflicts. Thus, there is no lack of logical reasons why nation-states can be drawn into conflicts in which IW will play anywhere from a minor to major role.

> *...in Europe and the Western Hemisphere, the governments, almost all of them, are democratic, and military conflicts between them are almost inconceivable. And the problems there are economic and social, except for some ethnic conflict at the fringes of Europe. But in Asia, the various states look at each other as political and geopolitical rivals. War is not impossible, and therefore in that region we have to pay more attention to balance of power and equilibrium. In the Middle East, the conflict is more ideological and religious, and there the conflicts take on a much more emotional character...[7]*

> —Dr. Henry Kissinger

Nation-states are run and controlled by governments. Governments are made up of various types of groups. Some are democratically based, in which the citizens of the nation-states elect representatives who, in turn, make up the government. Some consider these as individual representatives, while others consider them not as individual representatives but as part of a group that runs the government — some formal (e.g., the Republican Party, the Labour Party, the Christian Democrats), while others not as formal (e.g., the conservatives, the liberals, and the lobbyists). They also have some control over governmental employees and agencies, thus portions of the nation-state. Some governments are made up of religious groups such as Islamic fundamentalists and Catholics. Other government systems are based on the communist system; for example, China, where the selection of representatives of the people is significantly different. In addition, there are countries that are ruled by juntas and dictators.

Regardless of the type of people that control the government, thus the nation-state, they all want to retain their power over their nation-state and its citizens, but they may also want to increase their power and authority not only over the citizens of their nation-state, but also over the neighboring nation-states. In addition to those groups and individuals that can be readily identified as making up the "government" of a nation-state, there are literally millions of other "civil servants" who rely on the government in power for their livelihood. There are also the millions of those who rely on the government for support via some form of welfare aid. Another powerful group is corporations that rely on the government for support in global economic and business competitive warfare, as well as government contractors that make weapons systems and provide other services to the government and are usually handsomely paid for such services.

Add to all that those retired civil service workers, and possibly more importantly, the retired military personnel who also rely on the various nation-state governments for support in the form of retirement checks, medical benefits, etc. It is not inconceivable that in some nation-states, all these groups make up more than 50 percent of the population. Add to that the other nation-states, foreign nationals, and foreign corporations that also derive benefits from a specific nation-state government.

So, if one were to try to dissolve or overthrow a government — no matter how despotic and dictatorial it is — one would conceivably be fighting everyone that gained some type of support or advantage by being linked to the current government (e.g., the majority of its citizens backed by foreign support). By the way, this does not include all those who might fight for the government, even if they dislike it, under penalty of death, imprisonment, loss of jobs, and threats to family. Thus, in conflicts between nation-states, there are many who will support your adversary. In the case of information warfare, one would never know each and every entity that would be attacking them.

As nation-states develop their IW capabilities, there is the hope that there will be fewer deaths and less physical destruction. However, one should also consider the following.

- Will that then make it easier for those in power to more readily rationalize becoming involved in conflicts where IW tactics are used?

- Will it cause more IW-based conflicts to erupt between nation-states?
- If so, how will that impact neutral nation-states, and whose vital networks are integrated into the Internet and global information infrastructure and national information infrastructures of the warring nation-states?
- How will that affect the economic powers of not only the warring nation-states, but also their global corporations that rely on these global networks to conduct business?
- What about the effect on the world financial community when these global systems go down and traders cannot trade?

In February 2001, the U.S. Office of the Undersecretary of Defense for Acquisition, Technology, and Logistics in Washington, D.C., published a report, *Protecting the Homeland*, in which it said:

- The I Love You virus contaminated over one million computers in five hours.
- Viruses cost $1.5 trillion a year; the bill for large U.S. firms will be $266 billion (2.5 percent of U.S. GDP).
- At least 20 countries are developing tools to attack computer-based infrastructure; the Internet relies on 13 key nodes.
- More than 22,000 cyber "attacks" on DoD systems were reported to the Joint Task Force for Computer Network Defense in 1999.

This Office sees the issues as follows:

- Unconventional threats can act in concert (e.g., information warfare and biological warfare)
- Attribution can be a deterrent
- Perpetrators can be virtually "invisible" based on their scale, dual use tools
- Defense against attacks will require close cooperation between the public and private sectors and among countries, and such cooperation is controversial today
- The threat is evolving very rapidly; we must evolve even faster
- Government roles and government capabilities are not aligned
- Need national consensus that strenuous efforts are necessary
- To prepare for or defend against such attacks
- Time to reprioritize investment

Obviously, no one has all the answers to these issues. However, one can speculate on how much devastation can be caused by such information warfare-based conflicts.

Nation-States in Conflict

Information-dependent nation-states are more vulnerable to information warfare attacks, while at the same time, the agricultural and industrial period nation-states are less vulnerable to such attacks — for the obvious reasons: they have little in the way of information systems and less dependency on them.

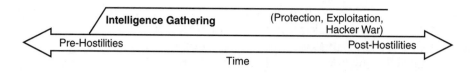

Exhibit 1.   Stage 1: Intelligence Gathering

It may behoove all information-dependent and information-based nation-states to make any potential adversary also information systems dependent. However, this is a two-edged sword. The adversary can then use more sophisticated weapons against you, while you are not prepared to defend against those that would otherwise not have been available to your adversary.

Wars between nation-states can be categorized in terms of their various phases. These phases are:

- Intelligence gathering
- Diplomatic pressure
- Economic pressure
- Military posturing
- Combat
- Reconstruction

Intelligence Gathering

During future wars in which IW plays a role, it will be important to develop and refine IW methodologies that can be used against an adversary. Many, possibly most if not all, gather intelligence on friends, foes, and potential foes alike. This is an ongoing task and has been ever since there has been the potential for conflicts.

Exhibit 1 depicts one of the first stages and an ever-present aspect of information warfare — intelligence gathering.

> *...Jay Valentine is president and CEO of Austin-based InfoGlide Corp., which makes powerful search software for such uses as insurance-fraud investigations. He will not license the technology to nine countries and three U.S. government agencies because of the potential for privacy abuse. That hasn't stopped at least one of those countries from trying. Two years ago, Valentine says, a company tried to buy rights to the technology. It turned out to be a front — for the Chinese government...*

—A Glimpse of Cyberwarfare:
Governments Ready Information-age
Tricks to Use against Their Adversaries
(By Warren P. Strobel with Richard J. Newman
World Report, March 13, 2000)

The use of spies, whether or not considered "gentlemanly" in the past, will continue to play a major role in information warfare. How will one know the power, defenses, and offensive capabilities of its potential adversaries if intelligence gathering does not take place? Nothing has changed because of high technology and the advent of information warfare except that it has gotten more sophisticated as high technology has become more sophisticated.

Today, because of the GII, the Internet in general, and the nation-states' NIIs, netspionage[8] (network-enabled espionage) is playing an ever increasing role. Take a look at an example of using information warfare methods for intelligence gathering:

> *By identifying an adversary's GII interface, NII, and DII in detail, mapping that environment against the strategies, and IW attack tools, techniques, and methodologies, one can prepare to successfully mount an IW offensive. During the Intelligence Collection Phase, one would collect and map the adversary's configurations, Internet nodes, links, groups, personnel, hardware, operating systems, and applications software being used by the adversary's networks. That information is analyzed to identify the vulnerabilities of the networks and what IW tools to be used against them. These tools can be used to collect and read messages being transmitted through the network for subsequent use in the other phases of national policy objectives implementation (e.g., send misinformation, take over the network, or just deny use of the network).*

Setting the Stage[9]

Netspionage, as we have defined it, is using networks and computers and associated capabilities to steal information of a nation-state or other entity which can be used to further the aims of another nation-state. It is a concept that is becoming increasingly important. Although it has not yet deeply penetrated the awareness of some nation-states' leaders, it is fairly well understood at the national governmental levels in the United States and other major global powers. Press reports[10] highlight the claim that the United States, in cooperation with its closest allies, the United Kingdom, Australia, and Canada, are suspected of creating a global monitoring and surveillance system. This system, reportedly code named ECHELON, is supposed to track all "telephone conversations, electronic mail messages, and faxes."[11] One can assume that this wholesale collection effort will undoubtedly include sensitive information being transmitted between nation-states, corporations, associations, and individuals.

In similar fashion, major European nations like France and Germany, as well as former Cold War enemies like Russia,[12] the People's Republic of China, and even Sweden have all been accused of implementing technical means of surveillance to deal with all forms of electronic communications. If any of these allegations are true, there is already a lot of netspionage being conducted by legions of techno-spies and netspionage agents.

Although a major concern is passive monitoring, no matter how pervasive, it depends on the negligence of the sender for success. Since quite often sensitive information is transmitted in an unprotected medium, such as an unencrypted e-mail message or a unencrypted cellular telephone call, it is vulnerable for collection by these organizations. The evolution of a "surveillance society" where every communication may be reviewed by faceless agents, competitive intelligence specialist(s), and others is very troubling.

Such measures include the systematic application of tools and techniques to achieve the classic objectives of espionage, acquisition of "secret" information by covert means. Such mechanisms have been used for centuries with varying degrees of success by nation-states' various governmental agencies. U.S. Central Intelligence Agency (CIA) Director Tenet said in testimony to the Senate Governmental Affairs Committee that 'potential attackers range from national intelligence and military corporations, terrorists, criminals, industrial competitors, hackers, and disgruntled or disloyal insiders.'[13] Tenet went on to say that the fundamental shift in the hacking challenge from that posed by individuals and terrorist groups to governments is already underway and that '...down the line we are going to encounter more [hacking] and it will be more organized.' By one estimate "23 countries are believed to have the capacity to engage in state-sponsored, surreptitious electronic raids. Among the most sophisticated: India, Syria, and Iran..."[13]

Monitoring, especially on a global basis has, to date, been the exclusive purview of large sophisticated governmental entities with vast budgets, huge numbers of people, and a large number of very fast and very powerful computers. In contrast, other techniques of netspionage may be engaged in by an individual or may involve the coordinated efforts of dozens to hundreds of collaborators distributed around the planet in a carefully orchestrated, yet "virtual" campaign to ferret out the digital treasures of a targeted nation-state.

Netspionage is made possible and indeed is inevitable due to the convergence of several major trends. The first, and most obvious of these is the increasing connectivity of all nation-states and corporations via the global Internet and the attendant evolution of electronic business and operations. This vast increase in the nature, extent, and duration of connectivity provide the potential access pathways for positive purposes, but also for abuse.

The second major trend has been the increasing digitalization of the most valuable assets of a nation-state or corporation. Nation-states and corporations are rapidly transitioning from the world where the greatest value has been created primarily by the physical contribution of people, manufacturing plants, and raw materials to a time where the acquisition, preservation, and exploitation of knowledge provides the primary competitive advantages. The products or services provided by most nation-states and corporations already reflect the increasingly important contribution of intangibles, such as the proprietary information and other intellectual properties created by the knowledge workers of the nation-state.

Whereas the first two trends are largely benign or at least neutral, the third trend is essentially negative. We have entered a time when there is widespread availability of both the software tools as well as the knowledge of how and why to exploit vulnerabilities in networks and computer systems. This is compounded by an increasing number of knowledgeable people who can operate these tools at the same time there is a general and rapid increase in the pool of people who have access to networks and computer systems as part of their work environment. The advent of what may be characterized as "electronic burglary tool kits" now allows nearly anyone with malice in their hearts to attempt to access or control the systems and networks that contain the valuable assets described above. As an ever-increasing number of the planet's population gain access to the global Internet, they have the potential of using these tools against nation-states and others entities in an attempt to gain personal benefit. The era of "security through obscurity" which assumed that no one knew where to go or what to do to hijack systems is now a fading memory. The barbarians are literally at the virtual gates, or more likely, on the payroll.

Given these trends, one might expect that nation-states would create strong protection programs to safeguard their sensitive information. Sadly, the protection of most nation-states' information assets is generally weak. Thus, they are vulnerable to those that operate in the "black zone" where legality and ethics are not considered.

Sounding the Alarm

The U.S. National Counterintelligence Center was one of the first agencies to alert businesses and other government agencies to the threat of netspionage as a tool for compromise of both classified (national security) information as well as proprietary and sensitive business or economic data. In an article in the July 1998 issue of the official publication *Counterintelligence News and Developments* entitled "Internet: The Fastest Growing Modus Operandi for Unsolicited Collection,"[14] the agency noted that:

> ...the Internet offers a variety of advantages to a foreign collector: it is simple, low cost, non-threatening, and relatively "risk free" for the foreign entity attempting to collect classified, proprietary, or sensitive information.

The article went on to describe how foreign intelligence collection was already exploiting the new access provided by the global Internet to obtain not just classified military information, but also economic and competitive information from nation-states and commercial business corporations. This demonstrates that professional intelligence specialists have already recognized that the Internet is revolutionizing their field. In the recent past, clandestine intelligence operatives would have likely been forced to engage in the risky business of recruiting foreigners to work on behalf of the spy agency in their plots to obtain information. Now, thanks to the ubiquitous Internet, collectors can safely sit at a keyboard in their most secure homeland site and execute a netspionage attack. Exploiting a variety of means (some of which are discussed in this book), they have the potential to engage in wholesale looting and plundering of the official economic and business secrets of their enemies.

The 1992 U.S. presidential campaign slogan, "It's the economy, stupid" captured a fundamental truth of life at the beginning of the new millennium.

Economic power, the ability to transform knowledge and information into goods and services, is now and will be for the foreseeable future the basis of national power. Most other nation-states understand very well that the United States is not only a military superpower, but also a technology and economic superpower. Otherwise "friendly" nations and their indigenous businesses have already directed efforts to penetrate the systems and networks that are the core strength of the economic and information infrastructures of the United States.

As global power is increasingly the result of the application of high technology to the national economy, one inevitable result of this will be that business corporations will become caught in the crossfire between those nation-states competing for global advantage. As incidents of network-enabled information theft proliferate, directed in some cases by competing national intelligence services, it is likely that business leaders have already concluded that the "rules" of the global business game have fundamentally changed. They will soon begin to use more aggressive tactics to survive and flourish and will use netspionage as a weapon in the global economic and business warfare arena.

There is precedent for this transition and how quickly attitudes can change. In the early Twentieth Century before World War I, a senior U.S. government official (Senator Henry Stimson) dismissed the idea of using spies to intercept official communiqués between foreign nations with the comment that, "Gentlemen don't read other gentlemen's mail." Such an idea now appears quaint and hopelessly idealistic in the Twenty-first Century as we review the crucial role that precisely such interception played in World War II and other conflicts in the past century. It is inevitable that legal and ethical constraints against aggressive techniques of netspionage by nation-states, businesses, and quasi-governmental entities will totally vanish over the next five to ten years as nation-states struggle to exert control and gain a dominant position in the global competition for survival and success. At some time early in this new century, such tools will be considered as fundamental to nation-states as any other weapon of warfare and, in particular, information warfare.

Netspionage is also gaining significance as a threat against other key industrial areas. According to Josef Karkowsky of the German Association for Economic Security:

> *The data net is not only used to make business contacts — it often also opens the way for modern spies. German companies in particular ignore security measures, making "self-service stores" for computer spies out of themselves.*[15]

Digital Tradecraft

The Internet, and specifically the Web, has now made available to increasing numbers of people with widely varying legal, ethical, and personal motivations and constraints the tools and technologies that are nearly ideal for the theft of the increasingly valuable digital assets of the typical nation-state. The global reach, now afforded to prospective techno-spies by the expanding connectivity provided by the Internet, has created a situation of unparalleled opportunity for anyone who is willing to go after a nation-state's secrets or set up a process to successfully attack their NII and other networks.

The pages that follow describe software programs and utilities that allow even unsophisticated attackers to steal information, then safely transfer these assets to others or successfully attack a nation-states' networks.

The focus in this section differs somewhat from a description of traditional hacker tools and how they have been used. Whereas a "hacker" may chose to engage in netspionage, it is also possible for a netspionage agent to commit netspionage without being an accomplished hacker. In fact, successful techno-spies may never need to engage some of the sophisticated technical tools, such as port scanners and attack simulators. Such powerful software requires some technical knowledge and tends to be the hacker's weapon of choice for neutralizing traditional information security technologies. Instead, we discuss an application of software and technical tools that can be described as "digital tradecraft."

Tradecraft is defined as "the technical skills used in espionage"[16] and typically might include knowledge of lock-picking, clandestine photography, secret writing, surveillance, and dead drops.[17] As one can deduce from such examples, a great deal of the traditional spy's life revolved around the means of acquiring information and then communicating it to the sponsoring agency for processing and analysis. Of course it is obvious that such means are intended to allow spies to operate in stealth and with anonymity, largely in hopes that they might continue to survive and perhaps even someday return safely to their homeland — but now they can even do it from their homeland!

In many ways, the objectives of netspionage are exactly the same as they have been for traditional espionage. The new tools are intended to allow the "virtual agents" operating against online nation-state government agencies to penetrate the internal systems of the target. Once successfully "inside," they can obtain information without detection, then communicate it so that the sponsoring nation-state can operate with the advantage of superior information while the attacked nation-state remains blissfully unaware of the nature and extent of its losses.

Take a closer look at some of the tools that comprise what can be characterized as "netspionage tradecraft."

Trojan Horse Software

There has been a literal plague of what has been described as "Trojan horse" programs. These software tools have been so named in honor of the famous beast of history that allowed the mighty walls of the city of Troy to be breached by cunning Greek warriors. Today's "Trojan" software is more likely to appear in the guise of a cute little attachment sent by what appears to be a friendly e-mail account. These programs represent a fundamental change in the threat matrix; they are particularly dangerous in less-sophisticated nation-states where there may not be a formal information security program, or in many modern nation-states where security rules are lacking or not followed.

The various Trojan software provides a common core of functions that typically includes the following (in a Windows operating system environment):

- Operate concealed, in "stealth mode" without any indication to the user of their presence. Nothing will be visible in the Windows system tray or will appear if the user activates the Close Program dialog box.
- Run programs already resident on the "target" system remotely without the user's intervention.
- Capture (log) user keystrokes without alerting the user
- Capture screen shots
- Reboot the computer
- Upload (and execute) programs to the "target" computer without the user's knowledge
- Operate microphones, Web cameras, modems, and other peripherals to gain information, including remotely turning on computers, downloading their contents, and then turning them off

BackOrifice2000

The program that probably best exemplifies the large number of new software tools that are magnifying the risks of network-enabled espionage (netspionage) is called BackOrifice (abbreviated as BO), which in its most current form is known as BackOrifice2000 (often abbreviated as BO2K). Named in a mocking double entendre to deride Microsoft's BackOffice with a teenager's sense of "potty humor" (the logo shows what one might infer to be a human buttocks), the software itself is no laughing matter. The original software was developed by a hacker who goes by the name Sir Dystic in what he or she has claimed is an effort to get Microsoft to improve the security features of the widely used Windows operating systems. The original version was released by the hacker group Cult of the Dead Cow at the annual DefCon hacker conference in Las Vegas in July 1998 and only operated under Windows 95. The BO2K version operates under Windows 95, 98, and NT 4.0 (sometimes). The most current version was released in July 1999, again at the DefCon conference.

The software is shareware, and is available for free to anyone, anywhere in the world with an Internet connection. By some estimates, several hundred thousand copies of this tool were downloaded in 1998 alone, and it continues to be very popular among hobbyists and hackers and others with an interest in computer and network security.

The program has a number of features that set the standard for other tools to follow and also has several unique features that distinguish it from the many copycat utilities. Some of the features included in the software itself and described in the documentation that are of greatest interest for the application of the product to netspionage include:

- Session and keystroke logging
- HTTP file system browsing and transfer
- Direct file browsing, transfer, and management

- Multimedia support for audio/video capture and audio playback
- NT registry passwords and Windows 9x screensaver password dumping

A computer user must be running either Windows 95 or 98 to allow this Trojan to automatically infect a target machine. The software will not install itself automatically on an NT system. The software comes in two parts: the client and the server. The server is installed on a target computer system and the client is used from another computer to gain access to the server and control it. The client connects to the server via a network.[18]

Once connectivity is established, it is possible to exercise considerable control over the server/target system. For example, the server (target) can be made to send an e-mail message to a designated address with key information about the system included in the message. It is also possible to connect to an Internet Relay Chat (IRC) server to inform all the users of a particular IRC channel that a specific computer is now available for remote operation and control.

What does this mean? If someone successfully installs BO (or one of the many imitators) on a target system, they have at least as much control as the assigned owner/user. Anything the user could do while sitting at their keyboard can probably be done by the techno-spy sitting at their own computer, which may be located in the same building as the target, or could possibly be located on the other side of the world. The techno-spy running the client software is able to search through the file listings of the target system; find any that are of interest; and copy, modify; or delete them as desired, or transmit them to another computer for future use. Cached passwords (passwords stored on the computer for example to log in to a remote system or Internet service provider) can likewise be copied and transmitted. The keystroke logging allows the client to capture any passwords entered by the user (e.g., those that have not been cached) and use them later, perhaps to impersonate the authorized user and gain access to an important database system.

The fact that the BackOrifice product accommodates plug-ins provides more reasons to be aware of this tool. As if the basic functionality was not dangerous enough, there are extensions that allow even unsophisticated users to package BO *into* another program. This allows BO to infect a target when the "doctored" program is executed. The plug-in called *SilkRope* also modifies BO so that it cannot be found with a common file scan. These tools are one reason why the various holiday executables can be a source of real danger. Although the "Dancing Santas" and "Happy Halloween ghosts" executables may be cute, it is a very simple matter to load BO into the file and send them out via e-mail to the desired target. When the target executes them, they enjoy the display, unaware of the infection and subsequent control over their system enjoyed by the operator of the client code.

These are features that are nearly ideal for the purposes of theft of sensitive information from computer and network systems. These tools are optimized for theft of passwords, documents, and other materials right out from under the noses of the often unsophisticated users (such as senior executives, managers, and other less technical staff), which means they are ideal tools for netspionage.

Other Trojan Software

For those netspionage agents operating in the "black zone" (illegal or unethical), BO is not the only sophisticated tool available. There are many others. An excellent listing of many additional "network Trojan" software programs can be found at http://xforce.iss.net/alerts/. The X-Force service, sponsored by ISS (Internet Security Systems), has documented more than 120 for the various versions of Windows. This site also provides a full technical description of how the software operates, as well as techniques for detecting and removing them from computers.

The following is a short list of some of the most common additional tools and the special features associated with each one. Note that the feature described is not the only function the software performs; most of them have the full complement of basic features similar to BO but also other unique features:

- *NetBus Pro*: presents itself as a remote administration and spy tool
- *NetSphere*: will operate the ICQ real-time messaging utility
- *SubSeven*: uses Internet Relay Chat or ICQ to inform the attacker when a target is infected

Digital Dead Drop

Imagine you are a techno-spy and you need to set up a secure place where you can "stash" copies of the critical information you have obtained from the penetrated nation-state. As good as you are, there is always the chance, no matter how small, that the authorities might raid your computer someday. If they find the copies of the stolen crown jewels on your computer system, you are going to be in big trouble. One way to avoid such professional embarrassment is to load a "digital dead drop" with the copies of the stolen valuables and get them out of your system as quickly as possible.

There is no need to purchase a server and set up an Internet connection as there are already many services offering 10, 20, 30 or more megabytes of online storage for free. The most generous provide 300 MB of personal "free disk space" on "secure servers."[19] All they require is some personal information about the subscriber, which of course could be completely fabricated, because the service apparently makes no effort to verify anything. The advantages to the techno-spies are obvious. They get a free online storage place that is accessible from anywhere on the Internet at any time of the day or night. Of course, if they are especially careful, they will protect the valuable stolen contents by using one or more methods of cryptography, perhaps even steganography (see below) to ensure that even an examination of the files deposited in the dead drop will be fruitless for investigators. Those that use such services should also be aware that the sites might be excellent targets for netspionage agents.

Even if the process is detected and investigated by the security group, they face an uphill battle. If the techno-spy uses digital dead drops properly, the "control agent" from their team or customer contact will be unloading the contents soon after the agent loads them. This downloading operation will be done using another expendable account, probably a new Web mail or front company address, for every transmission. This will be done from a safe location, probably outside the country. The contents will then be transferred to a safe location inside the sponsor's home nation-state, probably outside the target's homeland. Using a number of foreign locations for the transfer and processing of the stolen contents will make recovery more complicated and reduce the ability of the security officials to gain search warrants and execute them on a timely basis against multiple foreign locations and operations.

Steganography

Hiding information by embedding a file inside another seemingly innocent file is a technique known as steganography. It is most often used with graphics, sound, text, HTML, and PDF files. Stenography with digital files works by replacing the unused bytes of data in a computer file with bytes that contain concealed information.

Steganography (which translated from Greek means *covered writing*) has been in use since ancient times. One technique that was used in the distant past was to carve secret messages into wooden objects and then cover the etched words with colored wax to make them undetectable to an uninitiated observer. Another method was to tattoo a message onto the shaved messenger's head. Once the hair grew back, they were sent on their mission. Upon arrival, the head was shaved to reveal the message. The downside of this system was that you could only use a messenger once; and for the messenger, the major downside was that after reading the message, the carrier was normally killed to prevent anyone else from reading the message. The microdot, which reduced a page of text to the size of a typewriter's period so that it could be glued onto a postcard or letter and sent through the mail, is another example.[20]

Usually, two types of files are used when embedding data into an image. The innocent image that holds the hidden information is a "container." A "message" is the information to be hidden. A message may be plaintext, ciphertext, other images, or anything that can be embedded in the least significant bits (LSB) of an image.[21]

Steganographic software has some unique advantages as a tool for netspionage agents. First, if an agent uses regular cryptographic software on its computer systems, the files may not be accessible to investigators but they will be visible, and it will be obvious that the agent is hiding something. Steganographic software allows agents to "hide in plain sight" any valuable digital assets they may have obtained until they can transmit or transfer the files to a safe location or to their customer. As a second advantage, steganography can be used to conceal and transfer an encrypted document containing the acquired information to a digital dead drop. The agent could then provide the handler or customer with the password to unload the dead drop but not divulge the steganographic extraction phrase until payment is received or the agent is safely outside the target nation-state. As a final note, even when a file is known or suspected to contain information protected with steganographic software, it has been almost impossible to extract the information unless the passphrase has been obtained. (See Chapter 14 for further discussion on this topic.)

Computer Elicitation

Information requests via the Internet are likely to become a regular part of the techno-spies' arsenal. After all, e-mail communication using a sanitized account or free mail (see below) can be direct, quick, and inexpensive; and if it is successful, the requested information can be transmitted via return messages within minutes. Many people are now accustomed to receiving e-mail requests from people who they do not know, and few employees will take the time to verify the sender's identity and physical or Internet address when an unusual request is received. Even if a request is received from an unknown person, many people will likely just do their best to respond to the request without considering who is asking for what. The art of obtaining meaningful information from an unsuspecting source is known as *elicitation* in the intelligence community. It is really a skill to manipulate a conversation in such a way as to encourage a person to disclose sensitive information without raising suspicion. Because e-mail in many nation-states now takes the place of conversation, it offers the potential access to knowledgeable persons from whom information can be elicited.

The U.S. National Counterintelligence Center has advised that the following could be indicators of attempted collection efforts via computer elicitation:[22]

- The return e-mail address is from a foreign country.
- The recipient has never met the sender.
- The sender may claim to be a student or consultant working on a special project.
- The sender identifies his or her employer as a foreign government or claims the work is being done for a foreign government or program.
- The sender may advise the recipient to disregard the request if it causes any security problems or if it is for information the recipient cannot provide due to security classification, export controls, etc.

Network Mapping and Attack Tools

Software has become available that allows the techno-spy to probe even the largest networks for vulnerabilities. This probing is done in a way that creates the appearance of participating in a multi-nation attack force. This freeware allows the attacker to simulate any number of TCP/IP addresses as the originating points for these reconnaissance efforts. According to information security specialists, it is possible that by embedding attack scripts in the software, an attacker could exercise "one-button takeover of commercial or military computers."[22] It is now possible to execute very sophisticated scans for vulnerabilities and then to add the exploitation script for the specific vulnerabilities that have been detected. Once root/superuser access is obtained, the attacker will then have complete control of the system. *"Military and commercial espionage has never been so easy. Competitors inside or outside the country have little stopping them from closing down an enemy's electronic commerce and other network based services."*[29]

The tool is called NMAP;[24] it has a very good user interface, which has made it is accessible to a wide range of potential users. For the netspionage agent seeking an easy method of finding and exploiting vulnerabilities, NMAP has a lot to recommend it. Not only is it freeware, but it can be configured so as to mislead opponents into thinking people from an ostensibly hostile country are scanning them. Such misleading information can add a whole new level of protection to a collection effort. Imagine the excitement of a Republic of China (Free China on Taiwan) competitor with a good network security officer who detects a series of scans that appear to be originating from mainland China, or vice versa. As they leap to the conclusion that "hostile forces" are mounting an attack, the techno-spy could quietly complete his assignment. Once they provide the stolen information to their European or U.S. sponsor, they can have confidence that even if their efforts were detected, the company will be searching in the wrong direction.

Free Mail

One of the most important tools of choice for the successful netspionage agent is a Web-based e-mail service. There are several available and none that seem to have any particular advantages over the others. Two of the most commonly used are HotMail and Yahoo Mail; however, so many sites offer Web-hosted e-mail as a service that it is merely a matter of personal preference as to which one to use. These services are useful in part because, like so many other products and services on the Internet, they allow anyone to register and they do not appear to make an effort to verify any of the information provided by the registrant. E-mail services are accessible from any Internet access point.

When techno-spies are operating inside a target nation-state, there are several ways they could be tripped up. One of the most dangerous is to be caught in possession of documents or files they are not authorized to possess. They obviously need get the rid of the crown jewels quickly, and they need to get them to the customer. The longer they have it in their possession, the greater the chance that a search could discover the incriminating possession of critical information. If the knowledge worker has a Web browser on his desktop computer, then he can use Web mail to send the materials out of the nation-state. A couple of key caveats apply. First, the contents to be transmitted should not exceed the total bandwidth capacity to be sent. For most documents and files, this should not be a problem. Larger files may need to be compressed using zip or tar utilities, or perhaps subdivided into smaller components to avoid detection or to avoid crashing the mail gateway.

The reason Web mail is better for netspionage operations is because many companies have now begun monitoring outbound e-mail message traffic at the e-mail gateway. The SMTP gateway commonly has some type of filter, perhaps one created by the information technology group or maybe a commercial product. If the neophyte techno-spy is foolish enough to send a critical document containing keywords that trigger the filter, his career will likely be over. Of course, this includes the assumption that the keywords correspond to the current target. Because they will tend to change over time, it is possible that even a very good filter may not contain current priorities. The second is that someone notices and actually responds in a timely manner. Given the workload of most network and information systems and security personnel, it is very possible that even if the software is installed, they will not be detected. Reviewing the reports and logs is rarely a high priority and it is possible the techno-spy could perhaps complete the assignment and be gone before anyone had the opportunity to respond.

On the other hand, if a Web-based e-mail package is used, the techno-spy can transmit the stolen files to a safe holding position, perhaps one of the digital dead drops mentioned earlier. Because there is no practical limit to the number of free mail accounts or free drive server space to which they have access, the netspionage agent can operate without worrying too much about running out of space.

The only way to prevent a netspionage agent operating within the nation-state from using Web mail would be to deny all personnel at the location the option of using HTTP, which is the protocol used to surf the Web and engage in basic Internet research and searching. This means it is not likely that nation-states could take the steps to prevent netspionage agents from bypassing the company-provided (and monitored) electronic mail system in favor of the free mail access path that avoids detection.

Erasing Digital "Fingerprints"

Once stolen digital assets are transmitted to a safe location such as a digital dead drop, netspionage agents must then use a product that will help cover their trail and eliminate any digital evidence that could link them to the activity. There are a number of tools, such as East Tec Eraser,[25] that destroy much electronic evidence of the crime. These tools, often developed specifically to defeat computer forensic techniques used by law enforcement and security organizations, go far beyond the "delete" or "erase" functions provided by Windows and other operating systems. As almost everyone knows, files that are deleted or erased by earlier editions of the Windows operating system are not physically eliminated from the disk, but merely renamed and their assigned disk space opened to receive new data. Often, such files can be completely recovered by simple "Undelete" utility software. Forensic eliminators not only wipe (overwrite the clusters assigned to a specific file) the files that were sent, but also clean out the disk cache, file slack, all temporary files, and free space. Once these procedures are completed, they dramatically reduce the possibility that even a full computer forensic investigation of a computer system will uncover useful evidence.

Other Tools

The are many other tools that can be used as weapons in information warfare. Among them are:

- *WHOIS*: used to collect information from the InterNIC
- *DNSLOOKUP*: used to identify related networks
- *FINGER*: used to identify individual users and accounts
- *NetScan*: provides a group of tools to collect information
- *WhatsUp*: used to map a network and can be used for monitoring
- *Strobe*: used to automatically scan for ports[26]

Thus, one can see that the free arsenal of information warfare tools available to anyone makes it easy and cheap for even the poorest of nation-states to be a threat to an information-dependent nation-state. Add to this vast arsenal the offensive weapons that many of the more information-based nation-states are developing, and one can see that any all-out information warfare attacks against an information-dependent nation-state can be devastating.

Another Approach to Obtaining Information

An interview with Steve Lutz, President, WaySecure (see WaySecure.com), a highly specialized information systems security and high technology crime investigative company, (shown in Exhibit 2) in which he discusses obtaining information.

The Honey Pot and "Sexpionage"

In the old days of espionage before Internet and mass global communications, a well-known way of getting information from someone was to compromise them through sex-related blackmail. The adversaries provided members of the opposite sex or the same sex, depending on the individual target's profile and sexual preference. With the advent of mass global communications such as the Internet and World Wide Web, one can now find many "romance sites." Most of these sites are probably legitimate. For a price, they introduce two people to each other. However, these sites could also be ideal fronts for identifying new targets for exploitation. Also, the sites may be legitimate; however, one or more of the people listed on the site may be an agent for an adversary or competitor.

For example, on one site alone, 31 countries were listed with a total of more than 3528 women. Of those, 807 were Chinese from the People's Republic of

Exhibit 2.   Interview with Steve Lutz, President, WaySecure

I have been working on theoretical backdoors to operating systems and sub-systems (router boot images, special purpose devices, etc.) on and off for the last ten years or so. Anyone can do it with Trojan horse software; however, the manufacturer is in a supreme position to embed such a "feature" into the kernel as they own the source. If the source is open, then no problem. You just go through it line by line and make sure you understand everything it's doing. Microsoft is decidedly **not** open source. If you could get a copy of the source code for Windows (as was purported to have occurred several months ago by hackers that broke into MS http://www.zdnet.co.uk/news/2000/42/ns-18719.html) and examine it, you would find such a backdoor if it exists. Could there be a connection? Did the Bundeswehr get a hold of a copy and discover this? Possible. If so, the cat is out of the bag and we'll be hearing more about it soon. If not, then the Germans are just being paranoid. Or are they?

Looking at the problem, I see two main government strategies for inserting backdoors in an OS. The first is to have a way to remotely control the machines in question at some critical point in time, say in the event of hostilities, leaving the channel dormant until needed. Once activated, things would be bad for the target country for some time until they could firewall the systems and recover from whatever damage had been caused. This is a one-time use scenario that would be very effective. Once. Perhaps that would be enough in a war.

The second strategy is one that intrigues me more and is far more insidious. That is to implant a covert channel for intelligence collection purposes. This could be activated/ deactivated at any time using another covert channel, for example a broadcast (multicast [IGMP, mbone]) to a specific group of machines in a given country based on the version (German, French, Russian, etc.) being used. Obviously, if the target saw either inbound or outbound traffic on their network sniffers that looked unusual, they would be tipped off and possibly begin to investigate. This risk could be mitigated by using well-designed covert channels using steganography (no performance hit or discrete process to detect) and a slow leaking of the data embedded in, for example unused sections of TCP/IP packet headers of legitimate packets being passed through a given firewall. It could even be done by sending fragmented UDP/IP or multicast packets that may be passed by a firewall (outbound) and if detected would look like ordinary network noise. Even if analysis were done on the fragments or multicast packets, they would come up with seemingly random and useless patterns. What you would achieve as an attacker would be a slow, targeted, electronic intelligence feed that would be impossible to trace back to its source (remember, we're using multicast here). Pretty cool, huh?

So, given these and other scenarios that I'm sure we could cook up in a pub over a few pints, do you think it's wise for a government to process classified data on operating systems that they don't have the source to? My opinion is that either use open source OSs (after having carefully examined the code), commercial off-the-shelf software in stand-alone mode (no network connections outside the closed area), or build your own OS for classified data.

Check out these quotes from a MS tech bulletin. The bulletin talks about attacks against Windows machines using multicasting http://www.microsoft.com/technet/security/bulletin/ fq99-034.asp. However, they mention that fragmenting IGMP (multicast packets is OK and not to be concerned) may be caused by legitimate programs:

> *Windows 95, 98 and Windows NT 4.0 will operate correctly with any legitimate programs that may need to send fragmented IGMP packets. Translation: After you apply the patch, your machine is safe from inbound denial-of-service attacks so don't worry if your network people say you're generating or receiving fragmented multicast packets, that's normal.*

Exhibit 2.   Interview with Steve Lutz, President, WaySecure (Continued)

> *The vulnerability is unrelated to the IGMP functionality. Instead, it results because of the specific way that Windows 95, 98 and Windows NT handle fragmented packets. There is no inherent security risk associated with using IGMP."*

Translation: MS handles fragmented multicast packets in a unique way, but don't stop using IGMP because, once you apply the patch, you're secure. Perhaps you're secure from a denial-of-service attack, but not from what I outlined above.

Do you think I sound a little paranoid? Think again. Better yet, read this article by Craig Rowland "Covert Channels in the TCP/IP Protocol Suite" http://www.firstmonday.dk/issues/issue2_5/rowland/. In the Final Notes he concludes:

> *These methods could be incorporated into an operating system kernel or dae-mon that is set to automatically send a file when a particular site is contacted or whenever the system is used during normal operation. This method could turn a machine into a very stealthy transmission device that would appear to be working normally.*

He even includes a source code for constructing a covert channel on Linux machines. Even encrypted data packets have un-encrypted packet headers, otherwise there would be no way to route the packet. So if the target is using VPNs or other data encryption methods, it would not affect your covert channel.

And just when one thinks it is safe to go back online...

China and "overseas Chinese," 127 were from Russia, and 415 from the former Union of Soviet Socialist Republics (U.S.S.R.), excluding Russia itself.

Included on the Web site were color photographs and descriptions of the women, including their personal profile, their hobbies, as well as telephone numbers and e-mail addresses where they were available. Most of the women identified were professional, spoke at least some English, and many were college educated. One listed her occupation as "Chinese Army officer." Although it is not possible to know just by reading the sites whether any of the listings are in fact agents, it is likely that some foreign services are already exploiting this approach.

The use of men and women willing to exchange sexual favors for infor-mation is not new. However, in the pre-Internet days, such encounters were done person-to-person. Therefore, there was the risk of being caught by the intelligence agents of the targeted country. With the advent of the Internet "romance Web sites," this is no longer the case.

Take a look at an example of how such a system could be used. A country or foreign business, seeking to obtain sensitive information from an American business relating to the development of their new microprocessors, directs its netspionage agent to list himself or herself on one or more romance Web sites. The agent can be male or female; their background or description is not important. The netspionage agent creates profiles to attract the type of individuals being sought. The netspionage agent would of course get or "invent" photos of an attractive female or a handsome man (depending on the target) and ensure that the individual is fluent in the language of the targeted country where the target business is located, in this case English. That person could also be a college student or a college graduate majoring in computer systems or electrical engineering. Thus, they would naturally attract potential netspionage targets with similar interests.

Now along comes Fred, an American at work surfing the Internet out of boredom or looking for the one person to "light up his life." So, this American computer engineer from Silicon Valley, who is considered an extremely introverted nerd, makes contact with "Natasha." She appears to be the girl of his dreams and another computer nerd. They share so many common interests. Fred, of course, has never been briefed on the dangers of such contacts and what to watch for, or totally ignores the warnings that have been provided. So Fred sends off an e-mail to request a password to register for the site, or he logs into a URL to begin the registration.

Immediately, the site collection begins. From either Fred's e-mail header information or from the details provided by his Web browser, the adversary can determine the primary domain server associated with Fred's employer. From Consulting Internic or equivalent commercial registry services, they can learn the nation-state name and street address associated with the IP address. They will also learn the telephone numbers and fax numbers associated with the nation-state, as well as a system administrator and technician's telephone, fax, and e-mail information.

Once Fred completes the registration form, the adversary will have some additional information. Because this is a romance Web site, they may ask questions that are very personal and which could open the person to blackmail. The answers on the registration form can also be used to create a psychological profile of Fred so that he can be more effectively manipulated. Although Fred is under no obligation to answer truthfully or completely, the odds are good that he will give up some very useful information that helps the psychologists broaden his profile.

Because Fred works for a targeted company and is in one of the right groups with access to targeted information, a prompt response will be received from his new friend, Natasha. Over time, many conversations via e-mail would take place and this new "digital pen-pal" would gradually gain the trust of the targeted engineer who would of course open up and begin to share information as to what he is working on. It may be that the netspionage agent will be able to gain valuable information without ever actually meeting the targeted engineer. And if discovered, what is the result? There are no laws that can be enforced to prosecute that agent, especially not while he or she remains safely in Russia, China, or another country.

A variant on the above scenario would be to identify a specific business and identify individuals within that business to whom e-mails can be sent to develop a relationship over time. The netspionage agent interested in biotechnology, for example, would target biotechnology firms. Once the individual was identified, the netspionage agent would send an unsolicited e-mail: "Hello, you don't know me but I heard your lecture at the conference last week (or read your paper in the proceedings or...). I am also working on research in the field for a similar corporation in Peking. As a fellow scientist, I would like to share information on topics of mutual interest for the benefit of humanity. We are currently working on developing a prototype of XXX. What are you working on?"

There are so many variations that one can think of to begin the contact that may ultimately lead to compromise of sensitive business information. Remember, especially in nations like China and Russia, Internet e-mails, telephone calls, and other communications activity are often monitored. This means that if an innocent relationship did develop through the Internet, the netspionage agents of that nation could use the opportunity to require their citizens to gather information — and in countries like China, you do not say no to the government.

Potential Targets for Netspionage

The most important step in determining what assets to protect in a nation-state is to first determine what may be valuable and important, and then to think like the opponent, the techno-spies themselves. Because the global Internet allows the netspionage agents easy access to networked systems anywhere in the world, it is vital that the government agencies' managers understand they will inevitably be dealing with citizens from other countries, even other continents, as well as those of their homeland.

Access to Other Networks and Systems

Perhaps one of the least obvious and potentially most significant reasons to "target" a specific business or government agency is the particular locus it occupies on the Internet. To managers accustomed to thinking of their government agency as a purely physical entity, it may be difficult to accept that the Web of electronic connections that constitutes the Internet and electronic business may result in undesired attention. Even if the government agency itself has nothing of special value, the Internet connections it maintains and the information that flows between the nation-state and others may be cause for undesirable attention from other nation-states with an adversarial relationship with them, techno-spies, and others.

"Safe" Harbor and Staging Base

In some circumstances, inadequate security at a location will make it a convenient staging area for operations against another ultimate target located in another nation-state. Such situations are more likely if the intervening nation-state has extensive Internet connectivity with large or high-profile nation-states with valuable digital assets. This is a serious risk for nation-states extending their networks to vendors in the form of "extranets." Unless security standards are enforced for all participants in the "extranet," they may create an unmarked back door that can allow the techno-spies direct access to the nation-state's most valuable digital assets with no need to first circumvent existing security systems.

Vendors

The existing relationship between a vendor and its government agency customers may provide a convenient way to gain information concerning a primary target. Vendors' Internet connections create the risk that outsiders and others may gain unauthorized access to the nation-state's sensitive information. Another possibility is that the netspionage agents might use Internet connections to access sensitive information concerning the targeted nation-state stored *at* the vendor's location. The techno-spies might also gain initial entry to that vendor's network by invading others connected to that vendor's network. This "transitive property" of vulnerability can dramatically extend the risk to information far beyond the visible relationships of traditional business operations.

Nation-State or Corporate Dead Drop

If a nation-state's systems are poorly protected, they may find that although they are not the direct target of a netspionage agent, their servers can be used as a convenient place where the fruits of a crime are stored.[27] For example, the servers might be used to temporarily store source code or other digital property.

Intelligence collection by nation-states of other nation-states, businesses, associations, and even individuals is an ongoing threat. As an adversarial relationship begins to form between two nation-states, this information collection will undoubtedly intensify. One never has enough information about a potential adversary.

Intelligence collection is at the heart of all nation-states' information warfare plans. For those relying on their democratic form of government to protect them from such collections, think again. Even in the United States, information can be gathered with or without search warrants and other normal safeguards of a democracy. Using the United States — one of the world's freest forms of democratic government — as an example, there are several ways to covertly collect information without going through the usual safeguards:

- Government agents would determine that it was in the interest of national security to do so. After all, the survival of the nation-state is always paramount, regardless of any rules of law. However, the government agents could also go to a judge that has been identified to handle such matters and who has been given national security clearance to hear such cases.
- Government agents could rely on their allies, such as the British, Australians, and others, to collect the information and give it to the U.S. government agents. That way, the agents are not violating the laws by spying on their own citizens.

- The President of the United States could issue an executive order allowing the collection of that information. After all, was not the U.S. National Security Agency (NSA) established by a classified executive order?

Without current and accurate intelligence, one cannot adequately plan to conduct information warfare operations. The following are some examples of nation-states information collection methods:[28]

- *United States and allies.* Is the super-secret National Security Agency, working with its counterpart agencies in England, Canada, Australia, and New Zealand, eavesdropping on private communications from around the world? Credible reports suggest that a global electronic surveillance system — known by the code name ECHELON — is indeed capturing satellite, microwave, cellular, and fiber-optic communications worldwide. According to the American Civil Liberties Union, Echelon's constituents are:
  - United States: National Security Agency (NSA)
  - United Kingdom: Government Communications Headquarters (GCHQ)
  - Canada: Communications Security Establishment (CSE)
  - Australia: Defense Signals Directorate (DSD)
  - New Zealand: Government Communications Security Bureau
- *United States.* Carnivore is an Internet surveillance program currently being used by the U.S. government, and is somewhat similar to ECHELON. Carnivore is indeed capable of collecting all communications over the segment of the network being surveilled: "The results show that all TCP communications on the network segment being sniffed were captured by Carnivore." Moreover, the default configuration is to do just that: "When turning on TCP full mode collection and not selecting any port, the default is to collect traffic from all TCP ports." Carnivore is now being replaced by an even more powerful system, known as DCS 1000 or Enhanced Carnivore, which reportedly has a higher capacity to deal with speedier broadband networks.
- *United States.* U.S. intelligence officials have developed two programs that many experts believe may be used to enhance ECHELON's capabilities. One of these programs, Oasis, automatically creates machine-readable transcripts from television and audio broadcasts. Reports indicate that Oasis can also distinguish individual speakers and detect personal characteristics (such as gender), then denote these characteristics in the transcripts it creates. The other program, FLUENT, allows English-language keyword searches of non-English materials. This data mining tool not only finds pertinent documents, but also translates them, although the number of languages that can currently be translated is apparently limited (Russian, Chinese, Portuguese, Serbo-Croatian, Korean, and Ukrainian). In addition, FLUENT displays the frequency with which a given word is used in a document and can handle alternate search term spellings.
- *United States.* The Communications Assistance for Law Enforcement Act (CALEA) generally requires telecommunications carriers both to modify their existing networks and to design and deploy new generations of equipment (including software), all to ensure that carriers can meet certain specified "capability" and "capacity" requirements related to the ability of authorized government agencies to engage in wiretapping.

- *Russia*. SORM: System for Ensuring Investigated Activity is used by Russian agents to "monitor Internet transmissions coming in and out of Russia." In addition, the Russian Federal Agency for Government Communications and Information (Federal *'noye agentstvo paravitel'stvennykh svyazi i informatsu,* or FAPSI) has the "technical capabilities for monitoring communications and gathering intelligence, including monitoring of private networks. It too has its own troops (estimated at 54,000)."

- *China*. The Ministry of State Security, a special Internet police agency that was started in 1998, has launched several cyber-surveillance initiatives to track dissidents as well as conduct espionage on foreigners. In addition, Chinese authorities are planning to build more advanced Internet monitoring systems that rival those of the West, including the United States' Carnivore.

- *Germany*. Germany's Bundesnachrichtendienst (BND) has been engaged in intelligence gathering for nearly 50 years. Their official Web site is in German.

- *Israel*. Israel actually has at least three official intelligence-gathering organizations, commonly known as Mossad, Shin Bet, and Aman. Mossad handles surveillance outside of Israel, while Shin Bet conducts surveillance inside the country. Aman is charged with military intelligence.

- *France*. Reports indicate that two subdivisions of France's SGDN (Secretariat General de la Defense Nationale), the DRM (Direction du Renseignement Militaire) and the DGSE (Direction Generale de la Securite Exterieure), have created a French equivalent of ECHELON, or "Frenchelon." Some observers have charged that this system not only conducts surveillance, but also passes pertinent information along to French private companies. These allegations are documented in an article from ZDNet France. Further information in French (Francais) is available in an article from *Le Nouvel Observateur*.

- *India*. India's Central Bureau of Investigation (CBI) is tasked with "preservation of values in public life" as well as "ensuring the health of the national economy." In addition to national security matters, CBI also coordinates investigations with Infopol.

- *Other*. TEMPEST deals with the emanations from such things as computers, printers, CRTs, and other components. With the proper equipment and within the range of the equipment's emanations, the information being transmitted can be viewed on other equipment without the systems owner's knowledge. (See http://www.aclu.org/echelonwatch/networks.html for more information about this and the above topics.)

Diplomatic Pressure

Exhibit 3 depicts Stage 2 in information warfare — diplomatic pressure.

As the intensity of the relationship between two nation-states or groups of nation-states becomes more adversarial, there may be various forms of diplomatic pressure brought to bear on one or both nation-states who are slowly progressing toward more intense conflict. The diplomatic pressure may be brought to bear on the nation-states from various allies and associations (e.g.,

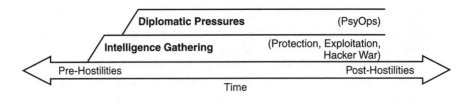

Exhibit 3. Stage 2: Diplomatic Pressure

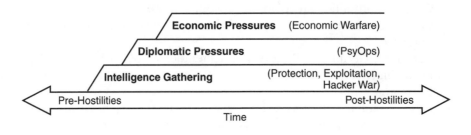

Exhibit 4. Stage 3: Economic Pressure

ASEAN and the UN); a call for calm and negotiations is usually heard. Whether or not either nation-state is in a "talking mood," they sure would love to know what the other is thinking and saying. During this phase of a conflict, there may be some forms of covert information warfare tactics used by a nation-state in gaining an advantage in the negotiations. For example:

- The adversary's network could be monitored to determine the reaction of those on the network to the diplomatic pressure.
- A nation-state could possibly read e-mail of the diplomats of its adversary, and others.
- A nation-state could increase its e-mail to others of influence, stating their position and asking for support; requesting that pressure be brought to bear on the other nation-state; why it is the right position; as well as post their information on their various government agencies' Web sites.
- Any information collected could be used as part of the psychological warfare part of diplomacy.
- Misinformation could be inserted so that the adversary's allies become confused or take exception to the adversary's diplomatic communiqués that in reality are placed there by the opposition.

Economic Pressure

Exhibit 4 depicts Stage 3 in information warfare — economic pressure.

As the intensity of disagreements escalates between two nation-states, one of them may take economic action and the other will undoubtedly retaliate in kind if able to do so. Because we are in a global economy where events in some distant place often adversely affect other nation-states and businesses, this build-up toward a heated conflict may draw others into the arena. One just has to look at what has happened in terms of recession in such nation-states as Japan, Indonesia, Malaysia, and their effects on the financial markets of the world. As we know, the economic power of a nation-state is equating more and more to global power. Therefore, anything that impacts a nation-state's economy will also impact its ability to project power and influence in the world. So, a nation-state might covertly attack the credibility, reliability, and stability of another nation-state by using information warfare tactics against the adversary. For example:

- Use the network to collect information, as was used during the diplomatic phase
- Release false economic-related information on targeted networks relating to the adversary
- Change economic information of the adversary
- Change the program used to compute interest and exchange rates, thus causing confusion for the global financial community and the adversary
- Using covert e-mail channels and Web sites, start adverse rumors relating to the stability of the adversary's government and its economy

One can probably think of many more devious information warfare tactics that can be employed. As the affected nation-state learns of the news being circulated around the world relative to its economic state of health, it will undoubtedly deny such rumors — usually taken with caution by those affected but not involved in the growing conflict. As the affected nation-state begins to suspect its adversary of using the above and other information warfare tactics against it, it will probably respond in kind (if possible), thus escalating the conflict.

When this occurs, one begins to see how quickly a disagreement can escalate. Because information warfare tactics at this stage generally do not cause the death or physical destruction of an adversary, while also providing some degree of anonymity or plausible deniability (e.g., "It's not me, must have been some hacker."), the nation-states may be more apt to escalate their conflicts. The conflicts may have emanated from disagreements as to who owns what territories, as noted in the CIA *World Factbook* appendices. It may begin to put pressure on its alliances formed in the various associations such as OPEC, NATO, or ASEAN to support its position. Thus, again, the conflict may begin to escalate. As the intensity increases, both sides, and possibly their allies, will begin their military posturing.

Military Posturing

Exhibit 5 depicts Stage 4 in information warfare — military posturing.

As military posturing develops, there may be a call to arms, a heightened alert status of the military, an increase in propaganda, and even probes of the adversary nation-state's network defenses using information warfare tools noted earlier. These would be preliminary "scouting" expeditions prior to actual combat. These would be probes looking for weaknesses in the information warfare

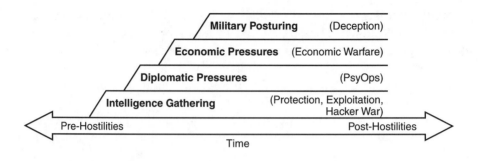

Exhibit 5.   Stage 4: Military Posturing

defenses of the adversary. Because most nation-states today run COTS, UNIX, Windows, and other software packages, their defenses will surely be porous.

Any of the above could also be used as part of military posturing. Other tactics include:

- The network or selected portions thereof could be prepared for attacks as part of denial-of-service attacks by embedding malicious code programs that could be activated later.
- Prepare misinformation relating to military actions to be inserted to get the adversary to act in a manner conducive to your objectives.
- Send out false messages over compromised links that would have the adversary maneuver its military forces in a manner conducive to the opposition.
- Prepare one's own information warfare defenses — remember that what goes around comes around.

As previously stated, these are just some examples of what tactics can be used by one nation-state against another. Once conflict breaks out, information warfare tactics take on both extreme defensive and offensive roles.

Combat

Exhibit 6 depicts Stage 5 in information warfare — combat.

By this phase of an adversarial relationship, a nation-state will be attacking its foe — both covertly and overtly. In information warfare, a nation-state may not know it has been attacked until it is too late. If the telephones are no longer working, or if the logistics system sends the wrong supplies to the wrong units, then one will be pretty much assured that the adversary has been successful in their information warfare attacks. Other tactics include, but are not limited to:

- Initiating the information warfare operational plan (e.g., send out misinformation; spoof and read the adversary's transmissions)
- Denying the adversary the use of their systems through the unleashing of denial-of-service attacks, viruses, and other malicious codes

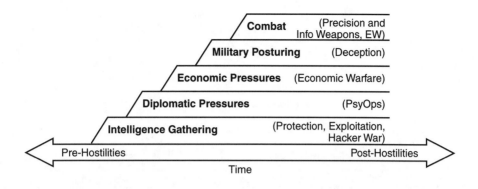

Exhibit 6.   Stage 5: Combat

- Breaking into the adversary's Web sites and posting misinformation and photographs of soldiers slaughtering women and children (these can easily be done using today's digitized software)
- Taking over the radio and television stations of the adversary and broadcast false information
- Sending e-mails to the adversary's allies, providing them with information that can be used to pressure that adversary to stop the conflict (of course at an advantage to the nation-state sending the information)
- Sending misinformation e-mails that appear to come from trusted sources of the adversary

These are just a few examples of what can be done to prepare and prosecute a war against an adversary and its allies. Think of unique ways in which you would prepare for and prosecute an information-based war against an adversary. Also, remember that what you are attempting to inflict on your adversary, they are also trying to use information warfare tactics to inflict the same or maybe better IW attacks against you. So, remember that a good offense starts with a good defense. In this case, it is very dangerous to think and implement information warfare operations based on a good defense starts with a good offense because one can never eliminate or totally mitigate a nation-state's information warfare capability. For example, the information warfare attacks against you may even come from within your own nation-state, or from a neutral or allied nation-state. One cannot take on the entire world and that is what one must be able to successfully do to avoid any successful information warfare attacks against oneself. It may only take that one individual sitting on an island with a notebook computer with Internet access to take down one of your most valuable networks.

Reconstruction

Exhibit 7 depicts Stage 6 in information warfare — reconstruction.

After a conflict prosecuted primarily based on information warfare tactics, some form of truce, peace treaty, or the like will generally take place. If the

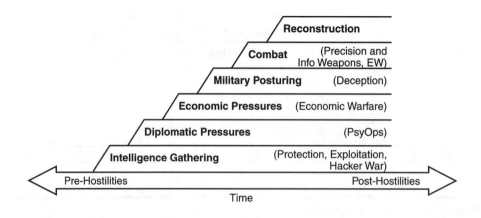

Exhibit 7.   Stage 6: Reconstruction

Western powers are successful, then the reconstruction phase will begin. Western powers are mentioned because history has shown that they invariably help a former adversary rebuild their infrastructure. This may or may not be the case if some other nation-state (e.g., China) were to win a war prosecuted mainly by using information warfare techniques.

The rebuilding phase of warfare, after hostilities have ended and a peace treaty signed, will be much faster and probably less costly than that of previous wars. That is because there will be less physical damage, and thus less costly to rebuild — unlike the World War II Marshal Plan. There will also be less loss of lives. Repairs and rebuilding will be primarily related to the national information infrastructure of a nation-state. The physical costs will primarily be in terms of hardware and software. Thus, a cheaper and more rapid recovery will take place.

Current Conflicts and Potential Conflicts

There are many conflicts and potential conflicts among the nation-states and other entities around the world. In fact, the U.S. Central Intelligence Agency (CIA) lists a total of 258 entities in which there may be a potential for more than 154 conflicts between nation-states.

All these nation-states and other entities will use all means at their disposal to successfully prosecute a war, "police action" — or whatever term is used — to win the dispute or conflict. If nothing else, they will use all means at their disposal to be able to negotiate from a position of strength and therefore win some concessions.

One should keep in mind that:

- Nation-states will continue to play a dominant role in information warfare.
- Nation-states will continue to fight over age-old issues.
- The information period will vary in degree for each nation-state; therefore, their degree of information warfare tactics will also vary in sophistication.

- Information warfare will provide the smaller Third World nation-states with the opportunity to conduct wars with larger nation-states on a more than equal battlefield.

The Pentagon identified 56 nations that went after controlled U.S. weapons technology in 1999. The findings are contained in the Defense Security Service's annual report, "Technology Collection Trends in the U.S. Defense Industry." In 1997, 37 countries were detected and 47 were identified in 1998.[28] The annual rates of increase were 27 and 19 percent, respectively, and the total increase was 51 percent in just two years. That large increase strains budgets and capabilities, sometimes beyond the breaking point.

The National Counterintelligence Center and the Central Intelligence Agency cited more than 23 countries with the capacity to engage in computer network attacks. Some of those countries include Israel, South Africa, the People's Republic of China, India, Taiwan, Syria, and Iran. Examples of other countries abound. The Indonesian government appeared to be behind a sustained attack on an Irish Internet service provider hosting a Web site supporting independence for East Timor. The Russian government targeted Pentagon computer networks in search of naval codes and missile information. Chinese hackers attacked U.S. government Web sites after a U.S. bomb struck the Chinese Embassy in Belgrade during the Kosovo Air Campaign. The attacks may have done the Chinese more harm than good because as a result of the cyber clean-up after the attacks, it was discovered there were well over 3000 back doors into U.S. computer systems allegedly put in place by China, and the possibility that there may be tens of thousands of other back doors. This is one of many actions supporting evidence of China's new asymmetric "unrestricted warfare" strategy detailed in Chapter 8.

The Information Warfare Process

Exhibit 8 depicts a general process used as part of an information warfare attack.

The basic, general process for information warfare attacks is as follows:

1. Identify the target (nation-state).
2. Identify its GII and NII interfaces.
3. Research the nation-state's systems.
4. Gather intelligence information.
5. Identify critical elements.
6. Identify its network vulnerabilities.
7. Covertly probe and document the results of the probes.
8. Once inside, find and transmit targeted, sensitive information.
9. Once inside, probe for other systems on the network and other networks.
10. Erase evidence of intrusion.
11. Search for additional networks and systems of the nation-state and repeat the process.

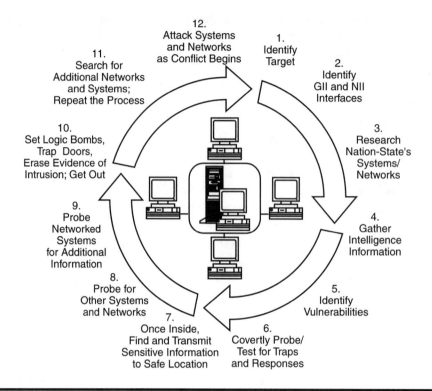

Exhibit 8. General Process Used as Part of an Information Warfare Attack

12. Attack the networks as conflict erupts, both overtly (e.g., denial-of-service attacks) and covertly, when it is in your interest to do so (allows the adversary to depend on compromised systems and networks).

Summary

The amount of information warfare tactics and methods used by a nation-state will of course vary with their ability to prosecute the conflict using information warfare weapons and also based on the vulnerability of their adversary to information warfare attacks. Needless to say that, as a minimum, the use of the Internet to get one's point of view across to the world via the Web sites and e-mails will be a basic information warfare tactic; as well as probable attacks against the Web sites of the adversary. After all, if it can be done by a teenager with a cheap personal computer, it can be done by someone working for the nation-state conducting the attacking — and this has been done on many occasions.

In discussing information warfare, one can see that it is not just a theory or something to be concerned about at some time in the future. It is taking place now and it appears that its use will intensify in the future. In fact, as you read this, sophisticated information warfare plans and scenarios are being developed by many nation-states around the world, and some are also being implemented.

There are literally hundreds of cases of information warfare attacks that have taken place or are taking place. There are also hundreds of examples of nation-states trying to defend against such attacks. This is because the defensive information warfare measures are categorized as information systems security, information operations, information assurance, and the like. Information warfare defensive mechanisms are in place, albeit not as in-depth and sophisticated as they should be; however, the real story is the offensive side because one must first understand the threats and information warfare offensive tactics before one can mount an adequate information warfare defense.

Notes

1. See James Dale Davidson and Lord William Rees-Mogg, *The Sovereign Individual: Mastering the Transition to the Information Age,* Touchstone, New York, 1999, p. 132.
2. *Encarta World English Dictionary* © & (P) 1999, Microsoft Corporation. All rights reserved. Developed for Microsoft by Bloomsbury Publishing Plc.
3. The U.S. Central Intelligence Agency's *The World Factbook* (see http://www.cia.gov) is a good source of information about "entities." Any individual, group, corporation, or government agency wanting to know about their adversaries will find the *Factbook* an excellent, basic source of information.
4. See the United States' CIA Web site for a detailed listing: http://www.cia.gov.
5. http://www.un.org/Overview/brief.html.
6. http://www.cnn.com/TECH/computing/9909/02/chinahack.idg/index.html.
7. http://www.cnn.com/2001/COMMUNITY/07/31/kissinger.cnna/index.html.
8. To learn more about netspionage, see *Netspionage: The Global Threat to Information,* published by Butterworth Heinemann, 2000.
9. This section is excerpted from *Netspionage: The Global Threat to Information,* published by Butterworth-Heinemann, 2000, and reprinted with permission.
10. *Secret Power— New Zealand's Role in the International Spy Network,* by Nicky Hager, Craig Potton Publishing, ISBN: 0-908802-35-8.
11. ECHELON Examined, article by Barbara Starr, ABCNEWS.com, November 22, 1999.
12. Russia establishes Internet surveillance network 08/12/99: http://www.zdnet.co.uk/news/1999/48/ns-12023.html.
13. *The Boston Globe,* June 25, 1998, CIA Chief Warns of Computer Warfare.
14. The complete report is available at the NACIC Web site: www.NACIC.gov.
15. http://www.berliner-morgenpost.de/bm/international/inhalt/2897/tunnel_2.html, In the Sights of the Spies.
16. Ibid, p. 161.
17. This information is derived from BackOrifice documentation.
18. http://www.freediskspace.com.
19. http://webopedia.internet.com/TERM/s/steganography.html.
20. http://www.jjtc.com/stegdoc/.
21. http://www.nacic.gov, Counterintelligence News and Developments, November 1996.
22. *The CIO Update* published by the SANS Institute, Vol. 2 No. 3, March 8, 1999.
23. Cracking Tools Get Smarter, *Wired News,* March 19, 1999.
24. http://www.east-tec.com.
25. Taken from *IA Newsletter* magazine, Vol. 2 (No. 3), Winter 1998–1999, p. 6; article entitled "Information Systems Security: The New Arms Race for the Information Age," page 6 published by the Information Assurance Technology Center, United States Department of Defense.

26. See Mitnick's alleged use of accounts on the Web to store the source code taken from Shimomura's systems.

27. See http://www.aclu.org/action/echelon107.html for more information.

28. Inside the Ring: Stealing Know-How, Bill Gertz and Rowan Scarborough, *Washington Times,* October 20, 2000.

Chapter 8

Information Warfare: Asian Nation-States

...in Asia, the various states look at each other as political and geopolitical rivals. War is not impossible, and therefore in that region we have to pay more attention to balance of power and equilibrium.[1]

—Dr. Henry Kissinger

This chapter discusses information warfare as it relates to nation-states in Asia. An Asian overview is presented with more details and discussions of selected nation-states that are involved or that may soon be involved in information warfare.

Asia

Exhibit 1 provides a map of Asia. Asia is rapidly growing into a powerful economic sphere of global influence. This is predicted to increase well into the Twenty-first Century. The two most economically powerful nation-states in that region are China and Japan. However, the Southeast Asian nation-states when considered as a group, as well as India and Pakistan (India and Pakistan from a military standpoint because they have nuclear weapons), are also growing economic and/or military powers. Although Australia and New Zealand may not often consider themselves a part of Asia, for our purposes, they are included in this chapter.

Asia is an area of the world where high technology is rapidly advancing and, in fact, often surpassing the rest of the world in its use. This is logical. Asian nation-states have in some instances gone from the agricultural period in human history directly into the information period. They may not have the older industrialized infrastructure to modify or replace. For example:

Exhibit 1.   Map of Asia (and Australia and New Zealand)

- The Republic of Indonesia is made up of more than 17,000 islands. To establish landline telephone communications throughout the country would be extremely expensive, and that it is not within the budgetary reach of this country. However, with the use of cellular telephone systems, many people in this country are able to communicate, and do so rather inexpensively.

- According to a report by Jupiter Media Metrix, Internet users in Asia will outpace the United States in less than five years and they will comprise about one-third of all the Internet users by 2005.[2]

Asia Overtaking Europe in E-Commerce
inforwar.com, August 18, 2000
...four years Australasia, Singapore and Korea will be seeing far more intensive E-commerce activity than Europe and, possibly, the USA... Research shows that, by 2004, six of the top seven most intensive E-commerce countries in the world will be in Asia led by Singapore (see also www.worldcom.org).

Among Asia's most interesting developments to watch are those between Singapore, China, and Taiwan. Previously, one would add Hong Kong; however, that is now part of China. What do they have in common? The following:

- Each nation-state is predominantly Chinese.
- China has cheap, abundant labor and natural resources — and with Hong Kong, an international financial center.
- Taiwan has highly educated citizens and high-technology knowledge and manufacturing.
- Singapore has highly educated citizens, a high technology base, and a financial center probably second only to China in the region.

If these nation-states continue to grow closer together for economic development, they will form a massive global economic powerhouse that will rival the United States and the European Union. The Chinese have a reputation for being shrewd business people and with economic power driving the power and influence of the nation-states, they may form a global entity that will lead the world in this century. According to John Naisbitt,[3] as far back as 1996 there were over 6000 Chinese "clan networks" sharing information, all through personal computers. Composed primarily of overseas Chinese spread throughout the world, they are quietly coming together and maybe becoming an economic force that cannot be ignored.

While nation-states develop their high technology-based infrastructure to form their national information infrastructure (NII) and integrate it into the global information infrastructure (GII), they make themselves more vulnerable to information warfare (IW) attacks by their adversaries or potential adversaries. The Asian nation-states are no exception to this rule.

Japan

As we all know, Japan in the past has been an economic giant. It was not that long ago that there was some fear that it would soon "rule the economic world." However, that has not occurred and, in fact, Japan has been mired in an economic slowdown and recession for some time. In Asia, Japan remains a powerful economic force to be dealt with. It is so intertwined with other nation-states in Asia that when the economy of Japan began to slide, it took much of Asia with it.

So, if one would want to become the Asian economic leader, one must first replace Japan. It would seem that the only nation that appears to want to take that position, or is capable of replacing Japan in that position, is China. China over the centuries has fought, and lost, several wars with Japan. There is no love lost between the two, nor with Japan and other Asian nation-states that she attacked and occupied during World War II and other wars. In addition, Japan is constantly meeting with Russia to gain back its northern islands lost to the Russians during the last days of World War II. Japan is also involved in some other disputes as noted in the CIA *World Factbook*[4] and other sources:

- Japan. Islands of Etorofu, Kunashiri, Shikotan, and the Habomai group occupied by the Soviet Union in 1945, now administered by Russia, claimed by Japan; Liancourt Rocks (Takeshima/Tokdo) disputed with South Korea; Senkaku-shoto (Senkaku Islands) claimed by China and Taiwan.[4]
- The *Taipei Times* newspaper, on January 14, 2001 stated:

  > *Hatred toward Japan should end...the people who immigrated to Taiwan with the KMT after the war and their descendants, view Japan as an enemy nation. Their hatred stems from the brutality that Japan waged against China during World War II....*

- On February 15, 2000, Reuters in Beijing reported that Japanese government Web sites had been attacked by Chinese hackers and derogatory information posted. China stated there was no way to confirm the attackers. *Note:* Is the Chinese government so overwhelmed by high technology and the Internet that it has no way of tracing the hackers operating in their nation-state and attacking Japan with impunity, or is it under the sponsorship of the Chinese government or even agents of the Chinese government that these attacks have occurred? Remember that there is still a deep-seated hatred by the Chinese because of past wars with Japan.
- On September 5, 2000, the *Mainichi Daily* News reported:

  > *...Putin refused a Japanese proposal to resolve a key territorial dispute.... The move would effectively return the four disputed Kuril islands, known in Japan as the Northern Territories, to Japanese sovereignty.... "It is important to resolve the dispute... before concluding the peace treaty," Mori was quoted by the officials as saying.*

  Note: Japan and Russia do not have a formal World War II peace treaty.

- Toward Security Independence; Japan Acquiring Ability to Project Air Power

  > *Corey Walrod, Special to ABCNEWS.com, December 29, 2000 Japan is taking its first step toward independence from the United States for regional security — undertaking a new defense program that is expected to allow its air force to project power beyond its borders...*

- According to John Quinn, retired CIA Officer, consultant, and writer for the Japan Forum for Strategic Studies: There is substantial interest in controlling and countering cyber terrorism (and cyber crime). Crime they can probably handle but external, foreign sponsored full-scale cyber war...would probably fry their shaky national infrastructure royally. Even Theater Missile Defense (TMD) is a situation they are concerned about.

Thus, Japan has several reasons why it may want to attack Russia, China, or even South Korea or Taiwan. However, it would probably be in its own best interest to do so covertly and, at the same time, do so under the guise of some global hacker or as coming from another nation-state. Consider the following:

> *If Japan were to attack Russia for failure to give back the Northern Territories, it may do so by compromising and then using some Chinese-based systems. That way, as Russia looked into the matter, they would trace it back to China. In conducting such a covert attack, Japan might wait until there is some breakdown in Russia–China diplomacy, or it may even make available through chat rooms, e-mail, and others' Web sites, some information that would cause some animosity between China and Russia. In that way, it would make it appear that the covert attacks were indeed coming from China. Japan could then do the same to China through Russia and then sit back and "enjoy the show."*

China's goal may be to replace Japan as the economic power in Asia. Consider the following scenario:

- China would identify Japan's networks, their vulnerabilities, and their systems administrators.
- China would then prioritizes attacks based on economic impact to Japan and itself.
- China blames attacks on global hackers and Taiwan because the attacks, although covertly done, are done on the anniversary of the Japanese invasion of Taiwan prior to World War II.
- Maybe China will wait for the breakdown of talks between Japan and Russia over the Northern Territories.
- China may conduct the covert and subtle attacks through Taiwan or Russian systems.
- Japan's networks begin to gradually slow down, possibly due to the use of a sophisticated version of the Code Red worm.
- Attacks appear to be coming from either Russia or Taiwan. Japan blames one or both.
- Japan's networks become unreliable and cause financial confusion; the economy begins to slow down and enter a deeper recession.
- Coupled with other factors, China replaces the void caused by an economically weakened Japan.

There are several other scenarios that one can think of between Japan and other nation-states. Put yourself in the place of Japan and consider whom you would want to covertly attack, why, when, and how. Do you think Russia would attack Japan? If so, when, why, and how? Here are some actual cases to help:

> ***Japanese Government on Alert for Chinese Computer Hackers***
> *infowar.com, February 24, 2000*
> *The Japanese government…issued an alert warning them of a possible attack on their Web sites by computer hackers critical of Japan's handling of historical issues with China…. The warning, issued in the name of "Hongke Federation," also cited the "current situation" in Japan in its call for a one-week campaign to disrupt Japanese Web sites.*

> ***Hackers Find Easy Target in Japanese Web Sites***
> *Kathryn Tolbert, Washington Post, TOKYO, JAPAN, February 2, 2000*
> *For eight days, a frenzy of computer hacking has swept through Japanese government Web sites, adding derogatory comments to home pages, wiping out census and personnel data, and alerting officials to the government's lax computer security…. In other attacks, data including the national census were erased from the Management and Coordination Agency's statistics bureau and 3500 files were deleted from a home page of the National Personnel Authority…*

North Korea and South Korea

One may not think that a small nation-state such as South Korea would be very active in the use of personal computers and accessing the Internet; however, the following news report may be surprising to many:

> ***Koreans Most Active Net Users In The World — Nielsen***
> *Adam Creed, Newsbytes, Seoul, South Korea, March 13, 2001*
> *Korean Internet users lead the world in many key Internet measurement metrics, according to a study released today by Nielsen/NetRatings. The study found that Koreans lead the world in certain areas — visiting the Web the largest number of times, visiting the most unique sites, generating the most page views, and time spent online per session and per month.*

When comparing the information warfare capabilities and vulnerabilities, one can quickly deduce that South Korea would be more vulnerable than North Korea to information warfare attacks.

There is no doubt that conflict between these two nation-states is always a possibility. North Korea seems to be constantly is the midst of a famine; and based on it being primarily an agriculture country, this does not bode well for its economy. South Korea, on the other hand, although as with other Asian nation-states facing some difficult economic times, is more of an industrial–information period nation-state. It is the case of the "haves" (South Korea) versus the "have-nots" (North Korea). However, because South Korea is so information systems dependent, North Korea may have the edge on the information warfare battlefield. After all, there is not much in the way of information systems for South Korea to attack in North Korea. Consequently, North Korea is less vulnerable to information warfare offensive weapons.

North Korea may want to continue to show its good side to the world community to ensure continued aid in the form of food donations. However, because of the continuing famine, the North Korean government may want to divert its people's minds to other matters, for example, South Korea as the enemy and the one to blame for North Korea's problems.

China, an old ally of North Korea (remember the Chinese "volunteer" army fighting UN forces during the Korean War?), may want to covertly test its arsenal of information warfare weapons such as sophisticated worms, viruses, and other forms of malicious code.

Remember that according to the CIA *World Factbook*:[4]

- *Korea, North:* 33-km section of boundary with China in the Paektu-san (mountain) area is indefinite; Demarcation Line with South Korea
- *Korea, South:* Demarcation Line with North Korea; Liancourt Rocks (Takeshima/Tokdo) claimed by Japan

Is this possible? Yes. Is it likely to occur? Maybe it already has: the South Korean Ministry of Information and Communication, which has received almost 2000 complaints, about a year ago said that South Korean firms had been attacked by the Chernobyl computer virus, wiping out hard disks at hundreds of companies. A ministry official stated that in 1999, the outbreak of the virus affected up to 300,000 computers; larger companies took the brunt of the damage. The year 2000 damage was estimated to be about five percent of that.

So, an information warfare scenario may go something like this:

- Famine increases in North Korea.
- South Korea refuses further aid without some concessions, such as families being able to visit each other as often as they wish.
- North Korea says no and considers it blackmail; seeks aid from China.
- China covertly assists North Korea by providing training on the use of information warfare weapons given to them by China.
- North Korea releases network viruses on South Korea via Japan (an added benefit to South Korea by "getting even" for the Liancourt Rocks dispute with Japan).
- North Korea places unfavorable economic rumors on the Internet chat rooms (covertly) concerning South Korea allowing old people and children to die of starvation.
- False information is embedded in South Korean financial systems.
- South Korea complains to the UN for help; North Korea denies any wrongdoing.
- Talks begin between North and South Korea and, with United States support, agrees to aid North Korea in exchange for Korean War missing-in-action information.

In addition to nation-state conflicts with other nation-states, there appears to be an almost global conflict within many nation-states between the government and its citizens. As noted previously when discussing the demise of nation-states and more power to the "sovereign individual," it appears the citizens want more power and that power is at the expense of the nation-state's power. One example of that is in South Korea:

> **Netizens Protest Government Media Repression**
> *infowar.com, August 26, 2001*
> *Some 30 members of the Democratic Participating Netizen Solidarity congregated at Tapgol Park, Jongno-gu at 2:00 p.m. Sunday and demanded that President Kim Dae-jung's administration must end its suppression of critical newspapers. Also, demonstrators insisted that the government should humbly accept globally rising criticism on its persecution of the media.[5]*

With the communications power of the GII, South Korea's NII, and the Internet in general, the South Korean "netizens" can reach the global community and solicit their assistance in bringing political and economic pressure to bear on the government of South Korea in order for the citizens to be given more freedom and human rights, such as the right to criticize their government.

South Korea is not the only nation-state that continues to see an increase in this type of activity. Information warfare tactics by nation-state citizens are being used as a powerful tool to gain more freedom of speech and the like.

Malaysia, Indonesia, Brunei, and Singapore

The nation-states of Malaysia, Indonesia, Brunei, and Singapore are all Southeast Asian nation-states and all members of ASEAN and the UN. Even so, the CIA *World Factbook*[4] indicates they have disputes among themselves and others:

- *Indonesia:* Sipadan and Ligitan Islands in dispute with Malaysia
- *Brunei:* Possibly involved in a complex dispute over the Spratly Islands with China, Malaysia, Philippines, Taiwan, and Vietnam; in 1984, Brunei established an exclusive fishing zone that encompasses Louisa Reef in the southern Spratly Islands, but has not publicly claimed the island
- *Malaysia:* Involved in a complex dispute over the Spratly Islands with China, Philippines, Taiwan, Vietnam, and possibly Brunei; Philippines have not fully revoked claim to Sabah State; two islands in dispute with Singapore; Sipadan and Ligitan Islands in dispute with Indonesia
- *Singapore:* Two islands in dispute with Malaysia

Exhibit 2 provides a map of Southeast Asia.

What is of interest is that of these four nation-states, all but Singapore pretty much share a common language, culture, and are predominantly members of the Moslem religion. Singapore, on the other hand, is predominantly

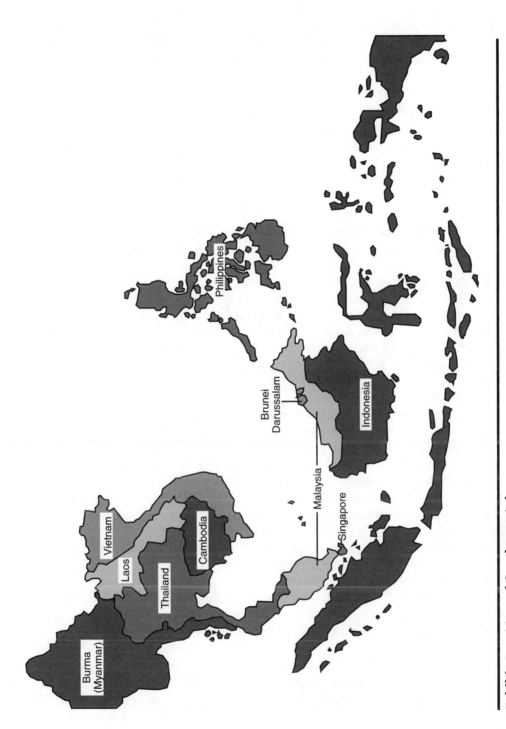

Exhibit 2. Map of Southeast Asia

Chinese, with a Chinese culture, language, along with English as its other
language, and predominantly Buddhists.

Of these nation-states, Singapore is probably the most advanced and most
vulnerable to information warfare attacks. In the event there is a conflict
between Singapore and one of the other three nation-states, Malaysia, Indonesia, and Brunei would be more likely to come to the aid of each other than
to Singapore.

In addition to the potential for conflict stated above, conflicts might erupt
because of Singapore not liking Malaysia opening a new, modern international
airport so close to its international airport. Singapore may also feel threatened
by Malaysia because of Malaysia's threat to restrict the supply of water to
Singapore.

Consider a possible scenario that may cause an information warfare-based
conflict to occur:

- Singapore is angered by increased traffic through Malaysia's new interna-
  tional airport that is taking tourist and airport revenues away from Sin-
  gapore.
- Malaysia threatens to shut off water to Singapore.
- Singapore plans military action to secure its water supply.
- Singapore gains global Chinese sympathy and China volunteers aid.
- Indonesia (a haven for anti-Chinese sentiment), in return for at least one
  of the disputed islands with Malaysia (see above), offers military support.
- Brunei says it will also support Malaysia.
- All boycott Singapore imports and block exports to Singapore.
- All economies begin to suffer.
- China begins aiding Singapore militarily and with oil.
- Mideast nation-states, to include the OPEC members, support Malaysia.
- Chinese businessmen develop a covert plan to destabilize the economies
  and governments of Malaysia, Indonesia, Brunei, and other nation-states
  who support Malaysia (*Note:* In information warfare, anyone can be a
  player as the battlefield is global because networks have global access).
- Each side begins information warfare attacks to include misinformation via
  Internet, Web sites, viruses, and denial-of-service attacks.
- Australia and New Zealand also side with Singapore (based on their dislike
  for Indonesia due to East Timor).
- Other ASEAN nation-states supported by the United States get warring
  parties to agree to a cease-fire and further talks under the auspices of the
  UN.
- UN troops are sent to Malaysia and Singapore as peacekeepers.

Farfetched? Remember the history and cultural backgrounds of each of these
nation-states: the issue of East Timor, and the killings and anti-Chinese sentiment in Indonesia due primarily to the 1960s Chinese attempt to overthrow
the Indonesian government to form a Communist government. Remember also
that Singapore is an island nation-state with little, if any, natural resources.

What may also help Malaysia's adversary are human errors such as the
following:

Malaysian ISP Admits 'Human Error' In Security Fiasco
Julian Matthews, Newsbytes, Kuala Lumpur, Malaysia, May 28, 2000
Malaysian Government-funded research corporation and Internet ser-
vice operator Mimos Berhad admits that a staff member carelessly placed
a large number of confidential company files on a publicly accessible
Internet server.... "One of our staff created a directory on a server and
accidentally made it publicly accessible...."

In addition to the human errors, nation-states with an adversarial relationship with Singapore can also capitalize on Singapore's relatively heavy-handed
approach against its own citizens, similar to that of South Korea. These nation-
states can "show" the world community via Web sites, e-mails, and chat rooms
that Singapore is not a true democratic nation-state and use examples such
as the following:

Singapore Community Web Site To Shut Down
Adam Creed, Newsbytes, Singapore, August 16, 2001
...owner of Sintercom, a community Web site in Singapore, has decided
to close the site. Sintercom is a non-profit site.... The Singapore Broad-
casting Authority (SBA), which is responsible for Internet content regu-
lation in Singapore, said registration was necessary as the site "engages
in the propagation, promotion, and discussion of political issues relating
to Singapore."

Singapore's government agencies, as with other nation-states, have a difficult
time accepting criticism. One way to mitigate this is to control and put pressure
on information "broadcast" through e-mail, Web sites, and the Internet in general.
It is quite probable that such pressures for more freedoms on nation-states such
as South Korea, Singapore, and others will continue. This will cause more conflict
between the nation-states leaders and its people. Such information wars are
expected to continue, increase, and grow in intensity in the future.

Singapore Military To Tackle New Threats, "Cyber-Attacks"
The Associated Press, March 14, 2000
Singapore's military will step up efforts to deal with new, 21st-century
threats such as "cyber-attacks," the country's defense minister said...
armed forces and government will "enhance Singapore's capabilities to
deal with a range of non-conventional threats, such as terrorism,
piracy and cyber-attacks," Defense Minister Tony Tan said.... Tan also
said that such areas as the Taiwan Straits, the Korean peninsula and
conflicting claims (see land disputes above) are "destabilizing issues"
in the region.

Malaysia: Sites Vulnerable To Hacker Attacks
Yu Wui Wui, New Straits Times Press (Malaysia), December 2, 1999
At least 65.4 percent of 7058 operational Web sites with .my domain
are easy preys for hackers, according to a survey by Infinitum (S) Pte
Ltd., an information technology (IT) security consultancy firm. Most of

> *these Web sites are using versions of Web servers which have security*
> *flaws that allows easy intrusion. Microsoft IIS version 4 has at least two*
> *bugs that people can exploit. There are about 1555 Web sites with .my*
> *domain using this version.*

Indonesia, Australia, and Portugal over East Timor

East Timor, until recently, was claimed by Indonesia. However, that claim was
not recognized by the UN and nation-states such as Portugal and Australia.
For several years, those that sided with East Timor for independence began
attacking the networks of the Indonesian government, as well as defacing and
changing their Web sites.

In fact, through the "underground hacker global communication system,"
some hackers had called on all the hackers around the world to attack the
Web sites and systems of the Indonesian government. The attacks were carried
out by hackers in Portugal, the United States, Australia, and it is assumed by
others in Europe and Asia. The actual scenario went something like this:

- In 1979, Portugal leaves East Timor
- Indonesia annexes East Timor
- Nation-states object
- The issue goes before the UN
- The UN does not recognize the Indonesian claim
- As high technology progresses, Indonesian Web sites and systems are attacked
- A "call to arms" goes out to global hackers to attack the Indonesian systems *en masse*
- Estimated that between 24 and 47 Indonesian Web sites and systems go offline in one day
- After more than 20 years, East Timor is set free

Some examples of the several defacements of the Indonesian Web sites
included: Indonesian Soldiers with Weapons at the Ready. On the top of that
Web page were the words, "Professional Killers" and on the bottom it said
"Made in Indonesia." Another defaced Indonesian Web site stated: ".id Portuguese Hackers against Indonesia" in the foreground and "East Timor Campaign" in the background. All the words were in English.

No claim is being made here that East Timor won its freedom due to the
attacks. However, Web sites and chat rooms filled the air with information
about Indonesian atrocities against the East Timor people. This, in fact, helped
keep the issue in the public's eye and probably did contribute in some part
to the freedom of East Timor people.

Did the Indonesian government fight back against these attacks? Yes, but
it did not have the tools or the level of sophistication to completely defend
its systems. Did the Indonesian government use offensive information warfare
tactics against those attacking them? There is some indication that it did to
some small degree.

However, the lesson to be learned from this is that information warfare tactics have and are being used to gain a public relations and political advantage over one's adversary or potential adversary. In this instance, the coordinated effort of hackers around the world should be a good lesson for all the nation-states that world opinion considers to have taken actions that are unacceptable. As nation-states try to advance into the information period, they cannot do so and still control the information that their citizens are able to access. For in doing so, they will remain a Third World nation-state with little influence on world affairs.

Information warfare is no longer new, and this is especially true in Asia. The following news articles explain one form of information warfare against Indonesia:

> ### East Timor Cyberwar On New Global Battlefield: The Internet
> *infowar.com, August 30, 1999*
> *JAKARTA (JP): Indonesia may be on the brink of a cyberwar.… The battlefield of such a war is the Internet, with bullets the words, and soldiers the hackers.… The global communication system…in which our increasingly cybernetic society…now wages war. It is becoming a conceptual space or site of political struggle, and a global high-tech battleground.…*

> ### East Timor-Cyberwar
> *Grant Peck. Bangkok, Thailand, August 18, 1999*
> *An international squad of computer hackers will wreak electronic mayhem on Indonesia if the country hampers voting in East Timor's independence referendum,.… More than 100 computer wizards, mostly teenagers in Portugal, Spain, Ireland, Belgium, Brazil, the U.S. and Canada, …are targeting the entire computer network of the Indonesian government, army, banking and finance institutions to create chaos,…*

> ### Cyber Raiders
> *Scripps Howard News Service, Washington, date unknown*
> *…So far, as many as 23 countries are believed to have the capacity to engage in state-sponsored, surreptitious electronic raids.… Indonesia:…was identified as being behind a coordinated assault on Ireland's Internet service provider, which hosted a Web site advocating independence for the province of East Timor.*

The Philippines and China

Is it possible that a conflict in which information warfare tactics are used might break out between the Philippines and China? Quite possible because, according to the U.S. CIA *World Factbook*,[4] China is in dispute with many nation-states in the region, and so is the Philippines:

- *China:* Boundary with India in dispute; dispute over at least two small sections of the boundary with Russia remain to be settled, despite 1997 boundary agreement; portions of the boundary with Tajikistan are indefinite; 33-km section of boundary with North Korea in the Paektu-san (mountain) area is indefinite; involved in a complex dispute over the Spratly Islands with Malaysia, Philippines, Taiwan, Vietnam, and possibly Brunei; maritime boundary dispute with Vietnam in the Gulf of Tonkin; Paracel Islands occupied by China, but claimed by Vietnam and Taiwan; claims Japanese-administered Senkaku-shoto (Senkaku Islands/Diaoyu Tai), as does Taiwan; agreement on land border with Vietnam was signed in December 1999, but details of alignment have not yet been made public
- *Philippines:* Involved in a complex dispute over the Spratly Islands with China, Malaysia, Taiwan, Vietnam, and possibly Brunei; claim to Malaysia's Sabah State has not been fully revoked

There is a reason for a conflict to occur if one or the other wanted to press the issue as to their rights to territory. However, there may be other reasons such as China possibly aiding the rebels in the Philippines and also an adversarial relationship may erupt over trade or competition for the placement of global corporations' manufacturing plants in their nation-states. In addition, the Philippines have Moslem rebels fighting the government. Is it possible that they are being supported by other Moslem nation-states, especially those in their area where they also have territorial disputes, such as Malaysia?

If such an information warfare-related conflict were to occur, it would seem that between the Philippines and China, China would be the most vulnerable to information warfare offensive weapons simply because it is more information-based, thus probably giving the advantage to the Philippines. After all, the Philippines do allegedly have the Love Bug virus developer.

China would also have more to lose in that they are currently trying to balance their communist ideology where information is controlled and manipulated by the government with that of a modern information period nation-state that requires its citizens to have access to information on a global scale. They are also trying to balance all that with some semblance of a capitalistic business system. Any degrading of their information systems and networks can help open the floodgates to calls for more freedom and blaming the control of the government for any economic collapse. Furthermore, now that the Chinese on the mainland have had a taste of "Western culture," it would be nearly impossible for the Chinese people to accept the "Second Cultural Revolution" and bringing back the Maoist ways of life.

India and Pakistan

Anyone reading the newspaper over the past several years knows that the conflicts between India and Pakistan have erupted more than once. Again, citing the U.S. CIA *World Factbook*:[4]

- *India:* Boundary with China in dispute; status of Kashmir with Pakistan; water-sharing problems with Pakistan over the Indus River (Wular Barrage); a portion of the boundary with Bangladesh is indefinite; dispute with Bangladesh over New Moore/South Talpatty Island
- *Pakistan:* Status of Kashmir with India; water-sharing problems with India over the Indus River (Wular Barrage)
- They also have differences in culture, religion, as well as vying for the title as "the toughest kid on the block"

Both nation-states appear to have limited information warfare capabilities. However, as with other developing nation-states, this too is changing. An interesting position for China, with which India has boundary disputes, is to covertly attack India's growing application software development systems and then blame it on Pakistan. China may want to do this at a time when India is again involved in a hot war with Pakistan. It may do this to retaliate against India because India harbors the Dalai Lama, who has always been a thorn in the side of the mainland Chinese government. In addition, it could help ensure that India would not be in a position to exercise its claim on the land that they dispute.

India is not immune to information warfare or other forms of attacks, as noted in Exhibit 3.

Vietnam and China

Border disputes between Vietnam and China and its other neighbors seem to heat up periodically. Previously presented were the disputes related to China. The following, according to the *World Factbook,*[4] are the disputes involving Vietnam:

- *Vietnam:* Maritime boundary with Cambodia not defined; involved in a complex dispute over the Spratly Islands with China, Malaysia, Philippines, Taiwan, and possibly Brunei; maritime boundary with Thailand resolved, August 1997; maritime boundary dispute with China in the Gulf of Tonkin; Paracel Islands occupied by China but claimed by Vietnam and Taiwan; offshore islands and sections of boundary with Cambodia are in dispute; agreement on land border with China was signed in December 1999, but details of alignment have not been made public

It is interesting that for such a small country, Vietnam has so many territorial claims. As with the potential conflict between China and the Philippines, Vietnam is probably in a better position than China when it comes to the potential damage that can be inflicted on an adversary, with China again in the worst position. This goes back to previous comments made about the more advanced a nation-state is in the use of high technology, information-based systems, and the like, the more vulnerable they are to action by nation-states that are not so heavily reliant on such high technology to the same extent as an adversary.

Exhibit 3.　Information Warfare on India

India: Lack of Transparency on Internet Leads to Increased Cybercrimes
Shilpa Garg, New Delhi, Business Standard Ltd., May 9, 2000
…From among the companies surveyed, 380 were attacked by virus, 353 by insider abuse of Net access, 297 were attacked on laptops, 108 suffered systems penetration, 75 faced telecom fraud, 68 financial fraud and 82 theft of proprietary information…

India — News Site's Cameras Secretly Record Army Sleaze
CT Mahabharat, Newsbytes, New Delhi, August 24, 2001
India news portal Tehelka, which exposed bribery and influence peddling relating to defense deals and rocked the Indian government the Web site released several hours of video showing a string of politicians, military officials and bureaucrats apparently accepting money…

Chinese Hackers Protest India's Missile-Defense Support?
CT Mahabharat, Newsbytes, New Delhi, India, May 7, 2001
Apparently, retaliating against India's support of the U.S. national missile defense plans, Chinese hackers are believed responsible for defacing a Web site belonging to India's CMC Ltd., a state-owned software company.

Indians, Pakistanis Play Patriotic Games on Net
http://www.timesofindia.com/060101/06home6.htm
…Indians and Pakistanis have found a virtually new battleground — cyberspace. And the first round of this war has gone to Pakistan…more than 500 Indian sites were defaced by Pakistani and other unidentified anti-India hackers last year.

Indians, Pakistanis Play Patriotic Games on Net
Times of India, Kaajal Wallia, January 6, 2001
Border skirmishes and the nuclear arms posturing between India and Pakistan have caused national pride in both countries to be an issue. Those with computer skills found a virtual battleground: cyberspace. The first round has gone to Pakistan. According to cyber experts, Pakistani and other unidentified anti-India hackers defaced over 500 (more than 635, according to India Express in the January 11, 2001 edition) Indian sites in 2000. Despite its numerous IT whiz kids, India has managed only one "victory": a warning on the Pakistani government site.

Hack the Hackers
Ravi Visvesvaraya Prasad, December 19, 2000
Pakistani hackers have regularly attacked Web sites of Indian organizations. The home pages of the Prime Minister's Office, the Bhabha Atomic Research Center, the Ministry of Information Technology, and Videsh Sanchar Nigam were hacked into and defaced with anti-India obscenities.

If the border disputes continue, Vietnam may seek covert assistance from Russia (as they did during the Vietnam War). What if China finds out about the request or, worse yet, that Russia was aiding Vietnam through an information warfare weapons assistance program? What information warfare offensive weapons can China use against Vietnam? What information warfare weapons can Vietnam use against China? Without Russian assistance, the answer is that China in all probability has more, and more devastating, information warfare weapons. However, the information infrastructure of Vietnam pales in comparison to that of China. Furthermore, Vietnam relies on it less than do the Chinese.

From its viewpoint, Vietnam probably has far fewer information warfare weapons. However, based on China's increasing dependence on its systems and networks with a more advanced information infrastructure, Vietnam could probably cause more damage to China than vice versa (using only information warfare weapons and tactics).

ASEAN and IMF

We previously discussed the potential information warfare conflict between ASEAN and the IMF. This is one of the more interesting scenarios. It is not about a nation-state against a nation-state, but about an association of nation-states in conflict with an "international financial institution" that for all practical purposes has the power to dictate economic policy to an association of nation-states, as a group or individually, that want international financial assistance.

So, why and how would such an information warfare conflict take place? Consider the following:

- ASEAN nation-states continue their economic slide
- IMF demands stronger economic action by the ASEAN nation-states
- ASEAN nation-states decline and say IMF is muddling in internal affairs of the countries
- IMF stops funding until its demands are met
- ASEAN nation-states seize the assets of foreign corporations in their countries that can influence the IMF; electronically transfers the foreign corporations' funds to the financial institutions of the ASEAN nation-states
- IMF requests and receives assistance from IMF-backed nation-states who then penetrate ASEAN nation-states financial institutions; transfers all funds to the IMF; and denies all electronic access to all worldwide ASEAN-associated government accounts
- ASEAN nation-states and IMF agree on economic talks under the auspices of the UN
- Funds are released by both sides; however, the foreign corporations are held until final talks and agreements are reached

This scenario is just another example of the use of information warfare tactics in today's information and information systems-dependent environment. The example brings together international associations, foreign corporations, and nation-states throughout the world. Can this happen? Probably. Will this happen some day? Maybe. One interesting thing: a foreign corporation may become involved in an information warfare conflict without ever having a clue that something like this could ever happen. The lesson here is for corporations to think of such scenarios as they develop their emergency, contingency, and disaster recovery plans.

Exhibit 4.   Burma's Information Warfare

Myanmar's Rulers Block Internet
Matthew Pennington, Associated Press Writer, 4/24/00, YANGON, Myanmar (AP) April 23, 2002
Dozens of key-tapping students stare intently at computer screens in a cramped classroom three floors up a crumbling colonial terrace in downtown Yangon.... But there's one glaring gap in the curriculum: the Internet.
http://dailynews.yahoo.com/h/ap/20000423/tc/internet_dreams_1.html.

Internet Attacks Burma!
http://www.zolatimes.com/v2.8/burma.html, April 24, 2000
Americans, Burmese and others are waging an elaborate Internet assault against Burma's (Myanmar's) military regime, which has retaliated by e-mail and a Web site to deflect charges of mass executions, torture and other abuses.... Another site, alerting Internet websurfers to the dangers of "disinformation," "propaganda" and the "New World Order," details "The Subjugation of Burma" under the military regime....

Burma Has Engaged in Extensive Monitoring of Communications in Thailand
Micool Brooke, Bangkok Post, August 8, 1999
The SIGINT station in the Burmese embassy in Bangkok is very active. From Burma itself, the Tatmadaw (Burmese military) has intercepted Thai government broadcasts as well as Thai Armed Forces radio traffic, the latter mostly in connection with insurgency activities along the Thai–Burmese border.... An extensive array of SIGINT equipment has been acquired, mostly from China, but also evidently from Singapore and Israel....

China's New War Fighting Skills Emerging Threats to the U.S., India, Taiwan and the Asia/Pacific Region
China Reform Monitor Special Issue, American Foreign Policy Council, Washington, D.C., http://www.afpc.org, undated
...Nyunt is the closest Burmese ally of Beijing. Nyunt is dependent on the Wa and the drug trade to pay for weapons from.... In addition, the Wa army, which was formerly the militant arm of the Beijing-backed Burmese Communist Party, has been commanded by Han Chinese PLA military officers since the 1950s.

Burma (Myanmar) and "Freedom-Fighters"

Burma is being controlled by an unelected government that renamed the country "Myanmar" when it took control, having ignored the election victory by the political party led by Aung San Suu Kyi. Burma, or as the despots that control this nation-state like to call it, Myanmar, is fighting its own information warfare by trying to control access by its citizens to information, especially access to the Internet (see Exhibit 4).

Thailand and Drug Traffickers

Thailand is also involved in information warfare, both with another nation-state, Burma (Myanmar), and also against drug dealers. As reported by infowar.com on August 20, 1999:

> *Thailand is fighting a full-scale war in cyber-space with the United Wa*
> *State Army (UWSA) in an effort to intercept drug caravans infiltrating*
> *northern.... Thailand was fighting a high-tech battle in cyber-space to*
> *win the war against drug traffickers...both sides were eavesdropping on*
> *each other's radio transmissions in spite of modern channel-hopping*
> *and encryption devices...*

This type of information warfare is not a nation-state against a nation-state; it is against a group of organized drug traffickers — or is it? Such shipments cannot easily be transported across nation-state boundaries without the nation-state tacit or implied permissions. Can it be that such shipments are being supported by some regional nation-states that make up the "Golden Triangle?" If so, conflicts between nation-states may increase; however, if so, they are expected to be fought the "old-fashioned way," by soldiers and supported with limited information warfare weapons and tactics.

China and Taiwan

If one wants to see actual information warfare tactics being used in Asia, one just has to look at China (officially known as the People's Republic of China — PRC) and Taiwan (officially known as the Republic of China — ROC).

Taiwan has been occupied by the Dutch, the Japanese, and the Chinese. It is an interesting predicament. It is inhabited by a minority (aboriginal) people, a majority who call themselves Taiwanese, and also by Chinese. The Taiwanese have a separate language and do not necessarily agree that they are part of China just because China and the United States said they were. The Chinese are those, and their descendants, who primarily came to Taiwan with Chiang Kai-shek's Nationalist government (Kuomintang party — KMT) and army after their withdrawal from mainland China in 1949, having been defeated by Mao Tse-tung's military forces.

Both China and Taiwan, for the most part, share a common culture and an official national language — Chinese Mandarin. Over the years, Taiwan has gradually undergone many changes. One of the most important is their shift from being controlled by the KMT to one where free elections are held. Taiwan's past two presidents, both freely elected by the people of Taiwan, are Taiwanese, a shift from the former presidents, the Chinese from the mainland led by Chiang Kai-shek, his sons, members of the KMT — the Nationalists. The PRC says that Taiwan is a "renegade province" and part of China. Under the KMT regime, they also considered Taiwan part of China; however, they believe the legitimate Chinese government is the ROC government on Taiwan.

Taiwan is now one of China's biggest supporters when it comes to economic development. There are many Taiwan-owned and backed businesses now operating in China. While the business aspect of the relationship is booming, the political aspects are not. For example:

- How many times have ROC government officials urged the PRC to resume cross-strait talks? Total: 21 times.
- Beijing claims that "The Taiwan question is purely an internal affair." Total 69 cases.[6]

Now that the ROC is led by Taiwanese, they have not committed to being a part of China. In fact, there appears to be growing consideration in Taiwan that it should be an independent nation-state (i.e., the Republic of Taiwan). The official government position does not take a firm stance on the issue. However, it is almost a certainty that China, if for no other reason than to save face, would not tolerate such an independent act. Therefore, if such a declaration was made by the ROC, even if voted on and accepted by the majority of Taiwan citizens, war between China and Taiwan would be all but a certainty. It is also interesting that the United States and others agree with China that there is only one China and that Taiwan is part of China. However, now that there is a freely elected government by the people of Taiwan, what would happen if this freely elected government supported by the majority of its people declared its independence from China? Would the United States and other democratically elected governments support such independence, or would they instead support a communist-controlled government and be neutral in any conflict between China and Taiwan? It is an interesting dilemma for democratic nation-states. The U.S. corporations and others doing business in China would probably lobby heavily for neutrality or for China. After all, when it comes to money versus human rights, money seems to win every time.

If the United States and others came to the defense of Taiwan, would anyone come to the defense of China? Yes, if you believe that the following is not in itself an information warfare tactic (misinformation):

> **Russian Fleet Will Intercept U.S. 7th Fleet's Intervention in Cross-Strait Russian Navy Reportedly Instructed to Stop U.S. Involvement in Taiwan Strait**
> *Li Nien-ting, Hong Kong Sing Tao Jih Pao (Internet Version-WWW)*
> *(in Chinese) July 8, 2000*
> *...Putin gave a special instruction to the Russian military that in case the Taiwan situation deteriorates and the U.S. military attempts to become involved in the situation, Russia will dispatch its Pacific Fleet to check the route of the Seventh Fleet of the U.S. Navy, to keep the latter far away from the Taiwan Strait...*

Both China and Taiwan are high-technology and information systems dependent; therefore, the use of information warfare tactics supporting industrial period-based warfare of planes, ships, missiles, and soldiers is a certainty. In fact, both sides are using information warfare tactics as you read this book. The *Taipei Times*, in its January 3, 2001, issue stated as follows:

> *...First information warfare group put into service; the military inaugurated its first information warfare (IW) force on Monday as part of*

its efforts to handle military threats that may appear in the new century, the Ministry of National Defense announced yesterday. The IW force was established mainly to cope with potential threats from China in the field.[7]

Consider the following possible information warfare scenario between China and Taiwan:

- China aggressively tries to isolate Taiwan.
- Taiwan declares to be the Republic of Taiwan, an independent nation-state and asks for UN and world recognition as such.
- China declares the action illegal, goes on military alert, and begins military buildup on its southeastern coast.
- The United States begins defense buildup around Taiwan, based on China's military threats and buildup.
- China begins propaganda and misinformation portion of information warfare campaign against Taiwan through its Web sites, chat groups, and e-mails.
- China covertly uses information warfare tactics to attack Taiwan's information infrastructure and U.S. fleet and air operations in the region.
- U.S. military information flow is impeded in the Pacific.
- Taiwan's power, water distribution, and telecommunication systems begin to fail.
- China initiates information deception and concealment by use of multi-mode, multi-path, and multi-frequency networks.
- Russian fleet moves into the area and attempts to jam, spoof U.S. military telecommunications traffic.
- The United States takes over China's primary Internet and internal networks or denies their use to Chinese military and government agencies.
- Taiwan and the United States begin PsyOps, impersonate PRC officials, and redirect Chinese military messages and activities.
- The United States electronically freezes all possible financial assets of China.
- United States NII begins to deteriorate.
- United States–China–Taiwan–Russian talks begin as a cease-fire is implemented.

There are more details and many conclusions that can be reached beginning with the above scenario. However, this scenario is only presented as a "what-if." There is no attempt to write a detailed Taiwan–China information warfare operations plan here. Suffice it to say that one can easily see the "hows and whys" of an information warfare conflict between these two entities; but there is more, as shown in Exhibit 5.

The letter shown in Exhibit 6, received by Dr. Kovacich, indicates how Taiwan is using some information warfare tactics and the Internet for political purposes.

Exhibit 5. Conflict between Taiwan and China

China Defends Military Buildup
Beijing (AP), October 16, 2000
China has said that tensions with Taiwan and bullying by major powers — an apparent reference to the United States — are forcing it to beef up its armed forces. In a lengthy policy paper Monday, China responded to foreign concerns about its growing military might and reiterated its threat to use force against Taiwan if it seeks independence.

Taiwan Prepares for Possible Chinese Cyber Attacks
Taipei (AP) — Dow Jones News, November 2, 1999
China could be able in five years to use computer viruses, hackers and other types of cyber warfare to quickly break down Taiwan's defenses and prepare for an invasion, the Taiwanese military...said Tuesday. Taiwan's economy, government and military are highly dependent on computers and could be vulnerable to such high-tech weapons,.... The high-tech weapons could quickly take out their targets without much expense and loss of life,...destroy public morale, spread disinformation and cause instability.

China Gears Army for Cyber-war
Lynne O'Donnell (China correspondent), Australia, October 11, 1999
China is planning to pour billions of dollars into a high-tech upgrade of its army to prepare to fight a future war in which software beats manpower.... With their efforts devoted to the improvement of electronic warfare capability, the Chinese communists are expected to pose a threat to Taiwan in 2005,.... We could wake up one morning and find a city, or the country or a section of the country without power because of a surprise electronic warfare attack.... Political and economic chaos could be created by hacking into or destroying computer systems with viruses, with terrorist acts or through biochemical warfare, the colonels say. Military strategists have also outlined plans to use the Internet and the global financial system as weapons of disruption against the U.S., western Europe and Japan. Taiwan plans to retaliate with computer viruses.

Taiwan Categorized Viruses for Attack on China
infowar.com, August 10, 1999
...The head of the Defence Ministry's Information and Communications Bureau, Lieutenant Lin Chin-ching, told the BBC that his officers had categorised about 1,000 different computer viruses, which could be used to fight back in the event of a Chinese electronic onslaught.... He said Beijing had not yet managed to penetrate the Taiwan military's computer network, although several government Internet sites were the target of mainland Chinese hackers last year...

Officials React to Hacking of Control Yuan
Central News Agency, Taipei, August 9, 1999
The Central Bank of China's (CBC) information system is secure from computer hacker invasions, the bank's Information Office Director Hsieh Chien-hsing said Monday [9th August]. A hacker broke into the Internet Web site of the Control Yuan and wiped out all the information on Sunday, rocking the whole island...

Taiwan-China Hackers' War Erupts
infowar.com, Taipei, Taiwan (AP), August 9, 1999
A cyberwar has erupted between Taiwanese and Chinese computer hackers lending support to their governments' battle for sovereignty over Taiwan. A Taiwanese hacked into a Chinese high-tech Internet site on Monday, planting on its webpage a red and blue Taiwanese national flag as well as an anti-Communist slogan: "Reconquer, Reconquer, Reconquer the Mainland."
...A Chinese railroad Web site and a securities Web site were hacked into in a similar way.

Exhibit 5. Conflict between Taiwan and China (Continued)

Taiwan: Cyber-hackers Strike Back at China

infowar.com, Reuters, August 10, 1999

Taiwan may be dwarfed by its sabre-rattling rival, mainland China, but it has shown it is not to be trifled with on at least one battleground — cyberspace. Hackers from the computer-savvy island have inserted pro-Taiwan messages into several Communist Chinese government Web sites in retaliation for a similar attack on Taiwan government sites by a mainland Chinese hacker. The Web attacks sparked concern from military authorities who said an Internet war could add to already simmering tension over Taiwan's drive for equal status with the mainland.

Taiwan: Pro-China Hacker Attacks Taiwan Govt Web Sites

infowar.com, Reuters, August 10, 1999

A person claiming to be from mainland China hacked into several Taiwan government Internet sites to insert pro-China messages amid a heated row between the two sides over Taiwan's political status. "Only one China exists and only one China is needed," read a message inserted on Sunday into the Web site of the Control Yuan — Taiwan's highest watchdog agency.

Taiwan: Prosecutors Probe Web Site Intrusion by Mainland Hacker

infowar.com, August 10, 1999

Following a series of intrusions by mainland Chinese hackers into a dozen government-run Internet Web sites over the weekend, prosecution and police authorities on Monday [9th August] launched an investigation into the incidents hackers may have invaded the Taiwan Web sites through Web sites backed by mainland Chinese authorities, or through U.S.-based Web sites run by mainlanders, as links between mainland and foreign Web sites are closely monitored by the authorities in Beijing.

China: Cyber 'War' May Lead to Real Thing

Straits Times, August 13, 1999

The war being waged between China and Taiwan in cyberspace, which led to government Web sites on both sides being hacked, could well lead to a real war, some analysts have warned. The attacks on the government Web sites had prompted concerns about Taiwan's ability to protect its electronic information infrastructure.

China: Troops Being Trained for Electronic Warfare

Zhongguo Tongxun She news agency, Hong Kong, August 9, 1999

China's armed forces are gearing up for the electronic warfare of the future. Computers and other types of electronic technology are already in place in communications, command and control and weapons systems. Training is being directed towards handling and applying advanced information technology in all types of military activity.

Taiwan on Guard for Cyberwar

http://news.bbc.co.uk/hi/english/world/asia-pacific/newsid_965000/965344.stm, October 10, 2000

Chinese and Taiwanese hackers have already crossed cyber-swords. As Taiwan celebrates its National Day on Tuesday, Web site operators on the island have been warned to prepare for a resumption of a cross-Straits cyberwar with China. The warning follows speculation that hackers in mainland China may be preparing to bombard Taiwanese sites with anti-independence messages. Major government Web sites and those belonging to associates of President Chen Shui-bien — who Beijing sees as a supporter of Taiwanese independence — are thought to be among the principal targets.

Exhibit 5.   Conflict between Taiwan and China (Continued)

China War Plan against Taiwan and U.S.

J. Michael Waller, Australian, September 21, 1999

The People's Republic of China is actively planning a military invasion of Taiwan and is preparing to wage war against the United States — including firing its small arsenal of strategic nuclear missiles on the territory of the United States — if Washington attempts to defend the island. In an internal document from the Chinese Communist Party's Central Military Commission to all its regional commanders, Beijing says it hopes to absorb Taiwan through nonviolent means but warns of an "increased possibility for a military solution," arguing: "It is better to fight now than [in the] future — the earlier, the better."

ROC Defense Ministry Sets Up Information Warfare Committee

Victor Lai, Taipei, (CNA), August 16, 1999

The Republic of China's Ministry of National Defense (MND)...said it had established a committee to deal with information warfare. A MND official disclosed the information at a public hearing on the protection of Taiwan's computer systems from mainland Chinese intrusion.... "we are able to defend ourselves in an information war, but we will not initiate an offensive." According to the official, in 1985 Beijing had already developed its plans for information warfare, and actual implication of the plan started in 1995...

Taiwan Military to Show off Computer Virus Capability

Report Unaccredited, China Times Interactive, August 8, 2000,
http://www.chinatimes.com.tw//english/epolitic/89080708.htm

Taiwan is not about to let its sovereignty be threatened by Chinese IW. The Liberty Times newspaper quoted a top defense ministry official as saying, "For the first time in the coming Han Kuang war games, computer viruses will be used to attack each other's information network." The paper said military authorities have worked out some 2,000 types of computer viruses, and the anti-virus capability of the military units has been upgraded.

Taiwan Establishes Info Warfare Center

USdefense.com, July 7, 2000

To develop doctrine and strategy, the Taiwanese military established an office to examine the concept of IW in response to China's. According to Taiwan's Central News Agency, a special center for study of strategic and tactical IW was established "in the face of Beijing's all-out efforts to develop advanced arms and surveillance equipment."

China and Russia

When dealing with information warfare, one cannot avoid addressing the possibility of information warfare between Russia and China. This is because the potential for conflict has existed and will continue to exist well into the future because:

- There continue to be border disputes between these two nation-states.
- There are political differences between them — now more than ever.
- China and Russia have a history of mistrust.
- China is very fearful of losing its communist party control, as Russia has.
- Russia has been a supporter of Vietnam, which has had border clashes with China.

Exhibit 6. Taiwan's Information Warfare Tactics

Dear Netizen,

Earlier this year you participated in the ROC Government Information Office's (GIO) slogan-writing competition, and we believe you may also be interested in our new Internet-based activity, which focuses on Taiwan's President Chen Shui-bian and his winning of the 2001 Prize For Freedom.

Earlier this year, Liberal International (LI) announced that President Chen would receive the prize for his leading role in Taiwan's democratization.

The prize is scheduled to be presented in Copenhagen, Denmark, this November.

President Chen has expressed his desire to travel to Denmark and receive the prize in person. He may not be able to do so, however, because China consistently opposes visits by Taiwan's leaders to countries, such as Denmark, that recognize Beijing.

To publicize LI's decision to honor President Chen, and raise global awareness of the difficulties Taiwan's democratically elected leader faces when traveling overseas in any capacity, the GIO in Taipei has prepared a special Internet-based activity at http://th.gio.gov.tw/freedom.cfm?Web=112 including background information and a message board where Netizens can post their comments.

Yours sincerely,

Internet Team
ROC Government Information Office
Taipei, Taiwan
www.gio.gov.tw

There are many scenarios that can be imagined, based on the above or other reasons, as to why China and Russia would end up in a conflict in which information warfare played a role. It is suggested that the reader develop some likely scenarios taking both sides of the conflict — as a Russian and as a Chinese. In doing so, it is suggested that the U.S. CIA *World Factbook*[4] relative to Russia and China be researched, as well as doing an online search of both nation-states to determine their information infrastructures and other vulnerabilities.

The People's Republic of China (PRC)

In the near future, information warfare will control the form and future of war. We recognize this developmental trend of information warfare and see it as a driving force in the modernization of China's military and combat readiness. This trend will be highly critical to achieving victory in future wars.

—Major General Wang Pufeng
Former Director of the Strategy Department
Academy of Military Science, Beijing
Excerpted from *China Military Science* (Spring 1995)

China is the home of a very old civilization. According to Siming Wang, professor and director, Institute of Agricultural History in China, China has a history of agriculture of 8000 to 10,000 years. China has either been at war with its neighbors most of that time, absorbed them, or took tributes from them. Sun Tzu's ancient book entitled *The Art of War* is mandatory reading for anyone interested in warfare, conflicts of any type, and also provides a key as to how China approaches warfare — even information warfare.

China is *the* major political entity in Asia, as well as a major economic power. Thus, it should not come as a surprise that China would have an interest in information warfare. With China's recent record as an advocate of information warfare tactics, it is believed that additional information about China's information warfare thoughts and tactics should be presented. Obviously, it is impossible and beyond the scope of this book to present an all-encompassing view of China's information warfare position, capabilities, etc. However, the information provided here should give the reader an overview of China's information warfare philosophies, policies, and actions:

- "China realized from the outcome of the Gulf War...that air raids and precision strikes from long distances are decision factors in the outcome of wars...information warfare and electronic warfare are of key importance...," Jun Jui-Wen in *Latest Trends in China's Military Revolution*.

- "Before a battle begins (sometimes dozens of hours in advance)...commanders will first use offensive information-war means (precision-guided weapons, electronic jamming, electromagnetic pulse weapons and computer viruses) to attack enemy information systems..." Lt. Gen. Huai Guomo, PLA, in *On Meeting the Challenge of the New Military Revolution*.

- *China's people's war scenario:* The enemy — the United States, Russia, or Japan — will invade and seek to subjugate China; the war will last many years; China's leaders will move to alternative national capitals during the war; China's defense industrial base will arm millions of militia in protracted war until the enemy can be defeated by the main army.[8]

- *China's local war scenarios:* The opponent will not be a superpower; the war will be near China's border; the war will not be a deep invasion; China will seek a quick military decision; rapid reaction forces will defeat the local forces of Japan, Vietnam, India, Central Asia, Taiwan, the Philippines, Malaysia, or Indonesia.[8]

- *China's revolution in military affairs scenarios:* Close an information gap; network all forces; attack the enemy command, control, computers, and intelligence (C3I) to paralyze its operations; preempt enemy attacks.

- Use directed energy weapons; use computer viruses; use submarine-launched munitions; use anti-satellite weapons; use forces to prevent a logistics buildup; use special operations raids.[8]

- "Whom does the Communist Party of China regard as its international archenemy? It is the United States." (From a 1993 report of senior PLA officers and high-level diplomats.)

- "(As for the United States) for a relatively long time it will be absolutely necessary that we quietly nurse our sense of vengeance.... We must conceal our abilities and bide our time." (LTG Mi Zhenyu, Vice Commandant, Academy of Military Sciences, Beijing, 1996.)

The inferior can defeat the superior...

—Fu Quanyou, Chief of Staff,
People's Liberation Army[8]

Some possible Chinese information warfare tactics related to the Internet, and the reason why, include:

- The Chinese believe that all its citizens are soldiers in time of need.
- They will apply current controls and espionage methods through networks.
- China has "Orwellian" plans for the networks.
- More capitalism has brought more Western support in the form of technology.
- More businesses and more citizens are beginning to own computers (e.g., over 35 million systems).
- They will concentrate monitoring, filtering, and auditing of incoming and outgoing Internet. If they cannot monitor it, they will shut it down. Examples of sites include:
  - CHINANET (www.cnc.ac.cn)
  - CERNET (China Education and Research Network) (www.net.edu.cn)
  - STNET (www.stn.sh.cn)
  - EIN (www.cein.cn)
  - Eight regional networks and subnets (e.g., *Shanghai Post,* Telephone and Telegraph PTT) (www.sta.net.cn)

As China continues to try to control its information infrastructure and those of others in its nation-state, one can envision the potential to network computers throughout this nation-state to attack specific information warfare-related problems. For example, how much computer power will it take to break encrypted messages if the power of more than 50 million Pentium-plus computers were attacking the problem (e.g., breaking foreign encrypted messages in almost real-time)?

One might say that this would not be possible; however, there are attempts at networking such systems:

> **Internet Tapped for 'Parasitic Computing'**
> *Richard Stenger; CNN Science and Technology, August 29, 2001*
> *Siphoning the computational power of the Internet, U.S. scientists have figured out a way to induce unwitting Web servers across the world to perform mathematical calculations...with the unauthorized help of computers in North America, Europe and Asia...using a remote server... The bits were then hidden inside components of the standard transmission control protocol of the Internet, and sent...*

It is interesting to note that the remote machine was used to force *unaware target computers* to solve a computational problem by engaging in standard communication. Does this mean that in information warfare, one will soon be able to read the adversary's encrypted communication in almost real-time? Ironically, that may not only be possible, but the adversary's own computers may be involuntarily used in portions of the decrypting.

There is so much information on China being reported these days that can be considered relevant to information warfare; however, it is impossible to keep up with all of it. Make no mistake about it, China is:

- Serious about pursuing information warfare strategies, policies, tactics, and weapons
- Gearing up for information warfare
- Testing offensive information warfare weapons
- Using information warfare weapons against Taiwan, the United States, and any others whose policies or political positions they do not like (e.g., Falun Gong)
- Confident that nation-states will do nothing but protest while China pleads ignorance and its confidence grows

Note: One finds it interesting that anyone using the Internet within China and calling for more freedoms in China, via e-mails and Web sites, appears to be quickly hunted down and prosecuted by Chinese government officials. While at the same time, when attacks against other nation-states' Web sites and networks are traced back to China, the Chinese government officials cannot seem to find the guilty parties — not even when the attacks are traced back to Chinese government systems.

The articles cited in Exhibit 7 are some examples of information warfare related to China.

Exhibit 7.   China's Information Warfare

China: Foreigners Must Disclose Internet Secrets to China Soon
infowar.com, Beijing, January 25, 2000
The Chinese government is about to require foreign firms to reveal one of their deepest secrets — the type of software used to protect sensitive data transfers over the Internet... foreign and Chinese companies must register the type of commercial encryption software they use. Such software makes it more difficult for hackers — or governments — to eavesdrop on electronic messages...

China: Beijing Developing Internet Control Device
Hong Kong iMail Web site/BBC Monitoring Media, March 21, 2001
...Beijing is working on developing a system similar to "black box" flight-data recorders capable of monitoring Internet traffic as it seeks to tighten its surveillance of cyberworld activities.

China Government Issues Internet News Regulations
Ian Stokell, Newsbytes, San Francisco, California, November 7, 2000
The Chinese government has issued regulations designed to control the distribution of news via Web sites in China, as well as chatroom content on the Internet. Reuters reported that Web sites are forbidden from reporting or writing news themselves, and must therefore rely on state media with whom they have signed contracts, but there is flexibility in that a clear

Exhibit 7.   China's Information Warfare (Continued)

definition of what constitutes "news" has been omitted. Analysts are hoping that an upcoming document defining "news" will allow for a variety of topics to exist outside of the regulations, such as sports, entertainment or financial stories.

China on Defensive?
Aug 9, 2000, BEIJING (Reuters)
...China issued a clarion call on Wednesday to Communist Party media to build up their Web sites for a propaganda fight against what it said were enemy forces at home and abroad.

China: Army Publishes Book on Information Warfare
Jiefangjun Bao, Beijing, in Chinese, p. 6, December 7, 1999
...At a time when the whole army is studying in depth "Outline of People's Liberation Army [PLA] Combined Operations," the book "Introduction to Information Warfare [IW]," a military studies project of the state social science fund, compiled by Dai Qingmin (2071 3237 3046) of the PLA Electronic Engineering Academy, has been published and distributed by the PLA Publishing House after appraisal by the Electronic Countermeasures Department of the General Staff Department.

China: Army Paper on Need for Offensive, Defensive Internet Warfare
Jiefangjun Bao, Beijing, in Chinese, p. 7, November 11, 1999
Internet warfare is quietly approaching in the world today.... Today finance, commerce, communications, telecommunications, military affairs and so on all depend on this mysterious space, and Internet order has become an important hallmark of a country's economic operations and development; countries have enormous economic and political interests on the net, and similarly exercise their powers and interests in cyberspace; the Internet is more and more displaying its economic, political, and military significance equivalent to land, sea, and air power. Violation of cyberspace is similarly violation of national sovereignty, and its seriousness is in no way inferior to violation of territory, territorial waters, or airspace.

Report: Net Risks on Rise in China
infowar.com., by Dan Verton, IDG, January 23, 2001
Intelligence and security experts are warning foreign firms in China of a growing threat of Internet-related crimes, government surveillance and loss of proprietary data. But some U.S. companies said they view those threats as exaggerated....the government-controlled Internet environment in China could put the integrity of their networks at risk. "The most important consideration is that, in one way or another, the government is involved in the operation, regulation and monitoring of the country's networks...

China Dragon Bares Its Claws for Cyberwar
NewsMax.com, November 17, 1999
Communist China is preparing to wage an "all-conquering" war in cyberspace, on a scale with conventional combat operations, by seizing "the Internet command power." Its plan to attack a technically sophisticated nation such as the United States through the worldwide Internet of computers...

U.S., Chinese Officials Discuss High-Tech Cooperation
David McGuire, Newsbytes, Washington, D.C., August 12, 2000
Looking for ways to bridge the gap between the U.S. and Chinese high-tech markets, officials from both countries are meeting here this week to discuss joint proposals in a number of technology-related areas.

Exhibit 7.   China's Information Warfare (Continued)

China Is Staging a 24-hour Nationwide Internet Security Surveillance
Hong Kong, China, source and date unknown
...as a Sino-U.S. cyber war intensifies. China's National Computer Network and Information Security Administration Center has issued a warning to all computer system administrators to watch out for major cyber assaults, according to state-run Xinhua News Agency. The chief of the center in Beijing was quoted as saying that hundreds of Chinese Web sites have been hacked into in April, with more than 100 attacks per day. The cyber war between China and the U.S. erupted after a U.S. Navy spy plane collided with a Chinese fighter jet over international waters in the South China Sea on April 1.

China's Military Plots 'Dirty War' against the West
David Harrison and Damien Mcelroy, Telegraph, October 17, 1999
...As President Jiang Zemin prepares for his state visit to Britain this week, details have emerged of a bizarre Chinese plan to destroy the West's financial institutions in the event of a major conflict breaking out. Senior members of the People's Liberation Army are openly urging the Beijing government to abandon conventional defence strategies and prepare a "dirty war." They advocate terrorism, biochemical warfare, environmental damage and computer viruses as a means to pitch the West into political and economic crisis.

Chinese Military Calls for Special Hacker Force
Hong Kong, China, By Staff, Newbsytes, April 8, 1999
The Chinese military is looking for a few good men — to be trained in the art of government sponsored hacking over the Internet. According to a report by asia.internet.com, The Liberation Army Daily newspaper, a "mouthpiece" of China's Peoples Liberation Army, recently called for the development of a hacking (more commonly called "cracking") capability made up of civilian experts and specially trained military personnel that could engage in online and Internet warfare.

PLA "Acupuncture" Info-War Targets U.S. Military/Civilian Strengths; Beijing Protests Cancellation of U.S.-China Satellite Deal
China Reform Monitor No. 175, March 3, 1999
In the wake of Congressional findings that commercial satellite deals with China have threatened U.S. national security, the Clinton administration has rejected a $450 million Hughes satellite deal with China, the New York Times reports. The Pentagon and State Department overruled the Commerce Department to deny permission for the deal, between Hughes and the Chinese-Singaporean Asia-Pacific Mobile Telecommunications Company, due to fears that sensitive U.S. military-related technology would be obtained by the PLA.

China: Air Force Publishes First Information Warfare Teaching Aid
infowar.com, date unknown
Text of report by Tang Baiyun, Chen Kecheng entitled: "Chinese air force publishes 'Information Warfare,' its first teaching material for signal corps," carried by Chinese news agency Zhongguo Xinwen She.
Information Warfare, a teaching aid compiled by a signal unit of the Guangzhou air force, was recently published. This is the first teaching aid on information warfare published by the Signal Corps of the Chinese air force.

Exhibit 7. China's Information Warfare (Continued)

China: Army Trains Information Warfare Personnel

Tai Yang Pao, Hong Kong, in Chinese 15/9/99 p A17; Text of report by Hong Kong newspaper 'Tai Yang Pao' on September 15, 1999

The Chinese military recently established a university of science and technology to train technological personnel for information warfare and to study information warfare theories and related technology. The People's Liberation Army [PLA] University of Science and Engineering was established after a merge of the Communication Engineering Institute, the Engineering College for Engineering Corps, the Air Force College of Meteorology, and 63 research institutes of the General Staff Headquarters.

China Sets Up Office to Regulate Internet News

Matt Pottinger, Beijing, Reuters, April 21, 2000

China has created an office to regulate news on the Internet and to help state media spice up their Web sites so they may compete in the booming market, officials and newspapers said Friday. Executives at privately owned Web sites said they were worried by the ambiguous mandate of the Internet Information Management Bureau, which includes countering the "infiltration of harmful information on the Internet."

Chinese Web Site Sina.com Suffers Hacker Attack

Beijing (Reuters), February 17, 2000

Sina.com, a top Chinese Internet portal, suffered a hacker attack around the same time several popular U.S. Web sites were crippled by online raids, a Sina.com executive said on Thursday. "The hackers — I don't know if there was one or several of them — successfully collapsed our e-mail service," said Wang Yan, general manager of Sina.com's China operations.

China-Taiwan Hacker Wars

janes.com, Foreign Report, Volume 000/2565, and infowar.com, October 21, 1999

…recent comment by President Lee Teng-hui that ties between China and Taiwan should be characterised as special state-to-state relations provoked a rash of attacks by hackers on each other's Web sites and computers,…. Taiwan's National Security Bureau estimates that between August and September Chinese hackers broke into Taiwanese computer networks 165 times. Taiwan's cyber-wizards are said to have hacked their way into the Web site of the Chinese State Tax Authority and the ministry of railways. Much more is being prepared. The first stage was an attempt by the two antagonists to improve their ability to take part in 'information warfare'. More recently, our informants have detected signs that they have begun to assemble teams of hackers to attack the civilian information infrastructure of a chosen country. How? By shutting down electricity grids, banks, stock markets, air-traffic control, telecommunications, and other vital parts of a country's economy that are run by computers.

How to Insert a Virus — Take China

infowar.com, date unknown

Following the 1990–91 Gulf War, its leaders first became aware of the immense techno-logical superiority of the United States. In response…China spent a great deal of time and money on research into ways of inserting viruses into foreign computer networks. According to sources from Taiwan's defence ministry, Chinese strategists began talking about incorpo-rating computers into their battle plans in 1985; laid concrete plans 10 years later; and subsequently carried out exercises aimed at interrupting, paralysing or destroying enemy broadcasting and military communications systems using computer viruses.

Exhibit 7.   China's Information Warfare (Continued)

China Threatens 'Electronic Pearl Harbor' Attack on U.S.
NewsMax.com, October 11, 2000
Nations lacking military muscle could create an "electronic Pearl Harbor" that could defeat the U.S. by using electronic warfare to cripple America's high-tech-dependent armed forces, an official Chinese report claims. In the report, "The U.S. Military's Soft Ribs and Strategic Weaknesses," which analyzed America's military doctrines, the strategies and tactics that Beijing could use against the U.S. were revealed by the official Chinese Xinhua news agency. According to the authoritative American Foreign Policy Council (AFPC), the Beijing document explained how China or any other country could use electronic warfare against the U.S....

Chinese Crackers Call Off Crusade
Robert MacMillan, Newsbytes, Washington, D.C., September 5, 2001
Saying it has reached its goal of hacking into and in some cases defacing 1,000 U.S. Web sites, a Chinese hacking group officially called off its online battle with American hackers. The Chinese group said it would not be responsible for any subsequent attacks on U.S. Web sites. One of the most high-profile hacks occurred on May 4 when the White House's Web site at http://www.whitehouse.gov fell victim to a denial-of-service attack that apparently originated in China.

Chinese E-commerce Site Attacked
infowar.com, date unknown
The hacker attack is likely to ignite the debate on internet security in China. A leading E-commerce Web site in China has come under attack from hackers.... "The damage on our Web site by hackers is huge, page contents files have been entirely destroyed and databases were damaged to varying degrees," the Beijing Youth Daily newspaper quoted the company's managing director Chen Yongjian as saying. As with several attacks on well-known sites in the United States last month, IT163.com was bombarded with so many messages at the same time that the system could not cope. Three successive days of attacks by hackers in the U.S. shut down Yahoo!, eBay, CNN.com, Amazon, Buy.com and E*Trade for varying lengths of time.

Chinese Hackers Step Up Attacks On U.S. Web Sites
Brian Krebs, Newsbytes, Washington, D.C., April 5, 2001
Pro-Chinese hackers have dramatically escalated attacks on U.S. commercial and government Web sites over the past 48 hours, shifting from mere Web page defacements to deleting important data from targeted sites, security experts said Thursday. The increasing ferocity of the attacks comes as several Chinese hacker outfits appear to have joined forces under the banner of "Project China," according to Web security analysts at iDefense. Project China, along with another large group of pro-Chinese hackers calling themselves the "Hackers Union of China (HUC)," have been responsible for 10 attacks on U.S. government and commercial sites in the last 12 hours, and more than two-dozen assaults in the past 24 hours, according to Web page defacement mirrors at http://www.attrition.org, and http://www.alldas.de.

Chinese Hacktivist Threat Continues to Build
Steve Gold, Newsbytes, Parsippany, New Jersey, April 30, 2001
Ongoing research compiled over the weekend suggests that Chinese threats of a hacker attack on U.S. Web sites is continuing to build..., Vigilinx, which was one of the first firms to report on the impending online skirmish between Chinese hackers and U.S. Web site operators, says that the hacker war now has a name: "The Sixth Network War of National

Exhibit 7. China's Information Warfare (Continued)

Defense." Citing reports on the Thousand Dragon (Qianlong) News Service, carried by China.com, the IT security firm says that the Chinese hack is actually the sixth campaign of its kind. "At least two sources (Qianlong and ChinaByte) have tied the threatened hacktivist action to the Hackers Union of China (aka Honkers Union of China, Red Guests and Red Guest Union), and various reports say the action will be called the '51 Network War of National defense' (wu yi wei guo wang zhan)," said the firm in an advisory to its customers overnight....

Chinese Hackers Launch Retaliatory Web-Site Attacks
Brian Krebs and Steve Gold, Newsbytes, Washington, D.C., April 30, 2001
Chinese hackers launched a highly coordinated volley of attacks on U.S. governmental and commercial Web sites early this morning, security experts report. The much-anticipated campaign was in apparent retaliation for nearly 100 U.S. hacker attacks on Chinese state-run sites following the collision of a U.S. spy plane and a Chinese jet last month. Last week, a U.S. government-funded security watchdog, the National Infrastructure Protection Center, warned the campaign could continue unabated until at least May 7, the two-year anniversary of the NATO bombing of a Chinese embassy in Belgrade.

Cyberattacks against NATO Traced to China
IDG, date unknown
Hackers with Chinese Internet addresses launched coordinated cyberattacks against the United States and allied forces during the air war against Yugoslavia this spring, the Air Force's top network communicator confirmed today. Lt. Gen. William Donahue, director of communications and information for the Air Force, said that during the 78-day air war, called Operation Allied Force, hackers "came at us daily, hell-bent on taking down NATO networks." Donahue, speaking here at the annual Air Force Information Technology Conference, said the cyberattacks emanated from the Serbs, what he called "Serb sympathizers" and from "people who came at us with an [Internet Protocol] address that resolved to China." He added that the U.S. military traced the attacks back to more than one Chinese IP address.

FBI Widens China Spy Investigation
infowar.com, Washington (AP), date unknown
New evidence widens the FBI's investigation into spying allegations and suggests China may have stolen information about America's most advanced nuclear warhead from one of the weapon's contractors or from the Navy, The Washington Post reported today.

China Orders Web Sites to Guard against Secrecy Leaks
infowar.com, date unknown
Chinese authorities have imposed security checks on all Web sites in an attempt to protect state secrets. The move is thought to be the first time a government has moved to restrict the availability of free information on the Internet. Chinese journalists found guilty of publishing state secrets now risk imprisonment or could even face the death penalty.

China Ponders New Rules of 'Unrestricted War'
John Pomfret, Washington Post Foreign Service, p. A01, August 8, 1999
In 1996, colonels Qiao Liang and Wang Xiangsui were in Fujian province for military exercises aimed at threatening the island of Taiwan. As Chinese M9 intermediate-range missiles splashed into waters off two main southern Taiwanese ports, the United States dispatched two aircraft carrier battle groups to the region. Like most Chinese officers, the

Exhibit 7.   China's Information Warfare (Continued)

colonels were furious at the U.S. move, seeing it as another sign of American interference in China's internal affairs. But to Qiao and Wang, the first crisis in the Taiwan Strait was also a lesson. "We realized that if China's military was to face off against the United States, we would not be sufficient," said Wang, an air force colonel in the Guangzhou military district's political department. "So we realized that China needs a new strategy to right the balance of power." Their response was to write a book called "Unrestricted War," (8/17/99).

Chinese Hacker Plants Flag on Web Page of Taiwanese Spy Agency
infowar.com, August 17, 1999
In an escalating cyberwar between Taiwan and China, a Chinese hacker got into a Taiwanese spy agency's Web site and planted a Chinese flag, officials said Tuesday. China has stepped up its saber-rattling against Taiwan and repeated its threat to use force against the island if it ever declares formal independence.

China — Huawei's Massive U.S. Connections
infowar.com, March 20, 2001
One of China's largest telecommunications companies, Huawei Technologies, is at the centre of a controversy over the sale and installation of fibre-optic cables by Chinese firms in Iraq...

Group Calls for Protest at Chinese Internet Detention
Adam Creed, Newsbytes, New York, June 28, 2000
New York-based Human Rights Watch has called on global companies involved in China's nascent Internet market to get involved with the case of a man who has been detained by the People's Republic of China Government. Huang Qi has been in detention since June 3 over a Web site he maintained that discussed human rights abuses in China.

Teacher Jailed for Anti-Beijing Banner on Net
Hong Kong (Reuters), March 12, 2001
China has sentenced a teacher to two years' jail for posting a banner critical of the ruling Communist Party on the Internet, a Hong Kong human rights group said on Monday. The Nanchong City Intermediate People's Court in central China's Sichuan province imposed the sentence on Jiang Xihua for "inciting to subvert state power," the Information Center for Human Rights & Democracy said in a statement. Jiang, 27 and a computer teacher, posted the banner that read "Down with the Communist Party" on an Internet chatroom last August. He was now appealing against the sentence which was delivered last December.

China's New War Fighting Skills Emerging Threats to the U.S., India, Taiwan and the Asia/Pacific Region
China Reform Monitor Special Issue, American Foreign Policy Council, Washington, D.C., http://www.afpc.org, undated
During July and August, China's People's Liberation Army [PLA] conducted large-scale joint military exercises in the Nanjing Military Region on the coast of the Taiwan Strait, demonstrating significant new fire power coordination and command-and-control capabilities. The PLA's modernization and joint war fighting capabilities are developing at a rate far ahead of the Pentagon's previous predictions. The "assymetrical" multi-dimensional threat posed by the Beijing-Islamabad-SLORC alliance to Asian democracies, such as India, Thailand and the Philippines is further demonstrated by their direct or indirect support for military aggression, terrorism and narcotics trafficking across southern Asia.

Exhibit 7.   China's Information Warfare (Continued)

China Police Want to Beef up Online Powers
infowar.com, date unknown
Investigators must be given power to demand secret digital keys from suspects to search encrypted online documents and transactions for criminal evidence, police specialists said yesterday. Hilton Chan Kwok-hung, chief inspector and head of the computer-crime section, said the Electronic Transaction Bill, now before the Legislative Council, should include powers to seize suspect electronic documents. "Criminals can use encryption to send messages for crimes from money laundering to sending coded triad messages," he said after an Internet security conference. "Without the digital keys, such documents cannot be decoded for evidence."

China Police Develop Software to 'Purify' Net
source unknown
The Ministry of Public Security has released new software designed to keep "cults, sex and violence" off the Internet in China, a police official said. "The software, Internet Police 110, was released yesterday. It will prevent users from getting unhealthy information from foreign and domestic Web sites," he said. "It was designed to block information of cults, sex and violence on the Internet," he said, without making clear whether installation was mandatory. "I believe it will help purify China's Internet service," the official said of the software, named for the emergency police telephone number. The software — which comes in three versions for households, Internet cafes and schools — can also monitor Web traffic and delete or block messages from sources deemed offensive…. But groups including dissidents and Falun Gong — banned in China as an "evil cult" — have used proxy addresses and other sophisticated methods to overcome Internet site blocks.

China Exacts Computer-Virus Samples
Wall Street Journal, March 3, 2001
The Chinese government has been requiring several anti-virus software to exchange samples of malicious viruses for access to the Chinese market…. Other officials were greatly upset by the news, saying the Chinese government could be stockpiling viruses for some sort of information warfare…

A Glimpse of Cyberwarfare, Governments Ready Information-Age Tricks to Use Against Their Adversaries
Warren P. Strobel with Richard J. Newman, World Report, March 13, 2000
At first, the urgent phone call from the U.S. Transportation Department confounded Cheng Wang, a Long Island-based webmaster for Falun Gong, the spiritual movement that has unnerved Chinese authorities. Why did the department think his computers were attacking theirs? The answer turned out to be startling. The electronic blitz hadn't come, as it seemed, from various Falun Gong Internet sites. Rather, someone had lifted their electronic identities. Computer sleuths followed a trail back to the XinAn Information Service Center in Beijing — where an operator identified it as part of the Ministry of Public Security, China's secret police.

ISPs Accuse China of Infowar
infowar.com. 30.Jul.99.PDT
Two Canadian ISPs said Friday that their networks were attacked this week by Chinese government crackers with a political agenda. "The hack attempts I could trace [originated with] Chinese government offices in Beijing," said Eric Weigel, director of Bestnet Internet, a Hamilton, Ontario-based ISP. Weigel said he suspected that the "denial-of-service" attack,

Exhibit 7. China's Information Warfare (Continued)

which ended at 4 a.m. EST Friday, was motivated by his organization's hosting a Web site for a religious group outlawed in China.

Chinese Military Calls for Special Hacker Force
Staff, Newbsytes, Hong Kong, August 4, 1999
Chinese military is looking for a few good men — to be trained in the art of government sponsored hacking over the Internet…called for the development of a hacking…capability made up of civilian experts and specially trained military personnel that could engage in online and Internet warfare.

Comments:
The rationale that Network Associates, Symantec, and Trend-Micro, according to the *Wall Street Journal* article, make up about 75 percent of the estimated $1.2 billion market of anti-virus products. Again, according to the article, these companies met with the Chinese and believed the Chinese explanation that they needed the software to run tests on software before it was marketed. Most information security, information warfare, and related professionals probably are saying: "So what else is new — money over national security and information security." One wonders if the company executives ever heard the communist doctrine that says something to the effect: "To lie in the furtherance of Communism is moral"? If so, would they care or continue to rationalize giving information and tools to an adversary to possibly use against us? Cold War rhetoric? The above information on China should put that question to rest.

China is perhaps the best example of a country that has embraced IW. To prevent external influences from undermining its implementation of Communism, China has tried unsuccessfully to isolate itself as much as possible from the outside world. Still, math, electronics, computing, and the laws of nature are the same on either side of its borders. It is the cultural difference that is important. An argument can be made that China is the first to codify IW. Sun Tzu explained many IW principles in his writings over 2500 years ago. Two important points come out of this. One is that IW principles have survived the test of time. The other is that IW does not require electronics and computers.

Over the past decade, China's IW thoughts and capabilities have caught the attention of the West. Examples of modern IW philosophical insight and intellectual understanding are expressed in two books. A 360-page book published in 1991 by two experts in China's intelligence community details how China's government acquires U.S. national security technology and information. *Sources and Techniques of Obtaining National Defense Science and Technology Intelligence (Guofang Keji Qingbaoyuan ji Huoqu Jishu)*, written by Huo Zhongwen and Wang Zongxiao for Kexue Jishu Wenxuan Publishing Company in Beijing, explains not just no, low, and high technology means of gathering intelligence, but also how the collected intelligence can be analyzed and shared for political and diplomatic, economic, and military uses.

Unrestricted Warfare was written by Qiao Liang and Wang Xiangsui for PLA Literature and Arts Publishing House in Beijing and published in February 1999. Their book offers their country's leaders a method for fighting war against the United States. In 1996, China backed down in its continued dispute over Taiwan to a show of military force by the United States. Liang's and Xiangsui's view of warfare makes Nazi Germany's unrestricted submarine warfare look tame. They realized that a military force-on-force war with the United States was unwinnable. Giving credit to Sun Tzu and the lessons of history, they detailed 24 types of warfare, ideally conducted simultaneously to achieve maximum effect, to defeat the United States. "Unrestricted war is a war that surpasses all boundaries and restrictions," they write. "It takes nonmilitary forms and military forms, and creates a war on many fronts.

Exhibit 7.   China's Information Warfare (Continued)

It is the war of the future."[9] If the combination of terrorist attacks on the Pentagon and World Trade Center, followed by the anthrax scare, is any indication, the two authors may well have a winning formula.

Unrestricted warfare is guerilla warfare, also known as asynchronous warfare, using the full spectrum of no, low, and high technology capabilities. Guerilla warfare uses a martial arts dictum of striking against weakness. This concept is also embodied in the principles of war used by many countries. One Chinese expectation may be that the United States will spend money on defensive measures, intelligence, personnel, training, economic analyses, and other areas, all the time diverting money away from offensive systems that could be used against China, and allowing China time to close the military gap. This is an excellent example of IW.

To back their rhetoric, the Chinese are fielding IW forces. Taiwan and U.S. defense officials have noticed that the Chinese government is considering creating a fourth branch of its armed services devoted just to information warfare.[10] Taiwan's Defense Minister Tang Fei said to Parliament, "Efforts by the Chinese Communists on computer viruses and magnetic pulses are shifting into high gear. They are planning to develop a computer virus that would be able to paralyze the rival's command and telecommunications systems." Tang went on to say that he expected Beijing to gain an overwhelming advantage over Taipei in electronic and information warfare by 2010.[11] These comments are supported by exhortations made by Chinese military leaders. "We can make the enemy's command centers not work by changing their data system," wrote Maj. Gen. Pan Junfeng. "We can cause the enemy's headquarters to make incorrect judgments by sending disinformation."[12] Incorrect judgments could be sending 155-mm artillery shells to 105-mm units because a database was altered, bombing friendly forces because false imagery was inserted in the reconnaissance downlink, or radioing troops to go to a wrong location to engage the enemy, permitting enemy forces to penetrate unscathed.

The Chinese Army's official newspaper, *The Liberation Army Daily,* wrote "Essential to an all-conquering offense is the development of software and technology for Internet offensives so as to be able to launch attacks and countermeasures on the Internet, including information paralyzing software, blocking software, and deception software. Weapons such as electronic bombs to saturate the enemy's cyberspace and software to scan the Internet, break codes, steal data, and enable anti-follow up measures will deny command and control. To ensure Internet warfare can play the maximum role in war, it is essential to integrate it with other combat actions. Modern warfare cannot win without the Internet, nor can it be won just on the Internet. In the future there must be coordinated land, sea, air, space, electronic, and Internet warfare. An Internet Force is very likely to become another military branch following the army, air force and navy, and it will shoulder the formidable task of protecting Internet sovereignty and engaging in Internet warfare."[13]

Focusing on "electronic bombs," which fall in the high-energy radiation frequency (HERF) category, the former Soviet Union and the United Kingdom are reputed to have such weapons. India fears Pakistan will use E-bombs on Bangalore and their other Silicon Valley-like cities. A rudimentary model, admittedly not military grade, can be built for as little as $400, as demonstrated at InfoWarCon in 2000 and the Army Research Laboratory in 2001. Although there are defenses, at this time they are expensive (e.g., special grounding and shielding, such as a Faraday cage). If deployed, the E-bombs and HERF weapons will have a direct negative effect on commerce, communications, power, transportation, and other vital aspects of the military and economy. If triggered intermittently and in a broad spread over a several-month period, it would be very difficult to conduct a meaningful response to aggression and to sustain a recovery.

Exhibit 7.   China's Information Warfare (Continued)

High-ranking officers in U.S Pacific Command confirmed that the People's Liberation Army (PLA) is pursuing IW. Since 1997, U.S. intelligence reported the PLA simulating computer attacks on key U.S. command and control systems, logistical choke points, critical communications nodes, and other sites.[14] Although striking military targets is a given, the military relies heavily on commercial systems, such as communications, power, transportation, and financial systems. All the critical infrastructures have deep cascading effects and are interdependent. The PLA knows this and no doubt would attack not in a one-time strike, but in a series of targeted attacks over several weeks or months. The combination of military and commercial strikes would ideally be designed to prevent the U.S. from being able to mount any meaningful kind of response. China is not the only country capable of this level of attack, so a sense of urgency should be associated with critical infrastructure protection.

The Chinese (and Asian) perspective of time is to allow events to unfold over years to attain desired goals and objectives. This is in juxtaposition to the United States and some other Western countries that are more focused on near-term actions. By sacrificing the long-term for the short, a degree of strategic perspective is given up. In this regard, it is best to learn vicariously and meld the best approaches from all sources. China represents an outstanding source.

New Zealand

New Zealand is one of those island nation-states where one seldom hears of any problems that others are facing, vis-à-vis computer attacks and information warfare. However, in today's globally connected nation-states, no one is immune to attacks by hackers, activists, and other nation-states. However, New Zealand is not one of those nation-states involved in a litany of conflicts and potential conflicts. Its only territorial claim is in Antarctica (Ross Dependency). However, unlike the other nation-states, there does not appear to be a real potential for hostility between New Zealand and other nation-states over that territory.

According to the *World Factbook*,[4] the British colony of New Zealand became an independent dominion in 1907 and supported the United Kingdom militarily in both World Wars. New Zealand withdrew from a number of defense alliances during the 1970s and 1980s. In recent years, the government has sought to address longstanding native Maori grievances.

It is logical to assume that if a nation-state such as New Zealand is not involved in any territorial or trade conflicts with other nation-states, then information warfare in the form we have been discussing thus far is pretty remote. So that leaves attacks by terrorists, activists, cyber-fraudsters, and global hackers. Of the four, the only real threats as of this date appear to come from the fraudsters and global hackers. Therefore, New Zealand, like other nation-states, is setting up a "cyber threat center":

> ***New Zealand Center To Combat Cyber Threats***
> *Adam Creed, Newsbytes, Wellington, New Zealand, August 8, 2001*
> *New Zealand government…will set up a government unit dedicated solely to protecting the nation's critical infrastructure from cyber threats by Internet hackers or computer viruses. "…distance from the large population centers of the world has provided us with some protection against threats to our infrastructure…. Those days are past."*

These comments, made by New Zealand's State Services Minister Trevor Mallard, hold true for many nation-states besides New Zealand. According to the news article, this Centre for Critical Infrastructure Protection (CCIP), to be established in 2002, will not only be responsible for protecting the nation-state's critical infrastructure (e.g., transportation, financial sectors, energy, and telecommunication systems) and its legal systems, but also as a coordinating center between the public and private sectors. It appears to be similar to what other nation-states are doing at the government level to begin to protect their critical infrastructures, of course including the information infrastructure. Although no mention is made of information warfare, that may be for several reasons, including:

- It may be that it is understood to have that responsibility.
- It may be that the news article just did not mention it.
- It may be that it is considered a national security matter, and thus not discussed.
- It may that such responsibility will fall on the military.

Because the CCIP will be under the Government Communications Security Bureau, it is assumed that there will be at least some coordination between the military and this organization. This points to an interesting predicament that many nation-states are facing and that is: who is responsible for the protection of the national infrastructure of a nation-state? Of course, it depends on the type of government in power to some extent. However, in those nation-states in which the infrastructure is owned by private businesses, does the government have the right, duty, and/or responsibility to dictate the infrastructure protection requirements to these private enterprises? Most think they do not. There are probably several reasons for this:

- The government agencies have done a poor job of protecting their own information infrastructure; therefore, they are in no position to tell others how they must do it.
- Protection costs money, which takes away from profits. Government agencies are not profit motivated; therefore, it is quite likely their approach would not consider costs a high priority when dictating protection requirements.
- Government involvement in private enterprise through regulations and laws has already been shown to be bureaucratic, complicated, and slow to change — changing slowly in today's rapid pace, high technology-driven world does not work.

Although New Zealand is an island, also "down under" in cyberspace, one cannot hide:

Japan Implicated in Kiwi Hacking Probe
John Leyden, The Register, August 13, 2001
Japanese government agency has been implicated in attempts to hack into a medical research institute in New Zealand. Kiwi news service NZOOM cites security consultant Philip Whitmore from Pricewaterhouse-Coopers to support its allegations of state sponsored espionage against New Zealand's private sector.

> **Chinese Spies Active in NZ Claims Web Site**
> *Phil Taylor, Sunday Star Times (NZ), January 22, 2001*
> *Chinese spies are operating undetected in the Pacific under the noses of New Zealand intelligence agencies, according to a Web site specialising in intelligence matters. An article on the Web site cryptome.org claims Chinese intelligence agents have worked in the area for years, working against Taiwanese efforts to gain diplomatic recognition with smaller Pacific nations.*

Australia

Australia is a nation-state that quietly supports the efforts of other friendly nation-states in their information warfare-related activities such as ECHELON. Australia has an active cyber-crime law enforcement program because it also has an active hacker population.

According to the *World Factbook*,[4] Australia became a commonwealth of the British Empire in 1901. It was able to take advantage of its natural resources to rapidly develop its agricultural and manufacturing industries and to make a major contribution to the British effort in World Wars I and II. Long-term concerns include pollution, particularly depletion of the ozone layer; and management and conservation of coastal areas, especially the Great Barrier Reef. A referendum to change Australia's status, from a commonwealth headed by the British monarch to an independent republic, was defeated in 1999. Disputes: international; territorial claim in Antarctica (Australian Antarctic Territory)

Summary

As pointed out in the quote of Dr. Henry Kissinger,

> *...But in Asia, the various states look at each other as political and geopolitical rivals. War is not impossible, and therefore in that region we have to pay more attention to balance of power and equilibrium....*
> *As Asia grows in its development, both economically and militarily, they will continue to take advantage of high technology as it relates to information warfare.*

The primary hotbeds of conflict continue to be South Korea versus North Korea, and China versus Taiwan. As we have shown, these conflicts will undoubtedly draw in the major world powers of the United States and Russia. They will bring with them their sophisticated information warfare weapons. Once these weapons are unleashed, they will affect the GII and NIIs throughout the world. Global financial systems will also suffer, and thus the world economy as malicious codes ricochet throughout networks damaging non-combatant and neutral networks. Containment will be difficult and may even result in global information warfare of offensive and defensive battles. As everyone tries to defend its systems from attacks, the warring parties will continue to unleash their industrial and information period weapons. In these wars, information warfare will play more than a supportive role.

Notes

1. http://www.cnn.com/2001/COMMUNITY/07/31/kissinger.cnna/index.html.
2. http:www.infowar.com, January 14, 2001.
3. Naisbitt, John. *Megatrends Asia: Eight Asian Megatrends that Are Reshaping Our World,* Simon and Shuster, New York, 1996.
4. CIA *World Factbook* (http://www.cia.gov/cia/publications/factbook/index.html).
5. http://www.chosun.com/w21data/html/news/200108/200108260133.html.
6. This and further information can be found at http://www.mac.gov.tw/english/MacPolicy.
7. http://www.taipeitimes.com/news/2001/01/03/story/0000068206.
8. From Michael Pillsbury's book, *China Debates the Future Security Environment,* National Defense University Press, Washington, D.C., 2000.
9. China Ponders New Rules of 'Unrestricted War', http://www.washingtonpost.com/wp-srv/inatl/longterm/china/china.htm, John Pomfret, The Washington Post Foreign Service, August 8, 1999.
10. Cyber-War Fears Continue, Emergency Response & Research Institute, March 5, 2000.
11. Taiwan behind China in Information Warfare, Defence Minister Warns, *Asia Free Press,* Taipei, May 5, 1999.
12. Pentagon Study Finds China Preparing for War with U.S., Bill Gertz, *The Washington Times,* February 2, 2000.
13. Bringing Internet Warfare into the Military System Is of Equal Significance with Land, Sea, and Air Power, *The Liberation Army Daily,* November 17, 1999.
14. China Focuses on Asymmetric Warfare, Michael Buonagurio, AntiOnline, March 12, 1999 (http://www.AntiOnline.com).

Chapter 9

Information Warfare: Middle East Nation-States

> *In the Middle East, the conflict is more ideological and religious, and there the conflicts take on a much more emotional character...*[1]

—Dr. Henry Kissinger

This chapter discusses information warfare (IW) as it relates to Middle East nation-states. A general overview is presented along with more details and discussions of selected nation-states. For our purposes, the Middle East will cover some nation-states that may be considered by some as being in the Middle East, Asia, or even Europe (e.g., Turkey).

The Middle East

When speaking of the Middle East, we are talking about the land occupied by the following 21 nation-states[2]: Algeria, Bahrain, Egypt, Iran, Iraq, Israel, Jordan, Kuwait, Lebanon, Libya, Morocco, Oman, Palestine, Qatar, Saudi Arabia, Sudan, Syria, Tunisia, Turkey, United Arab Emirates (UAE), and Yemen.

Exhibit 1 provides a map of the Middle East.

When we in the West think of the Middle East, we generally think of three things:

- Oil
- Islam
- Terrorism

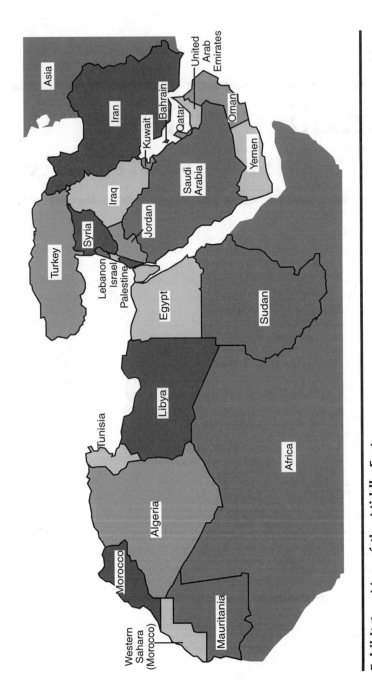

Exhibit 1.   Map of the Middle East

It is unfortunate that terrorism, as well as regional and global conflicts, sprout from that region of the world. The Middle East is home to some of the world's oldest cultures, religions, arts, sciences, and a truly great religion, Islam. Unfortunately, such violence overshadows the many wonders of the region.

Other than the use of Web sites, chat rooms, and e-mails to air their views, there is little in the way of actual offensive information warfare attacks other than some unsophisticated denial-of-service, Web site defacement, and virus attacks. The conflicts and terrorist acts are generally of the Industrial Age variety and little in the way of information warfare is practiced in the region, and for good reason. The nation-states of the region do not have a well-developed NII (national information infrastructure), and lag behind the West in becoming information-based and information-dependent nation-states.

A report conducted by McConnell International, in conjunction with the World Information Technology and Services Alliance, looked at the "E-readiness" of nation-states. Those in the Middle East were listed as lacking "the basic infrastructure needed for an online economy to flourish." Their measurements considered Internet access, online security, business climate, and workforce skills. Those specifically noted in the Middle East were Egypt and Saudi Arabia. The other nation-states in this region are probably less ready for E-commerce because they also lack the necessary infrastructure. However, they do have the one very important natural resource — oil. Because of oil, they are able to maintain their independence to a very large extent. Therefore, there is little incentive for the governments of the region to develop their NIIs. After all, who needs E-commerce when one has oil gushing out of the ground?

There may be other reasons for their lack of an NII. These may include:

- Wanting to remain independent of foreign assistance and influence
- Not being able to afford such a massive undertaking
- Being fearful of what the Internet might do to their culture, religion, and citizens
- The terrain and distribution of population
- A combination of, or all of the above

If they lack the necessary infrastructure for E-commerce at a time when economic power is very important in the world of nation-states for projecting power and influence, then they are not as susceptible to information warfare. The fact is that they may not be prepared for E-commerce does not mean that they are not prepared nor involved in information warfare. Countries of the Middle East have for some time used IW, albeit through the use of tactics that do no require high technology. One just has to look at the Web sites of the Arabs and the Israelis to see that they are involved in IW on several fronts. Among these is the establishment of Web sites expressing their side of the conflict to win the support of world opinion. Web sites on both sides of the conflict have been periodically attacked.

Areas of Disagreement That Have and Could Continue to Lead to Warfare

The more well-known conflicts in the Middle East that have joined what seems to be the continuous Israel–Palestinian–Arab conflicts are the Iran–Iraq war, and the Kuwait Coalition conflict known as Desert Shield/Desert Storm or the Gulf War. Because the Arab nation-states, and even Israel to a certain extent, do not have a sophisticated NII, the use of IW weapons to attack an adversary's NII is somewhat limited. What are not limited are the reasons for conflict as noted in the United States Central Intelligence Agency's *World Factbook*:[3]

- *Algeria:* part of southeastern region claimed by Libya
- *Bahrain:* the territorial dispute with Qatar over the Hawar Islands and the maritime boundary dispute with Qatar are currently before the International Court of Justice (ICJ)
- *Egypt:* asserts its claim to the "Hala'ib Triangle," a barren area of 20,580 square kilometers under partial Sudanese administration that is defined by an administrative boundary which supersedes the treaty boundary of 1899
- *Iran:* Iran and Iraq restored diplomatic relations in 1990 but are still trying to work out written agreements settling outstanding disputes from their eight-year war concerning border demarcation, prisoners of war, and freedom of navigation and sovereignty over the Shatt al Arab waterway; Iran occupies two islands in the Persian Gulf claimed by the UAE: Lesser Tunb (called *Tunb as Sughra* in Arabic by UAE and *Jazireh-ye Tonb-e Kuchek* in Persian by Iran) and Greater Tunb (called *Tunb al Kubra* in Arabic by UAE and *Jazireh-ye Tonb-e Bozorg* in Persian by Iran); Iran jointly administers with the UAE an island in the Persian Gulf claimed by the UAE (called *Abu Musa* in Arabic by UAE and *Jazireh-ye Abu Musa* in Persian by Iran) — over which Iran has taken steps to exert unilateral control since 1992, including access restrictions and a military buildup on the island; the UAE has garnered significant diplomatic support in the region in protesting these Iranian actions; Caspian Sea boundaries are not yet determined among Azerbaijan, Iran, Kazakhstan, Russia, and Turkmenistan
- *Iraq:* Iran and Iraq restored diplomatic relations in 1990 but are still trying to work out written agreements settling outstanding disputes from their eight-year war concerning border demarcation, prisoners of war, and freedom of navigation and sovereignty over the Shatt al Arab waterway; in November 1994, Iraq formally accepted the UN-demarcated border with Kuwait, which was spelled out in Security Council Resolutions 687 (1991), 773 (1993), and 883 (1993); this formally ends earlier claims to Kuwait and to the Bubiyan and Warbah islands, although the government continues periodic rhetorical challenges; dispute over water development plans by Turkey for the Tigris and Euphrates rivers
- *Israel:* the West Bank and Gaza Strip are Israeli-occupied, with current status subject to the Israeli–Palestinian Interim Agreement — permanent status to be determined through further negotiation; Golan Heights is Israeli-occupied; Israeli troops in southern Lebanon since June 1982

- *Jordan:* none
- *Kuwait:* in November 1994, Iraq formally accepted the UN-demarcated border with Kuwait, which was spelled out in Security Council Resolutions 687 (1991), 773 (1993), and 883 (1993); this formally ends earlier claims to Kuwait and to the Bubiyan and Warbah islands; ownership of Qaruh and Umm al Maradim islands disputed by Saudi Arabia
- *Lebanon:* Israeli troops in southern Lebanon since June 1982; Syrian troops in northern, central, and eastern Lebanon since October 1976
- *Libya:* maritime boundary dispute with Tunisia; Libya claims about 19,400 square kilometers in northern Niger and part of southeastern Algeria
- *Morocco:* claims and administers Western Sahara, but sovereignty is unresolved and the UN is attempting to hold a referendum on the issue; the UN-administered cease-fire has been in effect since September 1991; Spain controls five places of sovereignty (*plazas de soberania*) on and off the coast of Morocco — the coastal enclaves of Ceuta and Melilla that Morocco contests, as well as the islands of Penon de Alhucemas, Penon de Velez de la Gomera, and Islas Chafarinas
- *Palestine:* Although not listed in the CIA *World Factbook*,[3] it is listed here for information purposes. The exact, final borders of this potential nation-state has yet to be resolved, and may never be resolved based on ongoing conflicts. It is listed here because it is nation looking for nation-state status.
- *Oman:* northern boundary with the UAE has not been bilaterally defined; northern section in the Musandam Peninsula is an administrative boundary
- *Qatar:* the territorial dispute with Bahrain over the Hawar Islands and the maritime boundary dispute with Bahrain are currently before the International Court of Justice (ICJ); June 1999 agreement has furthered the goal of definitively establishing the border with Saudi Arabia
- *Saudi Arabia:* large section of boundary with Yemen not defined; location and status of boundary with UAE is not final, *de facto* boundary reflects 1974 agreement; Kuwaiti ownership of Qaruh and Umm al Maradim islands is disputed by Saudi Arabia; June 1999 agreement has furthered the goal of definitively establishing the border with Qatar
- *Sudan:* administrative boundary with Kenya does not coincide with international boundary; Egypt asserts its claim to the "Hala'ib Triangle," a barren area of 20,580 square kilometers under partial Sudanese administration that is defined by an administrative boundary which supersedes the treaty boundary of 1899
- *Syria:* Golan Heights is Israeli-occupied; dispute with upstream riparian Turkey over Turkish water development plans for the Tigris and Euphrates rivers; Syrian troops in northern, central, and eastern Lebanon since October 1976
- *Tunisia:* maritime boundary dispute with Libya; Malta and Tunisia are discussing the commercial exploitation of the continental shelf between their countries, particularly for oil exploration
- *Turkey:* complex maritime, air, and territorial disputes with Greece in Aegean Sea; Cyprus question with Greece; dispute with downstream riparian states (Syria and Iraq) over water development plans for the Tigris and Euphrates rivers; traditional demands regarding former Armenian lands in Turkey have subsided

- *United Arab Emirates (UAE):* location and status of boundary with Saudi Arabia is not final, *de facto* boundary reflects 1974 agreement; no defined boundary with most of Oman, but Administrative Line in far north; claims two islands in the Persian Gulf occupied by Iran: Lesser Tunb (called *Tunb as Sughra* in Arabic by UAE and *Jazireh-ye Tonb-e Kuchek* in Persian by Iran) and Greater Tunb (called *Tunb al Kubra* in Arabic by UAE and *Jazireh-ye Tonb-e Bozorg* in Persian by Iran); claims island in the Persian Gulf jointly administered with Iran (called *Abu Musa* in Arabic by UAE and *Jazireh-ye Abu Musa* in Persian by Iran) — over which Iran has taken steps to exert unilateral control since 1992, including access restrictions and a military buildup on the island; the UAE has garnered significant diplomatic support in the region in protesting these Iranian actions
- *Yemen:* a large section of boundary with Saudi Arabia is not defined

As one can see, there are many reasons for conflicts in this part of the world, and they do not even include the reasons for possible conflicts with the West and other nation-states around the world.

The Iran–Iraq War

The Iran–Iraq war was an "old-fashioned" industrial period type of war with massive killings, other physical destruction, guns, soldiers, bombs, rockets, ships, and aircraft all playing a part. Information warfare was basically limited to the propaganda tools of radio, television, and newspapers.

Therefore, the fact that people and nation-states in the Middle East share many things in common, such as language, religion, and culture, does not mean that they will not have conflicts. Do they not see the possibilities and benefits of joining together with other Arab nation-states to form a powerful Arab Union? This may some day be possible; however, there are differences that keep them apart. Because they do not have democratically elected governments, their government's leaders' thirst for power and control put themselves ahead of the good of their people and their nation-state. The combined resources of Iraq and Iran would make a formidable power block in the Middle East.

OPEC (Oil and Petroleum Exporting Countries) has surprisingly been able to survive such conflicts. One supposes that this is because each of these nation-states' leaders sees that this would be in their own best interest. This is logical because they share the wealth gained by being part of OPEC, with of course handouts doled to their loyal followers, with some trickle-down benefits to keep the masses in place.

Desert Shield/Desert Storm: The Gulf War

Although Desert Shield/Desert Storm was not officially a declared war, a political trend that began with the "police action" of Korea, it was in fact a typical Industrial Age conflict of bombs, tanks, ships, and infantry supported by some information warfare tactics and weapons systems. However, make no mistake about it, it was *not* an IW-based war. During this conflict, it was alleged that a group of Dutch hackers had volunteered to assist the Iraqi government by attacking the logistics computers systems of the U.S. military. Of course, the hackers denied that they had volunteered to support the Iraqis in the conflict. Ironically, it has also been alleged that the Iraqi government turned them down because the government thought it was a trick.

For the sake of discussion, assume that the Dutch hackers had indeed volunteered to assist Iraq in time of war against the coalition of forces amassed against Iraq. Also assume that the Iraqi government accepted their offer to use IW tactics and offensive weapons. It does not matter that the hackers were paid (in the form of money, computers, new cars) or not paid. They had now become the allies of Iraq. As they say, "You are either for me or against me." Well, in this scenario, it is quite clear that an alliance had been formed and that alliance, using IW concepts, is no different than any other wartime alliance. That being the case, the hackers were then "fair game" as information warfare soldiers for an adversary. Thus, could they not then be legally (using warfare rules of conflict) attacked, to the point of their equipment being destroyed and face charges for aiding the enemy?

Could the coalition forces not have sent a combat team into the homes of the hackers? Why not? The hackers were actively working for the enemy, were armed with their chosen weapons of warfare, and could have been denying the coalition forces the use of logistics systems that were vital to the shipment of soldiers, weapons, ammunition, and other supplies to the war zone. For example, what if there were delays in the shipment of ammunition to the infantry soldiers, a major battle ensued, the soldiers ran out of ammunition, and the soldiers were either killed or captured? Thus, the Dutch hackers would have been responsible for the death of coalition soldiers in the battlefield. What would have happened if hackers had broken into the military database and changed target coordinates that caused weapons to kill friendly forces? What would have happened if they had broken into the military blood database and mixed up the blood types in the soldiers' records? The blood transfusions given to wounded soldiers would have caused their deaths. Maybe as one perfects the control and direction of power surges, a "limited surgical" strike — in this case maybe to send a power surge to the hackers' CRT, causing it to explode and thus "fragging" the hacker — would be appropriate.

This is but one example of the use of IW tactics in wars between nation-states where individuals or groups can be drawn into the conflict from anywhere in the world. Has this changed from warfare in the industrial period of our nation-states? No, not really. The change is the environment and the choice of weapons. However, the reasons for the conflicts and all the basic human elements still hold true.

Two IW tactics were used by at least one of the coalition forces (the United States) and against U.S. citizens:

- Historically, the United States has sent military forces into conflicts to "make the world safe for democracy." What is also interesting about the Gulf War is that there were no political messages being given to the citizens of the coalition nation-states saying that the home forces were going to free a democratic Kuwait. Kuwait is a "nominal constitutional monarchy."[3] Military forces were used and lives lost, with apparently no thought of supporting the people of Kuwait in forming a freely elected, democratic form of government. Did not the coalition forces led by the United States in fact lead the effort to put the same people in power that were there prior to the Iraqi invasion?

- In the war in South Vietnam, the military forces learned a valuable lesson: to keep the news media "under control" by controlling their access to information. Some may refute this; however, one only has to look at how the media was handled by the United States during Desert Shield/Desert Storm. The briefings were of course conducted and the daily ritual showed smart weapons apparently hitting their target every time. Later, it was learned that the smart weapons were not always that smart. The same is true for the war in Afghanistan.

Both of these examples show an old and basic form of IW: control information and, as part of that control, limit information, divert attention away from some information, and do not mention other information that may adversely impact the policy of the nation-state.

The View of the Israeli Military Intelligence Services (AMAN)

According to janes.com,[4] they have seen a top-secret report of Israel's military intelligence service (AMAN) that provides some insight into Israel's thoughts which impact how they would look and prosecute a war, including:

- AMAN believes that peace agreements, such as with Palestinians and Syrians, will come with increases in terrorism.
- The Arab world is split.
- The United States is considered a great power in the region.
- The United States is allied with Egypt, Jordan, Saudi Arabia, and Turkey.
- Iraq is isolated and no longer militarily strong.
- Arabs are pursuing a policy of peace based, at least in part, on Israel's superior armed forces.
- Arabs citizens have put economic priorities over conflict with Israel.
- If the Palestinians' high expectations are not realized soon, they will be tempted to resume terrorism and rioting (this has already occurred!).
- Israel's greatest threat in the future will be the possession of nuclear weapons by states such as Iran and Iraq.
- Although Israel sees no immediate threat of war with Egypt, there is concern that the Egyptians have not come to terms with having peace with Israel.
- Possibly one of the most serious potential issues for future conflict with Israel, as they see it, is: The most difficult problem with which the regimes in the region are contending is economic distress. The birth rate is high; and in the Arab countries, a generation of despairing young people has grown up. Arab governments are not training staff for new industries. Israel and its GNP are a constant reminder of Arab shortcomings.

It is believed that Israel is continuing to improve its NII. As it does, and conflicts continue with its Arab neighbors, it will become more of a target in its conflicts with Arab nation-states. While at the same time, the NIIs of the region's nation-states will improve at a much slower rate and thus provide less of an IW target in time of war. Consequently, in any hot conflict between Israel and one or more Arab nation-states, it is expected that the Arab nation-states may be able to win the conflict on the IW battlefield. However, thus far, the IW conflicts have been relegated to Web sites expounding their positions. For example:

- For a Palestinian view of the situation:
    - http://www.palestinehistory.com/
    - http://www.intefada.com/
    - http://www.palestine-info.net/arabic/index.shtml
    - http://www.sabiroon.org/
    - http://archives.star.arabia.com/001019/TE1.html
- For an Israeli view of the situation:
    - http://www.israel.org/mfa/home.asp
    - http://www.aipac.org/
    - http://www.jpost.com/

Prelude to Attacking a Middle East Nation-State: Methodology for Analysis of Middle East Nation-State Internet Sites

The "hottest spot" in the world these days is the Middle East. Therefore, it was chosen for a case study in an intelligence collection project of a potential adversary as a prelude to conflict where information warfare would play a role — not a major role because of reasons stated earlier, but a role nonetheless. The information provided is factual; however, the names of companies, individuals, and even the Middle East nation-states involved are not provided. This was done, based on today's charged atmosphere, due to the latest terrorist attacks in the United States; it is not our intention to imply that there is any wrongdoing by anyone noted during this information warfare case study.

Description

This project was implemented to test the methodology that can be used to locate, review, research, and analyze "The Middle East Nation-State" (TMENS) oriented Internet sites as possible IW targets.

Locate

The primary method used to locate sites was to use Internet search engines, such as Yahoo, Lycos, Alta Vista, with keywords of TMENS (TMENS is used to represent "The Middle East Nation-State"). This yielded approximately 11,800 different results (Lycos). Search results that were obviously (to the search engines) related to news organizations (e.g., CNN and Time), book reviews, historical accounts of the Gulf War, and veteran issues were filtered, resulting in approximately 3500 (Alta Vista) remaining search results. Foreign language sites were also excluded for technical and linguistic reasons, such as the ability to properly display foreign language fonts and the inability to correctly translate a site's contents.

Note: In an actual case, one would of course want language specialists to translate the information in Arabic and Hebrew as they would probably provide more important information than would sites that used the English language, for instance.

Review

Of the remaining sites, those ranked (by the search engines) as being most relevant were reviewed. Most sites were personal home pages or TMENS "hate" pages. Furthermore, the manner in which sites were ranked by the search engines was based on the number of keyword occurrences rather than the actual content value of each site.

Research

Three characteristics of each site were researched: purpose, provider, and links.

Purpose

It was assumed that each site on the Internet was designed with a specific purpose in mind. By analyzing the content of each site, it is possible to determine the purpose or motivation for a site's existence (and determine if they are "friend" or "foe"). In addition to analyzing the content of each site, attention was paid to the overall architecture of the site design — keeping in mind that a site can be used for intelligence collection (e.g., honey pot) as well as dissemination.

Provider

The identity of the providers that hosted each site was also researched. The purpose of this research was to determine if there were any formal or casual relationships between a particular provider and any of the sites that it hosted.

Provider identification was accomplished by matching either the domain name or the IP address of a particular site against the registration databases of the InterNIC or RIPE. Once a provider was identified, its corresponding database entry was recursively queried so that possible relationships to other individuals, sites, and providers could be discovered.

The World Wide Web (WWW) home page of each provider was also analyzed to determine if the provider tended to solicit particular kinds of site hosting. Additionally, search engines were used to determine the kinds of sites hosted by each provider and if any formal or casual relationship existed between them.

Links

Often, a site would contain external links to other sites whose contents were related to that of the original site. External WWW links to other sites were investigated to determine if their destinations should also be researched. Links between similar sites were also analyzed to determine if any relationship existed. For example, research found links to Web sites of numerous groups that some have identified as terrorist groups. Also, the links provided a "flavor" of the political views of those managing the site (e.g., pro-Iraq, anti-Israel, and anti-United States).

The Case Study: TMENS Top-Level Domain (TLD)

The TLD for TMENS (.ns) was found in the InterNic database as entry "IX14-DOM." It is currently registered to ABC Corp., in Somewhere, U.S.A. The administrative contact for the domain was identified, along with his company name and location, which physically was in the Middle East nation of TMENS. The technical contact was also identified along with his company name and location. The primary Domain Name Server (DNS) is identified as a machine call "Righteous" on the RIGHTEOUS.NET network.

ABC Corp., according to its Web home page, is a combination of Internet service provider (ISP) and high-technology service provider. In particular, it resells access to the "MiddleEastSat" system.

However, research also revealed that ABC Corp. has only one ISP customer, XYZ Corp. of Small Town, U.S.A., an Islamic high-technology corporation that also has ties to other Middle East nation-states.

Research of the RIGHTEOUS.NET domain revealed that it is registered to Good Corp., which coincidentally has the same address as ABC Corp. XYZ Corp. did not have a home page, but research revealed that it hosts many Islamic-oriented sites that might be considered by some as "pro-terrorist."

During the research of XYZ Corp., one could search for default configuration pages. Assume one is discovered on one of its servers. The page may indicate that the system was XXXX based. Using anonymous FTP and the default log-in password, it may be possible to log in to the DNS host for RIGHTEOUS.NET. The password file may be located and it may even be determined when it was last modified. If the password file is discovered to be "world readable," it could be downloaded for analysis and later use. If not, password search tools could be used to break in and obtain the password file — of course, then leaving a trap door so one could get back in if the password files were ever updated.

An analysis of the password file could possibly determine that some of the users were individuals, while others were organizations. One may even find that the XXXX Software entry in the password file is a disabled vendor maintenance account.

UUCP (UNIX to UNIX Copy) is a method used to network computers over telephone lines. It was the primary transport for the Internet prior to the construction of today's TCP/IP networks, and is still being used to connect remote systems to the main Internet backbone. The RIGHTEOUS.NET system, according to its password file, supports three active UUCP accounts: two of them belong to named users (SA Systems and Anon Mous) and the other one is set up for anonymous system log-in.

A check of the InterNIC database revealed that the "SA System.com" SLD (Second-Level Domain) is configured to route its traffic to the UUCP account at ABC Corp. This is despite the fact that OMENS (Other Middle East Nation-State) using SA System.com has been directly connected to the Internet since 1996.

Anon Mous does not appear to be the name of an organization. It is most likely the name of an individual with a dial-up UUCP node that peers to ABC Corp. The anonymous UUCP log-in can support many dial-ups. It was not possible to determine the number or types of systems using that particular account without first compromising the security of the XYZ Corp. system. This could be easily done using information warfare offensive tools (in this case, simple hacker tools found on the Internet).

It is speculated that through this account TMENS TLD traffic passes on to its final destination. Monitoring such a system may provide valuable information. Furthermore, the compromise of the password file would allow misinformation to be sent under valid users' account names. This would be especially useful because in the event of conflict, the adversary's command and control and news media sources would have been disabled.

From what was learned, ABC Corp. is the primary provider for the TMENS TLD. It also hosts many other Arabic and Islamic sites. This could be a possible "choke point." If so, disabling or controlling this network may have an exponential adverse impact on the adversary.

Traffic for the TMENS TLD appears to be configured for UUCP transport mode over telephone lines. Monitoring of the UUCP traffic should be possible. Setup of the TMENS TLD has been within the last four months. Perhaps we are seeing the start of TMENS' attempt to gain Internet connectivity. No passwords on the RIGHTEOUS.NET server have been changed since June 1998, and the system home page has not been configured. This indicates that vulnerabilities may exist that would allow a compromise of system security.

Additionally, a block of Class C IP addressees have been allocated to SA Systems by the InterNIC. Attempts to scan for active IP ports within the block (123.4.56.7–123.4.56.255) have yielded no results of interest. It is suspected that this block is currently inactive, but that it is technically feasible to assign TMENS SLD to this block (as in Kabul.ns). This is because once a person has both a TLD and an IP address block, they can internally assign whatever SLD names they want to the IP address. Therefore, if 123.4.56.100 is assigned to "Kabul.ns," to the outside world it looks like another system connected at SA Systems.

Results of the Research Project

By taking some logical steps, one can learn a great deal of valuable information about one's potential adversary's systems and networks. By mapping out the connections of systems and the networks, identifying the types of hardware and software in use, individuals, and interfaces, one can then prepare to incorporate the information into IW operations plans for future use.

Note: Ironically, the actual IW project previously conducted prior to 9/11 provided information linked to an FBI "war on terrorism" investigation that was conducted in Texas. Results of the project were provided to the FBI.

Cyber War in the Middle East[5]

Introduction

Late September 2000 brought an escalation of the Arab–Israeli conflict, marked by an unreserved effort by self-proclaimed cyber-soldiers in support of information warfare. These attacks are not the first to be associated with a multi-national conflict. Tensions between China and Taiwan, and between India and Pakistan, have sparked some low-level computer attacks on the part of freelancing hackers, but not at the level seen during the latest Middle East conflict.

None of the perpetrators have publicly claimed an association with the governments involved. They appear to be acting on their own or in concert with other hackers or hacker groups.

Middle East Intifada

According to press articles, the Arab–Israeli cyber conflict is a cyber-Jihad, or a cyber holy war. This is not entirely true. This cyber activity is not a war, but more resembles an Intifada in which both sides throw rocks or fire small arms at the opponent. The hackers involved are lobbing E-munition (electronic ammunition) in the form of simple Web defacements, denial-of-service attacks, and the deliberate introduction of false information to computers belonging to the opposition to affect popular opinion. In some cases, the victim computers or attackers are resident in a country other than those involved in the conflict.

The Attacks

The first network attacks most likely occurred on September 29, 2000, when two Lebanese Web sites (www.hizbollah.org and www.hizballah.org) were flooded with electronic mail messages. Shortly thereafter, according to Beirut's The Daily Star, the Hizbullah's Al-Manar Television Web page was manipulated by an Israeli hacker identified as "Nir M.," who replaced the site with an animated Israeli flag and pictures of the violence. The Hizbullah Web master is quoted as saying, "This is part of the war."

Apparent Palestinian or Islamic hackers retaliated with a denial-of-service attack on the Israeli Foreign Ministry Web site hosted by NetVision, an Israeli Internet service provider (ISP). This action led the Israeli Defense Forces to host the site with AT&T for higher bandwidth. Gilad Rabinovich, NetVision ISP president, charged that the hackers were "trying to infect the Internet backbone" of Israel and claimed NetVision can reciprocate by destroying any hostile Web site within hours.

Other Israeli Web site cyber-attack victims include those registered to the Mahba University computing center, the Office of the Prime Minister, the Treasury, Israel's Ministry of Defense, the Ministry of Foreign Affairs, and the Knesset, which believes its site was penetrated by hackers in Saudi Arabia. The Director General of the Internet at King Abdulaziz City for Science and Technology, Dr. Fahad Al-Hewmani, called on Saudi hackers not to attack Israeli and Jewish Web sites, according to the Jedda Saudi Gazette.

The Jerusalem Middle East Newsline reported that Arab and Israeli hackers continued their E-intifada into late October and early November, attacking Web sites registered in Egypt, Israel, Lebanon, Jordan, and Palestine. Muslim supporters in Pakistan allegedly attacked Israeli Web sites, and Israeli hackers responded with two attacks of their own. A Tehran Times article claims Iran's Hyundai Company's Web site was attacked by the "Zionist regime."

The Amman (Jordan) newspaper Al-Sabil claimed its Web site had been the target of several hacking attempts and was inaccessible for several hours on October 25th. A U.S. ISP hosts the Web site. Another pro-Israeli Web site hosted in the United States, belonging to the American Israel Public Affairs Committee, fell victim to an attack that reportedly resulted in the theft of 700 credit card numbers and 3500 e-mail addresses. A third U.S.-based Web site, allegedly created by "extremist Palestinian elements," cloned the Israeli Defense Forces' official site, adding "insults against IDF soldiers and horror pictures," according to the Jerusalem Voice of Israel.

Several threatening messages circulated on the Internet. One, sent to Beirut's The Daily Star newspaper as well as AT&T, Verizon, British Telecom, French Telecom, Tel-Mex, and Tel Espana, called for punishment against The Daily Star for "promoting hacker attacks on Israeli and Jewish Web sites." Another message, sent to approximately 7000 subscribers of an Israeli cellular telephone service, read "Your telephone will explode in another five minutes."

Israel has one of the most computer literate populations and more Internet connections than all the Arab countries combined. This advantage became a disadvantage because Israel had many more targets the Palestinians and fellow Arabs could strike. Arab "hacktivists" launched attacks on Jewish Web sites, both in Israel and in the United States. Then Pakistani-based hackers attacked a U.S. Web site belonging to a powerful pro-Israel lobby, stealing credit card numbers and member records.[6] This is an example of how a regional conflict can easily escalate into a global imbroglio.

The Arabs were able to maintain a ten-to-one (10:1) advantage in striking sites and servers, a trend opposite to that seen in the war on the streets.[7] This became a symbolic source of pride, and the Arabs nicknamed their actions the

"E-Intifada." Of course, the quantity and quality of the sites attacked are two different things. The Bank of Israel and the Tel Aviv Stock Exchange were struck, but so were strategic Palestinian assets.

Media exposure is essential to most causes. Hizbullah maintains a high-standard, multilingual Web site, so it became a strategic target. The Israelis claimed to take down the main Hizbullah Web site, the alternate site, and Web sites related to Hamas and the Palestinian Authority.[8]

U.S. Cyber Defenses

The threat of attacks against U.S.-based Web sites related to Israel and pro-Palestinian organizations led the National Infrastructure Protection Center (NIPC) to issue an assessment[9] in late October 2001. The NIPC recommended that Web hosting enterprises "remain vigilant to the possibility that there could be some spill-over activity and that U.S. sites could become targeted." A later NIPC advisory[10] provides additional information, including an analysis of the difference between the attack types.

Of the eleven NIPC advisories issued during 2000, nine are technology related (viruses and vulnerabilities) and two pertain to hacktivity and related socially based cyber-actions. Vulnerability information can be obtained from a good number of sources, but only makes you aware of a general problem. Hacking threat intelligence collected by the NIPC and distributed in the form of an advisory or assessment can hold significance if it reports on likely imminent activity against your networks.

Identifying the Hackers and Hacker Groups

The London Al-Sharq al-Awsat published an interview with a young Saudi hacker who was not identified in the article. The hacker disclosed that he is associated with a "club" whose members in Saudi Arabia, the United Arab Emirates, Egypt, Kuwait, and several other Arab countries exchange ideas and programs related to hacking.

According to the unidentified hacker, most of the programs that are used are obtained from the Internet, and some are able to "override the security programs" of the targeted computers. However, the hacker displayed an apparent lack of knowledge by mentioning that "the number for Saudi Arabia on the Internet is 212," and by bragging that the number of members who reach technically advanced levels of ability (defined as being able to intrude via Telnet and FTP) is small.

The Israeli Internet Underground (IIU) hacker group states it is "dedicated to the Israeli spirit and united to protect Israel on the Internet against any kind of attacks from malicious hacking groups." This is interesting in light of the fact that one IIU alumnus is Ehud Tenenbaum, aka "Analyzer," who in 1998 was found to have mentored the two Cloverdale (California) teenagers who had penetrated numerous computers in the United States, including DoD computers operated by several universities, ISPs, and corporations.

Tenenbaum is now the Chief Technology Officer at the Israeli computer security firm 2XS, which claims to have collaborated with the IIU and helped hundreds of companies to enhance their computer security posture since the cyber-conflict began. Tenenbaum was fined approximately $18,000, sentenced in Israel to six months of community service, one year of probation, and a two-year suspended prison sentence that can be enforced if he commits another computer crime within three years.

Forecast

The malicious cyber-activities in the Middle East have slowed. While it is unknown if more Web hacks and denial-of-service attacks are lurking, the attackers' Internet activities seem to have legitimized computer network attack and defense (CNA/D) in conflict.

The technology and capability are still nascent and there is disagreement in international military doctrine as to the scope of involvement of CNA/D in a war. However, the cyber-activity seen thus far has been perpetrated by freelancing amateurs and there are no indications of government sponsorship. Future conflicts are likely to include this same type of low-level nonprofessional activity and, given the successes and wide press of the recent attacks, will likely be seen with more frequency and on a wider scale.

Summary

The Middle East will continue to be an area in conflict for the foreseeable future. However, the use of information warfare (IW) will for some time be limited to denial-of-service attacks; Web site defacements; and using chat rooms, e-mails, and Web sites to provide their side of the issues and conflicts. However, there will be an increasing number of successes at network penetrations as these information warriors gain experience and learn from those experiences.

However, if the conflicts continue as the nation-states' NIIs are developed, they will provide tempting targets for both sides, targets that — if history is any indication — will not be ignored.

Notes

1. http://www.cnn.com/2001/COMMUNITY/07/31/kissinger.cnna/index.html.
2. The countries listed were taken from http://www.middleeastinfo.com/countries.html. See that site for additional information.
3. See the CIA.org *World Factbook.*
4. See As Israel's Military Spies See It (http://www.janes.com).
5. Chip Seymour provided the information noted in this section. He kindly consented to its publication here. Seymour has been working with computer security issues since 1984. He serves on the Board of Directors for the Boston chapter of InfraGard. We thank him for his valuable contribution to this section of our book.

6. NIPC Assessment 00-057.

7. Hacktivists Target U.S. Pro-Israel Site, The Associated Press, CNET News.com, November 6, 2000.

8. Bytes Without the Blood in Mideast: The Internet Blossoms as a Battleground in the Conflict, http://www.msnbc.com/news/509349.asp, Hanson R. Hosein, NBC News, January 8, 2001.

9. Israeli-Palestine Cyber Conflict Report v2.0PR, analysis@iDefense.com, January 3, 2001.

10. NIPC Assessment 00-058.

Chapter 10

Information Warfare: European Nation-States

...in Europe and the Western Hemisphere, the governments, almost all of them are democratic, and military conflicts between them are almost inconceivable. And the problems there are economic and social, except for some ethnic conflict at the fringes of Europe...

—Dr. Henry Kissinger

This chapter discusses information warfare (IW) as it relates to European nation-states. However, only a European overview is presented, along with more details and discussions of selected nation-states.

Introduction to Information Warfare: Europe

Europe contains some of the world's oldest nation-states and is also where wars have started and encompassed the world on more than one occasion. It has matured, as Dr. Henry Kissinger has stated, to the point where at least warfare of the industrialized period is no longer probable. However, this only applies to the "old" European Union nation-states of Austria, Belgium, Denmark, Finland, France, Germany, Greece, Ireland, Italy, Luxembourg, the Netherlands, Portugal, Spain, Sweden, and the United Kingdom. In the other European nation-states, such as those of the former East European-controlled bloc of the former Union of Soviet Socialist Republics, that is not the case. The recent wars surrounding the demise of the former Yugoslavia is a prime example.

However, in the more "mature" European nation-states, economic and business aspects of information warfare, where a nation-state is actively involved in supporting its businesses in the world of global competition is, in fact, increasing. This is not a type of warfare that is limited to the European powers. It may very readily become more global as other nation-states request support from these more prosperous European nation-states and find to their chagrin that support and aid are not forthcoming. One of these nation-states that has asked for aid which is not forthcoming can easily rationalize using IW tactics to upset the economic and financial forces of these wealthier nation-states. If one of these "Third World" powers is trying but just cannot get out of its economic quagmire without outside support, it may see no prosperity in its future if economic superpowers decline to help. Consequently, it might use covert IW tactics to degrade one or more of the nation-states' economies. Why not? It has nothing to lose.

The Eastern Europe nation-states, now set free from the yoke of Communism, are restructuring their national information infrastructures (NIIs), which, as is repeatedly pointed out in this book, makes them more vulnerable to IW tactics.

> ### Some Countries Being Left Behind in E-Commerce
> *infowar.com, August 8, 2000*
> *In Europe, Estonia, benefiting from a close relationship with its Scandinavian neighbors...human capital and E-business climate. Nearly a third of Estonia's population is connected to the Internet, and nearly 90 percent of public employees work in a computerized environment. Hungary, Italy, Lithuania, Portugal, and Spain are not far behind Estonia.*

The high-technology environments in which today's modern European nation-states reside use common hardware, firmware, and software. To communicate globally, these networks must be compatible. Such products as Microsoft software, Cisco routers, and the like are used throughout Europe — and in fact the world. They all have the same inherent vulnerabilities and defensive mechanisms that can be applied. When it comes to warfare practiced over the ages by nation-states, we know from history that they also have a commonality (e.g., machine guns, aircraft, artillery, ships, and tactics).

The United States is one of the world leaders in the research and implementation of IW policies, strategies, concepts, and related areas. It is the home of Microsoft, Cisco, and many other high-technology products used throughout the world. The United States is the most powerful and high-technology-driven military force in the world today. Therefore, it is not surprising that European nation-states pattern their IW thinking after that of the United States and then modify it to meet their individual nation-state needs. This is also necessary as the European nation-states, in some instances through bilateral or multilateral relationships with the United States, share common defensive and offensive alliances, such as NATO. NATO forces have interchangeable and complementary military weapons systems, such as nine-millimeter (9-mm) handgun ammunition; their information warfare strategy and operations also have commonality.

So, when discussing IW in the European theater of operations, the individual nation-states appear to have more in common in IW strategies, policies, and tactics than not. Protecting their individual NIIs is a prime example.

What Is Europe?

The first question is: what do we really mean when we talk about Europe? In terms of the map showing the countries of Europe, it is just that. It is a geographical area, but the perception of its boundaries varies. Take a quick look at the history of the past 10 years to put this in perspective:

- Europe has a unique and colorful history that is characterized by:
  - Conflict
  - Changing alliances and borders
  - Multiple religions and ethnic backgrounds
  - Cultural diversity
  - Multiple languages (the primary four being English, German, French, and Spanish)
  - A history rich in conquest and colonization (England, the Netherlands, Spain, Portugal, and France)
  - Multiple overlapping alliances[1]
- Europe during the past ten-plus years has seen:
  - Fall of the Berlin Wall
  - Gulf War
  - Baader Meinhoff (German terrorists)
  - 1969–1998: IRA/UFF (United Kingdom terrorists)
  - To Present Date: ETA (Basque terrorists)
  - To Present Date: October 17 (Greek terrorists)
  - From 1992: Former Republic of Yugoslavia conflicts
  - NATO Enlargement and Partnership for Peace
  - 1996: Northern Ireland peace talks commence
  - 1998: Northern Ireland peace agreements signed
  - 2001: Global terrorist war

Exhibit 1 provides a map of the nation-states that are commonly referred to as Europe.

The European Union: Laws

As the European Union (EU) continues to mature (hopefully), there are attempts being made to harmonize its laws with regard to the cyber-environment. Following are some of the U.K. laws that must be considered to achieve this. (Remember that the laws that are listed are those within the United Kingdom; imagine the problem of harmonizing this throughout the more than 30 countries that currently make up the geographical area of Europe):

Exhibit 1.   Map of Europe

- Computer Misuse Act (1990)
- Telecommunications Act (1984)
- Telecommunications (Fraud) Act (1987)
- Obscene Publications Act (1959/1964)
- Protection of Children Act (1978)
- Criminal Justice Act (1988)
- Criminal Justice and Public Order Act (1994)
- Data Protection Acts (1984/1998)
- Theft Acts (1968/1978)
- Forgery and Counterfeiting Act (1981)
- Copyright Design and Patents Act (1988)

- Interception of Communications Act (1985)
- Regulation of Investigatory Powers Bill (2000)

As these types of laws and others are agreed on by the European nation-states, they will have an impact on the "rules of information warfare" as seen by the European Union (EU).

In addition, the European Union's commission plans to force manufacturers and operators of Internet service providers (ISPs) to build in "interception interfaces" to the Internet and "all future digital communications systems."[2] This is part of an overall strategy of about 20 nation-states that have usually met in secret. They include Hong Kong, Australia, the United Kingdom, the United States, Canada, Germany, and New Zealand. Others may include Ireland, Austria, Italy, Spain, and probably most of the major nation-states of the world whose NIIs and networks are being attacked by various groups, for various reasons, from all over the globe.

European Organizations and Information Warfare

There are many European nation-states who are intertwined through the various European organizations. In fact, there are 54 of them. These nation-states belong to the following main European organizations.

Organization for Security and Cooperation in Europe (OSCE)

The OSCE was formerly the Conference on Security and Cooperation in Europe (CSCE). It was established in 1995. Headquartered in Vienna, Austria, its goal is:

> *...to foster the implementation of human rights, fundamental freedoms, democracy, and the rule of law; to act as an instrument of early warning, conflict prevention and crisis management; and to serve as a framework for conventional arms control and confidence building measures.[3]*

The OSCE has the following 55 members: Albania, Andorra, Armenia, Austria, Azerbaijan, Belarus, Belgium, Bosnia and Herzegovina, Bulgaria, Canada, Croatia, Cyprus, the Czech Republic, Denmark, Estonia, Finland, France, Georgia, Germany, Greece, Holy See, Hungary, Iceland, Ireland, Italy, Kazakhstan, Kyrgyzstan, Latvia, Liechtenstein, Lithuania, Luxembourg, the Former Yugoslav Republic of Macedonia, Malta, Moldova, Monaco, the Netherlands, Norway, Poland, Portugal, Romania, Russia, San Marino, Slovakia, Slovenia, Spain, Sweden, Switzerland, Tajikistan, Turkey, Turkmenistan, Ukraine, the United Kingdom, the United States, Uzbekistan, and Yugoslavia (suspended).

European Union (EU)

The EU evolved from the European Community (EC). Headquartered in Brussels, Belgium, it was established in February 1992 with an effective date of November 1993. Its goal is:

> *...to coordinate policy among the 15 members in three fields: economics, building on the European Economic Community's (EEC) efforts to establish a common market and eventually a common currency to be called the "euro," which will supersede the EU's accounting unit, the ECU; defense, within the concept of a Common Foreign and Security Policy (CFSP); and justice and home affairs, including immigration, drugs, terrorism, and improved living and working conditions.[3]*

Its 15 members include: Austria, Belgium, Denmark, Finland, France, Germany, Greece, Ireland, Italy, Luxembourg, the Netherlands, Portugal, Spain, Sweden, and the United Kingdom. Membership applicants (13) are Bulgaria, Cyprus, Czech Republic, Estonia, Hungary, Latvia, Lithuania, Malta, Poland, Romania, Slovakia, Slovenia, and Turkey.

Western European Union (WEU)

The WEU, also headquartered in Brussels, Belgium, was established in October 1954 with an effective date of May 1955. Its purpose is to *"provide mutual defense and to move toward political unification."[3]*

There are ten members: Belgium, France, Germany, Greece, Italy, Luxembourg, the Netherlands, Portugal, Spain, and the United Kingdom; six associate members: the Czech Republic, Hungary, Iceland, Norway, Poland, and Turkey; seven associate partners: Bulgaria, Estonia, Latvia, Lithuania, Romania, Slovakia, and Slovenia; and five observers: Austria, Denmark, Finland, Ireland, and Sweden.

North Atlantic Treaty Organization (NATO)

NATO, also headquartered in Brussels, Belgium, was established in April 1949 to "promote mutual defense and cooperation."[3] It has 19 members: Belgium, Canada, the Czech Republic, Denmark, France, Germany, Greece, Hungary, Iceland, Italy, Luxembourg, the Netherlands, Norway, Poland, Portugal, Spain, Turkey, the United Kingdom, and the United States. Because NATO is currently the most active military force in Europe, its views on information are important, especially in light of its recent campaigns in the Balkans.

At a recent conference of NATO representatives, the definition of information warfare was discussed. It was determined that there were 17 different definitions. Albeit all not that much different, but different nonetheless. That being the case, how easy is it to establish a coordinated offensive and defensive information warfare doctrine if the definitions cannot even be agreed to?

Furthermore, since the fall of the Berlin Wall and the breakup of the former Soviet Union, NATO has been in discussion with a number of former Eastern European states that have borders with its member states who have expressed an interest in joining the organization. This has caused a number of issues to be addressed, including:

- Former "enemies" who were trained and equipped by the Soviet Union are now friends, but the investment in restructuring and reequipping their armed forces to be compatible with the NATO forces is high.
- How do we integrate them into NATO without threatening Russia?
- Can we trust former "enemies"?

Partnership for Peace (PFP)

The PFP operates under the auspices of NATO and is headquartered, as one would expect, in Brussels, Belgium. It was established in January 1994, with its goal:

> ...to expand and intensify political and military cooperation throughout Europe, increase stability, diminish threats to peace, and build relationships by promoting the spirit of practical cooperation and commitment to democratic principles that underpin NATO; program under the auspices of NATO.[3]

It has 27 members: Albania, Armenia, Austria, Azerbaijan, Belarus, Bulgaria, the Czech Republic, Estonia, Finland, Georgia, Hungary, Kazakhstan, Kyrgyzstan, Latvia, Lithuania, the Former Yugoslav Republic of Macedonia, Moldova, Poland, Romania, Russia, Slovakia, Slovenia, Sweden, Switzerland, Turkmenistan, Ukraine, and Uzbekistan.

Information Warfare: A More European Definition

Information warfare has many definitions. As discussed in Chapter 1, some believe that it can be categorized as:

- Class 1: personal information warfare
- Class 2: corporate information warfare
- Class 3: global information warfare

Some information warfare experts in Europe, and possibly elsewhere, disagree with this categorization. One European information warfare expert, who wished to remain anonymous, put it this way:

> *Warfare is not about individuals or commercial organizations — I would define an attack on an individual as criminal or a civil litigation issue. An attack on a company by another company is industrial espionage or industrial sabotage, or in many cases, normal aggressive competition, although an attack on a company by a government, terrorist group or pressure group may in fact be information warfare. The only group where I would mainly agree with the three categorizations is in Class 3 IW — warfare is about affecting nation states. I define warfare as noted in many dictionaries. Warfare is basically:*

- The waging of war against an enemy, as in armed conflict. In this case the enemy is considered a nation-state or group of nation-states. The armed conflict can include or be limited to information warfare weapons such as malicious code.
- Military operations marked by a specific characteristic, e.g., guerrilla warfare; chemical warfare — and now information warfare; and/or
- Acts undertaken to destroy or undermine the strength of another nation-state's government, e.g., political warfare where information warfare tactics and weapons may be used.

In June 2000, the United Kingdom, one of the European leaders when it comes to understanding and applying information warfare operations to conflicts, defined information warfare as:

> *Integrated actions undertaken to influence decision makers in support of political and military objectives by affecting others' information, information based processes, C2 systems and CIS while exploiting and protecting one's own information and/or information systems.*[5]

Many in Europe see information warfare playing a role in the move away from platform-centric warfare sponsored by nation-states and:

- Trend toward more intra-state conflict and operations other than warfare
- Growth in asymmetric adversaries and non-state actors (empowered small agents)
- Globalization of information systems, financial markets and economies, and the media
- Perceptions will play a more vital and important role because information warfare tactics and subsequent results are not always visible — until it is too late.

> *Within Belgrade, Serbian Officials are more concerned about stopping the <u>coverage</u> of the protest rallies against Milosevic than they are about stopping the actual rallies themselves.*
>
> —BBC News 24, June 29, 1999

> *If one doesn't talk about a thing, it has never happened. It is expression that gives reality to things.*
>
> —Oscar Wilde, *Portrait of Dorian Gray*

European Nation-States' Potential Conflicts

Although the organizations described above are basically for defense and mutual protection of the members of the various organizations, it is not surprising that many of the conflicts arise between the members, leading one to question the ability of its members to maintain peace among themselves. Among some of the disagreements and potential areas of conflict are:[3]

- *Albania:* The Albanian government supports protection of the rights of ethnic Albanians outside its borders but has downplayed them to further its primary foreign policy goal of regional cooperation; Albanian majority in Kosovo seeks independence from Serbian Republic; Albanians in the Former Yugoslav Republic of Macedonia claim discrimination in education, access to public-sector jobs, and representation in government
- *Armenia:* Armenia supports ethnic Armenians in the Nagorno-Karabakh region of Azerbaijan in the longstanding separatist conflict against the Azerbaijani government; traditional demands regarding former Armenian lands in Turkey have subsided
- *Azerbaijan:* Armenia supports ethnic Armenians in the Nagorno-Karabakh region of Azerbaijan in the longstanding separatist conflict against the Azerbaijani government; Caspian Sea boundaries are not yet determined among Azerbaijan, Iran, Kazakhstan, Russia, and Turkmenistan
- *Bosnia and Herzegovina:* Disputes with Serbia over Serbian populated areas of Bosnia and Herzegovina
- *Croatia:* Eastern Slavonia, which was held by ethnic Serbs during the ethnic conflict between the Croats and the Serbs, was returned to Croatian control by the UN Transitional Administration for Eastern Slavonia on January 15, 1998; Croatia and Italy made progress toward resolving a bilateral issue dating from World War II over property and ethnic minority rights; significant progress has been made with Slovenia toward resolving a maritime border dispute over direct access to the sea in the Adriatic; Serbia and Montenegro are disputing Croatia's claim to the Prevlaka Peninsula in southern Croatia because it controls the entrance to Boka Kotorska in Montenegro; Prevlaka is currently under observation by the UN Military Observer Mission in Prevlaka (UNMOP)
- *Cyprus:* 1974 hostilities divided the island into two *de facto* autonomous areas, a Greek Cypriot area controlled by the internationally recognized Cypriot government (59 percent of the island's land area) and a Turkish-Cypriot area (37 percent of the island), that are separated by a UN buffer zone (4 percent of the island); there are two U.K. sovereign base areas mostly within the Greek Cypriot portion of the island
- *Czech Republic:* Liechtenstein claims restitution for 1600 sq km of land in the Czech Republic confiscated from its royal family in 1918; the Czech Republic insists that restitution does not go back before February 1948, when the Communists seized power; individual Sudeten German claims for restitution of property confiscated in connection with their expulsion after World War II; agreement with Slovakia signed November 24, 1998 resolves issues of redistribution of former Czechoslovak federal land — approval by both parliaments is expected in 2000
- *Denmark:* Rockall continental shelf dispute involving Iceland, Ireland, and the United Kingdom (Ireland and the United Kingdom have signed a boundary agreement in the Rockall area)
- *Estonia:* Estonian and Russian negotiators reached a technical border agreement in December 1996 that has not been signed or ratified

- *Germany:* Remaining legal issues (restitution) arising from World War II and its aftermath
- *Gibraltar:* Source of friction between Spain and the United Kingdom
- *Glorioso Islands:* Claimed by Madagascar
- *Greece:* Complex maritime, air, and territorial disputes with Turkey in Aegean Sea; Cyprus question with Turkey; dispute with the Former Yugoslav Republic of Macedonia over its name
- *Hungary:* Ongoing Gabcikovo Dam dispute with Slovakia
- *Iceland:* Rockall continental shelf dispute involving Denmark, Ireland, and the United Kingdom
- *Ireland:* Northern Ireland issue with the United Kingdom (historic peace agreement signed April 10, 1998); Rockall continental shelf dispute involving Denmark, Iceland, and the United Kingdom (Ireland and the United Kingdom have signed a boundary agreement in the Rockall area)
- *Italy:* Italy and Slovenia made progress in resolving bilateral issues; Croatia and Italy made progress toward resolving a bilateral issue dating from World War II over property and ethnic minority rights
- *Kazakhstan:* Caspian Sea boundaries are not yet determined among Azerbaijan, Iran, Kazakhstan, Russia, and Turkmenistan; Russia leases approximately 6000 sq km of territory enclosing the Baykonur Cosmodrome
- *Latvia:* Draft treaty delimiting the boundary with Russia has not been signed; ongoing talks over maritime boundary dispute with Lithuania (primary concern is oil exploration rights)
- *Liechtenstein:* Claims 1600 sq km of land in the Czech Republic confiscated from its royal family in 1918; the Czech Republic insists that restitution does not go back before February 1948, when the Communists seized power
- *Lithuania:* Ongoing talks over maritime boundary dispute with Latvia (primary concern is oil exploration rights); 1997 border agreement with Russia not yet ratified
- *Macedonia, The Former Yugoslav Republic of:* Dispute with Greece over its name; the border commission formed by the Former Yugoslav Republic of Macedonia and Serbia and Montenegro in April 1996 to resolve differences in delineation of their mutual border has made no progress thus far; Albanians in F.Y.R.O.M. claim discrimination in education, access to public-sector jobs, and representation in government
- *Madagascar:* Claims Bassas da India, Europa Island, Glorioso Islands, Juan de Nova Island, and Tromelin Island (all administered by France)
- *Norway:* Territorial claim in Antarctica (Queen Maud Land)
- *Romania:* Dispute with Ukraine over continental shelf of the Black Sea under which significant gas and oil deposits may exist; agreed in 1997 to two-year negotiating period, after which either party can refer dispute to the ICJ
- *Serbia and Montenegro:* Disputes with Bosnia and Herzegovina over Serbian populated areas; Albanian majority in Kosovo seeks independence from Serbian republic; Serbia and Montenegro are disputing Croatia's claim to the Prevlaka Peninsula in southern Croatia because it controls the entrance to Boka Kotorska in Montenegro; Prevlaka is currently under

observation by the UN Military Observer Mission in Prevlaka (UNMOP); the border commission formed by the Former Yugoslav Republic of Macedonia and Serbia and Montenegro in April 1996 to resolve differences in delineation of their border has made no progress thus far

- *Slovakia:* Ongoing Gabcikovo Dam dispute with Hungary; agreement with Czech Republic (signed November 24, 1998) resolves issues of redistribution of former Czechoslovak federal property — approval by both parliaments is expected in 2000
- *Slovenia:* Significant progress has been made with Croatia toward resolving a maritime border dispute over direct access to the sea in the Adriatic; Italy and Slovenia made progress in resolving bilateral issues
- *Spain:* Gibraltar issue with United Kingdom; Spain controls five places of sovereignty (*plazas de soberania*) on and off the coast of Morocco — the coastal enclaves of Ceuta and Melilla, which Morocco contests, as well as the islands of Penon de Alhucemas, Penon de Velez de la Gomera, and Islas Chafarinas
- *Tajikistan:* Portions of the boundary with China are indefinite; territorial dispute with Kyrgyzstan on northern boundary in Isfara Valley area
- *Ukraine:* Dispute with Romania over continental shelf of the Black Sea under which significant gas and oil deposits may exist; agreed in 1997 to two-year negotiating period, after which either party can refer dispute to the ICJ; has made no territorial claim in Antarctica (but has reserved the right to do so) and does not recognize the claims of any other nation

Individual European Nation-States and Information Warfare

Any of the major European nation-states that have both a sufficiently developed NII to be information and information systems dependent, or that are vulnerable to information warfare attacks because of that, were selected for additional comment including:

- Sweden
- France
- Germany
- Russia
- United Kingdom

Sweden

Sweden is an interesting nation-state when it comes to information warfare. One would expect a great deal of IW activities, such as research and development, and integrating IW tactics into its military plans, from such nation-states as France, Germany, and Russia. However, Sweden also appears to have a very active IW program.

The subject of IW is being addressed by the Academy of War Sciences and government agencies, and is integrated into its military plans as noted by the

Swedish newspaper, *Svenska Dagbladet,* as far back as December 1999. In its December 29, 1999, articles, *Svenska* reported that the Swedish government was creating[6] an information warfare unit known as NTIC. The unit would be part of its armed forces. The mission was reported to be:

> *...not only defensive (to protect the governmental information processing systems of external attacks) but also offensive ("to destroy enemy servers remotely...").*

Sweden views IW as the use of technology to influence, change, or destroy information; as well as using IW tactics to influence the attitudes and information content of an adversary. They view the main military component of IW as the command and control warfare (C2W) dealing with protection. According to one source:

> *The Swedish taxonomy concerning IO/IW is about to adapt to the international taxonomy with Information Operations as the overarching term and with Information Assurance as the main defensive component. Our Swedish total defence system must within the widened view on threats and risks to society pay attention to the new questions which the threat from different forms of Information Operations Warfare imply...[7]*

The referenced Web site page went on to discuss the complications that all modern nation-states are trying to deal with and that is NII protection and the fact that its protection is divided among different government and private agencies. Based on these issues, it appears that the Swedish government is trying to develop a holistic approach and also increase research and studies in this topic. With this in mind, as far back as 1996, the Swedish Parliament's "1996 Defence Decision" was passed.

Sweden has a long history of defensive military operations, to include many sophisticated defenses such as protecting its military aircraft in caves. The aircraft were then able to use the adjoining roads to launch the aircraft. In addition, they were concerned with the protection needed in case of a nuclear bomb detonation.

Sweden has now progressed to the point of looking at electromagnetic pulse (EMP) weapons and their impact on computer systems. Although Sweden has in the past appeared to be more defense oriented than offense oriented, IW weapons of computers, networks, malicious codes, and related areas now make this nation-state as potent an information warfare offensive nation-state as any other.

Although Sweden does not have any major conflicts with other nation-states, one must always be prepared and, apparently, Sweden is quietly getting prepared. Manuel W. Wik, Chief Engineer and Strategic Specialist on Future Defence Science and Technology Programs in Sweden, has some very relevant views on information warfare. The reader is referred to the Appendices for his paper on the topic.[8]

France

France is an interesting nation-state when it comes to information warfare. It is an old country with a history of involvement in most of the major wars of the past century. It also has a history of attempting to maintain a viable, modern military force. So, it is no surprise that France would be very interested and active in IW tactics, operations, and weapons.

According to the CIA *World Factbook,*[3] France does have some territorial disagreements, including:

- *France:* Madagascar claims Bassas da India, Europa Island, Glorioso Islands, Juan de Nova Island, and Tromelin Island; Comoros claims Mayotte; Mauritius claims Tromelin Island; territorial dispute between Suriname and French Guiana; territorial claim in Antarctica (Adelie Land); Matthew and Hunter Islands east of New Caledonia claimed by France and Vanuatu
- *French Guiana:* Suriname claims area between Riviere Litani and Riviere Marouini (both headwaters of the Lawa)
- *French Southern and Antarctic Lands:* "Adelie Land" claim in Antarctica is not recognized by the United States

However, none of these disagreements would appear to be of such major importance as to lead to war. But if the disagreements became more intense, one could not rule out the use of information warfare tactics. After all — and as we have previously discussed — information warfare tactics generally do not cause loss of life or physical damage, so it is very tempting to use them against an adversary. In fact, because many of the attacks could be disguised as some hackers operating from within another nation-state, there is little risk of detecting or, if detected, being identified.

It would appear that France divides IW into military and civil issues. France has always had a rather interesting view of civil and military warfare. In military warfare, France appears to believe in the "concept of war of information," in which:

- It is prevalent in conflicts of low intensity.
- Information warfare is combined with electronic warfare.
- The conflicts are generally within the auspices of the UN or NATO.
- They believe that on the ground, the ally can be an enemy.
- Peace treaties are signed between winners and are overcome.
- Peace prevails in the international relations of the industrialized countries.

In the "civil war of information," France believes that:

- There is a diversity of conflicts.
- The adversarial relationships in economic warfare have a "richness of cases higher than the traditional military confrontations."
- "On the ground, an ally can be an enemy."
- Trade talks are compromises, unlike military warfare where the victor dictates the rules.

■ "Economic peace does not exist as much as competitors clash to conquer markets."

This view, although it may be held by many other nation-states, has never been so clearly defined as France has defined these two types of warfare. For example, the United States has been struggling for some time as to whether or not to collect business intelligence information and to pass that on to U.S. corporations. In France, there is little doubt that this is done. France over the years has grown infamous in its use of espionage for business reasons.

France learned long before other nation-states that economic power is more important than military power as a global concept of nation-state power. From its alleged bugging of business class compartments on its airlines to bugging hotel rooms of foreign business travelers, France has treated the collection of such information as vital to its national security and has dedicated a great deal of resources to the information collection efforts using information warfare tactics.

Is it any wonder then that France places a high regard on the use of information warfare tactics to gain an economic advantage against the competition — other nation-states and foreign corporations?

> *It would not be normal to spy on the United States in political matters or military matters. But in economic competition, in technical competition, we are competitors. We are not allied.*

> —Pierre Marion
> Former Director
> France's Direction Generale de la Securite Exterieure (DGSE)[9]

It has recently been alleged that some of the hotels in France where foreign business travelers stay have paper shredders conveniently located in the rooms for the use of security-minded travelers. However, it has been rumored that the shredders, in fact, digitally copy the information going into the shredder before the documents are shredded, thus allowing French government agents a rather unique and very useful information collection tool. The information gleaned from such a system can have an enormous impact in the global economic and business wars taking place. It is a good example of why information warfare tactics are used to further a nation-state's influence in the world.

And there is "Frenchelon." This information warfare tool allows the collection of vast amounts of information for France. As stated in an online article posted on zdnet.fr on June 21, 2000, a French government official affirmed the existence of Frenchelon and said:

> *...we do not have anything to envy Americans. We have our equivalent of level in France. In particular within a station of listening of the Paris area, with engines of semantic analysis to sort information...*

While at the same time:

French Prosecutor Starts Probe into U.S. Spy System
July 4, 2000, Paris (Reuters)
A French state prosecutor has launched a preliminary judicial investi-gation into the workings of the United States' Echelon spy system of satellites and listening posts.... Coincidentally, the European Parliament is due to decide in Strasbourg on Wednesday whether to set up a commission to investigate whether Echelon infringes the rights of Euro-pean citizens and industries...

Are government agents using malicious codes to penetrate the networks of foreign corporations and nation-states under the guise of "just another hacker"? Why not, when the benefits are great and the risks extremely low? France's track record of using sophisticated and not-so-sophisticated methods to steal information has been well-documented. Therefore, its covert use to obtain business and economic information cannot be ruled out. While its use of information warfare tactics to obtain military information of an ally would appear to be minimal, other allies of the United States, such as Israel, do not have the same ethical approach in nation-state relationships (research the Pollard espionage case for further details).

Germany

German Markus Hess leads a group of hackers to electronically break into several U.S. military bases to steal classified information. They sell the information to the KGB in exchange for cash and drugs.

—*Daily Telegraph*, August 20, 1996, p. 8

According to the CIA *World Factbook*,[3] Germany's "remaining legal issues (restitution) arise from World War II and its aftermath." Obviously, after two World Wars involving German aggression, there are no conflicts concerning territory since Germany's boundaries were dictated to them by the Allied Powers after World War II. Restitution issues do not at this time appear to be to the point of boiling over into some type of information warfare conflict. If that did occur, Germany, with her dependency on financial networks and other information infrastructure systems, could be targeted for an information warfare attack if some nation-state or other entity felt strongly enough that Germany was not meeting its obligations, vis-à-vis restitution.

The German military has been well-known over the past centuries to put great emphasis on having one of the world's best military forces. Although the World Wars and the Cold War of the past century are beginning to fade into memory of our younger generations, the world's political leaders have kept Germany's military forces at a level reasonable for defensive purposes, as has been done in Japan. However, IW for Germany is, as with all modern nation-states, a consideration in time of war. For example, the German Army has looked at the concept of information warfare in the space of battle in the year 2020.

As with most nation-states involved in IW, the Germans view it from both an offensive and defensive standpoint. They believe IW can be used to gain knowledge about an adversary, to destroy or modify an adversary's information systems while at the same time protect its own from malicious codes and attacks. The purpose of IW is to gain an advantage over an opponent.

So, there is nothing surprising or significant in such a description of IW or in fact that there is no unique definition found among the modern European nation-states. They are all interested in modifying and destroying an adversary's information systems while protecting their own. And for all European nation-states, it is a natural progression from the weapons of industrial period war to that of the information period.

They believe that information warfare compromises all measures that:

- Assist in gaining information superiority
- Deny the opponent the use of the information or communications systems
- Ensure that the information and communications are protected against manipulation

They view the elements of offensive information warfare as:

- Information gathering, generally from open sources
- Electronic warfare
- Malicious codes such as viruses, hacking, and logic bombs
- Destructive attacks against information nodes
- Psychological operations

Again, nothing surprising here from a nation-state point of view. Information from open sources such as those found on the GII, NIIs of nation-states (e.g., Web pages) offer an abundance of information of interest to an adversary. Add to that the information posted on the Web sites of businesses around the world, coupled with mass news media Web sites and communications via e-mail. These allow the use of various netspionage methods to obtain information useful to an adversary. The use of electronic warfare as part of information warfare, or information warfare as part of electronic warfare, has been the subject of debate in several nation-states. In Germany, however, it appears that information warfare includes electronic warfare — and not the other way around.

Germans view the elements of defensive information warfare as:

- Information technology security
- Communications security
- Information management
- Perception management and control
- Media management

Once again, not a unique view of defensive information warfare elements; however, an area not often discussed and not systems based in perception management and control. This is a very powerful and possibly the most powerful tool in the defensive information warfare toolkit. For example, if one can make any adversary perceive matters in a manner conducive to one's own success, one will have a great advantage. Even something as "minor" as having an adversary perceive that your systems have been rendered useless while they covertly continue to function would be of great advantage. Also, ISPs have been requested to install monitoring equipment:

German Carriers Told to Install Cyber-Snooping Tech
Steve Gold, Newsbytes, Berlin, Germany, October 25, 2001
The German government has rushed through proposals forcing telcos to install cyber-snooping technology that would give police and security agencies access to most German communications.... Surveillance of telco links has been legal — with a court order — for many years in Germany. The new legislation seeks to automate the process, and move the costs of surveillance from the government to the carriers, officials said.

Germany has also looked at the economic impact of information warfare and believes that:

- The economy is highly dependent on information and information systems.
- The economy is both a driver and goal of information warfare.
- The economy is the target of military/political information warfare intentions.
- Future international conflicts involving the military will affect other areas.

Germany has also defined categories of information warfare damage, to include:

- Loss of data
- Illicit exploitation of capacity
- Data/process manipulation
- Manipulation of software, and its functions and execution

Germany views information warfare "players" as a multitude of organizations and individuals, including:

- The national government
- Administrative authorities
- Military forces
- Communications services
- Police/customs
- Political organizations
- Corporations that impact the economy
- Church and religious groups and sects
- Environmentalists
- Humanitarian organizations
- Individuals

They view their threat agents as:

- Organized crime groups such as the Triad and Mafia
- Secret societies
- Criminal hackers
- Terrorists, both national and international

They view threats as coming from:

- Insiders working alone
- Outsiders working alone
- Those with grudges
- Contractors
- Religious/nationalist fanatics
- Criminals
- Secret services and foreign agents

They view the motives to use information warfare tactics as follows:

- Gain competitive advantage
- Illegal gains
- Revenge
- Sect membership
- Terrorists aims
- Political and military aims

Although there are hackers operating globally from their homes in Germany, there does not appear to have been any reported incidents in which the government of Germany was behind the attacks. Due to the nature of information warfare, that possibility can never be ruled out. However, there was the documented incident several years ago of the German hackers working for the old KGB, as reported by Cliff Stoll in his book, *The Cuckoo's Egg*. With the demise of the old U.S.S.R., has the "new Russia" and its "new KGB" denounced such activities? As we shall see, not very likely!

Russia

> *I cannot forecast to you the action of Russia. It is a riddle wrapped in a mystery inside an enigma.*

> —Winston Churchill, Radio broadcast, October 1, 1939[10]

The first question you might ask: should Russia be considered in this chapter, or should it be dealt with as a part of Asia? In reality, since the break-up of the former Soviet Union, the country of Russia is more European than Asian and, as a result, is dealt with here. Russia, like many other nation-states, is charging headlong into the Information Age. Russia has more than 18,000 Web sites registered in the .ru domain and, as with other developing nation-states, the Internet is a rapidly growing phenomenon. There are numerous chat rooms dealing with political, social, and other issues.[11] Some of the Web sites of interest include:

- www.lenta.ru
- www.gazeta.ru
- www.polit.ru
- www.ripn.net
- www.deadline.ru
- www.ripe.net
- www.vesti.ru
- www.smi.ru
- www.rbc.ru
- www.apn.ru
- www.svoboda.org

Like many other nation-states, Russia is struggling with issues such as how to control its citizens' access to information. Projects such as SORM and SORM2 are being used by Russian government agencies and are also meeting with resistance from those who now have had a taste of human rights and individual freedoms. Thus, some of the IW tactics that Russia is developing may in fact also be used against its own citizens.

Add to that the potential and actual conflicts that confront Russia and one can see her government agencies aggressively developing information warfare weapons and tactics. Russia's conflicts and potential conflicts include:

- Russia: Dispute over at least two small sections of the boundary with China remain to be settled, despite 1997 boundary agreement; islands of Etorofu, Kunashiri, and Shikotan and the Habomai group occupied by the Soviet Union in 1945, now administered by Russia, claimed by Japan; Caspian Sea boundaries are not yet determined among Azerbaijan, Iran, Kazakhstan, Russia, and Turkmenistan; Estonian and Russian negotiators reached a technical border agreement in December 1996 but has not been ratified; draft treaty delimiting the boundary with Latvia has not been signed; has made no territorial claim in Antarctica (but has reserved the right to do so) and does not recognize the claims of any other nation; 1997 border agreement with Lithuania not yet ratified.[3]

We addressed the Chinese, the major information warfare "player" in Asia, views and information warfare actions in Chapter 8. In this chapter, it is important to discuss the major European "player" in information warfare, and that is Russia.

The Union of Soviet Socialists Republics (U.S.S.R.) is no more; however, that does not mean that the former U.S.S.R. portion now known as Russia can be completely trusted. However, it cannot be abandoned or ignored either. Its fledging democracy must be supported and aid given in situations where the aid can be verified and assist in promoting democracy, human rights, and capitalism in that nation-state.

As with China, Russia was surprised and maybe even shocked that the Coalition forces were able to apparently defeat Iraq and in so short a period of time. And like China, it provided Russia with a sense of urgency to modernize and prioritize its forces, to include a major emphasis on information warfare doctrine, tactics, and weapons. After all, Iraq had the world's fourth largest military force and was alleged to have used Russian military tactics — oops! So, one can see why Russia is interested in information warfare. After all, the United States and China, probably in that order, are still considered Russia's main potential adversaries.

How the Russians View Information Warfare[12]

Many of the Russian views of information warfare, like those of other nation-states (e.g., China), are public knowledge. The following are some of their published views:

> *...the danger of an information war breaking out is coming to the fore, and information warfare will soon rank second only to thermo-nuclear war by its consequences.*

> —Vladimir Markonenko
> Russian First Deputy General-Director
> Federal Government Communications and Information Agency (FAPSI)

> *...a Military point of view, the use of Information Warfare means against Russia or its armed forces will categorically not be considered a non-military phase of a conflict whether there were casualties or not... considering the possible catastrophic use of strategic information warfare means by an enemy, whether on economic or state command and control systems, or on the combat potential of the armed forces, ... Russia retains the right to use nuclear weapons first against the means and forces of information warfare, and then against the aggressor state itself.*

> —Professor V. I. Tsymbal
> "Concept of Information War" 1995

> *Information Warfare is a way of resolving a conflict between opposing sides. The goal is for one side to gain and hold an information advantage over the other. This is achieved by exerting a specific information/psychological and information/technical influence on a nation's decision making system, on the nation's populace and on its information resources structures, as well as by defeating the enemy's control system and his information resource structures with the help of additional means...*

> —Grau and Thomas
> *The Journal of Slavic Military Studies,* September 1996

> *The danger of unsanctioned intervention in the operations of information systems has increased abruptly.*

> —Vladimir Markonenko
> Russian First Deputy General-Director
> Federal Government Communications and Information Agency (FAPSI)

> *After the surfacing of hostilities, combat viruses and other information related weapons can be used as powerful force multipliers by their synergistic or mutually deprecating effects from multiple weapon types in proximity to one another.*

—Author unknown
Information taken from "National Security in the Information Age"
handout at an IW security conference

The Russians believe that electronic (information) warfare is never declared; it never stops; it is conducted covertly and knows no boundaries in space or time. They further believe that:

- The struggle has begun for control over computer networks.
- Information strikes are becoming increasingly dangerous because their effectiveness is growing very fast.
- Nation-states with a well-developed information science sphere are preparing for a computer war and developing and testing methods of affecting computer systems.

Russian Honey Pots?

The Russians have many Web sites, as do many other nation-states. If you find one relative to Russian military; published and produced by Russians; its contributors Russian military personnel, experts, and scholars; high-quality translation and grammatically perfect English; and contains scores of articles about the Russian military; is the site controlled by Russian intelligence services? One never knows, but if so, are the Russians learning more about the readers than the readers are learning about the Russians? For example, at a minimum the following is known of each visitor:

- Registered Internet host IP address
- Specific areas of interest by visitor

In addition, one may find that some ask you if you want to be placed on their e-mail listing for new information. Sounds harmless enough, but after providing name, rank (yes, one had a box to insert your military rank), organization, telephone number, e-mail address, snail mail address, etc., one can quickly see how this may prove useful in identifying potential human intelligence targets. Furthermore, by tracking your "travels" through their Web site, they can determine where your main interests lie. Coupling your corporation or government agency, such as a military department in the Pentagon, one can begin to piece together valuable information on a potential adversary.

From the above information, the Russians can gain more intelligence about its viewers. By checking the InterNIC Registry, they can learn:

- The organization and street address (for the IP Address), telephone, and fax
- Names of system administrators and technicians, their telephone and fax numbers, and e-mail addresses

The aggregate data, over time, can be used to identify worldwide interest by country or organization. Think about it. If you were responsible for collecting information on potential adversaries and also developing an information operations plan in the event you must use IW tactics against an adversary, what could you do with the information you are collecting? [By the way, was it coincidence that our firewall alerted us to a possible intruder when we accessed one particular Web site? Maybe — but then again...]

Russian View of Computer Viruses

Like China and many other nation-states, Russia has an interest in computer viruses. They believe that one can expect the appearance of a so-called remote virus weapon against command and control computers. These software inserts will be introduced via radio channels and laser communications links.

Who potentially can create and plant bombs into the software components of military systems? The personnel directly involved in their development and those who are well acquainted with software development technology and the combat employment of the system. The Russians are involved in developing application programs for nation-states that they may consider an adversary or potential adversary (e.g., the United States).

There are many Russians who now work and live in the United States and other Western nation-states. Many find work as application programmers for corporations in their newfound county. With their skills, were they allowed to leave Russia as an ordinary immigrant, or were they directed to leave and find work as programmers in specific corporations or types of corporations?

Some years ago, several security managers working for corporations in Silicon Valley (an area south of San Francisco, California, in the United States known for its many high-technology companies) were discussing personnel security. One remarked about the unusual number of Russian immigrants applying for jobs with his company. Another stated that he too was seeing an increase. Because both companies required a minimum background check prior to hiring, they decided to compare notes. It was found that several had attended the same Russian university at the same time and had graduated with computer science degrees. They arrived in the area at about the same time, lived at the same address, but they each applied to different corporations for employment at the same time. Perhaps a "divide and conquer" approach?[13]

> *All warfare is based on deception.*
>
> —Sun Tzu

Were these individuals allowed to leave Russia with no strings attached? Are the applications programmers installing trap doors, logic bombs, or other malicious codes in the software they are developing. If so, is anyone checking their programs in sufficient detail to find such codes? Should one be concerned, or is this concern outdated and part of the Cold War era? Of course, this does not only apply to Russians. There are numerous nation-states with foreign nationals working on sensitive application programs in their corporations. Is that a concern? Over the years, many nation-states have found that their own citizens have been involved in working for an adversary. So, is it any more a concern that it would be for anyone else writing software programs? What do you think?

> **Soviets Sign Accord to Export Software to United States**
> *Orange County Register newspaper, United States, date unknown*
> *...Not only do the Soviets want to sell software in the United States, they want to guarantee that it will work on domestic platforms...*

> **America Taps into Russian Brainpower**
> *Washington Technology, September 29, 1994*
> *U.S. companies are luring Russian high-tech talent for more profit and a greater market share.... The trouble is, most of that talent went into the country's defense sector....*

>> *War to the hilt between Communism and Capitalism is inevitable. Today, of course we are not strong enough to attack. Our time will come in fifty to sixty years. To win, we shall need the element of surprise. The Western world will have to be put to sleep. So we shall begin by launching the most spectacular peace movement on record. There shall be electrifying overtures and unheard of concessions. The capitalist countries, stupid and decadent, will rejoice to cooperate to their own destruction. They will leap at another chance to be friends. As soon as their guard is down, we shall smash them with our clenched fist.*

>> —Alleged declaration by Dimitry Manuilski
>> Professor, Lenin School of Political Warfare
>> Moscow, 1930

Russian Case Studies

The articles cited in Exhibit 2 are just a few examples of information warfare activities involving the Russians.

Exhibit 2.   Information Warfare Activities Involving the Russians

Russian Hackers Quietly Invade U.S.
infowar.com, June 26, 2001
Government Struggles to Defend against Mysterious New Security Threat.... Russian hackers are invading the U.S. electronically, quietly downloading millions of pages of sensitive data and even one colonel's entire e-mail inbox.

KGB Successor Monitors Net Traffic
The Associated Press, Moscow, February 21, 2000
The successor to the KGB is now spying on the Internet, raising fears that the information it collects could be used for blackmail and business espionage. "The whole Federal Security

Exhibit 2.   Information Warfare Activities Involving the Russians (Continued)

Service will be crying tomorrow over your love letters," warns one of the banners angry Russian Web designers have posted on the Internet.

Networks Attack from Russia?
Reuters, Washington, October 6, 1999
A major effort to pierce U.S. government and private-sector computer networks seems to have originated in Russia,...intruders had stolen "unclassified but still-sensitive information about essentially defense technical research matters."

Russia, Hack Zone
The Straits Times' Moscow-based writer Michael Walker reports infowar.com, October 25, 1999
Hackzone is the term Russian hackers use to describe their homeland.... Hackzone.ru is not only the name of a popular Russian hacking Web site,...individuals and probably some governmental institutions use them for global hacking...hundreds of hackers use their knowledge to attack foreign and domestic Web sites to steal commercial secrets.... Russian hackers often target U.S. Web sites,...successfully attacked the White House official Web site on one occasion earlier this year, "in retaliation for the U.S. air attacks against the Serbs."

Cyber-war Russians Hack into Military Secrets
infowar.com, Newsweek magazine, September 13, 1999
Russian cyber-spies with funding from Moscow have hacked into the computer network of the U.S. Defense Department, extracting information that may include classified naval codes and data on missile-guidance systems. The breach of security, code-named Moonlight Maze, is so serious that all workers at the Department have had to change their passwords in an effort to fight off the threat.

Hacker Invades Web Site Via Russia
infowar.com, Washington (AP), August 10, 1999
In the high-tech equivalent of a thief burglarizing police headquarters, a hacker using an unusually clever tactic vandalized a prominent Internet site devoted to computer security. The electronic assault against the AntiOnline site last week occurred days after other hackers altered the Web site for Symantec Corp., whose software is used by millions of consumers to protect against viruses and electronic snoops. In the latest incident, a hacker using an Internet account in Russia successfully tricked the site's computer to load hidden software code from elsewhere onto one of its own Web pages, called "Eye on the Underground."

Russia: Computer Networks of the Cold War: Is It Possible to Enter a Military Computer Network that Has No Internet Entry?
inforwar.com, date unknown, by Andrei Soldatov and Andrei Bystrov (Translation)
US security services believe that Russia poses a threat to the American computer networks. Russian specialists believe vice versa.... The Russian intelligence services possess the largest opportunities of waging information warfare throughout the world. Russia is focusing its foreign intelligence operations on industrial espionage...

Russian Hackers for Hire — The Rise of the e-Mercenary
01/Jul/01 by Ruth Alvey, July 1, 2001, rferl.org
The underemployment of highly skilled Russian hackers has increased the danger that intelligence agencies or criminal organizations will employ them for more sinister activities. In April 2001, the Russian newspaper *Moskovsky Komsomolets* reported that the U.S. embassy in Moscow had attempted to recruit a Russian hacker, known as 'Verse', to secretly collaborate "in the interests of American intelligence services against the Russian Federation."

Exhibit 2. Information Warfare Activities Involving the Russians (Continued)

US Losing Top Secrets to Russian Hackers
infowar.com, date unknown
US investigators say Russia has been stealing weapons data…. US officials believe Russia may have stolen some of the nation's most sensitive military secrets, including weapons guidance systems and naval intelligence codes, in a concerted espionage offensive that investigators have called operation Moonlight Maze…involved computer hacking…was so sophisticated and well-coordinated that security experts trying to build ramparts against further incursions believed the U.S. might be losing the world's first "cyber war,"…

Cyber Raiders
Scripps Howard News Service, Washington, date unknown
So far, as many as 23 countries are believed to have the capacity to engage in state-sponsored, surreptitious electronic raids. Among the most sophisticated: India, Syria and Iran, experts say…. Russia: Hackers working for the Russian government targeted Pentagon computer networks between January and May, apparently in search of naval codes and missile guidance data….

Russia Spies on Net Traffic; Civilians Worried about Blackmail, KGB Tactics
Infowar.com, date unknown
The successor to the KGB is now also spying on the Internet, raising fears that the information it collects could be used for blackmail and business espionage…. Russian human-rights and free-speech advocates say the security service has already forced many of the country's 350 Internet service providers to install surveillance equipment.

Comments:
Read the full story at:
http://www.abcnews.go.com/sections/tech/DailyNews/russia_netspy000221.html

CIA Says Russia and China Gearing up Cyber Attack Potential
infowar.com, February 23, 2000
The Central Intelligence Agency said on February 23rd that it has received recent indicators that countries such as Russia and China are building tools capable of attacking commercial computer networks. Cyber warfare programs appear to be proliferating globally, directed primarily at national infrastructures, and often replacing the concept of conventional warfare.

Comments:
Of course, this is not surprising. The surprise would be if other nation-states were surprised. *The CIA statement on the subject given to Congress may be found at http://www.cia.gov/ cia/public_affairs/speeches/cyberthreats_022300.html*

United Kingdom

The United Kingdom (U.K.) is made up of England, Scotland, Wales, and Northern Ireland. Its primary conflicts are with terrorists in Northern Ireland, as noted in Chapter 14. However, the United Kingdom, like other nation-states, does have some potential for conflicts over landmasses, as noted in the CIA *World Factbook*:[3]

- *United Kingdom:* Northern Ireland issue with Ireland (historic peace agreement signed April 10, 1998); Gibraltar issue with Spain; Argentina claims Falkland Islands (Islas Malvinas); Argentina claims South Georgia and the South Sandwich Islands; Mauritius claims island of Diego Garcia in British Indian Ocean Territory; Rockall continental shelf dispute involving Denmark, Iceland, and Ireland (Ireland and the United Kingdom have signed a boundary agreement in the Rockall area); territorial claim in Antarctica (British Antarctic Territory); Seychelles claims Chagos Archipelago in British Indian Ocean Territory

The United Kingdom has a number of government agencies that contribute to the U.K.'s information warfare strategies, policies, plans, and weapons systems. Among the major contributors are:

- The Prime Minister's Office
- Cabinet Office (CO)
- Home Office (HO)
- Security Service (SS)
- Security Intelligence Service (SIS)
- Ministry of Defence (MoD)
- Government Communications Headquarters (GCHQ)

Most of these organizations have their own Web sites and these give a strong indication of roles that the individual organizations undertake.

The Web site of the Prime Minister's Office provides good insight into the very heart of the decision making machine and gives up-to-date information on the position of the government with regard to current events. The Web site at http://www.pm.gov.uk/ has good links to articles on current issues.

The Cabinet Office includes a number of functions, including the Cabinet Secretariat for defense and overseas affairs, economic and domestic affairs, legislation, and European Union issues, and would be involved in the process of policy and strategy. It is rumored that within the Cabinet Office is the COBRA (Cabinet Office Briefing Room A) Committee, which briefs the Prime Minister during world crises. The COBRA Committee is believed to consist of top ministers and senior military staff who meet regularly during any crisis.

The Security Service (MI5) Web site at http://www.mi5.gov.uk/ gives in-depth information as to its structure and role, but this is summed up in the statement

> *...to protect national security, to safeguard the economic well-being of the U.K. against threats from overseas, and to act in support of the police and other law enforcement agencies in the prevention and detection of serious crime.*

Interestingly, the Security Intelligence Service (SIS) does not have a Web page of its own, but an indication of their involvement in information warfare can be found in the 1998–99 Intelligence and Security Committee Annual Report,[14] which states that:

> *...SIS made a significant effort to provide the Government with secret intelligence in the confusing and difficult situation in the Former Yugoslavia, including in support of military operations. The efforts of SIS in obtaining sources to gather intelligence from Saddam Hussein's inner circle continued to be very important. Another significant area for SIS was international counter-terrorism work.*

The Ministry of Defence has an obvious role and defines that role as "to defend the United Kingdom and Overseas Territories, our people and interests and act as a force for good by strengthening international peace and security" on its Web page at http://www.mod.uk/.

The GCHQ Web site at http://www.gchq.gov.uk/index.html gives good insight into its role in intelligence gathering. It describes itself as follows: "In business, knowledge is power. For GCHQ it's vastly more important. Knowledge ensures our nation's security, economic well-being, and protection against..."

Within the United Kingdom there are the same tensions present as there are in any other developed country. On one side is the requirement of the individual for privacy and the right to go about their business, while on the other side there is the need for government to maintain control and to make the environment a safe place to live and to do business. At the same time, law enforcement needs to be able to have the power to gain access to information that has been stored in digital form and also to protect the individual from coming to harm or being exploited. In the middle of all this is industry and commerce. Business needs a sterile environment to work, digital tools, and a place to work from.

The news items detailed in Exhibit 3 reveal some of the attitudes toward the current and developing environment within the United Kingdom.

Possible European Information Warfare Incidents and Conflicts

When a nation-state's information systems or Web sites are attacked, it is not easy to determine if the attacker is a government agent of another nation-state, a third party hoping to cause friction between several nation-states, or some teenage hacker out for a cyber-joyride. It is difficult enough to make that determination with state-of-the-art and proactive information warfare defenses in place. It is almost impossible with today's lax information warfare defenses.

If one were to engage in all-out information warfare against another nation-state because of attacks against one's systems, one may be attacking a nation-state's systems that were only a conduit for the attack actually being made by someone else. How can one ever be actually sure of the attacker? You cannot be sure of the attacker by just tracing the attacks through computers. Some very good human intelligence is still the best way of identifying the attacker. Thus, many in the United Kingdom and other nation-sates believe that more surveillance and monitoring are required as a means of intelligence collection activities.

Exhibit 3.   The Current and Developing Environment within the United Kingdom

Warning over Wiretaps
Wednesday, 22 August, 2001, 07:39 GMT 08:39 UK
Mark Ward, BBC News Online technology correspondent,
http://news.bbc.co.uk/hi/english/sci/tech/newsid_1500000/1500889.stm
Laws designed to catch computer criminals could result in a huge increase in the amount of covert surveillance carried out on British citizens by the police and intelligence services. The controversial Regulation of Investigatory Powers Act requires many companies providing communication services to install technology that allows up to one in 10,000 of their customers to be watched at the same time.

'Big Brother' Watching? In Britain, Quite Likely
Mike Collett-White, London, Reuters, August 13, 2001
A government decision…to broaden the network of roadside speed cameras…has raised fresh concerns…. There are estimates of 1.5 to 2.5 million closed circuit television (CCTV) cameras in Britain,…. But it was another example of how technological advances have made snooping easier, from hidden cameras to mobile phones and the Internet, creating an Orwellian-like nightmare of "Big Brother is Watching You"…

Kashmiri Hackers Attack BBC Site; Corporation Takes Server Offline
Will Knight, ZDNet U.K., 5/1/2001; http://www.msnbc.com/news/566618.asp
BBC has confirmed that hackers supporting the liberation of Kashmir have attempted to break into a server used to provide information to BBC staff around the world…attack is thought to have been carried out by hackers supporting Kashmir independence who intended to post Kashmiri separatist…

Inquiries Begin into Hacking of NSW Opposition's Files
http://www.abc.net.au/news/2001/08/item20010805130216_1.htm
Separate inquiries have been launched into claims a computer belonging to a senior New South Wales Government MP has been used to hack into the Opposition's files. New South Wales Premier Bob Carr returns from overseas tomorrow morning and heads straight into the political controversy. Opposition leader Kerry Chickarovski says the New South Wales Police Commissioner has promised a full and thorough investigation into the matter…. They concern sensitive policy documents that she claims were hacked into.

Comment:
While the item above is reporting on an Australian event, the reader should remember that there are very close ties between the United Kingdom and Australia, and that the Queen of England is still the sovereign of Australia.

If government officials are actually doing such attacks against each other, would they really have any qualms of attacking other nation-states or individuals or groups?

__Berlin Cyberwar Conference — Most Hacking Hides Real Threats__
infowar.com, July 3, 2001
The high profile of such relatively inconsequential online political warfare as denial-of-service attacks and playful site defacement has the general public distracted from much graver risks…. "Do Europeans care about information warfare?" asks Christiane Schultzki-Haddouti, a German journalist who specializes in information warfare. "Not much. Compared to America, Europe is still sleeping."

Russian–Chechen Conflict

The Russians' previous hot conflicts were with the Chinese over border disputes and in Afghanistan. In both instances, the fighting was industrial period based warfare of guns, bombs, aircraft, and soldiers. The same is true in Chechnya; however, there is a small aspect of information warfare tactics and that is the use of Web sites to provide information (or propaganda as some see it).

For example, if one were to go to http://www.qoqaz.net/, one would find the Chechen version of the war in that disputed territory. The site is available in Arabic, German, Swedish, Bosnian, Turkish, Malay, Italian, French, Ukrainian, Dutch, Albanian, Somalian, Indonesian, Spanish, Russian, English, and Macedonian. Thus, the Chechens are able to state their side of the conflict and try to draw on the nation-states and people from all over the world for support. They hope that by doing so, sufficient pressure would be brought to bear on Russia to withdraw from the area.

If one were to do an InterNIC search, would one find this Web domain somewhere in the heart of Chechnya? Look it up and find out. The answer may surprise you. What is interesting is that many of the sites are registered with Web masters located in another part of the world. If the Russian military wanted to attack and silence this Web site, located in another nonbelligerent nation-state, would that be considered "collateral damage" against the nation-state, or would it be considered an act of war because the "cyberspace" was actually located within the boundaries of a neutral nation-state?

Using a search engine and searching for *Chechnya,* the engine had 385,000 hits. Under *Russian Chechnya,* there were 156,000 hits; and under *Chechnya war,* there were 137,000. What this dramatically points out is that information warfare using Web sites for propaganda purposes, to get one's views across, is being used on a global scale. Prior to today's Internet, news of such rebellions and incidents could be controlled by the major nation-state. However, this aspect of information warfare can be done quickly, cheaply, and reach throughout the world. However, with the events of September 11, 2001, it is believed that the Chechens have lost much of their Western support for their cause because some identified terrorists had fought for them in that region.

The Greater Balkans Area

This area of the world is expected to continue to be an area of friction and conflict for years to come just as it has for centuries. When Yugoslavia broke up, "new" nation-states were formed, including:

- Slovenia
- Croatia
- Bosnia–Herzegovina
- Serbia and Montenegro
- Macedonia

These nation-states are made up of several religious groups: Catholic, Ortho-dox, and Muslim.

Conflicts arose when the Serbs tried to form a new, broader nation-state. When that happened:

- Orthodox Greeks began supporting the Orthodox Serbs.
- Russians resisted attempts by Western nation-states to mitigate the Serbs' aggression.
- The United Nations tried to decree an arms embargo that appeared to be largely ignored by the warring parties.
- Muslims from the Middle East and other regions supported the Muslims fighting in the Bosnia–Herzegovina area.
- Greece became concerned that the newly formed Macedonia may threaten their northern Greek territory; while at the same time Greece was obtaining economic ties with Bulgaria and Romania.
- The Muslim Croats found allies in Turkey and Iran.

Exhibit 4 is a map of the Balkan region. As one can easily see, the Balkans can be a source of a much broader conflict at any given time. Besides these nation-states, add in the nation-states that send in troops under NATO and UN banners; as well as all the treaties and organizations to which they all belong, such as NATO. A strike against one is a strike against all. Thus, one can easily see how another "world war" can come again to Europe.

Exhibit 5 depicts some of the possible reasons for conflicts in the Balkans.

The nation-states involved in the Kosovo, Croatian–Serbian, and NATO–Serbian conflicts, just to name a few, have all used various forms of information warfare tactics. As usual, because of the low level of NII in those nation-states, excluding the NATO nation-states, much of the information warfare was through Web site, chat rooms, and e-mails denouncing the other side and promoting their own. There were, of course, the usual Web site defacements, denial-of-service attacks and virus attacks. However, none of these were sophisticated; but then again, they did not have to be, to be successful in the minds of the attackers. On the other hand, some NATO forces, like the United States, decided to try out some of their information warfare tactics and weapons on Milosevic and his supporters (see Exhibit 6).

Kosovo Air Campaign

The Kosovo Air Campaign is rich with IW examples and lessons learned. Perhaps the most important lesson learned, or rather relearned, in the coalition effort against Milosevic was the poor public relations campaign on the part of the Allies. Public relations is a polite way of saying perception management. NATO may have won militarily, but lost this round of the IW campaign. "The enemy deliberately and criminally killed innocents by the thousands, but no one saw it," stated Admiral Ellis, Commander, Joint Task Force NOBLE ANVIL during Operation ALLIED FORCE. "We accidentally killed innocents sometimes by the dozens and the world watched on the evening news. We were

Exhibit 4.   Map of the Balkan Region

continuously reacting, investigating, and trying to answer, 'How could this happen?'"[15] It happened because of an arrogant oversight that the Serbs would be effective at manipulating international opinion. Underestimating the competition is a critical planning failure.

The Serbs would shell people or shoot them and throw them in craters created by Allied bombs. CNN and other news services were permitted in the country and shown "the Allied atrocities." Not until Allied forces were able to view the sites for themselves and conduct chemical and other forensic analysis did they uncover Milosevic's scheme.

Another tactic used by the Serbs to force NATO into reactive mode was to go after NATO computer systems. These attacks marked the first time that NATO's computer systems had been attacked in wartime.[16] NATO used defensive measures to protect its e-mail and Web site systems against a well-thought-out propaganda campaign launched by the Serbs. The risk is that without a rapid solution, the hackers may move on to more damaging activities such as downloading press releases and overhead imagery available on the site, tampering with them, and then releasing them as official policy. "All of this is well prepared, and part of [Yugoslav President Slobodan] Milosevic's propaganda war," the source explained.[17]

Country	(Population)	Slovenians	Croats	Serbs	Muslims	Macedonians	Albanians	Greeks	Turks	Kurds	Catholic	Orthodox	Muslim
		Ethnicity									Religion		
Slovenia	(1.9M)	91%	3%	2%	1%						96%		1%
Croatia	(4.6M)		78%	12%	1%						75%	11%	1%
Bosnia and Herzegovina	(4.6M)		17%	31%	44%						15%	31%	40%
Serbia and Montenegro	(10.7M)			63%			14%					65%	19%
Macedonia	(2.2M)					65%	22%		4%			67%	30%
Albania	(3.3M)						95%	3%			10%	20%	70%
Greece	(10.5M)							98%				98%	
Turkey	(62.1M)								80%	20%			99%
Cyprus	(0.7M)							78%	18%			78%	18%

(Slovenia through Macedonia grouped under: **Yugoslavia**)

Internal Conflict

Outside Intervention

Exhibit 5.   Possible Reasons for Conflicts in the Balkans

U.S. Air Force Major General Ronald Keys, U.S. European Command's Director of Operations, portrayed U.S. military leaders as just beginning to grapple with the enormous potential — and limitations — of information as a tool of warfare. Keys noted that U.S. officials must begin to think broadly about IW. "This is not just STO [special technical operations] and computer attack."[18]

The Serbs understood asymmetrical warfare and employed the full spectrum of capabilities available to them. Non-military participants took part in the attack on NATO's computer systems, employing neither STO nor computer attack. A Yugoslav Internet provider and a peacetime graphic designer, Nenad Cosic, spent up to 18 hours a day compiling what amounted to an anti-NATO air defense early warning system, to include a blow-by-blow description of the strikes. "We can hear the idiots flying towards Yugoslavia," warned an e-mail sent from Slovenia hundreds of miles away. "Good luck, Yugoslavia!"[19] Could these civilians have been considered combatants, considering that the information was shared with the military, and been subjected to hostile Allied action? Probably not, according to the outstanding legal interpretations in *Cyber Space and the Use of Force,* written by Walter Gary Sharp, Sr. This book is the definitive think-piece in this murky legal area. Even if the civilians were considered to be combatants, they would not have been subjected to direct lethal action because of the negative public relations that would have ensued. Instead, the sites would have been subjected to electronic warfare or their power sources would have been destroyed.

Exhibit 6.   NATO Forces Try Out Information Warfare Tactics on Milosevic and His Supporters

Cyberwar? The U.S. Stands to Lose Experts Argue Plan to Raid Milosevic's Bank Accounts Would Do More Harm than Good
infowar.com, May 28, 2001
CIA was planning to tinker with international bank accounts full of Slobodan Milosevic's money — just another way of getting under the Yugoslav president's skin. Information warfare experts disagree about the feasibility of such a cyberattack.

Serb Computer Viruses Infected Western Firms
infowar.com, August 17, 1999
Pro-Serbian hackers attacked more than 170 organizations worldwide, including some in Britain, in a cyber war directed at shutting down key computer systems in NATO countries during the Kosovan conflict.... Black Hand, a notorious Serbian paramilitary group, and its sympathisers in Russia, Latvia, Lithuania and eastern Europe, targeted banks, Internet service providers and media organizations in revenge for NATO bombing campaigns.

Digital Bombing Technology
Howard Banks, Forbes, November 15, 1999, p. 94
During the Kosovo War, the air force aimed precision weapons through the drawing room windows of Slobodan Milosevic's official residence. This hit was made possible by a piece of software called RainDrop, developed by a firm far from Silicon Valley and almost entirely out of sight of Wall Street. The firm is Comptek Research, Inc. of Buffalo, N.Y.

The Other Kosovo War
William Arkin and Robert Windrem, MSNBC, August 30, 2001, http://www.msnbc.com/news/607032.asp?cp1=1
On the night of May 15, 1999 day 53 of the Kosovo War a flight of American B-2 bombers attacked two industrial facilities in eastern Serbia. Before the raid, under a covert operation dubbed Matrix, U.S. and British information warfare specialists used e-mail, faxes and cell phones to forewarn the plant owners of the attack. The warnings had nothing to do with limiting casualties, nor were the targets of great military value. Rather, the operation was designed to send a message to cronies of President Slobodan Milosevic enriching themselves through these factories: Prevail upon the Yugoslav leader to withdraw his forces from Kosovo or face further attacks on your sources of income.

Among the low-technology ways that the Serbian military sought to undo NATO's combat advantage was by intermittent use of its integrated air defense system. Serb forces were aware that the standard U.S. Air Force approach was to destroy an adversary's air defense system at the outset of a conflict, a factor toward achieving air supremacy for Allied planes. Leaving their mobile radars turned off for most of the war allowed Serb forces to maintain a constant, although severely hindered, threat against Allied aircraft. Fearing loss of public will and coalition cohesion if there were combat losses, Allied aircraft were restricted to flying above 15,000 feet for most of the air campaign. Due to these rules of engagement, hundreds of Serbian ground targets survived the conflict. The Serbs won that round because they used IW to influence NATO to take actions that benefited the Serbs.

Those Serbs with computer attack skills provided their services to their government. The Black Hand, a Serbian paramilitary group, and Russian, Latvian, Lithuanian, and Eastern European sympathizers went after more than 170 organizations worldwide, targeting banks, Internet service providers, and media organizations. More than 30 British companies suffered partial shutdowns or were denied access to their computer systems. For some, the disruption cost up to £25,000 per day.[20] Nationalistic pride in the form of revenge can be attributed to these actions, but so too can the hope that disruptions would slow the pace of NATO's military actions and also permit the Serbs more time to swing international opinion in their favor.

Another example of the Allied forces losing an IW round was in psychological operations (PSYOPS). The Allies dropped poorly translated leaflets on Serb locations, and radio and television transmissions from Allied aircraft and ground stations were focused on Serb population centers and military forces. Djordje, a mechanical engineering student at Belgrade University, said, "I cannot understand how such an organization with so many analysts and brains can transmit something like this to Serbs. They should at least broadcast our folk music. It is completely unconvincing."[21] Poorly executed PSYOPS will harden resolve against the attacking force.

The Allies did have IW successes but they were a struggle to achieve. The Commander, U.S. Air Forces Europe, and the U.S. Chairman of the Joint Chiefs of Staff confirmed successful IW attacks against the Serb's air defense system. The major problem was that it took too long for U.S. operators to find an offensive computer attack target and then get permission to strike.[22] As with any form of war in which decision makers are unfamiliar (in this regard, computer network attack) and the laws lagging technology, decisions are often slow due to conservatism and caution. Allied IW could have been far more effective — certainly the capabilities were available. The limiting factors were that laws, doctrine, and strategy were either nonexistent or immature and untested.

Admiral Ellis said, "Properly executed, IW could have *halved* the length of the campaign."[15] Proper execution means not just having capabilities, but also a framework of laws, strategy, and decision processes, a total package. Timothy Thomas, an analyst for the U.S. Army, articulates this well.[23] Information superiority is a relative state between actors, and is achieved when one can process information, execute faster, and achieve desired end states before the competition. While high technology is important, it is prudent not to be enamored with it because this leads to ignoring perhaps better no and low technology approaches. The Serbs won the IW campaign.

Many practical lessons for those not engaged in combat can be found in the Kosovo Air Campaign. Continuing the "from battlefield to the boardroom" theme, is there a difference between a public relations or advertising campaign and what Milosevic tried to do to sway international opinion to support his cause? He masterfully and persuasively orchestrated the press, television, and radio. He applied information and misinformation in a well-thought-out plan to grab the imagination and backing of people, some of whom were willing to give their lives. That is stronger branding than "Intel inside." Looking at the other side of the coin, what is the underlying reason for business decisions? Revenue and profit. Decision processes are fine-tuned to this. There is a keen understanding of the cost of a delayed decision, often articulated in the present and future value of money and opportunity costs. As business can learn from the military, so can the military learn from business.

Governments can afford sophisticated and complex research and development to produce the best IW technologies. As technologies evolve, what at one time was bleeding edge becomes outmoded. This now-"obsolete" technology often finds its way into industry — and into the hands of activists.

Azerbaijan and Armenia

In today's world, even the most basic of nation-state information warfare capabilities are being used as they would any other weapons in their arsenals. Even such nation-states as Azerbaijan have this capability. One wonders what will happen over the years as the NIIs of these nation-states are developed.

> **A Glimpse of Cyberwarfare: Governments Ready Information-Age Tricks to Use against Their Adversaries**
> *Warren P. Strobel with Richard J. Newman, World Report, March 13, 2000*
> *Hackers from Azerbaijan…tampered with dozens of Armenian-related Web sites, including host computers in the United States. Experts suspect involvement or support from the Azerbaijani government, which imposes tight controls over Internet use within its borders…*

Ukraine

In the Ukraine, a computer virus released classified documents into the wrong hands.

> **Web Site Says Virus Leaked Ukrainian Secret Documents**
> *infowar.com, August 2, 2001*
> *A Ukrainian Web site…received secret documents from the administration of President Leonid Kuchma due to a computer virus that infected government computers and e-mailed it the files. The Sircam virus, having infected the computers in the presidential administration, is bombarding our editorial department with their documents…*

This is an interesting incident because it is not just another virus attack or just another virus. The details of this matter, if ever fully known, may prove to be an interesting case study on how an information warrior can gain access to the secrets of a nation-state. One thing is certain: a basic information systems security principle was surely violated. The principle is never to have classified documents residing on a network that is not self-contained in a classified mode, residing in a classified environment — unless encrypted with the best encryption software available.

Summary

European nation-states are some of the oldest on Earth and, as such, have a long history of conflicts. The majority of these countries have been invaded a number of times and have learned from living under imposed foreign governments with alien cultures. At the same time, European countries are the very ones that invaded and colonized so many parts of the world and seeded their cultures and values in all continents. As they have modernized and become more and more information dependent, they have come to realize that their NIIs and other massive networks are vulnerable, as are those of their potential adversaries. Nation-states such as the United Kingdom, Sweden, France, Russia, and Germany are actively involved in both information warfare defensive and offensive strategies, policies, and tactics, often using the United States' approach as a baseline.

At the same time, many "Third-World" European nation-states (using their technology sophistication and dependency as a gauge) are beginning to use information warfare tactics as part of their overall warfare strategies. Even those European nation-states with little in the way of sophisticated information warfare tools are using the information warfare tactics and weapons that they have, albeit primitive, in their conflicts with their adversaries.

Notes

1. Go to cia.gov and read the *World Factbook* appendix relating to associations and such.
2. *GUARDIAN* newspaper, April 29, 1999.
3. United States Central Intelligence Agency *World Factbook*.
4. Winn Schwartau defined information warfare by these three classes.
5. C2 is the acronym for command and control; CIS is the acronym for computer information system.
6. See ZDNET-France article, entitled The Swedish Government Wants Its "Soldiers of Information," September 12, 1999.
7. See http://www.fhs.se.
8. The authors are grateful to Manuel Wik for his support of this project and for allowing us to publish his papers in this book.
9. These comments have been widely circulated but never verified as coming from Mr. Marion. However, they do represent what appears to be an accurate view by France of the world of civil and military information warfare.
10. *Encarta Book of Quotations,* Microsoft Corporation. All rights reserved. Developed for Microsoft by Bloomsbury Publishing Plc 1999.
11. Russian Authorities and the World Wide Web, by Floriana Fossato, March 2000, http://www.rferl.org.
12. A special thanks to Paul Zavidniak, Information Warfare Manager, Logicon Corporation.
13. We do not mean to single out Russians or any others as being the only ones who might use such tactics and we mean no offense to the honest, hardworking immigrants that have made our nation-states so great. However, when it comes to information warfare tactics and defenses, one must never take such things at face value.
14. See http://www.official-documents.co.uk/document/cm45/4532/4532-02.htm#gen1.

15. U.S. Commander in Kosovo Sees Low-Tech Threats to High-Tech Warfare, Elaine M. Grossman, *Inside The Pentagon,* September 9, 1999.

16. A Hacker Attack against NATO Spawns a War in Cyberspace, Frank Vizard, *Popular Science,* September 17, 1999.

17. NATO Reinforces against Net Attack from Serbs, Elizabeth de Bony, IDG News Service, April 2, 1999.

18. Air Force General Found Information Tools a Mixed Bag in Kosovo War, *Defense Information and Electronics Report,* January 7, 2000.

19. Web Site Warns of Attacks on Serbs, George Jahn, Associated Press, April 24, 1999.

20. Serb Computer Viruses Infected Western Firms, Maeve Sheehan, *The Sunday Times (Britain),* August 15, 1999.

21. Propaganda; NATO TV Is Sent to Serbs, Who Are Harsh Critics, G. Platt, *The New York Times,* May 26, 1999.

22. Telecom Links Provide Cyber-Attack Route, David A. Fulgham, Washington, November 17, 1999.

23. Kosovo and the Current Myth of Information Superiority, Timothy L. Thomas, Foreign Military Studies Office, Fort Leavenworth, Kansas, Parameters, Spring 2000, p. 13–29.

Chapter 11

Information Warfare:
Nation-States
of the Americas

Government is not reason, it is not eloquence, it is force; like fire, a troublesome servant and a fearful master. Never for a moment should it be left to irresponsible action.

—George Washington

This chapter discusses information warfare as it relates to the Americas.

Introduction to the Americas

When a person outside of the Americas thinks of America, they often think of the United States. However, the Americas are made up of some 35 nation-states, or entities. Some are independent nation-states, some are "owned" by other nation-states, and some are in between, as "protectorates."

The Americas primarily comprise the following nation-states and entities:

- *South America:* Argentina, Bolivia, Brazil, Chile, French Guiana, Guyana, Colombia, Ecuador, Paraguay, Peru, Suriname, Uruguay, Venezuela, and Falkland Islands
- *Central America:* Mexico, Belize, Costa Rica, El Salvador, Guatemala, Honduras, Nicaragua, and Panama
- *Caribbean:* Bahamas, Barbados, Bermuda, Cuba, Dominican Republic, Haiti, Jamaica, Puerto Rico, and Trinidad and Tobago
- *North America:* Greenland, Iceland,[1] Canada, and United States

With the exception of the North American grouping, many of these nation-states have been plagued, and many still are, by dictators, terrorists, and drug cartels, while others are governed by democracies. Cuba is the last holdout in the area that still represents a Communist nation-state. Its days of supporting Communist guerrilla forces in the region have waned over the years, such as Nicaragua and El Salvador. Other nation-states such as Columbia are known not for their culture, beautiful country, and friendly people, but for drug cartels and violence. Still other nation-states in the region are plagued by guerrilla forces that appear to be at least as interested in drug operations and other criminal activity as trying to replace a government. Some are supported by Caribbean entities whose banks launder drug and terrorist funds.

Many of these nation-states use information warfare tactics to their advantage as best they can. However, except for the more modern nation-states such as the United States and Canada, the information warfare tactics are about the same as previously described. That is, the use of Web sites to present their side of the conflicts, e-mails, and once in awhile some unleashing of viruses against an adversary or the defacement of Web sites of the adversary.

Associations

Like other nation-states of the world, a number of countries in this region also belong to nation-state associations, as noted in the CIA *World Factbook*. Some of those more closely related to the Americas include:[2]

- *Agency for the Prohibition of Nuclear Weapons in Latin America and the Caribbean (OPANAL):* Headquartered in Mexico City, Mexico, it was established in 1967 with the aim of "encouraging the peaceful uses of atomic energy and prohibit nuclear weapons." It consists of 32 members: Antigua and Barbuda, Argentina, the Bahamas, Barbados, Belize, Bolivia, Brazil, Chile, Colombia, Costa Rica, Dominica, the Dominican Republic, Ecuador, El Salvador, Grenada, Guatemala, Guyana, Haiti, Honduras, Jamaica, Mexico, Nicaragua, Panama, Paraguay, Peru, Saint Kitts and Nevis, Saint Lucia, Saint Vincent and the Grenadines, Suriname, Trinidad and Tobago, Uruguay, and Venezuela
- *Andean Community of Nations (ACN):* Formerly known as the Andean Group (AG), the Andean Parliament, and most recently the Andean Common Market (Ancom). It is headquartered in Lima, Peru. It was established in 1969 with the aim to "promote harmonious development through economic integration." It consists of five members — Bolivia, Colombia, Ecuador, Peru, Venezuela — with Panama as an associate member.
- *Asia-Pacific Economic Cooperation (APEC):* Headquartered in Singapore, this association was established in 1989 to "promote trade and investment in the Pacific basin." Its 21 members include: Australia, Brunei, Canada, Chile, China, Hong Kong, Indonesia, Japan, South Korea, Malaysia, Mexico, New Zealand, Papua New Guinea, Peru, Philippines, Russia, Singapore, Taiwan, Thailand, United States, Vietnam; and three observers (Association

of Southeast Asian Nations, Pacific Economic Cooperation Conference, South Pacific Forum).

- *Caribbean Community and Common Market (Caricom):* Headquartered in Georgetown, Guyana, and established in 1973, this association's aim is "to promote economic integration and development, especially among the less developed countries." It has 14 members: Antigua and Barbuda, the Bahamas, Barbados, Belize, Dominica, Grenada, Guyana, Jamaica, Montserrat, Saint Kitts and Nevis, Saint Lucia, Saint Vincent and the Grenadines, Suriname, Trinidad and Tobago; three associate members (Anguilla, the British Virgin Islands, and the Turks and Caicos Islands); and ten observers (Aruba, Bermuda, the Cayman Islands, Colombia, Dominican Republic, Haiti, Mexico, Netherlands Antilles, Puerto Rico, and Venezuela).

- *Caribbean Development Bank (CDB):* Headquartered in St. Michael, Barbados, this association became effective in 1970 with the aim to "promote economic development and cooperation." It has 20 members (Anguilla, Antigua and Barbuda, the Bahamas, Barbados, Belize, the British Virgin Islands, the Cayman Islands, Colombia, Dominica, Grenada, Guyana, Jamaica, Mexico, Montserrat, Saint Kitts and Nevis, Saint Lucia, Saint Vincent and the Grenadines, Trinidad and Tobago, the Turks and Caicos Islands, Venezuela) and also six non-regional members (Canada, China, France, Germany, Italy, the United Kingdom).

- *Central American Bank for Economic Integration (BCIE):* This association (acronym from Banco Centroamericano de Integracion Economico) was established in 1960 with the aim to "promote economic integration and development." It has five members (Costa Rica, El Salvador, Guatemala, Honduras, and Nicaragua) and four non-regional members (Argentina, Colombia, Mexico, and Taiwan).

- *Central American Common Market (CACM):* This association is headquartered in Guatemala; established in 1960; and later collapsed in 1969. It was reinstated in 1991. Its aim is to "promote establishment of a Central American Common Market." It consists of five members: Costa Rica, El Salvador, Guatemala, Honduras, Nicaragua; note: Panama, although not a member, pursues full regional cooperation.

- *Economic and Social Council (ECOSOC):* This organization is headquartered at the United Nations, New York City, was established in 1945 and is an example of one of the numerous organizations to which nation-states of the Americas belong. Thus, like most of the other regions of the world, they are closely tied to nation-states around the world. The aim of this organization is "to coordinate the economic and social work of the UN; includes five regional commissions (see Economic Commission for Africa, Economic Commission for Europe, Economic Commission for Latin America and the Caribbean, Economic and Social Commission for Asia and the Pacific, Economic and Social Commission for Western Asia) and 10 functional commissions (see Commission for Social Development, Commission on Human Rights, Commission on Narcotic Drugs, Commission on the Status of Women, Commission on Population and Development, Statistical Commission, Commission on Science and Technology for Development, Commission on Sustainable Development, and Commission on Crime Prevention and Criminal Justice). Members — (54) selected on a rotating basis from all regions."

- *Economic Commission for Latin America and the Caribbean (ECLAC):* Headquartered in Santiago, Chile, and established in 1948 as the Economic Commission for Latin America (ECLA), its aim is to "promote economic development as a regional commission of the UN's Economic and Social Council." It has 41 members: Antigua and Barbuda, Argentina, the Bahamas, Barbados, Belize, Bolivia, Brazil, Canada, Chile, Colombia, Costa Rica, Cuba, Dominica, the Dominican Republic, Ecuador, El Salvador, France, Grenada, Guatemala, Guyana, Haiti, Honduras, Italy, Jamaica, Mexico, the Netherlands, Nicaragua, Panama, Paraguay, Peru, Portugal, Saint Kitts and Nevis, Saint Lucia, Saint Vincent and the Grenadines, Spain, Suriname, Trinidad and Tobago, the United Kingdom, the United States, Uruguay, Venezuela; and seven associate members: Anguilla, Aruba, British Virgin Islands, Montserrat, Netherlands Antilles, Puerto Rico, and the Virgin Islands.

- *Organization of American States (OAS):* One of the few of the associations to be headquartered in the United States, excluding the UN associations, this organization is headquartered in Washington, D.C. It was established in "April 1890 as the International Union of American Republics; 30 April 1948 adopted present charter; effective — 13 December 1951. Its aim is to promote regional peace and security as well as economic and social development." It has 35 members: Antigua and Barbuda, Argentina, the Bahamas, Barbados, Belize, Bolivia, Brazil, Canada, Chile, Colombia, Costa Rica, Cuba (excluded from formal participation since 1962), Dominica, Dominican Republic, Ecuador, El Salvador, Grenada, Guatemala, Guyana, Haiti, Honduras, Jamaica, Mexico, Nicaragua, Panama, Paraguay, Peru, Saint Kitts and Nevis, Saint Lucia, Saint Vincent and the Grenadines, Suriname, Trinidad and Tobago, the United States, Uruguay, Venezuela; and 45 observers: Algeria, Angola, Austria, Belgium, Bosnia and Herzegovina, Bulgaria, Croatia, Cyprus, Czech Republic, Egypt, Equatorial Guinea, the European Union, Finland, France, Germany, Ghana, Greece, Holy See, Hungary, India, Israel, Italy, Japan, Kazakhstan, South Korea, Latvia, Lebanon, Morocco, the Netherlands, Pakistan, Poland, Portugal, Romania, Russia, Saudi Arabia, Spain, Sri Lanka, Sweden, Switzerland, Thailand, Tunisia, Turkey, Ukraine, the United Kingdom, and Yemen.

These are just a few of the more noteworthy organizations to which the nation-states of the Americas belong. One can see that there are some with local and some with global ties. Most of the associations are based on economic ties, or various other ties through the UN. As was mentioned earlier in this book, economic warfare, coupled with territorial conflicts, and political conflicts, or variations of these three, can also give rise to conflicts in which information warfare may play a role.

Territorial Conflicts and Potential Conflicts

When looking at the Americas, one finds that the nation-states and other entities of the region are no different than any other part of the world. Their conflicts and potential conflicts often bring in other nation-states of the world to support either side. The Falklands war between the United Kingdom and Argentina is but one example. In that case, the United States was a vocal supporter of the United Kingdom, although the United States did not provide direct military support in that conflict. The territorial conflicts in the region, according to the United States' Central Intelligence Agency's *World Factbook* are as follows:

- *Argentina:* Claims U.K.-administered Falkland Islands (Islas Malvinas); claims U.K.-administered South Georgia and the South Sandwich Islands; territorial claim in Antarctica
- *Belize:* Territory in Belize claimed by Guatemala; precise alignment of boundary in dispute
- *Bolivia:* Has wanted a sovereign corridor to the South Pacific Ocean since the Atacama area was lost to Chile in 1884; dispute with Chile over Rio Lauca water rights
- *Brazil:* Two short sections of boundary with Uruguay are in dispute — Arroio Invernada (Arroyo de la Invernada) area of the Rio Quarai (Rio Cuareim) and the islands at the confluence of the Rio Quarai and the Uruguay River
- *Canada:* Maritime boundary disputes with the United States (Dixon Entrance, Beaufort Sea, Strait of Juan de Fuca, Machias Seal Island)
- *Chile:* Bolivia has wanted a sovereign corridor to the South Pacific Ocean since the Atacama area was lost to Chile in 1884; dispute with Bolivia over Rio Lauca water rights; territorial claim in Antarctica (Chilean Antarctic Territory) partially overlaps Argentine and British claims
- *Colombia:* Maritime boundary dispute with Venezuela in the Gulf of Venezuela; territorial disputes with Nicaragua over Archipelago de San Andres y Providencia and Quita Sueno Bank
- *Cuba:* U.S. Naval Base at Guantanamo Bay is leased to the United States and only mutual agreement or United States abandonment of the area can terminate the lease
- *Ecuador:* Demarcation of the agreed-upon border with Peru was completed in May 1999
- *El Salvador:* The Honduras–El Salvador Border Protocol ratified by Honduras in May 1999 established a framework for a long-delayed border demarcation, which is currently underway; with respect to the maritime boundary in the Golfo de Fonseca, the ICJ referred to the line determined by the 1900 Honduras–Nicaragua Mixed Boundary Commission and advised that some tripartite resolution among El Salvador, Honduras, and Nicaragua likely would be required
- *Falkland Islands (Islas Malvinas):* Claimed by Argentina
- *France:* Madagascar claims Bassas da India, Europa Island, Glorioso Islands, Juan de Nova Island, and Tromelin Island; Comoros claims Mayotte; Mauritius claims Tromelin Island; territorial dispute between Suriname and French Guiana; territorial claim in Antarctica (Adelie Land); Matthew and Hunter Islands east of New Caledonia claimed by France and Vanuatu
- *French Guiana:* Suriname claims area between Riviere Litani and Riviere Marouini (both headwaters of the Lawa)
- *Guatemala:* Territory in Belize claimed by Guatemala; precise alignment of boundary in dispute

- *Guyana:* All of the area west of the Essequibo River claimed by Venezuela; Suriname claims area between New (Upper Courantyne) and Courantyne/ Kutari [Koetari] Rivers (all headwaters of the Courantyne)
- *Haiti:* Claims U.S.-administered Navassa Island
- *Honduras:* The Honduras–El Salvador Border Protocol ratified by Honduras in May 1999 established a framework for a long-delayed border demarcation, which is currently underway; with respect to the maritime boundary in the Golfo de Fonseca, the ICJ referred to the line determined by the 1900 Honduras–Nicaragua Mixed Boundary Commission and advised that some tripartite resolution among El Salvador, Honduras, and Nicaragua likely would be required; maritime boundary dispute with Nicaragua in the Caribbean Sea
- *Iceland:* Rockall continental shelf dispute involving Denmark, Ireland, and the United Kingdom (Ireland and the United Kingdom have signed a boundary agreement in the Rockall area)
- *Nicaragua:* Territorial disputes with Colombia over the Archipelago de San Andres y Providencia and Quita Sueno Bank; with respect to the maritime boundary question in the Golfo de Fonseca, the ICJ referred to the line determined by the 1900 Honduras–Nicaragua Mixed Boundary Commission and advised that some tripartite resolution among El Salvador, Honduras, and Nicaragua likely would be required; maritime boundary dispute with Honduras in the Caribbean Sea
- *Peru:* Demarcation of the agreed-upon border with Ecuador was completed in May 1999
- *Southern Ocean:* Antarctic Treaty defers claims (see Antarctic Treaty Summary in the Antarctica entry); sections (some overlapping) claimed by Argentina, Australia, Chile, France, New Zealand, Norway, and United Kingdom; the United States and most other nations do not recognize the maritime claims of other nations and have made no claims themselves (the United States reserves the right to do so); no formal claims have been made in the sector between 90 degrees west and 150 degrees west
- *Suriname:* Claims area in French Guiana between Riviere Litani and Riviere Marouini (both headwaters of the Lawa); claims area in Guyana between New (Upper Courantyne) and Courantyne/Koetari [Kutari] Rivers (all headwaters of the Courantyne)
- *United States:* Maritime boundary disputes with Canada (Dixon Entrance, Beaufort Sea, Strait of Juan de Fuca, Machias Seal Island); U.S. Naval Base at Guantanamo Bay is leased from Cuba and only mutual agreement or United States abandonment of the area can terminate the lease; Haiti claims Navassa Island; the United States has made no territorial claim in Antarctica (but has reserved the right to do so) and does not recognize the claims of any other nation; Marshall Islands claims Wake Island
- *Uruguay:* Two short sections of the boundary with Brazil are in dispute — Arroyo de la Invernada (Arroio Invernada) area of the Rio Cuareim (Rio Quarai) and the islands at the confluence of the Rio Cuareim (Rio Quarai) and the Uruguay River
- *Venezuela:* Claims all of Guyana west of the Essequibo River; maritime boundary dispute with Colombia in the Gulf of Venezuela

South America

South America is involved in information warfare activities; however, very little of it is related to nation-state conflicts. The majority appears to be conflicts due to drug cartels and terrorist activities in the region. For example, in the United States' "war on drugs," it is rumored that information warfare operations are taking place to read e-mails and other communications between drug cartels; as well as attacks against their computer systems and tracing the flow of drug funds through various laundering activities (e.g., Caribbean banks). Exhibit 1 is a map depicting the major South American nation-states and entities.

Argentine Hacker

An incident in South America drew the attention of both Argentine and U.S. officials. A 23-year-old Argentine man pleaded guilty to hacking into U.S. university and military computer networks where he reportedly obtained access to sensitive but unclassified research files on satellites, radiation, and energy-related engineering. Although he was sentenced under a plea agreement to three years of probation and a $5000 fine, the agreement allowed him to serve his probation in Argentina.

This case received significant media coverage when it occurred as a sensational example of a non-U.S. hacker penetrating primarily government systems. Although one lesson to be drawn from this case is that intruders are able to access systems with sensitive information from many places outside North America and Europe. The results of a successful prosecution are likely to be very disappointing. In this example, a national government was able to track down, with the aid of a court-ordered wiretap, an international hacker. What is also interesting is that the father of the hacker reportedly was an intelligence officer for the Argentine military. If this were true, is this the case of "plausible denial" using a hacker in the event the individual is caught? The ramifications of such action and the response by the United States, if tied directly to government officials, may have been quite different. Is there some covert plan related to this incident based on the United States siding with the United Kingdom over the Falkland Islands dispute with Argentina? One never knows; and in information warfare where the adversary can easily hide in cyberspace, one may never know.

Peru and Ecuador

When conflicts arise between nation-states (e.g., Peru and Ecuador) of South America, like other regions, they "exchange shots" over the Internet via Web sites and e-mail. However, as with other less-developed regions, their use of information warfare is relatively primitive compared to the capabilities of more modern nation-states.

Exhibit 1.   Map of South America

Central America

Mexico

Central America, like South America, is involved in the United States' "war on drugs." Mexico has become a major player in that war, both as a conduit for drugs and also possibly home to some of the drug cartels.

Exhibit 2 provides a map of Central America and the Caribbean, depicting the major nation-states and entities.

Mexico also has an interesting war on its hands and that is the war of words for the "hearts and minds" of the people of Mexico, specifically the Zapatistas. The Zapatistas have been in rebellion for some time. This is another indication of the information warfare tactics of groups being used against a more powerful nation-state government — and in doing so, giving the group the equivalent power of the nation-state, vis-à-vis information warfare tactics and strategies.

According to Major General Michael V. Hayden, who in 1997 was the Commander, United States Air Intelligence Agency, the Zapatista rebels used "information cover" and not air cover to stop attacks against them. Apparently, the Zapatistas were withdrawing from an area in Mexico and expected to be attacked by the Mexican Air Force. Consequently, they sent out e-mail and through their Web site made their case known to the world. The Mexican government was flooded with feedback from all over the world to such an extent that they thought the "world was watching" their actions. Consequently, the Mexican Air Force did not bomb the Zapatistas.

This is interesting in that if true, a "poor rebel group" using the Internet could gather sufficient world opinion as to impact the decisions of a nation-state's military. While some nation-states spend literally billions of dollars on air defense systems, the Zapatistas were able to gain the same effect with a computer, e-mails, and a Web site. Now, this may not always work for a group trying to stave off air attacks, but in this case at least, the defensive information warfare tactics appeared to provide air cover for their movements.

In another incident, Mexican rebels had hacked into the Mexican government's Web page (http://www.shcp.gob.mx) and defaced it with anti-government propaganda. The defacement included a statement saying: "We belong to no group, we do not belong to the Zapatista Army for National Liberation, but we are expressing our free expression as Mexicans...." The Zapatistas are credited with pioneering the use of the Internet to support their cause (their Web site is located at http://www.ezln.org).[3]

Another example of the use of the Internet for applying information warfare tactics against a nation-state is as follows:

Mass Virtual Direct Action (MVDA), Individual Virtual Direct Action (IVDA) And Cyber-wars
Tim Jordan, Open University, date unknown
MVDA emerged in 1998 with the call by the Electronic Disturbance Theatre (EDT) for mass participation in a denial-of-service assault on Mexican Government networks in protest against attacks on the Zapatistas, using a tool called Floodnet.[4]

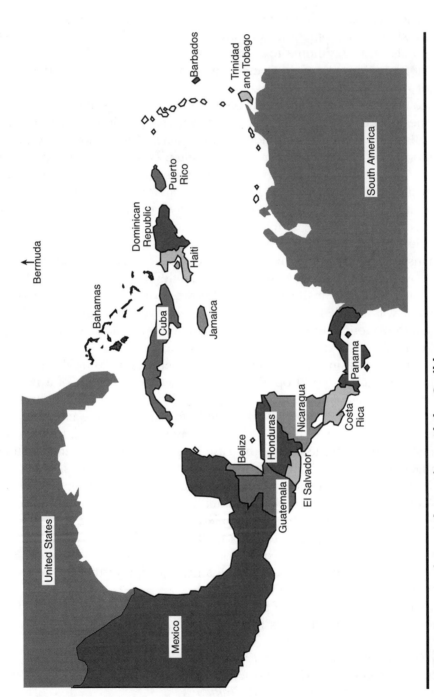

Exhibit 2.   **Map of Central America and the Caribbean**

Such attacks as these are common throughout the world. However, what it points out is the power of today's technology to rally forces around the world for a specific cause. This is somewhat similar to the use of information warfare tactics mentioned in a previous chapter covering East Timor.

Cuba

Cuba, one of the world's last communist-run nation-states has had Internet access since about 1996.[5] However, it is estimated that only about 60,000 Cubans out of a total population of about 11 million people have e-mail capabilities and of those, less than half can e-mail outside Cuba. As is always the case when a communist-run government takes charge, they quickly rush to control the media (e.g., newspapers, radio, and television).

As far as can be determined, there appears to be little in the way of information warfare tactics being used by Cuban exiles, or any type of ongoing information warfare between Cuba and those against the communist regime. This is somewhat surprising considering how effective the Zapatistas in Mexico have been in using the Internet against the Mexican government. The Cuban government does use its Web sites and e-mail for propaganda purposes. There have also been alleged attempts traced back to Cuba of unauthorized accesses to United States' computer networks. However, because of vulnerabilities that allow for the penetration of computers and their use as a proxy, one never knows if these attacks are actually from Cuba or they are from another nation-state.

Possibly, they are from other countries using Cuban systems because they were able to take them over, or to start a conflict between the United States and Cuba. There are many Cuban-related Web sites, as there are with every nation-state and those that oppose their governments. The reader can find them using the various search engines available on the Internet (three Web sites related to Cuba are provided as examples and identified in the notes at the end of this chapter[6]).

North America[7]

North America is dominated by two nation-states: Canada and the United States. They are both modern, information-based, and information-dependent nation-states. This section draws heavily from the Web sites of government agencies. Instead of paraphrasing their information warfare visions and missions, they are quoted in detail to offer the reader first-hand information relative to their information warfare views.

Exhibit 3 provides a map of North America depicting the major nation-states and entities.

Exhibit 3.   Map of North America

Canada

Canadian territorial "conflicts" are minor and appear at this time to only be with the United States, as noted above. Canada is a modern nation-state with a relatively mature and growing national information infrastructure (NII). Its military is relatively small and primarily based on homeland defense. However, it can be called on to supply some military forces outside its region if required.

Similar to most modern nation-states, hackers, Internet fraudsters, and the usual type of miscreants found on the global and national networks these days have plagued Canada. Canadian officials have been focusing on these type of crimes, as well as on terrorists. Canada's military is also modernizing to confront the information warfare threats of this century.

Canadian Companies Victimized by Hackers

In a report published in the *Counterintelligence News and Developments* (Volume 3, September 1998) the Canadian Secret Intelligence Service (their equivalent of the U.S. CIA) stated that an unnamed foreign government had ordered its intelligence service to obtain specific business intelligence. This intelligence service contacted computer hackers to help achieve the objective, which resulted in the loss and compromise of numerous computer systems, passwords, personnel, and research files of two companies.[8]

"Netspionage" (network-enabled espionage) by official government agencies may involve the use of common computer criminals/hackers. By using members of the cyber under-culture to achieve their objectives, the foreign government sponsors of a netspionage effort gain increased deniability in the event that the operation is ever discovered or publicized in the media. Although it may be extreme, to some another "advantage" of using hackers is that they may be terminated "with extreme prejudice" if necessary. As a chilling example, many believe that at least one of the German hackers who helped the Russian KGB search out "Star Wars" information in the early 1980s was allegedly killed by the KGB. This may have been done to prevent him from disclosing to Western investigators the full extent of the KGB's netspionage.[9]

Canada's "Information Warfare Plan" 2001

Canada's Defence Plan 2001[10] alludes to information warfare, but provides very little in the form of details (which is an example of good operations security, or OPSEC). It states:

> *...307. INFORMATION MANAGEMENT GROUP— ADM(IM)*
> *General.... The integrity of our information holdings must be assured while striving to exploit the use of information from many sources to achieve information superiority. Rapid advancements in technology necessitate an almost constant state of building and improving the information management infrastructure (IMI) with the ultimate goal being an Integrated Information Environment (IIE).... Mission. The mission of ADM(IM) is to provide leadership, products and services needed to manage information as a mission critical corporate resource.... ADM(IM) Tasks. ADM(IM) is assigned the following tasks in support of the Defence Objectives. The ADM(IM) business plan is to indicate the intended level of effort for these tasks:*
> *a. Defence Objective 1 (IMDO1): To provide strategic defence and security advice and information to the Government.*
> *b. Task. Provide the DM and CDS with advice on Information Management, Information Technology and Information Operations.*
> *c. Defence Objective 2 (IMDO2): To conduct surveillance and control of Canada's territory, airspace and maritime areas of jurisdiction.*
> *d. Tasks:*
>   *1. Maintain and improve the capability to conduct Information Operations (IO) from Canadian land based facilities and to co-ordinate IO on air, land and sea assets. Priority is to be given to defensive IO; and*
>   *2. Maintain and continuously operate the Information Management Infrastructure in support of Command, Control and Intelligence Systems (C²IS) and Departmental Command and Control.... ADM(IM)*

> *Goals. ADM(IM) is assigned the following goals to support change objectives. The ADM(IM) Business Plan is to indicate the intended level of effort for these goals: Change Objective 3 (IMC03): Modernize Goal. Active Information Protection Capability. Develop an active information protection capability with initial operational capability (IOC) by 31 March 2003.[11]*

A report in the *Montreal Gazette* on October 6, 2001, reported that although Canada has not acknowledged that it "employs hackers to break into computers...very likely Canadian spy agencies...have been hacking into computers for years." The report went on to say:

> *Canada's military says it wants to engage in hacking...want to include hacking in their military arsenal...military hackers would be trained to disable communication systems, destroy electronic information and plant destructive viruses....*
>
> *When we talk about information warfare, people don't see it applies to them. But it does. We've created this social space (on the Internet), and conflict is moving into it. Every decision you make is mediated by computer. In that sense, the computer layer becomes very powerful when you can manipulate it.*
>
> —Robert Garigue[12]
> retired Canadian Forces lieutenant-commander

With the close cooperation between the United States and Canada in military and political affairs, it is expected that the Canadian military forces will draw heavily on the United States' expertise in information warfare to upgrade its own information warfare defenses and offensive capabilities. This would include such operations as information war-gaming and training. Such assistance and support would not be unusual as the precedent was set years ago with both nation-states sharing industrial-age military information, holding joint training exercises and the like.

United States

The last of the America's nation-states being addressed is the United States. The United States, while probably the most information-dependent and information-driven nation-state, is also the leader in the high technology that drives information warfare. However, perhaps more importantly, the United States is the leader in terms of advances in information warfare strategies and tactics, and weapons systems, as well as vulnerabilities. One can write volumes as to what the United States is doing relative to information warfare, which is of course beyond the scope of this book. An overview is presented with Web sites cited so, if interested, the reader can obtain more detailed information about many aspects of what U.S. government agencies are doing concerning information warfare.

Many of the world's nation-states have taken information warfare developments in the United States and adopted and adapted them to meet their own information warfare needs. These include Germany, Sweden, the United Kingdom, Canada, and even China and Russia to a large extent.[13] So, the information presented here will give the reader some indication as to what other nation-states that follow the United States' lead are also doing.

Everyone knows that the United States' NII and other information infrastructures, like other nation-states around the world, are vulnerable to both Industrial-Age and Information-Age attacks. This, of course, is because most modern nation-states use the same hardware, firmware, and software as has been discussed (e.g., Chapter 6 on COTS). However, what seems to set the United States apart from the majority of the nation-states of the world is the fact that it is the biggest target for attacks by anyone in any part of the world.

To compound the U.S. information warfare vulnerabilities is the fact that the United States is relying heavily on high technology to meet the defensive and offensive needs of the U.S. military. Over the past decade or so, the United States has dispensed with:[14]

- 709,000 regular (active duty) personnel
- 293,000 reserve troops
- Eight standing army divisions
- 20 air force and navy air wings with:
  - 2000 combat aircraft
  - 232 strategic bombers
- 19 strategic ballistic missile submarines containing 3114 nuclear warheads on 232 missiles
- 500 ICBMs with 1950 warheads
- Four aircraft carriers and 121 surface combat ships and submarines...plus all the support bases, shipyards, and logistical assets needed to sustain such a force

It was assumed that, with the decreased U.S. defense budget, especially over the eight years of the Clinton Administration, that high technology would be used to make up the difference. However, as we have seen since the September 11, 2001 massacre, the United States must still rely on Industrial-Age warfare weapons, albeit "smart weapons," in its wars with today's adversaries (e.g., terrorists). After all, most of the adversaries of the United States are still in the agricultural or industrial periods of maturity as nation-states or more primitive as, for example, living in caves. If one were to rely solely on information warfare tactics against the Taliban, what would the United States attack?

In the following paragraphs, we will address:

- How the United States views information warfare
- How the United States is structured to support information warfare
- What the United States is doing to defend itself against successful attacks

- What the United States is doing to advance the art of information warfare through its military laboratories
- What the United States, and any other nation-states vulnerable to information warfare attack, should do now and in the future

U.S. Overview of Information Warfare Issues

The topic of information warfare (IW) in the United States is viewed from many perspectives, many depending on the viewers area of expertise, responsibilities, and personal agendas. However, most seem to agree that IW has one fundamental objective and that is to gain "information superiority" or "information dominance" over an adversary. The U.S. methodology is as follows:

- Affect (disrupt, exploit, corrupt) adversary information (and information systems)
- Protect U.S. information (and information systems)

The United States breaks down the information warfare domain into several sub-areas. For example, on offense:

- Targets:
  - Counter/disrupt adversary's command, control, and communications (C^3)
  - Disrupt, exploit, corrupt the adversary's information systems/networks (computers, data links, etc.)
  - Both military and civilian
- Techniques/methods of attack:
  - Electronic warfare (e.g., jamming)
  - Hacker-like intrusions into computer systems
    - Intelligence systems
    - Command and control (C^2)
    - Commercial systems
- Breaking cryptography
- Physical destruction
- Deception
- Psychological operations (leaflets, video, broadcasts, etc.)

Some argue from a macro systems-based approach, beginning with the requirements definition/architecture design/systems acquisition process. This construct postulates that the first step is to get one's own command, control, communications, computers, and intelligence (C^4I) and battle management right, after which one can execute many of the information warfare missions described above.

Others say that the high technology is changing too fast for a top-down approach to succeed, and that the United States' course must be to encourage user-driven solutions within a framework of standards that allows interoperability, broad access to data, etc.

Some claim that most of the United States' present approaches are aimed at re-fighting Desert Storm with new technologies; and to understand the impact of the information revolution on warfare, more effort should be put into understanding its impact on society at large, and so forth.

The United States has coined new terms to address the various aspects of information warfare. Some argue that the terminology was changed to make it more "politically correct." For example, there is less use of the term "information *warfare*." After all, the United States does not want to be seen as some big, war-like nation-state. Decades ago, the U.S. War Department was renamed the Department of Defense for that very same reason (e.g., "We do not make war, but only fight wars in self-defense"), and that sort of thing. Similar action has apparently been taken by the United Kingdom (e.g., one information warfare-related organization is now " a secure information" organization).

New terms have been added, primarily by the U.S. military, to describe the various components of information warfare — or as some argue, information warfare is a component of the larger picture (e.g., information operations or information superiority). Some of the changes are the result of a maturing of the concept of information warfare and its role in the total warfare environment.

Some of these components are called defensive information warfare (DIW), information operations (IO), information assurance (IA), and yes, good old information security (INFOSEC). Take a look at the defensive aspects of all this. There is a movement to integrate and have these various concepts work in a cohesive way to begin a series of steps toward true and total information protection and defense. True, the use of the word "total" seems unrealistic, but that is what one must strive to accomplish. So why not say so? If one just settles on "mitigating risks," that means "striving" for information protection mediocrity. If one is to have an information protection goal, one should strive for the ultimate and not settle for something short of that. Yes, one has a lot of walking to do before running and a lot of running to do before the goal is even realistically in sight. Concepts such as DIW, IO, and IA are bringing the United States much closer to that goal.

Before discussing the topics of IW, IO, etc., one point to keep in mind is that the U.S. military agencies first thought of some of these concepts and put them to use. They have followed a long history of technological tradition when it comes to the development of technology. From the computer to the hardware/software developed for space travel, the government's military agencies have been at the forefront. There are of course several good reasons for that — they have more money to spend on such things than a private business and the need for national defense made research and development in technology a necessity. However, those of you who are in the commercial world should not discard what the government has developed and is using to protect information — not that they are doing such as great job themselves. However, they have some good ideas. As we continue to have more common systems, common architectures, and common protection needs, the fact that it was developed by a government agency has no bearing on its usefulness to protect business information.

The United States understands that it is also important to remember that today's and tomorrow's adversaries will attack the private sector as quickly in the information environment that exists, as it would a military sector, if not more so. One just has to look at what factories (non-government) were bombed by the Allies in World War II and the Battle of Britain to clearly understand that today's commercial information systems are subject to attack from adversaries — both foreign and domestic. One also just has to recall September 11, 2001, to bring that point home.

The U.S. military, it is believed, first formalized the term "information warfare" (IW). They stated it involves actions taken to achieve information superiority by affecting adversary information, information-based processes, information systems, and computer-based networks while defending one's own information, information-based processes, information systems, and computer-based networks (U.S. DoD Directive 3600.1).

Then it follows that offensive IW is to deny, corrupt, destroy, or exploit another's information, or influence another's perception, while DIW is to safeguard one's information or systems from similar action. At the same time, one wants to exploit the available information in any combination that will succeed in enhancing one's decision/action cycle while disrupting another's cycle.

Information operations (IO) is another of those terms that have grown out of the U.S. information warfare concept. A U.S. DoD directive (IATAC TR-97-002) says "IO conducted during time of crisis or conflict to achieve or promote specific objectives over a specific adversary or adversaries."

> *...defining how joint forces use information operations (IO) to support our national military strategy. Our ability to conduct peacetime theatre engagement, to forestall to prevent crisis and conflict and to fight and win is critically dependent on effective IO at all levels of warfare and across the range of military operation...*

> —General Henry H. Shelton
> former Chairman of the Joint Chiefs of Staff

The U.S. military Joint Publication 3-13, Joint Doctrine for Information Operations, says the following:

> *Information operations (IO) involve actions taken to affect the adversary information and information systems while defending one's own information and information systems[15].... IO target information or information systems to affect the information-based process, whether human or automated.... Many different capabilities and activities must be integrated to achieve a coherent IO strategy.*

The U.S. military also believes that in IW conflicts, intelligence and communications support is critical to conducting offensive and defensive IO. The design and correct operation of information systems are fundamental to the overall conduct of IO. Additionally, to achieve success, IO must be integrated with other operations and contribute to national and military objectives.

In the United States today, many are tired of the global attacks on their networks and Web sites. Some in the United States are fighting back (is this offensive IO?). This is tricky. The old U.S. Wild West philosophy and vigilantism sounds like a just and noble cause. However, one must be sure of the true identity of the attacker and understand the ramifications of attacking the wrong system — not to mention the fact that it may just be against the law and one can be held civilly and criminally responsible for such activities.

When one is concerned with what appears to be attempted or actual unauthorized access, how does one know it is some form of malicious attack by an adversary as a prelude to or act of war, or an accident? One must consider the effects that human error and misconfiguration of systems may have on the issue. Although not an attack against a military information warfare target, the following actual example makes that point clear:

> *A large international corporation called in an information systems security consultant and explained that they were under attack by their competitor who was headquartered in another country. They were ready to file a lawsuit against that competitor who appeared to be accessing their systems without authorization and possibly also performing a denial-of-service attack against them. Furthermore, the attack appeared to be coming through a third party in a third country in an apparent attempt to hide the origin of the attacks. This of course is not unusual and was considered a further indication of the covert attacks by the competitor. The consultant traveled to both other countries, and talked to law enforcement authorities on the best course of action. The consultant also discussed the matter with the apparently unwitting third party whose systems were being used as the conduit for the attacks. While collecting audit trail evidence and conducting detailed analysis of the third party's system configuration, it was determined that the third party had a business relationship and was networked with both parties. It was further determined that the systems were misconfigured in such a manner as to cause the problems being experienced. Thus, it was not an attack at all! Can you imagine the consequences of a lawsuit if the "victim" corporation had hastily filed the lawsuit and the subsequent cause of the attacks was determined?*

One would not want to attack the wrong target and begin a war by accident. In cyberspace, the chances of that occurring are much greater than in the physical world.

Defensive IO is another concept that is an integral part of the entire realm of information warfare. Again quoting from the cited publication:

> *Defensive IO integrate and coordinate policies, procedures, operations, personnel, and technology to protect and defend information and information systems.... Offensive IO also can support defensive IO.... Defensive IO ensures the necessary protection and defense of informa-*

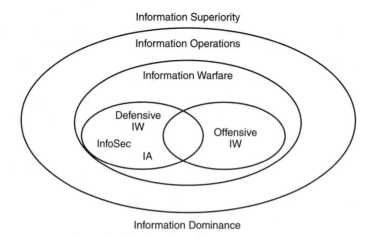

Information Superiority

Information Operations

Information Warfare

Defensive IW

Offensive IW

InfoSec

IA

Information Dominance

Exhibit 4.   Information Operations (IO)

> *tion and information systems on which joint forces depend to conduct operations and achieve objectives.... Four interrelated processes support defensive IO: information environment protection, attack detection, capability restoration and attack response.... Because they are so interrelated, full integration of the offensive and defensive components of IO is essential...should plan, exercise, and employ available IO capabilities and activities to support integrated defensive IO...*

Information superiority is also a useful concept to integrate into a holistic information protection program. Information superiority "is that degree of dominance in the information domain which permits the conduct of operations without effective opposition." (U.S. DoD Directive 3600.1)

Some would argue that information warfare is a subset of information operations (IO). IO occurs across the spectrum of conflict, whereas IW may be reserved for combat operations. This entire topic of information warfare, as seen in the Twenty-first Century, is considered by many to be part of the "revolution in military affairs" (RMA) that is occurring throughout many of the world's nation-states' military organizations. It is fairly new; therefore, there is some discussion and confusion as to how all these concepts are integrated into today's warfare environment. It is a "work in progress."

Exhibit 4 depicts one example of such a concept.

Information assurance (IA) is one of the latest hot topics related to information warfare. Is it INFOSEC by another name, a subset, or just the other way around? There is some argument as to the pecking order. Information assurance (IA) is information operations (IO) that

> *...protect and defend information and information systems by ensuring their availability, integrity, authentication, confidentiality and non-repudiation. This includes providing for restoration of information systems by incorporating protection, detection and reaction capabilities.*

—U.S. DoD 3600.1

For the purposes of this definition, the following meanings also apply:

- *IA authentication:* security measure designed to establish the validity of a transmission, message, or originator, or a means of verifying an individual's authorization to receive specific categories of information (National Telecommunications Information Systems Security Instructions (NSTISSI) 4009)
- *IA availability:* timely, reliable access to data and information services for authorized users (NSTISSI 4009)
- *IA confidentiality:* assurance that information is not disclosed to unauthorized persons, processes, or devices (NSTISSI 4009)
- *IA integrity:* protection against unauthorized modification or destruction of information (NSTISSI 4009)
- *IA non-repudiation:* assurance the sender of data is provided with proof of delivery and the recipient is provided with proof of the sender's identity, so neither can later deny having processed the data (NSTISSI 4009)

These are the basic concepts on which the United States is developing its information warfare defensive and offensive strategies. How they relate to one another may change over time. There also may be new concepts developed and some of the above integrated into them or renamed. Regardless, when one tries to understand U.S. information warfare strategies, policies, plans, and weapons, one should clearly first understand the above concepts.

U.S. Information Warfare-Related Agencies

The United States has many government agencies that contribute to its information warfare strategies, policies, plans, and weapons systems. Among the major contributors are:

- National Security Council (NSC)
- United States Congress (USC)
- Central Intelligence Agency (CIA)
- Department of Defense (DoD)
  - Joint Chiefs of Staff offices (JCS)
  - National Security Agency (NSA)
  - Defense Intelligence Agency (DIA)
  - Defense Information Systems Agency (DISA)
  - Department of the Air Force (USAF)
  - Department of the Army (USA)
  - Department of the Navy — includes the Marine Corps (USN and USMC)
  - Defense Advanced Research Projects Agency (DARPA)
  - Homeland Security

Exhibit 5 depicts the organizational structure that plays a major role in U.S. information warfare efforts.

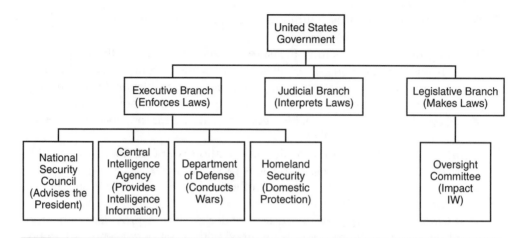

Exhibit 5.   Organizational Structure of U.S. Information Warfare Efforts

National Security Council (NSC)

The NSC was established by the National Security Act of 1947. The Council is in the Executive Office of the President, who is also Commander-in-Chief of the military forces of the United States. The NSC is chaired by the President. Its regular attendees (both statutory and non-statutory) are the:

- Vice President
- Secretary of State
- Secretary of the Treasury
- Secretary of Defense
- Assistant to the President for National Security Affairs

According to the Office of the President:

> *The Chairman of the Joint Chiefs of Staff is the statutory military advisor to the Council, and the Director of Central Intelligence is the intelligence advisor. The Chief of Staff to the President, Counsel to the President, and the Assistant to the President for Economic Policy are invited to attend any NSC meeting. The Attorney General and the Director of the Office of Management and Budget are invited to attend meetings pertaining to their responsibilities. The heads of other executive departments and agencies, as well as other senior officials, are invited to attend meetings of the NSC when appropriate.*
>
> *The National Security Council is the President's principal forum for considering national security and foreign policy matters with his senior national security advisors and cabinet officials. Since its inception under President Truman, the function of the Council has been to advise and assist the President on national security and foreign policies. The Council also serves as the President's principal arm for coordinating these policies among various government agencies.[16]*

When it comes to any type of conflict, the NSC is, as implied above, the group on which the U.S. President heavily relies for counsel. Information warfare, especially those related to offensive information warfare, would be debated and recommendations given to the President (e.g., attack and subvert the networks of China if China attacked Taiwan).

The use of any offensive information warfare weapons would of course be a topic for the NSC to consider and weighed against the potential of an adversary to wage information warfare against the United States using offensive information warfare weapons against the vulnerable military and NII networks. The United States may hold back on the use of such weapons to limit the war; however, that position implies that the adversary will play by the same rules — an unlikely event, especially if the adversary is losing the war. In the United States, information about offensive information warfare strategies, plans, and weapons is considered highly classified — in fact, similar to that of nuclear weapons because they both can be considered weapons of mass destruction, especially against an information-dependent and vulnerable nation-state such as the United States.

One might wonder if that level of detail (e.g., use of offensive information warfare weapons) would be discussed by the NSC or such matters as choice of weapons would be left to the military forces. Such matters would probably be discussed, but the level of detail may vary. Surely, the war plans would be discussed and debated by the NSC and those plans would include information warfare tactics. As history has shown, the President's staff has been involved in all conflicts; for example, World War II (drop or not drop the atomic bomb), Korean War (use nuclear weapons, cross the Yalu River), Vietnam (whether or not to bomb North Vietnam, a foot bridge, a village), and the Gulf War and War on Terrorism.

United States Congress (USC)[17]

The U.S. Congress, as it relates to information warfare, is responsible for initiating applicable legislation to include funding, as well as providing oversight (e.g., Armed Services, Appropriations, and Intelligence). The USC relies on the President to enforce the legislation passed and also to keep the USC apprised of developments vis-à-vis warfare.

One of the most lacking warfare-related tasks of the USC has been its reluctance to formally declare war on a nation-state, group, or other entity. The United States last formally declared war in December 1941. The Korean War was a "police action" with no formal declaration of war. The Vietnam War was not a formal war, nor was the Gulf War. However, in each of these conflicts, hundreds of thousands of military personnel and civilians were killed or wounded.

In this latest "War on Terrorism," no formal declaration of war was declared by the USC. Some have argued that such a declaration would be too restrictive on the President because of "legal" issues related to a formal declaration of war.

As it relates to information warfare, offensive operations would be primarily covert and at a level of detail not requiring any type of USC approval. However, the USC members may or may not be briefed in detail as to the information warfare offensive and defensive actions taken. As we have seen, members of the USC or their staffs often "leak" the classified information to the news media. This has led to the President sometimes restricting briefings to only key members of the USC. For some in the United States, it is difficult to differentiate between information leaks in peacetime versus information leaks in wartime which may get people killed. As they said in World War II: "Loose lips sink ships." The same concept applies today in information warfare.

The USC's recourse in the event of the use of any type of information warfare is either to agree with the President's actions or disagree. Either way, their power is limited to providing or withholding funding or passing legislation that would prohibit some action by the President. If the President ignored or in some way violated the USC, a Constitutional crisis would arise.

Central Intelligence Agency (CIA)

My greatest concern is that hackers, terrorists organizations, or other nations might use information warfare techniques as part of a coordinated attack designed to seriously disrupt infrastructures such as electric power, ATC, financial, international commerce, deployed military forces in time of peace and war.

—John M. Deutch, Former DCI[18]

The CIA, like the NSA, is a key agency for providing intelligence information to U.S. government officials. And as noted in earlier chapters, intelligence collections as part of any information warfare efforts are crucial. See Exhibit 6 for an excerpt from the CIA's Web site.[19]

The CIA often operates using open source information but also, obviously, covertly conducts much of its missions and projects. Of major importance to the CIA is its ability to collect information via technological means (e.g., spy satellites). Over the years, its ability to use people to collect information (HUMINT) has been severely limited by USC legislation. One can see many uses of the CIA as it relates to information warfare because intelligence collection is at the heart of any warfare strategy, as noted in Chapter 7.

As with the NSA, some of the collection means may take advantage of TEMPEST. Also, the CIA using HUMINT may determine the configuration of a potential adversary's networks, obtain copies of password files, and even the use of moles at key positions within an adversary's NII or DII. These individuals may collect information and be in a position to insert malicious codes in time of war. In addition, working with various software and hardware vendors, the CIA can learn what systems, hardware, and firmware are being installed within a potential adversary's nation-state. They may possibly, like the NSA may do, work with vendors to insert trap doors and the like that could be used in time of war. As stated in Chapter 6, COTS are used throughout the world. Potential adversaries of the United States, such as China, are undoubtedly concerned about such matters and are likely to be conducting various tests to ensure that no trap doors and the like are in their U.S. vendor-provided information systems products.

Exhibit 6.   Excerpt from the CIA's Web Site

The Central Intelligence Agency was created in 1947 with the signing of the National Security Act by President Truman. The National Security Act charged the Director of Central Intelligence (DCI) with coordinating the nation's intelligence activities and correlating, evaluating and disseminating intelligence which affects national security. The *Director of Central Intelligence (DCI)* serves as the head of the *United States Intelligence Community*, principal advisor to the President for intelligence matters related to national security, and head of the Central Intelligence Agency (CIA).

The CIA is an independent agency, responsible to the President through the DCI, and accountable to the American people through the intelligence oversight committees of the U.S. Congress. CIA's mission is to support the President, the National Security Council, and all officials who make and execute the U.S. national security policy by:

- Providing accurate, comprehensive, and timely foreign intelligence on national security topics.
- Conducting counterintelligence activities, special activities, and other functions related to foreign intelligence and national security, as directed by the President.

To accomplish its mission, the CIA engages in research, development, and deployment of high-leverage technology for intelligence purposes. As a separate agency, CIA serves as an independent source of analysis on topics of concern and also works closely with the other organizations in the Intelligence Community to ensure that the intelligence consumer — whether Washington policymaker or battlefield commander — receives the best intelligence possible.

As changing global realities have reordered the national security agenda, CIA has met these new challenges by:

- Creating special, multidisciplinary centers to address such high-priority issues such as nonproliferation, counterterrorism, counterintelligence, international organized crime and narcotics trafficking, environment, and arms control intelligence.
- Forging stronger partnerships between the several intelligence collection disciplines and all-source analysis.
- Taking an active part in Intelligence Community analytical efforts and producing all-source analysis on the full range of topics that affect national security.
- Contributing to the effectiveness of the overall Intelligence Community by managing services of common concern in imagery analysis and open-source collection and participating in partnerships with other intelligence agencies in the areas of research and development and technical collection.

The CIA is made up of the following directorates:

- The *Directorate of Intelligence*, the analytical branch of the CIA, is responsible for the production and dissemination of all-source intelligence analysis on key foreign issues.
- The *Directorate of Science and Technology* creates and applies innovative technology in support of the intelligence collection mission.
- The *Directorate of Operations* is responsible for the clandestine collection of foreign intelligence.
- The *Center for the Study of Intelligence* maintains the Agency's historical materials and promotes the study of intelligence as a legitimate and serious discipline.

Exhibit 6.   Excerpt from the CIA's Web Site (Continued)

- The *Office of General Counsel* advises the Director of Central Intelligence on all legal matters relating to his roles as head of the CIA and the Intelligence Community and is the principal source of legal counsel for the CIA and the DCI's *Community Management Staff.*
- The *Office of Public Affairs* advises the Director of Central Intelligence on all media, public policy, and employee communications issues relating to his role as head of the CIA and the Intelligence Community and is the CIA's principal communications focal point for the media, the general public and Agency employees.

Department of Defense (DoD)[20]

The DoD is headquartered in the Pentagon, Washington, D.C. The Pentagon was completed for the "War Department" in 1943. As the nerve center for DoD activities, it is the target of U.S. adversaries, as was seen on September 11, 2001. In addition, the computers and networks of the Pentagon are under almost constant attack by U.S. adversaries and various miscreants around the world. It is ironic that the DoD, responsible for the *defense* of the United States appears to continue to have some of the most vulnerable computers and networks of any nation-state. Information warfare is a real and constant threat to the DoD. In addition to the Pentagon computers and networks, those of the other military agencies are also under almost constant attack (e.g., at air bases, Army bases, and even those systems aboard naval vessels).

Exhibits 7 and 8 depict the organizational structure of the DoD as it relates to its major information warfare responsibilities.

Joint Chiefs of Staff Offices (JCS)[21]

The JCS is the focal point for the U.S. military services and all the armed forces are represented in the JCS. The JCS is also very involved in the development of offensive and defensive information warfare strategies, plans, tactics, and weapons systems. It has defined the information environment and information superiority, an obvious U.S. objective in a war where information warfare plays a role.[22]

- *Information environment*: the aggregate of individuals, organizations, and systems that collect, process, or disseminate information, including the information itself (JP1-02)
- *Information superiority*: the capability to collect, process, and disseminate an uninterrupted flow of information while exploiting or denying an adversary's ability to do the same (JP1-02); information superiority is achieved in a noncombat situation or one in which there are no clearly defined adversaries when friendly forces have the information necessary to achieve operational objectives.

The JCS view of information warfare is stated in Exhibit 9.

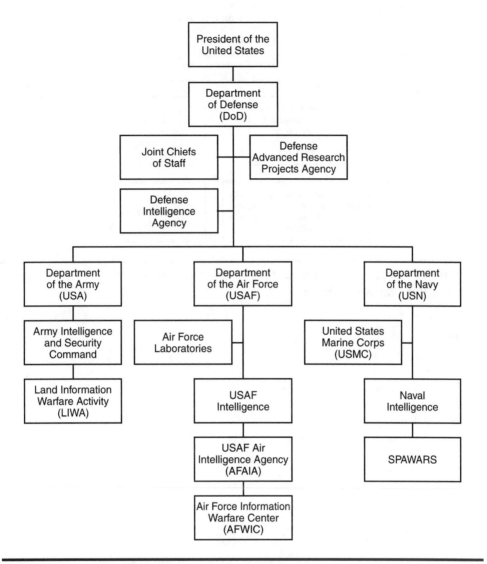

Exhibit 7. Organizational Structure of the DoD

The JCS operates the Joint Command & Control Warfare Center (JC2WC) whose mission it is to enable the "JCS and the combatant commanders to plan and execute C2W operations."[23] As part of their "mission execution," their activities center around:

- Commander-in-Chief support teams
- Protect and defend activities
- C2W modeling and simulation
- C2W information base coordination
- C2W doctrine development
- Technology

The JC2WC views information warfare target sets as follows:

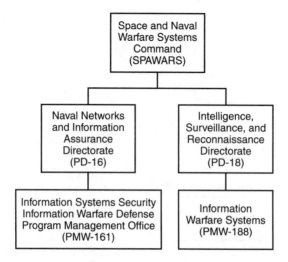

Exhibit 8.   Organizational Structure of the DoD as it Relates to its Major Information Warfare Responsibilities

- Leadership
  - Key personnel
  - ADP support
  - Strategic communications
  - Power base
- Civil infrastructure
  - Communications (links and nodes)
  - Industry
  - Financial
  - Power
  - Space
- Military infrastructures
  - Commanders
  - C2 communications links
  - C2 nodes
  - Troops
  - Intelligence collections
- Weapons systems
  - Planes
  - Ships
  - Artillery
  - Programmable guided munitions
  - Air defenses

The U.S. military still uses a hierarchical approach to warfare, to include information warfare. This top-down approach is based on pre-Information Age warfare concepts. It is in need of review; however, changes are difficult due to the many "turf battles" and "rice bowls" that must be eliminated. Thus far, any top-down reviews have resulted in little structural change in

Exhibit 9.   Joint Chiefs of Staff View of Information Warfare

Information, information processing, and communications networks are at the core of every military activity. Throughout history, military leaders have regarded information superiority as a key enabler of victory. However, the ongoing "information revolution" is creating not only a quantitative, but a qualitative change in the information environment that by 2020 will result in profound changes in the conduct of military operations. In fact, advances in information capabilities are proceeding so rapidly that there is a risk of outstripping our ability to capture ideas, formulate operational concepts, and develop the capacity to assess results. While the goal of achieving information superiority will not change, the nature, scope, and "rules" of the quest are changing radically.

The qualitative change in the information environment extends the conceptual underpin-nings of information superiority beyond the mere accumulation of more, or even better, information. The word "superiority" implies a state or condition of imbalance in one's favor. Information superiority is transitory in nature and must be created and sustained by the joint force through the conduct of information operations. However, the creation of infor-mation superiority is not an end in itself.

Information superiority provides the joint force a competitive advantage only when it is effectively translated into superior knowledge and decisions. The joint force must be able to take advantage of superior information converted to superior knowledge to achieve "decision superiority" — better decisions arrived at and implemented faster than an oppo-nent can react, or in a noncombat situation, at a tempo that allows the force to shape the situation or react to changes and accomplish its mission. Decision superiority does not automatically result from information superiority. Organizational and doctrinal adaptation, relevant training and experience, and the proper command and control mechanisms and tools are equally necessary.

The evolution of information technology will increasingly permit us to integrate the tradi-tional forms of information operations with sophisticated all-source intelligence, surveil-lance, and reconnaissance in a fully synchronized information campaign. The development of a concept labeled the global information grid will provide the network-centric environ-ment required to achieve this goal. The grid will be the globally interconnected, end-to-end set of information capabilities, associated processes, and people to manage and provide information on demand to warfighters, policy makers, and support personnel. It will enhance combat power and contribute to the success of noncombat military operations as well. Realization of the full potential of these changes requires not only technological improve-ments, but the continued evolution of organizations and doctrine and the development of relevant training to sustain a comparative advantage in the information environment.

We must also remember that information superiority neither equates to perfect information, nor does it mean the elimination of the fog of war. Information systems, processes, and operations add their own sources of friction and fog to the operational environment. Information superiority is fundamental to the transformation of the operational capabilities of the joint force. The joint force of 2020 will use superior information and knowledge to achieve decision superiority, to support advanced command and control capabilities, and to reach the full potential of dominant maneuver, precision engagement, full dimensional protection, and focused logistics. The breadth and pace of this evolution demands flexibility and a readiness to innovate.

the DoD bureaucracy. However, if one is to fight a Twenty-first Century war where information warfare concepts play a greater and greater role as United States' potential adversaries increasingly become dependent on information systems and their various NIIs and DIIs, it is believed some major changes are required.

The United States' Land Information Warfare organization views the JC2WC as follows:

> *The Joint Command and Control Warfare Center (JC2WC), under the operational control of the JCS, provides the combatant commands and teams of command and control warfare specialists. Each JC2WC team has a habitual relationship with a supported command. Teams provide technical and operational specialists to support IO planning, operations, and exercises. The JC2WC emerged from the former Joint Electronic Warfare Center (JEWC), transitioning from purely EW to encompass all elements of C2W. Field Support Teams (FSTs) will be deployed to support operations ranging from peace keeping to major regional conflicts. FSTs also support operational planning, wargames, exercises, and training programs. The FST is structured to fill gaps in the command's IO cell, provide connectivity to CONUS resident agencies and databases support-ing IO, and coordinate with the IO cells at the JTF or CINC level, as well as with the IW staff elements from other component commands in the operational area.*

National Security Agency (NSA)

The NSA's mission statement[24] is as follows:

> *The ability to understand the secret communications of our foreign adversaries while protecting our own communications— a capability in which the United States leads the world— gives our nation a unique advantage. The resources of NSA/CSS are organized for the accomplish-ment of two national missions:*
>
> > *The Information Assurance mission provides the solutions, products and services, and conducts defensive information operations, to achieve information assurance for information infrastructures criti-cal to U.S. national security interests. The foreign signals intelligence or SIGINT mission allows for an effective, unified organization and control of all the foreign signals collection and processing activities of the United States. NSA is authorized to produce SIGINT in accor-dance with objectives, requirements and priorities established by the Director of Central Intelligence with the advice of the National Foreign Intelligence Board.*

Its information warfare mission includes:

> *...protecting all classified and sensitive information that is stored or sent through U.S. Government equipment. INFOSEC professionals go to great*

lengths to make certain that Government systems remain impenetrable. This support spans from the highest levels of U.S. Government to the individual warfighter in the field.

The NSA is believed to have two primary information warfare research and development organizations: one that deals primarily with unclassified and classified defensive projects while the other deals with classified offensive projects. Some of the projects that they may be working on include:

- Sophisticated pings sent back to an attacker's origin. By measuring the ping's travel time, one has a reasonable chance of determining the location of the attacker.
- Attacker profile project or projects that gather information about an attacker and the attacker's methodology. By collecting that information on attackers, one may, over time, be able to identify the attackers, their native language, where they were trained, where they reside, and thus be in a better position to actively target them.

All that said, one must remember that the primary existence for the NSA is to collect information (e.g., ECHELON), analyze and interpret it as a primary information gathering source not only for the DoD but also the U.S. government in general. As with the CIA, the NSA has a major role to play in information warfare, primarily from an intelligence collection standpoint, but also has an information warfare defensive and offensive responsibility.

According to the Frequently Asked Questions section of the NSA Web site at http://www.nsa.gov/about_nsa/faqs_internet.html,[25] there are 13 federal organizations in the United States Intelligence Community, which it identifies as:

- National Security Agency/Central Security Service (NSA/CSS)
- Central Intelligence Agency (CIA)
- National Imagery and Mapping Agency (NIMA)
- Federal Bureau of Investigation (FBI)
- Defense Intelligence Agency (DIA)
- National Reconnaissance Office (NRO)
- Department of Energy (DoE)
- Army Intelligence
- Air Force Intelligence
- Navy Intelligence
- Marine Corps Intelligence
- Department of Treasury
- Department of State

Defense Intelligence Agency (DIA)[26]

The DIA is a Department of Defense combat support agency and an important member of the United States Intelligence Community. With over 7000 military and civilian employees worldwide, the DIA is a major producer and manager of foreign military intelligence. The DIA provides military intelligence to warfighters, defense policymakers, and force planners, in the Department of Defense and the Intelligence Community, in support of U.S. military planning and operations and weapon systems acquisition. The DIA is often overlooked by the NSA and CIA, as was the old Soviet GRU by the KGB.

The DIA states that:

> One of the key challenges of this thrust is to identify and define the challenge. Asymmetry— or the lack of symmetry— in reference to warfare is the use of power in unanticipated or nontraditional ways. For decades, the United States maintained a traditional mindset on the nature of its foe, developing "force multipliers" for ensuring the advantage over projected threats. This strategy led to the emergence and recognition of the United States as an overwhelming conventional military power. To challenge this resultant tenable power, state and non-state actors developed means for avoiding traditional confrontations with the United States. The intent of these "asymmetric" means is to project unexpected or unbalanced capability for reducing the conventional military superiority of the United States, render it irrelevant, or exploit perceived weaknesses. The asymmetric thrust challenges us not only to identify unanticipated threats, but also to develop a new mindset for assessing potential asymmetric threats and developing effective measures for combating them. The Asymmetric Threat Senior Steering Group, chaired by the Marine Corps Intelligence Activity, is developing a community approach to address the disparate nature of the asymmetric threat.

The DIA Web site[26] states the following:

- *Asymmetric approaches.* The variables involved in identifying an asymmetric threat make this problem extremely complex. Since the adversary, objectives, targets, means of attack, and context are all situation dependent, the challenge is like "wrestling with Jell-O." It is difficult to pin it down. To simplify its understanding, the intelligence and operational communities have defined asymmetric approaches to focus on strategic, operational and tactical targets.
- *Strategic asymmetries.* Strategic asymmetries attempt to preclude, deter, or degrade our ability to use military force by focusing on such things as our national will, public opinion, our national infrastructure, the highest-level civilian-military command and control, our overseas deployment flow, and our domestic mobilization capability
- *Operational asymmetries.* Operational asymmetries would work against our preferred warfighting concepts, as described in Joint Vision 2020, by attempting to undermine our ability to execute one or more key elements (dominant maneuver, precision engagement, focused logistics, full dimensional protection) of our operational plan. Examples might include chemical and biological warfare capabilities or computer network attacks that would render us ineffective.
- *Tactical asymmetries.* Tactical asymmetries focus on U.S. and allied forces already engaged in theater, such as attacks similar to the 1996 Khobar

Towers bombing. During Operation ALLIED FORCE, Serb forces implemented asymmetric warfare by widespread use of denial and deception measures in attempting to shoot down NATO aircraft and protect their deployed forces from NATO air attack. They also tried to sway international opinion by exaggerating or even fabricating collateral damage incidents.

- *The challenge.* The Intelligence Community's overarching challenge in shaping to meet this emerging threat is to develop an approach to the problem within the current system and its capabilities. There has not been a coherent plan for addressing the disparate nature of the asymmetric threat. Our strategy has been more reactive — allocating resources to meet individual, immediate issues. Whether it was drugs in the late 1980s, terrorism in the early 1990s, or information warfare most recently, new organizations, new products, and new systems were created to deal with the current mission. This approach is no longer expedient. The coordinated strategy of this thrust includes building the right skills mix, collection and analytic methods, and organizational linkages to deal with the intangible and "soft" data that characterize asymmetric issues. Resources will be allocated more effectively within this coherent approach.

The DIA is active in the information warfare arena for the U.S. government. Of course, the majority of its information warfare activities remain classified for national security reasons. However, the following gives some indication as to at least one of their areas of responsibility. It is also interesting to note the last sentence (italicized by the authors to highlight the DoD perceived problem).

Veridian Drafted into Pentagon Cyber Defense

The Department of Defense awarded Virginia-based security firm Veridian an exclusive contract to monitor and analyze foreign-borne cyber attacks against the Pentagon's IT assets. Under the auspices of the Defense Intelligence Agency, Veridian will monitor intrusion alerts, trace and log hostile IP addresses, and assess the threat capability of potential foreign adversaries. Intelligence sources told reporters that the program is geared toward assessing the cyber threat posed by potential adversaries, such as nation-states. Some question the wisdom and necessity of the government hiring a private contractor for intelligence work, saying the contract demonstrates the government's incapacity to protect its IT assets.[27]

It is not unusual to "outsource" information warfare work to government contractors. The rationale for that includes the fact that outsourcing is often cheaper; it is difficult to hire technically skilled employees at government wages. On the other side, one can argue that one cannot trust these contractor-provided employees as much as one can trust a government employee. That way of thinking has been ingrained in government minds for many years, especially in the minds of security professionals. Contractor employees are required to have polygraph examinations, for example, prior to having access to high levels of classified information while their government employee counterparts may not.

However, the facts do not support the security logic. One just has to look at the number of government employee and military personnel in the United States who have been imprisoned for spying for a United States' adversary. These employees worked for one of the military services, NSA, CIA, or the FBI, just to name a few relatively recent examples.

Defense Information Systems Agency (DISA)[28]

DISA is responsible for the networks and information systems of the DoD and as such defines its mission as:

> *To plan, engineer, develop, test, manage programs, acquire, implement, operate, and maintain information systems for C4I and mission support under all conditions of peace and war.*

The Defense Information Systems Agency is transforming the way Department of Defense users move, share, and use information. All DoD personnel need information whether they maintain or fly aircraft, operate a periscope, move a platoon, perform surgery, process transactions, or any of hundreds of other jobs supporting our country's defense. As the manager of the Global Information Grid (GIG), DISA is integrating hardware and software and constructing a common operating environment to sustain warfighters' need for information anytime, anywhere. The pillars of the GIG include the Defense Information System Network, the Defense Message System, the Global Command and Control System, and the Global Combat Support System. DISA is also helping protect against, detect, and react to threats to both its information infrastructure and information sources. Additionally, DISA is aggressively working with DoD agencies, military departments, other federal agencies, and industry.

DISA is on the leading edge of defensive information warfare. The DoD is a popular target for crackers, so DISA has the difficult task of balancing protecting the information environment (IE) while allowing access to the Internet. To do this, they are engaged in a number of activities. Some of these efforts include:

- Purchasing multiple anti-virus software products for Department-wide use
- Providing intrusion detection systems and firewall guidance
- Providing leadership and technical guidance for the Department's certification and accreditation of computer networks and systems
- Writing Security Technical Information Guides
- Conducting Security Readiness Reviews
- Providing Information Assurance Vulnerability Alerts
- Hosting DoD's Computer Emergency Response Team

DISA understands that protecting the IE requires the coordination of network operations and information assurance (IA). It has combined network operations with IA into the Global Network Operations and Security Center (GNOSC). Within budgetary and legal constraints, DISA does all that it can to ensure that the DoD's IE is as secure as possible.

Department of the Air Force (USAF)[29]

The USAF, of course, has the primary mission to fly and fight as the primary air combat arm of the United States. According to Air Force Chief of Staff Gen. John P. Jumper, the USAF is going through a transformation that:

> *will allow the Air Force to leverage the nation's technology and what the service brings to the fight — stealth, precision, standoff, information technology and space — to create asymmetrical advantages against the enemy.*

The USAF is very active in information warfare and was the first to establish an information warfare center and info-warrior organizations. It also has an active research and development (R&D) organization.

One of the leading DoD R&D laboratories is the USAF's Rome Laboratory (AFRL), whose mission is *"leading the discovery, development, and integration of affordable warfighting technologies for our aerospace forces."* The AFRL R&D projects include those relative to information warfare. For example, one of its projects that is unique has one foot in the industrial period of warfare (radar) and the other in the information period (computers). As we know, today's radar systems are computer supported. Because they are based on computers and we know computers are vulnerable to attack, is it possible to attack a computer supporting a radar system to the point that it may be spoofed so that a friend is viewed as an enemy and vice versa? What if the computer system is successfully attacked and the radar is not functional in real-time? One may never be able to release the defensive weapon or see the adversary's aircraft or missile coming. One project is to determine if one can quickly — for example, as close to real-time as possible (without physically going to backup files) — get the radar up and running before the next wave of the radar screen. If done successfully, one can see its use being ported over to the Federal Aviation Administration to support them in their tracking of commercial traffic.

Another agency of the USAF that plays a major role in information warfare is the Air Force Information Warfare Center (AFIWC). The AFIWC develops, maintains, and deploys Information Warfare/Command and Control Warfare (IW/C2W) capabilities in support of operations, campaign planning, acquisition, and testing. The AFIWC acts as a time-sensitive, single focal point for intelligence data and C2W services, providing technical expertise for computer and communications security. The Center is the Air Force focal point for Tactical Deception and Operations Security training.

The AFIWC is a subordinate organization of the USAF Air Intelligence Agency (AIA). The AIA is a field operating agency (FOA) under the Assistant Chief of Staff, Intelligence (ACS/I). It is located at Kelly Air Force Base, San Antonio, Texas. The Center:

- Develops, maintains, and deploys information warfare and command and control warfare (IW/C2W) capabilities in support of operations, campaign planning, acquisition, and testing

- Acts as a time-sensitive, single focal point for intelligence data and C2W services
- Provides technical expertise for computer and communications security
- Is the Air Force focal point for tactical deception and operations security

AFIWC responsibilities include:

- Satisfies other services, Department of Defense components, and U.S. government agencies with IW/C2W products and services in direct response to stated requirements
- Is a multi-service organization with Marine and National Security Agency participation; the AFIWC continues a steady growth with personnel from both organizations; the center also continues to expand its participation with the Army, Navy, and other U.S. agencies
- Communicates with supported commands, Air Force organizations, other services, and U.S. agencies on matters of direct operational impact to its mission areas and responsibilities

The AFIWC, as with other DoD agencies, is actively involved in R&D activities. It has a component that deals in the classified world of offensive information warfare. It is also a focal point for the collection and analyses of information on attacks against USAF computer systems (e.g., at USAF bases around the world). The information is collected and stored in a database. One R&D project is to establish a neural network, feed the system the information in the attacks database, and, using artificial intelligence, learn from that history. The information gleaned could be used to:

- Look for attack similarities to identify and profile attackers
- Predict, with some reasonable certainty, future attacks
- Automatically heighten defenses based on predicted future attacks

As with other R&D projects, such systems, once "perfected," could be extremely beneficial to not only other government agencies but also to the private sector. As such, should such defensive systems be limited to the U.S. government and U.S. corporations? If so, how could that be enforced? That would be extremely difficult. The problem with such defensive systems that are often outside the realm of the classified environment is that they can be easily obtained by a potential adversary and used by them against the United States, as has been the case for decades with U.S. technology.

As with many of the R&D projects, where one is developing an information warfare defensive capability, another project may be developing an offensive information warfare capability to defeat it. That may include placing back doors in the systems prior to releasing them to the general public, of course without the source code.

Department of the Army[36]

The United States Army (USA) vision is:

While aspiring to be the most esteemed institution in the Nation, we will remain the most respected Army in the world and the most feared ground force to those who would threaten the interests of the United States. Our commitment to meeting these challenges compels comprehensive transformation of the Army.

The USA also has a vital role in information warfare. Its primary thrust into the information warfare arena is through their Land Information Warfare Activity (LIWA) (see Exhibit 10).[31]

Exhibit 10.   Land Information Warfare Activity (LIWA)

LIWA is a Headquarters Department of the Army operations support activity assigned to the Intelligence and Security Command (INSCOM), provides multi-discipline Information Operations (IO) support to the U.S. Army's component and major commands. LIWA has broad authority to coordinate IO topics and establish contact with Army organizations, USN, USAF, and JCS IO Centers, and with DoD and National Agency IO elements.

IO Strategic Role

The strategic goal of IO is to promote freedom of action for U.S. Forces while hindering adversary efforts. U.S. Army IO integrate all aspects of information to support and enhance the elements of combat power, with the goal of dominating the battle space at the right time, at the right place, and with the right weapons or resources. Activities to support IO include acquiring, using, protecting, managing, exploiting, and denying information and information systems. The strategic purpose of IO is to secure peacetime national security objectives, deter conflict, protect DoD information and information systems, and to shape the information environment. If deterrence fails, IO seeks to achieve U.S. information dominance to attain specific objectives against potential adversaries in time of crisis or conflict. Information Operations focus on maximizing friendly information capabilities, while degrading the opponent's information capabilities.

Army component commands may perform strategic missions such as employment of deep strike weapons, special forces, and other special capabilities. Information operations broadens the scope of strategic and EAC military operations. Emerging high technology military capabilities may be employed independently as stand-alone actions supporting national security objectives. When these capabilities are employed in a military operation they become part of the IO planning strategy under the control of a Unified or Joint Task Force (JTF) commander. Coordination with U.S. Army intelligence and operational threat analysis activities is essential for IO planning and operations.

Operational commanders weigh the advantages to be gained by countering adversary C2 nodes against the potential loss of intelligence from enemy signatures, radiation, or emissions, and the need to protect intelligence sources and methods. In some cases, the decision authority to destroy or degrade an adversary's higher command echelons will be held at the national strategic level. Assistance in understanding an adversary's information system and his cycle of information processing is available through the Defense Intelligence Agency's (DIA) Tailored Analytical Intelligence Support to Individual Projects (TASIP).

The U.S. Army may be called on to assist with Information Operations of other services, joint commands, National agencies, or allied forces as authorized by CJCSI 3210.01, DoD 3600.1, and AR 525-20. The U.S. Army could be assigned a specific IO mission by the

Exhibit 10.   Land Information Warfare Activity (LIWA) (Continued)

National Command Authority (NCA), through the National Military Command Authority (NMCA), to an Army component of a unified command. The Joint term IW connotes the application of C2-Attack means to degrade or destroy an adversary's information system and to protect friendly command and control. Information Operations, unlike Information Warfare, are conducted continuously, e.g., defensive IO measures are applied routinely on a day-to-day basis. As a subset of IO, C2W is the application of IO strategy during military operations by engaging specific C2 targets. C2-Attack calls for the coordinated employment of destruction, deception, operations security, psychological warfare and electronic warfare, synchronized with the main operation.

The LIWA IO Role

LIWA teams support the Army Commander's goal of achieving Information dominance with the other JTF components or organizations. LIWA's purpose is to provide Army commands with technical expertise that is not resident on the command's general or special staff, and to exercise technical interfaces with other commands, service components, and National, DoD, and joint information centers. When deployed, LIWA FSTs become an integral part of the command's IO staff. To facilitate planning and execution of IO, LIWA provides IO/ C2W operational support to land component and separate Army commands, and reserve components commands as required.

LIWA Mission

The mission of the LIWA is to provide IW/IO support to the land component and major/ separate Army commands, active and reserve component (AC/RC), to facilitate planning and execution of information operations (IO).

LIWA Functions include:

- Act as the focal point for Land IO.
- Coordination with U.S. Army intelligence and operational threat analysis activities is essential for IO planning and operations.
- Coordinate, arrange for, and synchronize intelligence and counterintelligence support.
- Coordinate and deploy field support teams (FST) to assist and support the land component commands in C2-Protect and C2-Attack.
- Coordinate and deploy FSTs to provide battlefield deception support.
- Coordinate and assist TRADOC in the development and integration of doctrine, training, leader development, organization, materiel, and soldier requirements for IO.
- Act as the combat developer for C2-Attack and C2-Protect systems.
- Develop IO models and simulations in support of IO systems development, planning, training, and exercises.
- Assist in the development and integration of IO requirements in Army modernization strategy and policy scenarios, modeling, and simulations.
- Initiate and coordinate requirements for IO area studies.
- Assist in the development and evaluation of IO systems performance and operational employment tactics, techniques, and procedures in combat operations, operational tests, and training exercises.
- Identify technology for possible application to Army IO.
- Establish, develop, and promote IO interoperability with other services and allies.
- Assess IO force readiness and IO operational capabilities.
- Conduct IO vulnerability analyses of Army commands.

Exhibit 10.   Land Information Warfare Activity (LIWA) (Continued)

- Develop and sustain a rapid response capability to combat attempted penetrations of Army C2 systems and processes.
- Develop and coordinate requirements for operational IO from National and Defense reconnaissance.
- Identify and report changes in worldwide signature information that may require the software rapid reprogramming of Army Target Sensing Systems (ATSS), i.e., smart/ brilliant munitions, sensors, processors, and aviation electronic combat survivability equipment.

Field Support Teams

Field Support Teams (FST) normally augment the Army or Land Component Command with IO expertise similar to the way JC2WC teams support the JTF or CINC. FST will also support Army divisions and corps when needed to plan and implement information operations below the Army component command level. Team members consist of a need-driven mix of PSYOPS, deception, OPSEC, EW, C2-Protect, C2-Attack, and intelligence specialties. When deployed, the FST becomes an integral part of the supported command's IO cell. The FST is structured to fill gaps in the command's IO cell, provide connectivity to CONUS resident agencies and databases supporting IO, and coordinate with the IO cells at the JTF or CINC level, as well as with the IW staff elements from other component commands in the operational area.

LIWA Red Team

The LIWA Red Team provides an Information Operations Vulnerability Assessment capability and an independent opposing force (OPFOR) type of capability to the Army component commands, the Army acquisition community, and separate Army commands. The Red Team provides a capability to assess the vulnerability of U.S. information, information systems, and information infrastructure.

IOVAP

The Information Operations Vulnerability Assessments Program (IOVAP) provides the supported command a perspective of the command's susceptibility to an opponent's C2W operations. The IOVAP can be focused on garrison activities, field exercises, or both. In addition to isolating a command's vulnerabilities, the team recommends ways to reduce those vulnerabilities, allowing commanders to apply remedial action on the spot. In addition, the team will provide limited training to system managers on protection tools and procedures.

OPFOR

The Red Team has the capability of assembling an independent C2W opposing force (OPFOR) to support exercises and experiments. The size and composition of the OPFOR will vary by type of exercise, and by what must be learned about the command's vulnerability. Army warfighting experiments (AWE) involving brigade or division size elements may require high-technology intelligence systems and processors from the National level down to tactical Army systems, as well as systems from other services and agencies to provide the high-resolution information required. Other IO OPFOR operations may be successfully conducted using local collection systems.

Army Computer Emergency Response Team (ACERT)

ACERT's mission is to conduct Command and Control Protect (C2-Protect) operations in support of Army commanders worldwide. The objective is to ensure the availability, integrity and confidentiality of the information and information systems used in planning, coordinating,

Exhibit 10. Land Information Warfare Activity (LIWA) (Continued)

directing and controlling forces. ACERT supports systems administrators reporting suspicious activity on their computer networks. ACERT also has the responsibility of keeping Army leadership informed of incidents, and promulgating alerts and warnings based on information collected from a variety of sources.

Army Reprogramming and Analysis Team — Threat Analysis (ARAT-TA)

The ARAT- TA supports warfighters, the commodity commands' post-deployment software support (PDSS) centers, and combat and materiel developers. ARAT-TA identifies and reports changes in worldwide signature information requiring reprogramming of Army Target Sensing Systems (ATSS) software. Army Target Sensing Systems include smart and brilliant munitions, sensors, processors and aviation electronic combat survivability equipment. Identified threat signature changes are "flashed" to tactical units' subscribers over the ARAT Project Office electronic bulletin board.

Army Rapid Reprogramming Analysis Team Program Office (ARAT PO)

Established in 1994 with a charter through 1999, the ARAT PO acts as the technical intermediary between the CECOM Systems Engineering Center and ARAT-TA on matters related to rapid reprogramming of Army Target Sensing Systems (ATSS). ARAT PO developed the Memory Loader Verified (MLV) to reprogram the memory of ATSS when the threat changes, or when the Army deploys to an operational area with a threat array unlike the one the deploying unit TSS were programmed to handle.

Advanced Programs Division

The Advanced Programs Division leads LIWA in the innovation, development, and employment of advanced IO/C2W capabilities (C2-Protect and C2-Attack) using multi-disciplined approaches. The Division monitors technology advancements, looking for opportunities to advance the state of the art in C2-Attack and C2-Protect capabilities. Modeling and simulations are used extensively to support the combat development process. The Advanced Programs Division acts as the IO combat developer, in close coordination with TRADOC. The Division explores lethal, non-lethal, destructive, and nondestructive means to meet information dominance requirements in peacetime, conflict, war, and military operations other than war (MOOTW). The Advanced Programs Division is the focal point for technology transfer opportunities.

Support Division

The Support Division consolidates LIWA intelligence and support activities. The Division manages the overall support functions including security, intelligence, information management, and resource management. Members of the Support Division may augment other LIWA teams during deployments, as required.

LIWA considers its core competencies as:

- *IO Operational Expertise:* LIWA's Operations Division contains a mix of military and DA civilian personnel with a variety of skills including combat arms, special operations, aviation, communications and computer specialists, and intelligence analysts. Personnel with tactical and operational-level training and experience and are capable of operating in joint and combined operational environments. Contractor personnel with additional specialties augment the Operations Division as required. The IO operational expertise (C2-Attack, C2-Protect, and C2-Exploit) represented by this array of skills and experience is task organized on a mission-by-mission basis into teams, and deployed to support Army commands.

Exhibit 10. Land Information Warfare Activity (LIWA) (Continued)

- *IO Intelligence Support:* LIWA's structure contains a small intelligence organization designed to be the focal point for IO intelligence support. The value of this organization resides in its ability to respond rapidly to field-generated, IO-unique intelligence requirements, and to forward and track requests for IO intelligence support. A mix of intelligence specialties, supported by automation and connectivity to DIA, NSA, joint intelligence centers and IW cells of the other services, allows LIWA to request and receive IO specific data from multiple sources. In addition, LIWA provides liaison personnel to selected intelligence organizations, increasing their awareness of Army IO needs and facilitating the exchange of IO-related intelligence. LIWA intelligence analysts provide deployed teams with sharply focused IO area studies, IO targeting products, and quick-response one-of-a-kind reports designed to meet specific needs from the field.

- *Future IO Requirements and Capabilities:* LIWA conducts and participates in studies, war games, and exercises designed to identify future IO requirements and capabilities. Models and simulations are developed to support analysis and decision making. Working closely with government, industry, and academia, LIWA looks for opportunities to apply advanced technology against IO requirements using commercial, off-the-shelf hardware and software. The dynamic nature of C2-related technology, and RDA funding constraints places a premium on off-the-shelf applications, and low density procurements. Information Operations, directed against opponents employing advanced commercial C2 systems, may require state-of-the-art systems to effectively exploit and attack an opponent's C2 systems, or to protect friendly systems.

From http://www.fas.org/irp/doddir/army/fm34-37_97/3-chap.htm.

Department of the Navy (includes the Marine Corps) (USN and USMC)[32]

The USN's information warfare efforts are accomplished primarily through their Space and Naval Warfare Systems Command (SPAWAR). SPAWAR[33]

> *is one of the Department of the Navy's three major acquisition commands. SPAWAR's mission is to provide the warfighter with knowledge superiority by developing, delivering, and maintaining effective, capable and integrated command, control, communications, computer, intelligence and surveillance systems. And, while our name and organizational structure have changed several times over the years, our basic mission of helping the Navy communicate and share critical information has not. SPAWAR provides information technology and space systems for today's Navy and Defense Department activities while planning and designing for the future.*

One example of the USN's information warfare defensive approach is the Navy Marine Corps Intranet (NMCI). It is rumored that the NMCI, which includes a Security Operation Center, has over 440,000 users, literally worldwide. It uses COTS and is attempting to perform host-based intrusion detection on the client. It also uses INFOSEC features such as firewalls, VPNs, PKI, and malicious code detection and eradication software.

Defense Advanced Research Projects Agency (DARPA)[34]

The Defense Advanced Research Projects Agency (DARPA) is the central research and development organization for the Department of Defense (DoD). It manages and directs selected basic and applied research and development projects for the DoD, and pursues research and technology where risk and payoff are both very high and where success may provide dramatic advances for traditional military roles and missions.

The DARPA mission is to develop imaginative, innovative, and often high-risk research ideas offering a significant technological impact that will go well beyond the normal evolutionary developmental approaches, and to pursue these ideas from the demonstration of technical feasibility through the development of prototype systems.

Currently, DARPA's Advanced Technology Office (ATO) is involved in several projects that can be used as part of information warfare. According to the DARPA Web page (http://www.darpa.mil/ato/), they include:

- Airborne Communication Node (ACN), an autonomous communications infrastructure simultaneously providing in-theater assured communications and signals intelligence
- Antipersonnel Landmine Alternative (APLA)
- Buoyant Cable Array Antenna (BCAA), high data-rate, two-way submarine connectivity at speed and depth
- Center of Excellence for Research in Oceanographic Sciences (CEROS)
- Command Post of the Future, to shorten the commander's decision cycle to stay ahead of the adversary's ability to react
- Cyber panel
- Drag reduction program
- Dynamic coalitions
- Fault tolerant networks
- FCS command and control
- FCS communications
- Metal storm
- Micro-electronics and bio processes (MEB)
- Operational partners in experimentation
- Optical tags
- Project Genoa: research, develop, and deliver general-purpose tools for supporting early pre-crisis understanding and mitigation through collective reasoning
- Robust passive sonar (RPS)
- Small unit operations: Situational Awareness System (SUO SAS): Mobile communication system with high data-rate capacity that is optimized for restrictive terrain
- Submarine Payloads and Sensors Program (SPSP)
- Tactical mobile robotics
- Tactical sensors
- Totally Agile Sensor Systems (TASS)

- Undersea Littoral Warfare: Netted Search, Acquisition and Targeting (Net-SAT): provide real-time target localization and tracking for torpedo guidance in the presence of countermeasures
- Wolfpack: deny the enemy use of radio communications throughout the battlespace

DARPA has a long history of sponsoring and active involvement in projects and programs that further the state-of the-art in warfare. DARPA is also, as would be expected, involved in classified projects. This agency also works with various other DoD and governmental agencies relative to many of their projects. They are expected to continue to play a key role in activities related to information warfare.

Homeland Security

The United States has awakened to the threats to its NII and other networks. Over a decade of studies and countless testimonies by experts in the field and others did not awaken government officials to actively establish aggressive defensive measures to protect the nation. However, on September 11, 2001, all that changed. Unfortunately, it took another "Pearl Harbor" costing thousands of lives to do so. The President has established a cabinet post, Homeland Security, to be the focal point for such activities (see Exhibit 11).

Such a position will require budget and staff. It will take some time to determine exactly how this new bureaucracy will fit in with the others responsible for the defense of the United States. As part of that "Homeland Security," defensive information warfare will play a role, and hopefully a key role. The White House released the following on October 9, 2001:

> *Special Advisor to the President for Cyberspace Security: Richard Clarke*
> *The information technology revolution has changed the way business is transacted, government operates, and national defense is conducted. The United States now depends on a complex, interdependent network of critical infrastructure information systems that are essential to our national and economic security. These networks include information systems in the government, telecommunications, banking and finance, transportation, energy, manufacturing, water, health and emergency services networks.*
>
> *The United States must protect against the disruption of the operation of these systems. Any disruption that occurs must be infrequent, of minimal duration, manageable, and cause the least possible damage.*
>
> *The President's Special Advisor for Cyberspace Security will coordinate interagency efforts to secure information systems. In the event of a disruption, the Special Advisor will coordinate efforts to restore critical systems.*
>
> *The Special Advisor will work in close coordination and partnership with the private sector, which owns and operates the vast majority of America's critical infrastructure. The Special Advisor will be the President's principal advisor on matters related to cyberspace security and*

> *report to the Assistant to the President for Homeland Security and to the Assistant to the President for National Security Affairs.*
>
> *The Special Advisor will also serve as chairman of a government-wide board that will coordinate the protection of critical information systems (the President is expected to sign an Executive Order soon establishing the board).*

In this era where attacks can occur in a nanosecond and the requirement for aggressive defenses and nanosecond responses to attacks, it is hoped that the bureaucracy will not get to the level of detail and decision making that was used during the Vietnam War. It is hoped that strategies and plans with a complement of defensive and offensive information warfare systems will be aggressively installed and maintained in a current state of readiness.

The Twentieth Century, top-down bureaucratic approach is too slow and cumbersome to be of much use in the era of Twenty-first Century information warfare. However, the U.S. government's approach to any problem seems to be to legislate it away, for a committee to study the issue and form a department saddled with a bureaucracy, budget issues, political issues, and turf battles, to fix the problem. If Homeland Security meets that profile, millions of tax payers' money will again be wasted based on the probable "return on investments."

Exhibit 11.   Executive Order Establishing the Office of Homeland Security and the Homeland Security Council

By the authority vested in me as President by the Constitution and the laws of the United States of America, it is hereby ordered as follows:

Section 1. Establishment. I hereby establish within the Executive Office of the President an Office of Homeland Security (the "Office") to be headed by the Assistant to the President for Homeland Security.

Sec. 2. Mission. The mission of the Office shall be to develop and coordinate the implementation of a comprehensive national strategy to secure the United States from terrorist threats or attacks. The Office shall perform the functions necessary to carry out this mission, including the functions specified in Section 3 of this order.

Sec. 3. Functions. The functions of the Office shall be to coordinate the executive branch's efforts to detect, prepare for, prevent, protect against, respond to, and recover from terrorist attacks within the United States.

   a. *National Strategy.* The Office shall work with executive departments and agencies, State and local governments, and private entities to ensure the adequacy of the national strategy for detecting, preparing for, preventing, protecting against, responding to, and recovering from terrorist threats or attacks within the United States and shall periodically review and coordinate revisions to that strategy as necessary.
   b. *Detection.* The Office shall identify priorities and coordinate efforts for collection and analysis of information within the United States regarding threats of terrorism against the United States and activities of terrorists or terrorist groups within the United States. The Office also shall identify, in coordination with the Assistant to the President for

Exhibit 11. Executive Order Establishing the Office of Homeland Security and the Homeland Security Council (Continued)

National Security Affairs, priorities for collection of intelligence outside the United States regarding threats of terrorism within the United States.

  (i) In performing these functions, the Office shall work with Federal, State, and local agencies, as appropriate, to:

  (A) facilitate collection from State and local governments and private entities of information pertaining to terrorist threats or activities within the United States;

  (B) coordinate and prioritize the requirements for foreign intelligence relating to terrorism within the United States of executive departments and agencies responsible for homeland security and provide these requirements and priorities to the Director of Central Intelligence and other agencies responsible for collection of foreign intelligence;

  (C) coordinate efforts to ensure that all executive departments and agencies that have intelligence collection responsibilities have sufficient technological capabilities and resources to collect intelligence and data relating to terrorist activities or possible terrorist acts within the United States, working with the Assistant to the President for National Security Affairs, as appropriate;

  (D) coordinate development of monitoring protocols and equipment for use in detecting the release of biological, chemical, and radiological hazards; and

  (E) ensure that, to the extent permitted by law, all appropriate and necessary intelligence and law enforcement information relating to homeland security is disseminated to and exchanged among appropriate executive departments and agencies responsible for homeland security and, where appropriate for reasons of homeland security, promote exchange of such information with and among State and local governments and private entities.

  (ii) Executive departments and agencies shall, to the extent permitted by law, make available to the Office all information relating to terrorist threats and activities within the United States.

  c. *Preparedness.* The Office of Homeland Security shall coordinate national efforts to prepare for and mitigate the consequences of terrorist threats or attacks within the United States. In performing this function, the Office shall work with Federal, State, and local agencies, and private entities, as appropriate, to:

  (i) review and assess the adequacy of the portions of all Federal emergency response plans that pertain to terrorist threats or attacks within the United States;

  (ii) coordinate domestic exercises and simulations designed to assess and practice systems that would be called on to respond to a terrorist threat or attack within the United States and coordinate programs and activities for training Federal, State, and local employees who would be called on to respond to such a threat or attack;

  (iii) coordinate national efforts to ensure public health preparedness for a terrorist attack, including reviewing vaccination policies and reviewing the adequacy of and, if necessary, increasing vaccine and pharmaceutical stockpiles and hospital capacity;

  (iv) coordinate Federal assistance to State and local authorities and nongovernmental organizations to prepare for and respond to terrorist threats or attacks within the United States;

  (v) ensure that national preparedness programs and activities for terrorist threats or attacks are developed and are regularly evaluated under appropriate standards and that resources are allocated to improving and sustaining preparedness based on such evaluations; and

Exhibit 11.   Executive Order Establishing the Office of Homeland Security and the Homeland Security Council (Continued)

   (vi) ensure the readiness and coordinated deployment of Federal response teams to respond to terrorist threats or attacks, working with the Assistant to the President for National Security Affairs, when appropriate.
   d. *Prevention.* The Office shall coordinate efforts to prevent terrorist attacks within the United States. In performing this function, the Office shall work with Federal, State, and local agencies, and private entities, as appropriate, to:
   (i) facilitate the exchange of information among such agencies relating to immigration and visa matters and shipments of cargo; and, working with the Assistant to the President for National Security Affairs, ensure coordination among such agencies to prevent the entry of terrorists and terrorist materials and supplies into the United States and facilitate removal of such terrorists from the United States, when appropriate;
   (ii) coordinate efforts to investigate terrorist threats and attacks within the United States; and
   (iii) coordinate efforts to improve the security of United States borders, territorial waters, and airspace to prevent acts of terrorism within the United States, working with the Assistant to the President for National Security Affairs, when appropriate.
   e. *Protection.* The Office shall coordinate efforts to protect the United States and its critical infrastructure from the consequences of terrorist attacks. In performing this function, the Office shall work with Federal, State, and local agencies, and private entities, as appropriate, to:
   (i) strengthen measures for protecting energy production, transmission, and distribution services and critical facilities; other utilities; telecommunications; facilities that produce, use, store, or dispose of nuclear material; and other critical infrastructure services and critical facilities within the United States from terrorist attack;
   (ii) coordinate efforts to protect critical public and privately owned information systems within the United States from terrorist attack;
   (iii) develop criteria for reviewing whether appropriate security measures are in place at major public and privately owned facilities within the United States;
   (iv) coordinate domestic efforts to ensure that special events determined by appropriate senior officials to have national significance are protected from terrorist attack;
   (v) coordinate efforts to protect transportation systems within the United States, including railways, highways, shipping, ports and waterways, and airports and civilian aircraft, from terrorist attack;
   (vi) coordinate efforts to protect United States livestock, agriculture, and systems for the provision of water and food for human use and consumption from terrorist attack; and
   (vii) coordinate efforts to prevent unauthorized access to, development of, and unlawful importation into the United States of chemical, biological, radiological, nuclear, explosive, or other related materials that have the potential to be used in terrorist attacks.
   f. *Response and Recovery.* The Office shall coordinate efforts to respond to and promote recovery from terrorist threats or attacks within the United States. In performing this function, the Office shall work with Federal, State, and local agencies, and private entities, as appropriate, to:
   (i) coordinate efforts to ensure rapid restoration of transportation systems, energy production, transmission, and distribution systems; telecommunications; other utilities; and other critical infrastructure facilities after disruption by a terrorist threat or attack;

Exhibit 11.   Executive Order Establishing the Office of Homeland Security and the Homeland Security Council (Continued)

(ii) coordinate efforts to ensure rapid restoration of public and private critical information systems after disruption by a terrorist threat or attack;

(iii) work with the National Economic Council to coordinate efforts to stabilize United States financial markets after a terrorist threat or attack and manage the immediate economic and financial consequences of the incident;

(iv) coordinate Federal plans and programs to provide medical, financial, and other assistance to victims of terrorist attacks and their families; and

(v) coordinate containment and removal of biological, chemical, radiological, explosive, or other hazardous materials in the event of a terrorist threat or attack involving such hazards and coordinate efforts to mitigate the effects of such an attack.

g. *Incident Management.* The Assistant to the President for Homeland Security shall be the individual primarily responsible for coordinating the domestic response efforts of all departments and agencies in the event of an imminent terrorist threat and during and in the immediate aftermath of a terrorist attack within the United States and shall be the principal point of contact for and to the President with respect to coordination of such efforts. The Assistant to the President for Homeland Security shall coordinate with the Assistant to the President for National Security Affairs, as appropriate.

h. *Continuity of Government.* The Assistant to the President for Homeland Security, in coordination with the Assistant to the President for National Security Affairs, shall review plans and preparations for ensuring the continuity of the Federal Government in the event of a terrorist attack that threatens the safety and security of the United States Government or its leadership.

i. *Public Affairs.* The Office, subject to the direction of the White House Office of Communications, shall coordinate the strategy of the executive branch for communicating with the public in the event of a terrorist threat or attack within the United States. The Office also shall coordinate the development of programs for educating the public about the nature of terrorist threats and appropriate precautions and responses.

j. *Cooperation with State and Local Governments and Private Entities.* The Office shall encourage and invite the participation of State and local governments and private entities, as appropriate, in carrying out the Office's functions.

k. *Review of Legal Authorities and Development of Legislative Proposals.* The Office shall coordinate a periodic review and assessment of the legal authorities available to executive departments and agencies to permit them to perform the functions described in this order. When the Office determines that such legal authorities are inadequate, the Office shall develop, in consultation with executive departments and agencies, proposals for presidential action and legislative proposals for submission to the Office of Management and Budget to enhance the ability of executive departments and agencies to perform those functions. The Office shall work with State and local governments in assessing the adequacy of their legal authorities to permit them to detect, prepare for, prevent, protect against, and recover from terrorist threats and attacks.

l. *Budget Review.* The Assistant to the President for Homeland Security, in consultation with the Director of the Office of Management and Budget (the "Director") and the heads of executive departments and agencies, shall identify programs that contribute to the Administration's strategy for homeland security and, in the development of the President's annual budget submission, shall review and provide advice to the heads of departments and agencies for such programs. The Assistant to the President for Homeland Security shall provide advice to the Director on the level and use of funding

Exhibit 11.   Executive Order Establishing the Office of Homeland Security and the Homeland Security Council (Continued)

in departments and agencies for homeland security-related activities and, prior to the Director's forwarding of the proposed annual budget submission to the President for transmittal to the Congress, shall certify to the Director the funding levels that the Assistant to the President for Homeland Security believes are necessary and appropriate for the homeland security-related activities of the executive branch.

Sec. 4. Administration.

a. The Office of Homeland Security shall be directed by the Assistant to the President for Homeland Security.
b. The Office of Administration within the Executive Office of the President shall provide the Office of Homeland Security with such personnel, funding, and administrative support, to the extent permitted by law and subject to the availability of appropriations, as directed by the Chief of Staff to carry out the provisions of this order.
c. Heads of executive departments and agencies are authorized, to the extent permitted by law, to detail or assign personnel of such departments and agencies to the Office of Homeland Security upon request of the Assistant to the President for Homeland Security, subject to the approval of the Chief of Staff.

Sec. 5. Establishment of Homeland Security Council.

a. I hereby establish a Homeland Security Council (the "Council"), which shall be responsible for advising and assisting the President with respect to all aspects of homeland security. The Council shall serve as the mechanism for ensuring coordination of homeland security-related activities of executive departments and agencies and effective development and implementation of homeland security policies.
b. The Council shall have as its members the President, the Vice President, the Secretary of the Treasury, the Secretary of Defense, the Attorney General, the Secretary of Health and Human Services, the Secretary of Transportation, the Director of the Federal Emergency Management Agency, the Director of the Federal Bureau of Investigation, the Director of Central Intelligence, the Assistant to the President for Homeland Security, and such other officers of the executive branch as the President may from time to time designate. The Chief of Staff, the Chief of Staff to the Vice President, the Assistant to the President for National Security Affairs, the Counsel to the President, and the Director of the Office of Management and Budget also are invited to attend any Council meeting. The Secretary of State, the Secretary of Agriculture, the Secretary of the Interior, the Secretary of Energy, the Secretary of Labor, the Secretary of Commerce, the Secretary of Veterans Affairs, the Administrator of the Environmental Protection Agency, the Assistant to the President for Economic Policy, and the Assistant to the President for Domestic Policy shall be invited to attend meetings pertaining to their responsibilities. The heads of other executive departments and agencies and other senior officials shall be invited to attend Council meetings when appropriate.
c. The Council shall meet at the President's direction. When the President is absent from a meeting of the Council, at the President's direction the Vice President may preside. The Assistant to the President for Homeland Security shall be responsible, at the President's direction, for determining the agenda, ensuring that necessary papers are prepared, and recording Council actions and Presidential decisions.

Sec. 6. Original Classification Authority.
I hereby delegate the authority to classify information originally as Top Secret, in accordance with Executive Order 12958 or any successor Executive Order, to the Assistant to the President for Homeland Security.

Exhibit 11. Executive Order Establishing the Office of Homeland Security and the Homeland Security Council (Continued)

Sec. 7. Continuing Authorities. This order does not alter the existing authorities of United States Government departments and agencies. All executive departments and agencies are directed to assist the Council and the Assistant to the President for Homeland Security in carrying out the purposes of this order.

Sec. 8. General Provisions.
   a. This order does not create any right or benefit, substantive or procedural, enforceable at law or equity by a party against the United States, its departments, agencies or instrumentalities, its officers or employees, or any other person.
   b. References in this order to State and local governments shall be construed to include tribal governments and United States territories and other possessions.
   c. References to the "United States" shall be construed to include United States territories and possessions.

Sec. 9. Amendments to Executive Order 12656. Executive Order 12656 of November 18, 1988, as amended, is hereby further amended as follows:
   a. Section 101(a) is amended by adding at the end of the fourth sentence: ", except that the Homeland Security Council shall be responsible for administering such policy with respect to terrorist threats and attacks within the United States."
   b. Section 104(a) is amended by adding at the end: ", except that the Homeland Security Council is the principal forum for consideration of policy relating to terrorist threats and attacks within the United States."
   c. Section 104(b) is amended by inserting the words "and the Homeland Security Council" after the words "National Security Council."
   d. The first sentence of Section 104(c) is amended by inserting the words "and the Homeland Security Council" after the words "National Security Council."
   e. The second sentence of Section 104(c) is replaced with the following two sentences: "Pursuant to such procedures for the organization and management of the National Security Council and Homeland Security Council processes as the President may establish, the Director of the Federal Emergency Management Agency also shall assist in the implementation of and management of those processes as the President may establish. The Director of the Federal Emergency Management Agency also shall assist in the implementation of national security emergency preparedness policy by coordinating with the other Federal departments and agencies and with State and local governments, and by providing periodic reports to the National Security Council and the Homeland Security Council on implementation of national security emergency preparedness policy."
   f. Section 201(7) is amended by inserting the words "and the Homeland Security Council" after the words "National Security Council."
   g. Section 206 is amended by inserting the words "and the Homeland Security Council" after the words "National Security Council."
   h. Section 208 is amended by inserting the words "or the Homeland Security Council" after the words "National Security Council."

GEORGE W. BUSH
THE WHITE HOUSE,
October 8, 2001

An Information Warfare Concern

When one looks at the projects under DARPA, it is easy to see many of them applying to information warfare. These are just a sample of what is being considered. Many of the government agencies have projects related to information warfare. Some are classified and some are not. What is troublesome from an information warfare standpoint is the vast amount of information available at various U.S. government agency Web sites as noted in this chapter.

Information on these Web sites provides organizational charts; the names of agency leaders; e-mail addresses; wonderful graphics on how information warfare communications and weapons systems are to function; project schedules; project leaders, complete with names, telephone numbers, fax numbers, and e-mail addresses; and so much more. An adversary just has to surf the U.S. government agencies' Web sites to build an excellent profile of U.S. information warfare capabilities, strategies, plans, and such. One can argue that that is the price of living in a free society. Others might say that the general public is not interested in such information to the level of detail provided in many of these Web sites. Therefore, such details assist no one but the adversaries of the United States. One final example on this topic:

> **Military Secrets Posted on Internet**
> *Josh Gerstein, ABCNEWS.com, October 16, 2001,*
> *http://news.excite.com/news/abc/011015/17/military-secrets-posted*
> *Key details about secure bunkers used by President Bush and Vice President Cheney are available on the Internet...the locations and layout of presidential and military command centers— even information about their water supply— are accessible worldwide at the click of a mouse. Experts say some of the information should be classified.*

The article quoted former CIA Director James Woolsey as saying, "I had absolutely no idea they were on the Web — plans of facilities and the like. That's just crazy." The article went on to say that the Federation of American Scientists reportedly has removed approximately 200 Web pages that allegedly contained sensitive information about the White House and other facilities. However, there are literally thousands more out there that probably should not be, especially in time of war.

Case Studies

There are thousands of incidents that can be cited that relate to information warfare attacks against the United States, as well as some relating to U.S. efforts to defend against such attacks or mount information warfare attacks in self-defense. Details of such attacks and subsequent responses would fill volumes and are beyond the scope of this book. However, Exhibit 12 contains examples of what is occurring everyday in the United States.

Eligible Receiver 97-1 and Solar Sunrise

Eligible Receiver is a Chairman, Joint Chiefs of Staff exercise. Planning is as close-hold as possible to ensure the exercises are as realistic as possible so the Chairman can have an objective assessment of specific capability. Exercises have included logistics and non-combatant evacuation operations. Eligible Receiver 97-1 was not only DoD's, but also the U.S. government's, first information warfare exercise.

ER 97-1 involved a hypothetical belligerent nation-state that wanted to force the United States to change its aid policy toward it. This country conducted (in exercise time) two years of reconnaissance on U.S. military capabilities, U.S. critical infrastructures, and the Internet. Within exercise rules of engagement, they physically and virtually attacked military networks (they left exercise bombs) to cause delays in the deployment of U.S. forces, civilian critical infrastructures such as supervisory control and data acquisition systems (the control system underlying infrastructures) and 911 emergency systems, even used a media campaign to sway American public opinion. Who was the target? U.S. lawmakers who would modify laws to favor the belligerent nation-state. All the attacks were based on information and capabilities found on the Internet and through other public sources.

The Deputy Secretary of Defense and the lead IA lawyer for the Department reviewed "with a fine tooth comb" all the scenarios. Many events were either deleted or scaled back. Why? To ensure that real-world operations would not be adversely affected. Despite all the constraints, the results of the exercise were shocking. Every attack was successful, and the vast majority of computer virtual intrusions were undetected.

There were the usual finger-pointing and knee-jerk reactions that follow negative events such as this. A few people did not believe the results. After about six months, the sense of urgency had waned and the exercise was on its way to becoming a memory. There were some minor budget adjustments, a few programmatic changes, and the lengthy debates that go into changing laws, but nothing meaningful occurred during those six months.

Solar Sunrise occurred six months after ER 97-1. Details about this event are given in Chapter 14. The important point to note here is that the United States was gearing up to go back into Iraq. It took a painfully long time to ascertain that it was a few seriously misguided youths who attacked the DoD systems. If ER 97-1 had not happened six months earlier, finding the miscreants would have taken even longer. The Joint Staff Crisis Action Team (CAT) Leader said, "I don't know if we should give them medals or jail them." This was a humorous aside because during the time the CAT was assembled, the general consensus among those on the CAT was that more was done to advance protecting DoD's information environment in two weeks than would have occurred during two years of staff work.

Exhibit 12. Case Studies: The United States

U.S. Embassy Recruits Russian Hacker

In April 2001, the Russian newspaper *Moskovsky Komsomolets* reported that the U.S. embassy in Moscow had attempted to recruit a Russian hacker, known as 'Verse,' to secretly collaborate "in the interests of American intelligence services against the Russian Federation."

Comments:

This report is probably true and is the work of the CIA. No doubt that Russia is also attempting to do the same to the United States. Although the Cold War is officially over, the "games" will go on.

16-Year-Old Attacks Labs

One of the attacks in 1994 against Rome Laboratories, Goddard Space Flight Center, Jet Propulsion Laboratory, South Korean Atomic Research Institute, and other sites was conducted by a U.K. 16-year-old hacker known as "Datastream Cowboy," with the assistance of a 22-year-old hacker known as "Kuji."

Comments:

What is of special interest about this case is that such places as Rome Laboratories and the Jet Propulsion Laboratory are involved in information warfare-related R&D efforts. If a 16-year-old and a 22-year-old can, apparently with ease, successfully gain access to systems and networks where sensitive work is being done, what can an experienced netspionage agent accomplish?

U.S. Information Warfare Active for 10 Years

Christiane Schultzki-Haddouti, a German journalist who specializes in information warfare.... Up until 1995, for example, she traced 10 U.S. developments, including in 1992 the Pentagon's "first top secret directive TS-3600 on 'Information Warfare'" and the use of "computer network attacks" in the U.S. operation in Haiti in 1994 to return Bertrand Aristide to power.

Comments:

Although the source of the above is lost, it is placed here to show the reader that (1) the United States has been thinking of information warfare for more than ten years; and (2) the United States apparently used information warfare offensive tactics and weapons almost a decade ago.

Sandia Labs Target for Hackers

Sandia National Laboratory in New Mexico in the United States was attacked, and attacks continue to be attempted against this sensitive laboratory. Apparently, the attackers gained access to sensitive information and were thought to be foreign government agents. The United States intelligence community suspects Russia, China, Iraq, and North Korea.

Comments:

Such attacks are almost a daily occurrence. How does one know whether it is someone like Datastream Cowboy or Kuji, or a foreign Netspionage agent — or both? That is the challenge for all those responsible for defensive information warfare, information assurance, and INFOSEC.

Exhibit 12. Case Studies: The United States (Continued)

U.S. May Help Chinese Evade Net Censorship
Reuters, August 30, 2001,
http://dailynews.yahoo.com/h/nm/20010830/wr/tech_china_internet_report_dc_1.html
United States government agencies hope to finance an American-based computer network designed to thwart attempts by the Chinese government to censor the World Wide Web for users in China...

Comments:
The use of such information warfare offensive weapons (at least considered offensive by the Chinese government), if implemented, can be viewed by the Chinese as an attack against its NII and an act of war. How would China respond? At least with diplomatic protests to the United States. Even if the United States had a "plausible denial" at the ready, China would not believe it. Would China then become more aggressive in their information warfare operations against the United States? It is believed that right now, the Chinese are involved in an information warfare thrust toward the United States in intelligence gathering — netspionage and the like. This may change as China is very sensitive to the United States becoming involved in what China sees as its internal affairs. Furthermore, this may also be seen by China as an attempt to foment a revolution or rebellion in China with the objective of ending the power of the communist regime. On the other hand, if such a system could be implemented throughout the world where nation-states control access to the Internet, the ramifications could include more wars, revolutions, rebellions, but also more freedoms for the people of those nation-states, such as Cuba.

U.S. Embassy Web Site in China Hacked
Hong Kong, China, Newsbytes, Neil Taylor, IT Daily, September 8, 1999
The official Web site for the U.S. Embassy, at *http://www.usembassy-china.gov/* in China was brought down by hackers (crackers) on Tuesday...

Comments:
Attacks against the Web sites of the United States government agencies happen everyday and are no longer news. However, if the attacks are traced back to Chinese government agencies, China will of course deny they know anything about it nor could they determine who actually was the attacker. That has been their position in the past. One wonders how China could condone its networks being so defenseless that a Chinese security official could not trace such culprits. What if a Chinese internal hacker attacked a Chinese government site? One would believe that Chinese security officials could quickly hunt down the culprits and imprison them. Such attacks would lead one to believe that they are sponsored and even conducted by or with the approval of the Chinese government security officials. Are they probes as part of an information warfare intelligence collection effort? If the attacks can be traced directly to a Chinese government computer system, would this be considered an act of war against the United States? In today's political environment, not likely.

NIPC Warns China Hackers May Target U.S. Sites
Dick Kelsey, Newsbytes, Washington, D.C., April 26, 2001
An arm of the FBI that watches for cybercrime and online security threats today warned that Chinese hackers may escalate their attacks on U.S. Web sites and mail servers.

Comments:
The United States' National Infrastructure Protection Center (http://www.nipc.gov) provides warnings such as the above. Unfortunately, history has shown that those responsible for

Exhibit 12. Case Studies: The United States (Continued)

information warfare defense (e.g., information systems security, information assurance) pay little attention to such warnings. They often do not even have plans in place to heighten information protection, update their systems security software, patch security holes based on CERT advisories, and the like. In 1987, the United States Congress passed the Computer Security Act to provide some baseline of information systems protection. As of this date, one will find that on many government systems, even the most basic and prudent protection measures are not taken unless they are forced on the IT personnel responsible for such defenses. Will the new United States "cyber-security" czar make a difference? One hopes so, but based on previous track records of the government agencies, one cannot be too optimistic.

CIA: China, Russia Develop Cyberattack Capability
Jack McCarthy, IDG News Service February 24, 2000
The United States is vulnerable to cyberattacks from a growing list of terrorists and foreign countries, including Russia and China, a U.S. Central Intelligence Agency (CIA) official *told a congressional committee.*

Comments:
John Serabian, CIA's information operations issue manager, told a United States Congressional committee that: We are detecting, with increasing frequency, the appearance of doctrine and dedicated offensive cyberwarfare programs in other countries,…. We have identified several (countries), based on all-source intelligence information, that are pursuing government-sponsored offensive cyberprograms…. "Information warfare" is becoming a possible strategic alternative for countries that realize that, in conventional military confrontation with the United States, they will not prevail…

Hill Air Force Base Being Targeted in Ongoing Cyberspace War
Security News Portal, http://www.securitynewsportal.com/article.php?sid=1659, August 27, 2001
Utah's Hill Air Force Base is one battlefield in this digital war. About 1000 hackers a day attack Hill's computer system and its 64,000 dataports, reports John Gilchrist, Hill's chief of information assurance.

Comments:
If there were over 1000 attacks against this one United States Air Force Base imagine the attacks against United States government network and systems worldwide. One of the reasons that there are so many attacks is because the systems have been found to be so vulnerable to even 16-year-old kids. Once the United States has the will to defend itself and takes a more aggressive defensive and offensive posture, such attacks will be drastically reduced. Until then, they will continue almost unabated.

U.S. Railroad Regulator Hit by Hacker
Washington, Reuters, May 1, 2001
The railroad-regulating U.S. Surface Transportation Board said on Tuesday hackers had knocked out its computer system amid a threatened assault on U.S. Web sites by Chinese activists. Since Monday, when the threatened week-long blitz began, the board has been unable to post its decisions and unable to send or receive e-mail outside the organization,…

Exhibit 12.　Case Studies: The United States (Continued)

Comments:

This is an example of Chinese attacks against United States networks, done with impunity. As stated earlier, is this possible without Chinese government support? Unlikely, and yet the United States does little in the way of protesting to the Chinese government. This can do nothing but encourage them to continue and to increase such attacks. It took the attacks of September 11, 2001, for the United States to wake up to the threat of global terrorism after it found that lobbing a few cruise missiles into some Arab desert tents, and maybe killing a few camels, did nothing but encourage them further. Will the United States have to learn that same, painful lesson when it comes to information warfare attacks? It seems so.

Moonlight Maze (Renamed Storm Cloud)

Research and development laboratories and other sites were virtually bleeding intellectual property. None of the standard information assurance devices went into alarm mode. One way that the attacks were identified was using heuristics on intrusion detections system audit log data. Even with that knowledge, for over three years, Moonlight Maze, renamed Storm Cloud, stumped investigators. The leads ended at a Russian ISP in Moscow. The expected time it would take a computer to communicate with the ISP was calculated, the time was converted to distance, a circle using the distance as the radius was drawn around the ISP, and possible perpetrators within the circle were identified.

By mid-year 2001, the heist was the "equivalent to a stack of paper three times the height of the Washington Monument," said Air Force Maj. Gen. Bruce Wright, Commander of the Air Intelligence Agency. Officials found software sensors inside federal computers that modified a private Web site in Britain whenever new documents were available. The hackers would view the Web site to see if it had changed, thus limiting their exposure to detection. The FBI followed additional leads and connected to Russian computers overseas. They downloaded 781 megabytes, a portion of the purloined data.[35]

Summary

Nation-states will continue to play the dominant role in information warfare. Nation-states, such as those in the Americas, will continue to fight over age-old issues. The Information Age and information warfare will provide the smaller Third World nations and groups such as the Zapatistas with the opportunity to conduct wars with larger or more developed nation-states on a more than equal playing field.

Information warfare in the Americas, excluding the United States and Canada, is very primitive. However, as these nation-states increase their high-technology base, they will undoubtedly increase their information warfare expertise.

The United States is the leader in technology, information warfare strategies, concepts, tactics, and weapons systems. Information warfare activities are carried out by a wide variety of U.S. government agencies. The United States is also the number-one target for United States' adversaries' information warfare specialists.

Unless the United States wakes up to this ever-increasing threat, as it did relative to the global terrorists on September 11, 2001, well — "if it keeps doing what it is doing, it will keep getting what it is getting."

One final thought — after reading this chapter you will note the amount of information taken from U.S. government Web sites. From an IW viewpoint, a veritable gold mine of information is available for adversaries. That raises the question, Why is this posted? Who benefits more, the American people or their adversaries?

Notes

1. Iceland is somewhat unique in that it straddles North America and Europe in physical location but has strong ties with both; thus it is listed here.
2. As noted in other chapters about nation-states, these associations are listed to provide the reader with some insight into the relationship between nation-states as well as potential territorial conflicts. Thus, one can begin to see where and why conflicts might arise and draw in other entities. Coupled with the nation-states' technology level, one can establish the sophistication and amount of information warfare tactics that may play a role in conflicts between nation-states.
3. See http://www.yahoo.com/headlines/980205/wired/stories/mexico_1.html
4. Floodnet is a Java applet that, once the launching Web page had been called up, repeatedly loaded pages from targeted networks. If enough people participated, the targeted computer would be brought to a halt — bombarded by too many messages for it to handle. See http://www.yahoo.com/headlines/980205/wired/stories/mexico_1.html
5. Information on Cuba taken primarily from "The Internet and State Control in Authoritarian Regimes: China, Cuba, and the Counterrevolution," by Shanthi Kalathil and Taylor C. Boa, Global Policy Program, Number 21, July 2001, Carnegie Endowment.
6. See http://www.cubaweb.cu/esp/main.asp; http://www.cubasolidarity.net/cubalink.html; and http://www.state.gov/www/regions/wha/cuba/.
7. Only the United States and Canada have the information infrastructure, local and world presence, and history of conflicts in which information warfare has played a role. Greenland and Iceland are not in a similar position and therefore not discussed here.
8. http://www.nacic.gov/cind/Sep98.html, *Counterintelligence News and Developments,* September 1998.
9. For discussion of this matter, see *The Cuckoo's Egg* by Cliff Stohl.
10. See http://www.vcds.dnd.ca/dgsp/dplan/intro_e.asp for details.
11. The Canadian Defence Plan 2001 defines information operations as "Actions taken in support of national objectives which influence decision makers by affecting other's information while exploiting and protecting one's own information."
12. Also reported in the *Montreal Gazette* article on October 6, 2001.
13. As noted, the United States is at the forefront of information warfare strategies and tactics, as well as information warfare weapons and battles. Therefore, although some of the information presented was alluded to in previous chapters of this book (e.g., definitions), it is presented here because it is very relevant to the United States and it is hoped that it will help the reader obtain a better understanding of not only information warfare as seen by the United States, but also information warfare in general.
14. Source is Jerry Ervin, former INFOSEC and military specialist.

15. Sounds like IW, but perhaps more "politically correct" for the politicians and other bureaucrats.
16. This information was derived from http://www.whitehouse.gov/nsc/.
17. See http://www.access.gpo.gov/congress/.
18. Statement to U.S. Senate Committee on Government Affairs, Permanent Subcommittee on Investigations, June 25, 1996.
19. See http://www.cia.gov/cia/information/info.html.
20. See http://www.defenselink.mil/.
21. See http://www.dtic.mil/jcs/.
22. See http://www.dtic.mil/jv2020/.
23. From a briefing given by Colonel Frank Goral, at the time Vice Director, JC2WC.
24. See http://www.nsa.gov/ for additional information.
25. See http://www.nsa.gov/about_nsa/faqs_internet.html for additional details.
26. See http://www.dia.mil/.
27. See *Information Security* magazine, October 2001, page 30. Special thanks to Andy Briney for allowing the republication of this information.
28. See http://www.disa.mil/.
29. See http://www.af.mil/; http://www.afrl.af.mil/; http://www.fas.org/irp/agency/aia/afiwc/.
30. See http://www.army.mil/.
31. See http://www.fas.org/irp/doddir/army/fm34-37_97/3-chap.htm.
32. See http://www.navy.mil; http://www.usmc.mil/.
33. See http://enterprise.spawar.navy.mil/spawarpublicsite/aboutspawar/index.htm.
34. See http://www.darpa.mil/.
35. Net Espionage Rekindles Tensions as U.S. Tries to Identify Hackers, Ted Bridis, *The Wall Street Journal,* June 27, 2001.

Chapter 12

Information Warfare: African and Other Nation-States

Progress lies not in enhancing what is, but in advancing toward what will be.

—Kahlil Gibran, "A Handful of Sand on the Shore"

This chapter discusses information warfare as it relates to Africa and other entities not covered in previous chapters.

Introduction to African Nation-States

Africa is sometimes divided into four areas:

- East Africa
- Southern Africa
- West Central Africa
- North Africa

Africa is currently made up of the nation-states as listed in Exhibit 1.

Exhibit 1. African Nation-States

East Africa	Southern Africa	West Central Africa		North Africa
Burundi	Angola	Benin	Ivory Coast	Western Sahara
Eritrea	Botswana	Burkina Faso	Liberia	Morocco
Ethiopia	Malawi	Cameroon	Mali	Algeria
Kenya	Mozambique	Chad	Mauritania	Tunisia
Madagascar	Namibia	D.R. Congo	Niger	Libya
Rwanda	South Africa	R. Congo	Nigeria	Egypt
Somalia	Swaziland	Gabon	Senegal	
Sudan	Zambia	Gambia	Sierra Leone	
Tanzania	Zimbabwe	Ghana	Togo	
Uganda		Guinea		

However, many of these countries are usually associated with the Middle East by customs and religions. They are not addressed here, as their primary issues were addressed in Chapter 9.

Exhibit 2 provides a map of Africa depicting her nation-states.

African Associations

As with other regions, the African nation-states also belong to various associations. These associations, as with others between nation-states throughout the world, may draw African nation-states into conflicts not previously considered likely. For example, one would not have considered Africa to play a role in the United States' war on terrorists. However, with the bombings of U.S. embassies on August 7, 1998, in Kenya and Tanzania, African nation-states were drawn into that conflict.

These, according to the *World Factbook,*[1] include but are not limited to:

- *African, Caribbean, and Pacific Group of States (ACP Group)* located in Brussels, Belgium. Established in June 1975, its aim is to manage their preferential economic and aid relationships with the European Union. Members — 71 — include Angola, Antigua and Barbuda, the Bahamas, Barbados, Belize, Benin, Botswana, Burkina Faso, Burundi, Cameroon, Cape Verde, Central African Republic, Chad, Comoros, Democratic Republic of the Congo, Republic of the Congo, Cote d'Ivoire, Djibouti, Dominica, Dominican Republic, Equatorial Guinea, Eritrea, Ethiopia, Fiji, Gabon, the Gambia, Ghana, Grenada, Guinea, Guinea-Bissau, Guyana, Haiti, Jamaica, Kenya, Kiribati, Lesotho, Liberia, Madagascar, Malawi, Mali, Mauritania, Mauritius, Mozambique, Namibia, Niger, Nigeria, Papua New Guinea, Rwanda, Saint Kitts and Nevis, Saint Lucia, Saint Vincent and the Grenadines, Samoa, Sao Tome and Principe, Senegal, Seychelles, Sierra Leone, Solomon Islands, Somalia, South Africa, Sudan, Suriname, Swaziland, Tanzania, Togo, Tonga, Trinidad and Tobago, Tuvalu, Uganda, Vanuatu, Zambia, and Zimbabwe.

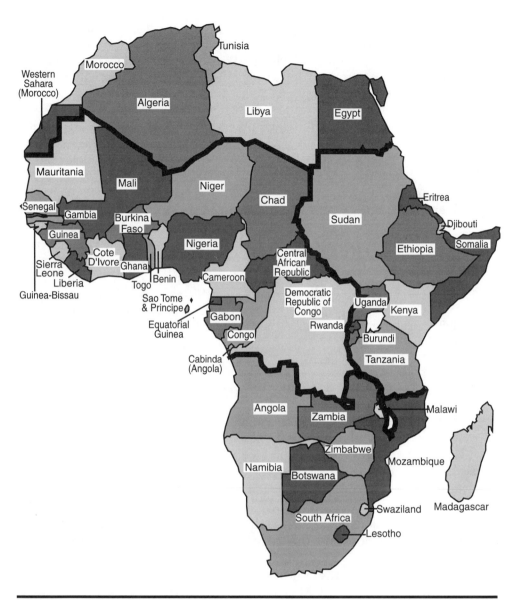

Exhibit 2.   Map of Africa

- *African Development Bank (AfDB),* also known as Banque Africaine de Developement (BAD), Abidjan 01, Cote d'Ivoire. It was established in August 1963 with the aim of promoting economic and social development. Regional members — 53 — include Algeria, Angola, Benin, Botswana, Burkina Faso, Burundi, Cameroon, Cape Verde, Central African Republic, Chad, Comoros, Democratic Republic of the Congo, Republic of the Congo, Cote d'Ivoire, Djibouti, Egypt, Equatorial Guinea, Eritrea, Ethiopia, Gabon, the Gambia, Ghana, Guinea, Guinea-Bissau, Kenya, Lesotho, Liberia, Libya, Madagascar, Malawi, Mali, Mauritania, Mauritius, Morocco, Mozambique, Namibia, Niger, Nigeria, Rwanda, Sao Tome and Principe, Senegal, Seychelles,

Sierra Leone, Somalia, South Africa, Sudan, Swaziland, Tanzania, Togo, Tunisia, Uganda, Zambia, and Zimbabwe. Non-regional members — 25 — include Argentina, Austria, Belgium, Brazil, Canada, China, Denmark, Finland, France, Germany, India, Italy, Japan, South Korea, Kuwait, the Netherlands, Norway, Portugal, Saudi Arabia, Spain, Sweden, Switzerland, the United Arab Emirates, the United Kingdom, and the United States.

■ *Agency for the French-Speaking Community (ACCT),* formerly Agency for Cultural and Technical Cooperation, Paris, France. It was established in March 1970, with its name changed in 1996. Its aim is to promote cultural and technical cooperation among French-speaking countries. Members — 41 — include Belgium, Benin, Bulgaria, Burkina Faso, Burundi, Cambodia, Cameroon, Canada, Cape Verde, Central African Republic, Chad, Comoros, Democratic Republic of the Congo, Republic of the Congo, Cote d'Ivoire, Djibouti, Dominica, Equatorial Guinea, France, Gabon, Guinea, Haiti, Laos, Lebanon, Luxembourg, Madagascar, Mali, Mauritius, Moldova, Monaco, Niger, Romania, Rwanda, Sao Tome and Principe, Senegal, Seychelles, Switzerland, Togo, Tunisia, Vanuatu, and Vietnam. Associate members — 5 — include Egypt, Guinea-Bissau, Mauritania, Morocco, and Saint Lucia. Participating governments — 2 — include New Brunswick (Canada) and Quebec (Canada).

African Conflicts and Potential Conflicts

As with other nation-states noted in earlier chapters, the African nation-states also have political and territorial conflicts. These are mentioned because, regardless of the means used to conduct war (e.g., bombs, infantry, computers), it is important to understand the reasons for these conflicts. For those involved in defensive information warfare (IW-D) strategies and tactics, it is vitally important to understand where some future conflicts may arise and why. This information is necessary so that IW-D professionals can prepare for such future conflicts now, in a proactive manner and not in a reactive manner. Being on the IW-D side of information warfare does not mean being "defensive" by only being in a position to react to conflicts after they occur.

The following is a list of African nation-states that have some territorial conflicts as reported by the United States Central Intelligence Agency in the *World Factbook*:[1]

■ *Bassas da India:* Claimed by Madagascar
■ *Botswana:* Dispute with Namibia over uninhabited Kasikili (Sidudu) Island in Linyanti (Chobe) River resolved by the ICJ in favor of Botswana (December 13, 1999); at least one other island in Linyanti River is contested
■ *Cameroon:* Delimitation of international boundaries in the vicinity of Lake Chad, the lack of which led to border incidents in the past, is complete and awaits ratification by Cameroon, Chad, Niger, and Nigeria; dispute with Nigeria over land and maritime boundaries around the Bakasi Peninsula and Lake Chad is currently before the ICJ, as is a dispute with Equatorial Guinea over the exclusive maritime economic zone

- *Chad:* Delimitation of international boundaries in the vicinity of Lake Chad, the lack of which led to border incidents in the past, has been completed and awaits ratification by Cameroon, Chad, Niger, and Nigeria
- *Congo, Democratic Republic of the:* The Democratic Republic of the Congo is in the grip of a civil war that has drawn in military forces from neighboring states, with Uganda and Rwanda supporting the rebel movement which occupies much of the eastern portion of the state; most of the Congo River boundary with the Republic of the Congo is indefinite (no agreement has been reached on the division of the river or its islands, except in the Pool Malebo/Stanley Pool area)
- *Equatorial Guinea:* Exclusive maritime economic zone boundary dispute with Cameroon is presently before the ICJ; maritime boundary dispute with Gabon because of disputed sovereignty over islands in Corisco Bay; maritime boundary dispute with Nigeria and Cameroon because of disputed jurisdiction over oil-rich areas in the Gulf of Guinea
- *Eritrea:* Dispute over alignment of boundary with Ethiopia led to armed conflict in 1998, which is still unresolved despite arbitration efforts
- *Ethiopia:* Most of the southern half of the boundary with Somalia is a Provisional Administrative Line; territorial dispute with Somalia over the Ogaden; dispute over alignment of boundary with Eritrea led to armed conflict in 1998, which is still unresolved despite arbitration efforts
- *Europa Island:* Claimed by Madagascar
- *Gabon:* Maritime boundary dispute with Equatorial Guinea because of disputed sovereignty over islands in Corisco Bay
- *Gambia, The:* Short section of boundary with Senegal is indefinite
- *Glorioso Islands:* Claimed by Madagascar
- *Juan de Nova Island:* Claimed by Madagascar
- *Kenya:* Administrative boundary with Sudan does not coincide with international boundary
- *Madagascar:* Claims Bassas da India, Europa Island, Glorioso Islands, Juan de Nova Island, and Tromelin Island (all administered by France)
- *Malawi:* Dispute with Tanzania over the boundary in Lake Nyasa (Lake Malawi)
- *Morocco:* Claims and administers Western Sahara, but sovereignty is unresolved and the UN is attempting to hold a referendum on the issue; the UN-administered cease-fire has been in effect since September 1991; Spain controls five places of sovereignty (*plazas de soberania*) on and off the coast of Morocco — the coastal enclaves of Ceuta and Melilla which Morocco contests, as well as the islands of Penon de Alhucemas, Penon de Velez de la Gomera, and Islas Chafarinas
- *Namibia:* Dispute with Botswana over uninhabited Kasikili (Sidudu) Island in Linyanti (Chobe) River resolved by the ICJ in favor of Botswana (December 13, 1999); at least one other island in Linyanti River is contested
- *Niger:* Libya claims about 19,400 sq km in northern Niger; delimitation of international boundaries in the vicinity of Lake Chad, the lack of which led to border incidents in the past, has been completed and awaits ratification by Cameroon, Chad, Niger, and Nigeria
- *Nigeria:* Delimitation of international boundaries in the vicinity of Lake Chad, the lack of which led to border incidents in the past, has been

completed and awaits ratification by Cameroon, Chad, Niger, and Nigeria; dispute with Cameroon over land and maritime boundaries around the Bakasi Peninsula is currently before the ICJ; maritime boundary dispute with Equatorial Guinea because of disputed jurisdiction over oil-rich areas in the Gulf of Guinea

- *Rwanda:* Rwandan military forces are supporting the rebel forces in the civil war in the Democratic Republic of the Congo
- *Senegal:* Short section of boundary with the Gambia is indefinite
- *Somalia:* Most of the southern half of the boundary with Ethiopia is a Provisional Administrative Line; territorial dispute with Ethiopia over the Ogaden
- *South Africa:* Swaziland has asked South Africa to open negotiations on reincorporating some nearby South African territories that are populated by ethnic Swazis or that were long ago part of the Swazi Kingdom
- *Sudan:* Administrative boundary with Kenya does not coincide with international boundary; Egypt asserts its claim to the "Hala'ib Triangle," a barren area of 20,580 sq km under partial Sudanese administration that is defined by an administrative boundary that supersedes the treaty boundary of 1899
- *Swaziland:* Swaziland has asked South Africa to open negotiations on reincorporating some nearby South African territories that are populated by ethnic Swazis or that were long ago part of the Swazi Kingdom
- *Tanzania:* Dispute with Malawi over the boundary in Lake Nyasa (Lake Malawi)
- *Tromelin Island:* Claimed by Madagascar and Mauritius
- *Tunisia:* Maritime boundary dispute with Libya; Malta and Tunisia are discussing the commercial exploitation of the continental shelf between their countries, particularly for oil exploration
- *Western Sahara:* Claimed and administered by Morocco, but sovereignty is unresolved and the UN is attempting to hold a referendum on the issue; the UN-administered cease-fire has been in effect since September 1991

African Nation-States and Technology

The nation-states of Africa, as one would expect, are far behind in developing a NII and using technology to assist in developing into information-based nation-states.

> ***Some Countries Being Left Behind in E-Commerce***
> *infowar.com, August 8, 2000*
> *Some nations are falling behind as the world becomes increasingly reliant on information technology and E-commerce.... African countries were described as needing "substantial improvement" to support new, electronic-based businesses...and Egypt, Ghana, Kenya, Nigeria, Saudi Arabia, South Africa, and Tanzania all have considerable room for improvement, the report said. However, the governments of Egypt, Ghana, and South Africa, in particular, have recognized the coming shift in economic focus, and are dedicated to preparing their countries for that change...*

This is expected when looking at their types of governments, their economies, especially their national and international debts. However, at the same time, it is believed that technology offers them an opportunity to progress as nation-states. While at the same time, the conflicts and coups of governments also make such technology either something to be feared and banned to their citizens, or used in information warfare of the future.

As Dana Ott, Program Analyst in the Africa Bureau, Office of Sustainable Development of the U.S. Agency for International Development in Washington, D.C., wrote in his article, "Power to the People: The Role of the Electronic Media in Promoting Democracy in Africa"[2]

> **Internet.** *It is an astounding fact that where the Internet barely existed in Africa even five years ago, today 44 of the 54 nations in Africa have some form of Internet access in their capital cities. But as Mike Jensen notes, "Of the countries with full Internet access, only Burkina Faso, Mauritius, Morocco, Senegal, South Africa, and Zimbabwe have pervasive local dialup facilities outside of the capital city, while Benin, Botswana, Egypt, and Kenya have services in the second major city"…access costs are generally higher in Africa than elsewhere,…this may be related to the profit margins of the telecommunications sector in African countries, where the revenue per subscriber line is twice as high as in Europe.… The two greatest impediments to the development of electronic communication in Africa are insufficient infrastructure and regulatory barriers. Infrastructure limitations include such problems as scarce and/ or poor quality telephone lines, unreliable power supplies, outdated equipment, and a lack of knowledge and training. Regulatory barriers include government monopolies on telecommunications, high access rates for telephone service, and legal disincentives to foreign investment.*

African Nation-States and Information Warfare

Because of the lack of a mature NII and currently not being dependent on information systems to conduct government, business, and social functions within the African nation-states, it stands to reason that mature information warfare strategies and tactics are also not in the military arsenal of the majority of these nation-states.

In Somalia, the conflicts rage between government forces and militia. Warlords such as Hassan Ali "Atto" and Hussein Mohamed Aidid continue to retain Somalia in the grip of war. They battle with the brutal weapons of the agricultural and industrial ages. Propaganda via radio or television is their primary use of information warfare tactics, with some minor use of the Internet for assisting in the propagation of the war.

The United States learned that valuable lesson when it sent troops into Somalia.[3] In fact, when such international networks such as CNN showed a U.S. soldier being dragged through the streets in Mogadishu, it quickly caused the United States to withdraw its military from Somalia. Yes, it was probably not the only reason, but it also made a strong case for such action. Thus, the use of such "primitive" information warfare weapons as international television newscasts can be a powerful weapon to be used by small Agricultural or Industrial Age nation-states against the more advanced and powerful Information Age nation-states. This is an example of how high technology is not a prerequisite for conducting information warfare. However, that said, one must realize that even today's "primitive" information systems of these nation-states do have at least some Internet access, e-mail, and Web sites (e.g., http://www.somalipress.com). This incident was also a salutary lesson for the United States and for all other countries in the use of weapons of information warfare. The first rule is that you must have a comprehensive understanding of your enemy and of the enemy's culture and values. It is clear that Aidid understood the psyche of the Western visitors far better than they understood that of the Somalis, which is not surprising because the West trained him.

Other conflicts — such as that which resulted from the apartheid between blacks and whites in South Africa, the conflict between Ethiopia and Eritrea, and tribal conflicts in other nation-states — are all examples of Industrial Age warfare based on guns and infantry. One must not forget Sudan with its alleged connections to Middle East terrorists, which caught the attention of the United States who bombed alleged terrorist support sites in the Sudan.

In the matter of the war between Ethiopia and Eritrea, like other nation-states addressed in previous chapters, even some primitive information warfare tactics are being used. According to a BBC news article,[4] both of these nation-states have a history of using radio, clandestine and unofficial, to get their messages across to the people of the region. It is interesting to note that three of these anti-Eritrea radio stations apparently are using the same transmitter in the Sudan, thus giving cause for the expansion of the conflict. The "war of words" has gone from the radio to the Internet. The BBC states:

> *The battle for the hearts, minds, and pockets of the large Ethiopian and Eritrea exile communities has also led to a spate of fake e-mails and postings aimed at spreading disinformation and sapping the opponent's morale.*

More information on this conflict can be found[5] at:

- www.africanews.org
- www.tidalwave.net/ethiopia
- www.primenet.com/ephrem
- www.monitor.bbc.co.uk

It appears that, like many other nation-state conflicts around the world, those in Africa have been going on for some time and there is no end in sight. As these nation-states slowly begin to enter the Information Age, there is little doubt that they will embrace the information warfare strategies and tactics now being used by other, more modern nation-states around the world. In fact, it is expected that the African nation-states will learn quickly from others around the world, and more readily and more quickly integrate information warfare strategies and tactics into their war plans. Until then, we may see more Industrial Age warfare in Africa (e.g., bombing of embassies).

Other Entities and Nation-States

There are many "entities" in the world that are territories claimed by one or more nation-states, as well as some smaller nation-states that are not active participants in the world's conflicts. However, such entities may at some point become embroiled in conflicts between nation-states. Therefore, they are listed here as sort of a "catch-all" so that the reader is familiar with these territorial conflicts and is better prepared to understand and address them.

The following is a list of other entities that have some territorial conflicts, as reported by the United States Central Intelligence Agency in the *World Factbook*:[1]

- *Antarctica:* Antarctic Treaty defers claims (see Antarctic Treaty Summary in Government type entry); sections (some overlapping) claimed by Argentina, Australia, Chile, France (Adelie Land), New Zealand (Ross Dependency), Norway (Queen Maud Land), and the United Kingdom; the United States and most other nations do not recognize the territorial claims of other nations and have made no claims themselves (the United States reserves the right to do so); no formal claims have been made in the sector between 90 degrees west and 150 degrees west
- *Arctic Ocean:* Some maritime disputes (see littoral states); Svalbard is the focus of a maritime boundary dispute between Norway and Russia
- *British Indian Ocean Territory:* The Chagos Archipelago is claimed by Mauritius and Seychelles
- *Comoros:* Claims French-administered Mayotte; the islands of Anjouan (Nzwani) and Moheli (Mwali) have moved to secede from Comoros
- *French Guiana:* Suriname claims area between Riviere Litani and Riviere Marouini (both headwaters of the Lawa)
- *French Southern and Antarctic Lands:* "Adelie Land" claim in Antarctica is not recognized by the United States
- *Marshall Islands:* Claims U.S. territory of Wake Island
- *Mauritius:* Claims the Chagos Archipelago in U.K.-administered British Indian Ocean Territory; claims French-administered Tromelin Island
- *Mayotte:* Claimed by Comoros
- *Seychelles:* Claims Chagos Archipelago in British Indian Ocean Territory
- *Southern Ocean:* Antarctic Treaty defers claims (see Antarctic Treaty Summary in the Antarctica entry); sections (some overlapping) claimed by Argentina, Australia, Chile, France, New Zealand, Norway, and the United Kingdom; the United States and most other nations do not recognize the maritime claims of other nations and have made no claims themselves (the United States reserves the right to do so); no formal claims have been made in the sector between 90 degrees west and 150 degrees west
- *Wake Island:* Claimed by Marshall Islands

Based on the nation-states that may become embroiled in a "hot war" over these territories or as part of a conflict that will draw these territories into an already wider conflict, information warfare strategies and tactics may play a role — even a vital role — in such conflicts. However, at this time, information warfare based on the territorial conflicts seems remote.

Summary

African nation-states and other entities have not sufficiently developed into the Information Age; therefore, their use of all but primitive information warfare strategies and tactics are remote at this time. However, as with other parts of the world, there are hot wars being fought by nation-states on this continent. These wars and the animosities that have built up over the years will not quickly go away — if ever. Therefore, as they progress into the Information Age, it is expected that they will quickly adopt information strategies and tactics and use them against their already-hated adversaries.

Notes

1. See http://www.cia.org.
2. The authors thank Mr. Ott for allowing the republication of his paper and also Edward J. Valauskas, Chief Editor, *First Monday* (see http://firstmonday.org), who first published his paper.
3. For an excellent read on that topic, see Mark Bowden's *Black Hawk Down: A Story of Modern War.*
4. See http://news.bbc.co.uk/hi/english/world/monitoring/newsid_280000/280680.stm, October 14, 2001.
5. Throughout this chapter, and the entire book for that matter, the references and Web site addresses "were there when last we looked". However, as all I-way surfers know, the Internet is a very dynamic environment. The means that Web sites come and go. So, if you cannot find a cited Web site, this is the reason for it.

Chapter 13

It's All about Profits: Information Warfare Tactics in Business

In every business, no matter how small or how large, someone is just around the corner forever trying to steal your ideas and build his success out of your imagination, struggling after that which you have toiled endless years to secure, striving to outdo you in each and every way. If such a competitor would work as hard to originate as he does to copy, he would much more quickly gain success.

—Alice Foote MacDougall (1867–1945)

This chapter discusses how businesses use offensive, exploitative, and defensive information warfare tactics to their advantage.

Introduction

Imagine presenting a business case to your company president. You know he is a "numbers man." More than once you have had to answer three questions: How much is it going to cost me? When am I going to see a return? What is my profit going to be? After what you think is a brilliant presentation, the president asks you, "Have we protected this project from an attack by competitors? Has any of this information been sent on the Internet? What have we found our competitors doing in this area?" Whoa! Although you answered the traditional questions, three new questions reflective of doing business on the Internet were posed. The president may not know much about information warfare technical details, but he does know that the marketplace is a dangerous place and that he must use every tool at his disposal that will give him a competitive advantage so that he can make a profit and deliver positive returns to shareholders.

There is much information citing the costs and losses of information warfare attacks. The cost of servicing cyber-warfare incidents worldwide exceeded U.S.$20 billion in 1999, said DK Matai, founder of mi2g software. When asked, "How does one plan for business continuity in the coming years?," he responded, "The answer lies in a properly funded bespoke security architecture to which the board of directors commits itself completely."[1] We can suspect that one of the president's top three issues is security because he is asking security-related questions.

Lapses in information and information systems (INFOSEC) security are costing companies up to seven percent of their annual revenues. Computer hackers cost European businesses U.S.$4.3 billion (£3 billion) in lost revenue in 2000, according to new research. A study of 3000 businesses worldwide found lapses in security cost companies between 5.7 and 7 percent of their annual revenue, or six cents for every dollar in sales.[2] These numbers very quickly grab the attention of senior management.

Coherent's Chief Operating Officer, John Ambroseo, said, "For a company like ours, there are two most valuable assets: our people and our proprietary information. That's what drives senior management and security here — protecting our information."[3] That is true at Coherent and hundreds of thousands of other companies. The 2001 Global Information System Security Survey by *InformationWeek* estimated the cost of security-related downtime to U.S. businesses in the 12 months covered by the survey at $273 billion. Worldwide, the tally was $1.39 trillion.[3] The survey did not address all information warfare areas we detailed in Chapter 1. We estimate the U.S. and worldwide costs for total IW losses are each an order of magnitude higher. Surveys of companies suffering Internet-based attacks, again not total IW, show positive responses in the 70 to 90 percent range. These numbers have increased every year. That is, of those who know they have been attacked (think about it — would a competitor want you to know an attack occurred?) *and* who are willing to report an attack. Not every business is willing to report an attack.

Computer criminals in the United Kingdom continue to get away with their dirty deeds because companies cannot afford the price of prosecution. The threshold for pursuing attackers appears to be more than £50,000 (U.S.$73,125) in damage or strong evidence publicity of the attack hurt a company's reputation, according to Nigel Layton, CEO of security consultancy Quest.[4] This finding is consistent with the KPMG electronic fraud survey that found 83 percent of the CEOs and CIOs it surveyed suffered attacks but decided against taking legal action. The reasons included difficulty in gathering sufficient evidence and the low conviction rate of hackers.[5] A manufacturer recently suspected former employees of stealing hundreds of thousands of dollars worth of proprietary information from its computer network. But rather than call the police, the manufacturer hired a private consultant to help settle the matter quietly, thus dodging the scrutiny of the courts, the glare of the media, and the attention of investors.[6] For every breach reported, there may be ten or twenty that have not been reported. Organizations do not like admitting they lost money or their systems were breached. If there is a serious breach, the directors could be held accountable if they have not taken security seriously. It is not a technical issue; it is a management issue that requires attention at the board level.[7] The last thing the corporate general counsel wants is a class-action lawsuit claiming the company failed to use best practices and due diligence to protect information, profits, and the shareholders' stock value.

This is a stunning paradox. Companies worldwide are losing over a trillion dollars, are not reporting the crimes to law enforcement, and are doing little to help themselves. *Network World* released the findings of its 2001 Network World 500 study. The survey was circulated to companies with more than 1000 employees, multiple sites with inter-networked local and wider area networks, and expenditures of more than $10 million. Of the network managers, 90.9 percent said security is their biggest challenge. Almost 70 percent say that the E-business activities they perform are vulnerable to security breaches.[8]

Corporate spies do not always go after exotic information. Seemingly boring stuff such as customer lists, employee forms, and supplier agreements are some of the most sought-after items for competitors to steal. Most often, victimized companies do not even know anything has happened, let alone take action. "Companies don't usually report information theft," FBI agent Thomas Purcell said.[9] Some companies are compelled to report. It is not unusual for the price of a stock to go up on the expectation of future business. When the expected business does not materialize, questions are asked. In September 2000, Irwin Jacobs, founder and chairman of Qualcomm Inc., had his laptop computer stolen after a speech. Jacobs told *The Wall Street Journal* the information on the hard drive would be very valuable to foreign governments. Qualcomm was then negotiating a deal for CDMA service across the People's Republic of China.[10]

Malicious acts can happen in both the physical and virtual environments to widen or narrow competitive advantage.

Overview of Techniques Used to Widen or Narrow Competitive Advantage

New technologies have affected competitive advantage by:

- Leveraging existing strategies or efforts
- Enabling new and unexpected strategic uses of existing technology
- Providing new capabilities
- Neutralizing or mitigating the effects of competitors' capabilities and strategies
- Providing or denying the element of surprise

Technology (high technology) enables the acquisition of information concerning the disposition, objectives, and vulnerabilities of the competition to gain a competitive advantage. Businesses use disciplines such as communications intelligence (COMINT), electronic warfare (EW), electronics intelligence (ELINT), imagery intelligence (IMINT), open source intelligence (OSINT), and signals intelligence (SIGINT). Sound militaristic? Maybe these are not the specific terms used, but these are the capabilities they address. Go to terraserver.homeadvisor.msn.com to obtain one-meter aerial and satellite imagery, for free, of competitors' properties and other sites. Other services and Web sites are available, some offering for a small fee even clearer imagery through computer enhancement.

Information warfare (IW) is the modern construct that embodies and demonstrates the dependency of modern business, especially on telecommunications and computers. Fundamentally, IW includes any activity that influences the production, modification, falsification, distribution, availability, or security of information. These activities may be wide-ranging, even low technology, as long as they influence the gathering, analysis, distribution, or implementation of useful information. Sabotage, physical destruction of communications infrastructure, radio frequency jamming, high-energy radio frequency weapons, and electromagnetic pulse generation are all examples of relevant, modern IW, as previously discussed.

Offensive, exploitative, and defensive IW implications of new technologies must constantly be assessed. Specifically, computer networking technologies are becoming more and more integrated into modern commerce and communications, and consequently, attacks on these computer networks must be integrated into offensive and defensive strategies. Relevant IW computer technologies include the full networking spectrum from small, hardened, independent, local area networks (LANs) and regionally distributed wide area networks (WANs) to the use of the global, publicly supported Internet.

Most bolt-out-of-the-blue attacks are not bolts out of the blue. While some hackers may attack only via the Internet, a sophisticated and persistent threat agent dedicated to compromising a computer system will attempt to surveil the system physically and electronically. Significant reconnaissance and probing are necessary, especially if non-attribution of the attack is a goal. Information gathered from conventional forms of surveillance and analysis is very effective in determining which type of intrusion will be the most successful. Insiders, of course, are the greatest threat to any computer system — they have authorized access.

If physical access is obtained, both information gathering and actual system compromise are significantly easier. Corporate spies may gain physical access to a company's computers through employment as a janitor or temporary secretary, or they may simply be a client or customer who is momentarily left alone near a computer. Once they gain physical access to a computer, hackers can immediately download or corrupt information, or install sniffer software to collect information. A sniffer is a program that runs in the background of the target machine, collecting information, such as passwords or credit card numbers, during normal operations. It generally requires a return (physical or virtual) visit to retrieve the collected information, but these programs are normally quite small and difficult to detect.

Finally, attackers can use social engineering techniques to learn information that compromises a computer system. Social engineering takes advantage of the fact that most people endeavor to be honest and helpful. The most obvious of these in businesses is the help desk or the reception area. Unless an enterprise has taken steps to educate its user base to the vulnerabilities represented by releasing seemingly innocuous information, social engineering gathers attack design information very effectively. Typically, a perpetrator will call on an overworked employee, either in person or by telephone, invent a plausible need-to-know excuse, and ask for relevant information. Another method that can be used is to offer a free magazine subscription or some other incentive in return for answering a few survey questions. Or, they may actually send free software (which contains malicious code) to try out on a computer. A skilled practitioner in social engineering will usually obtain at least unclassified system details, but often passwords and sensitive information can also be obtained.

Seemingly innocuous information can also be very useful, leading to ease of access through system configuration details, personnel information, or guessed passwords. Public records, such as a company's Web site, or public business relationships allow a significant amount of information to be collated for use against the target. This information may point to a vulnerable electronic interface or an insecure business partner with full access. These essential elements of friendly information (EEFIs) may be insignificant in isolation but can generate considerable weight when collected and pieced together.[11]

Gaining Situation Awareness Through Competitive Intelligence, Knowledge Management, and Espionage

A sample of 128 South African marketing decision makers was asked to indicate the importance of different types of information. The most important was competitor's strategies. However, this area also had the lowest rate of availability.[12] Leaders have to make decisions in the absence of complete information. What can be done to improve the information set with which they have to work? One way is to take full advantage of employees' knowledge and improving situational awareness.

"Why is Company X doing that? How did they know about...?" These are not friendly questions to have to answer at the morning meeting with The Boss. A more amiable discussion would be, "Good job finding that out about Company Y. Based on this information, put together a plan and brief me tomorrow morning on how we are going to take advantage of this situation." That plan will not include key information, and if implemented may fail, if it does not take full advantage of employees' knowledge.

In an intelligence, surveillance, and reconnaissance context, people and devices are collection sensors. Your employees are your best sensors because they have first-hand knowledge of customer needs, what the competition is doing, and what is good and bad about the products and services your company offers. That adds up to a lot of valuable information, and every employee has unique gold nuggets. Employees will not take time from their already-full schedules to input these into a knowledge management (KM) system. Capabilities need to be developed where these gold nuggets are routinely mined as a normal part of business. Clearly, this goes beyond electronic records management and data mining, and is the essence of KM. In addition to this knowledge, take advantage of the information the competition gives you — for free.

Open source intelligence (OSINT) is the art of legally finding out about the competition through public media. Competitive intelligence professionals emphasize that they can satisfy 95 percent of their executives' needs for information about competitors using open, public sources, and legal, ethical methods.[10] Radio, television, publications, brochures from trade fairs, conversations with sales representatives, dumpster diving (yes, they really do wear wet suits), public financial statements, reviewing Web sites, buying products to reverse-engineer hardware and decompose software, and any other creative approach are yours for the taking. Web sites can be very lucrative. The U.S. Department of Defense (DoD) established the Joint Web Risk Analysis Cell to ensure that what is posted on DoD Web sites will not harm military operations or national security. On the Web, one can find "how-to" tutorials to jam a satellite, break into a bank, own a Cisco router in one easy step, and hack into personal computers of any type. For your organization, establish a policy of what can be sent on the Internet and posted on your Web site so that nothing of value can be collected and used against you.

Integrating employee knowledge with OSINT will provide important insights into what, where, when, and how to market products and services to counter the competition, seize and maintain a competitive advantage, and maximize profits.

Why do all this? Sun Tzu, 2500 years ago, quite wisely wrote, "If you know the enemy and know yourself, you need not fear the result of a hundred battles. If you know yourself but not the enemy, for every victory gained you will also suffer a defeat. If you know neither the enemy nor yourself, you will succumb in every battle."

Knowing about your competition is good, but it is not enough. Situational awareness of the "big picture" is essential. Understanding not just the "local" marketplace, but the political, social, economic, and military factors affecting the business environment, is vital to making sound decisions. Situational awareness must be done in both the micro and macro sense. Examples of micro metrics are monthly and quarterly sales, and comparison with the competition's and your sales for last quarter and the same fiscal period last year. Macro metrics consider leading and lagging indicators for the sector, industry, and country; government policies, laws, and regulations; cultural shifts; national and international economic variances; and the military-industrial complex downsizing or gearing up.

This situational awareness permits you to understand what is influencing your company and what influence you may wield. It also helps bring into focus what information you need to go after, as well as the information environment (IE) you need to protect. Concentrate resources on the competition's essential elements of information (EEIs) and your essential elements of friendly information (EEFIs). This traditional operations security (OPSEC) approach requires a thorough understanding of both your competition's and your own organization's IE and processes.

What is an essential element? What is the intellectual property, infrastructure, or process you must have, or else sales will decline? What would the effect on sales be if the competition knew what you had in research and development (R&D) or knew the contents of the contracts with your customers? Change the questions to ask: What if you knew this information about your competition? Would your competitive advantage go up or down?

Every morning you need to ask, "What can I do to beat Company Z today?" Neither your competition nor technology will wait for you. The innocent PalmPilot, a personal digital assistant, is one example. It can be used as a cracking tool and at the same time is targeted for theft because of the valuable information that may be stored on it.

Sun Tzu wrote, "The acme of skill is to avoid conflict." To state the obvious, avoid conflict on your terms. That requires intelligence and counterintelligence (CI) functions. These are absolutely essential to gain and maintain a competitive advantage. Do not outsource these functions. Remember that every employee can offer valuable insights. The KM process discussed previously needs to directly feed the intelligence and CI functions.

Poorly executed intelligence and CI will result in successful attacks against your IE. The recent theft of credit card information, ransom requests for millions of dollars, and distributed denial-of-service (DDoS) attacks will pale in comparison to future attacks that affect your information. Denying or destroying information is not as insidious as diddling with the bits. An order from St. Louis destined for Chicago ends up in Kansas City, or the order does get to Chicago but it is acetone instead of acetylene. Imagine if this happened on the battlefield, and 155-mm shells were delivered to a 105-mm howitzer battery. Accounting ledgers are just as easily altered. It could be months before an audit catches the error.

The best way to protect the confidentiality of online information is via encryption. However, this neither stops attacks nor prevents the information from being altered or destroyed. To protect the integrity of online information, a defense in depth is necessary. Intrusion detection devices, filtering routers, identification and authentication, firewalls, anti-virus/malicious code software, and audit systems, as well as understanding out-of-norm events on the network management side, are some of the essential tools if you do business online. And do not forget to keep several backup copies of databases and other information off-site. Take this seriously. A system administrator recently said he was installing so many system, network, and security patches, upgrades, and new capabilities that he semi-jokingly remarked he would need all users to restart every four hours to keep pace with the *known* threats and to close vulnerabilities.[13]

To help achieve situational awareness, behavioral profiling of competitors is in order. Competitors' anticipated behavior is based on what they are willing to do and how they will act or react. Doing this leads to informed decisions, reduces risk, and improves competitive position.[14] Another tool to call upon is competitive intelligence (CI). CI should focus on (in rough order):[15]

- Monitoring current competitor activities and strategy
- Monitoring customers and vendors
- Operational benchmarking
- Strategic probabilities and possible futures
- Product/service sales and marketing support
- Internal knowledge management
- Intellectual property exploitation/protection
- Alliance and investment support
- Long-term market prospects
- Counterintelligence and information security
- Legislative/regulatory effect on business issues
- Decision support and consultative briefings

Alan Breakspear, a former officer in the Canadian Security Intelligence Service, said, "Information management helps us understand what has happened. CI helps us understand what is likely to happen externally and potential options. KM helps us change what is likely to happen internally and externally, through innovation, reinventing, and repositioning."[16] This nicely ties together the major components necessary to attain situation awareness.

An important, but not openly recognized means of gaining situation awareness is through spying. The terms "sharks," "wolves," and "predators" are used to describe those who are dangerous in the marketplace. How about "spies"? External and internal spies are the most insidious threat of all. Virtual spying (netspionage), of course, is the preferred means because it is easy to do, usually goes unnoticed, and when it is noticed there is little recourse for the harmed party.

In its 1995 unclassified "Annual Report to Congress on Foreign Economic Collection and Industrial Espionage," the U.S. National Counterintelligence Center stated a key method of intelligence gathering is "telecommunications targeting and intercept, and private-sector encryption weaknesses. These account for the largest portion of economic and industrial information lost by U.S. corporations."

Edwin Fraumann, a U.S. Federal Bureau of Investigation agent, stated in a 1998 article in *Public Administration Review* that a number of countries practice economic espionage against U.S. firms, including France, Germany, Israel, China, Russia, and South Korea. The Canadian Security and Intelligence Service has gone further, stating that Canada's aerospace, biotechnology, chemical, communications, and information technology companies are targets for economic spying from 25 foreign intelligence agencies.

Spies can use very sophisticated means of collecting information. One method is to capture compromising signals emanating via conduction on power lines and air conditioning ductwork, or by radio frequency signals radiating into the air. Examples of emanating devices are modems, fax machines, CD-ROM and hard disk drives, and speakerphones. Computer monitors with cathode ray guns are some of the worst culprits. Poorly designed ones are referred to as "screamers" by those in the collection field. The signals are isolated and captured with a directional antenna, amplified, and reconstructed to show precisely what is on the screen from as far away as several hundred meters. However, it may be far easier to pay an insider for the information than to mount this type of attack.

If spies gain access to a facility, they can plant a device in a computer or piece of office furniture that constantly or periodically transmits information. An active (as opposed to passive) electronic bug can be detected in a basic countermeasure sweep. To counter detection, infrared devices are used; even a PalmPilot is possible. For less than $1000, it is possible to build monitoring systems and even disguise them as common devices such as buttons, telephone plugs, pens, or power strips. Soviet spies hid a surveillance device in a conference room in the U.S. State Department by replacing a piece of wall molding with a look-alike piece containing an electronic bug, a government spokesman said.

To reduce the risk that is being taken by a business, it may be necessary to take a range of measures. A defense, albeit difficult with commercial off-the-shelf products, is to frequently change transmitting and refresh frequencies, but this will not stop an automated system. To prevent sophisticated detection, you need shielding as well as good design and operation practices. Specialty devices are expensive, costing four times that of an ordinary device. Companies sell devices such as portable tents to shield ordinary computer equipment. The tents are made of highly conductive fabric to significantly reduce the chance of having computer emissions intercepted; they start at $30,000. Secure compartmentalized information facilities are rooms wrapped in screening made of copper and other metals and cost orders of magnitude more.

Another defense is for security to authorize — and ensure there is — an escort for every visitor. When contract maintenance wants to run any program, it should first be checked for malicious code by the network security staff.

High-tech companies appear to be especially vulnerable to espionage. To attract employees, they promote a casual atmosphere and bypass security because it "hinders creativity." Guards detract from an open atmosphere and networks lack firewalls and intrusion detection safeguards — it is a "drain on profits." On the other hand, a company called Transmeta was careful. It had hundreds of millions of dollars of research at stake in its computer chip. Employees kept a careful watch on what, and who, went into trash bins and who attempted to gain access to the work areas. They seemed to have successfully kept their project a secret for five years.[17]

Methods Used to Slip by the Physical and Virtual Guards

Antisthenes said, "Pay attention to your enemies, for they are the first to discover your mistakes." Here are a number of clever ways competitors are getting the drop on businesses.

Web Bugs

A common way to get "back doored" is via the Internet. Web bugs are one-pixel image files almost too small to see on the monitor. Companies use Web bugs for profiling. These electronic beacons track a person's whereabouts in cyberspace. Anti-cookie filters do not detect Web bugs, which were developed to not let users know they are being tracked (thus the term "spyware"). Retailers use Web bugs to get the most bang for their advertising buck and to design more attractive sites and features. Real Networks, a Seattle-based company, was forced to admit it had used an identification code in its RealJukebox software to secretly profile the listening tastes of its users. If information is flowing freely out of your system, you are in trouble.[17]

Corporate spies are using covert JavaScript code within e-mail to track the contents of sensitive financial communications, warned managed service provider Activis. These Web bugs can be embedded into HTML-based e-mails, then act to covertly copy the original sender each time the e-mail is forwarded on within the recipient's system.[18]

The Privacy Foundation, which supports an Internet privacy research institute at the University of Denver, published a report demonstrating how Word documents can be planted with "Web bugs," and these can also be embedded in Excel spreadsheets and PowerPoint files.[19]

Low-Technology Methods

Do not overlook low-tech approaches for spies to "get the goods" on you. Corporate spies look for physical and virtual telephone books, customer lists, forms, personnel information, and contracts. "Cold" letters or e-mails, social engineering, and leads from corporate publications, Web sites, and trade shows are common means of obtaining information. A Web site might reveal sensitive projects by identifying openings for specialists with skills related to the project.

Collecting intelligence is a never-ending process. Use it to tailor your products and services, marketing, and perception management. Keep on top of your competition at home and in the countries in which you do business by collecting on political legislation, patents and trademarks filed, societal trends, and other factors to give you total situation awareness.

Take advantage of as many sources of information as possible. Do you debrief employees after they have had job interviews with the competition? How about after they have been to trade shows and conferences? Speak with them before they go and give them specific instructions as to what information to gather, with whom to speak, and to take advantage of targets of opportunity.

(Be sure you use legal means.) Debriefings should be formal so that subtle points are not overlooked.

All communications with customers should be consistent — salesmen, Web sites, public affairs releases, etc. Your Web site needs to be checked for information that could disclose sensitive information. Do you have multidisciplinary teams from across your organization analyze and offer options based on the gleaned information? Are these teams close-hold? Make sure your CI team looks for moles, people planted to siphon off information. Is information accounted for? What are the penalties for not properly handling information?[17]

Malicious Insiders

Perhaps the most difficult security problem with which to deal is a trusted insider gone bad. There may be no outward signs that an employee will harm the company. The best approach is to be vigilant and to withdraw access to computer systems as soon as practical.

A computer genius with grievance against his former employer hacked into the employer's Vodaphone message network, sent a message to 32,000 international subscribers that they had won a Peugeot 106, and that they must phone the employer to claim it. The firm's business came to a standstill, resulting in lost business of approximately £10,000.[20]

Three Internet-only radio stations went off the air after they were removed from the computer server they were hosted on by a "disgruntled former employee." The three electronic music stations, E101, Pro G, and Trance Invasion, are operated by EbandMedia, a start-up company owned by Internet incubator iWeb Corp.[21]

A Fortune 1000 company IT employee knew he was about to be fired. He rigged the payroll server to send a virus network-wide on the first payday the server did not cut him a check. He brought "the whole corporation to a screeching halt," said Bill Hancock, executive vice president and CTO at Network-1 Security System.[22]

Reverse-Engineering

Reverse-engineering is the process of taking a piece of software or hardware, for example, analyzing its functions and information flow, and then translating these processes into a human-readable format. The goal is often to duplicate or improve upon the original item's functionality.[23]

The Soviet Union confiscated one of the most advanced aircraft in America's World War II fleet, took it apart, then replicated it in just two years, historians say. Soviet leader Joseph Stalin ordered the bomber copied. Soviet engineers copied the B-29 Superfortress and renamed it the Tu-4 bomber.[24]

What about the well-funded competition? They can overtly and covertly acquire products, then reverse-engineer hardware and decompose software to understand how they work and to identify vulnerabilities. Tools are then developed to exploit the vulnerabilities with the purpose of narrowing or widening the competitive edge.

Attacks on Wireless Systems

Research at Berkeley and the University of Maryland shows how easy it is for a determined adversary to defeat the "secure" IEEE Standard 802-11b network.

Analyst firm Gartner predicts that by the end of 2002, nearly one in three U.S. businesses will have come under attack by hackers exploiting security weaknesses in wireless LANs (WLANs). Firms must take security of their WLANs seriously.[25] Further, Gartner estimates that at least 20 percent of organizations already have "rogue" WLANs attached to their corporate networks. Hackers can easily break the over-the-air security built into today's 802-11b WLANs. According to Gartner, this is a primary risk associated with WLANs. Few WLAN installations operate with even a minimal level of protection. "Wireless LANs are broadcasting secrets of enterprises that have spent millions on Internet security," said John Pescatore, a research director at Gartner.[26]

Wireless keyboards are insecure and hackers can sniff out every password you type on them.[27] This is one of many reasons companies are not rushing into wireless.[28] They are concerned about the security weakness in the Wireless Application Protocol standard. Data being carried over a wireless network using the standard Transport Layer Security protocol must be decrypted at a carrier's WAP gateway and then re-encrypted using the Wireless Transport Layer encryption protocol to be delivered to a WAP device. It is that point between encryption and re-encryption that concerns some enterprises.[29] Any time information is "in the clear" it can be intercepted and read. Excerpted information can also be easily intercepted but not easily read.

Shipley and Peterson say it is not necessary to be close to a network to listen in. They plan to head for the hills above San Francisco, where they will use a special amplifier to pick up networks in downtown office buildings many miles away. Says Peterson, "That ought to really scare people."

Computer-Based Attacks

Because we have cited many computer-based attacks throughout this book, only two will be mentioned here. The purpose is to use them to establish specific points.

A Netherlands-based Internet company was back online Wednesday after 11 days during which executives say their service was held hostage by a skilled European hacker making political and monetary demands.[30] *Eleven* days? If a $10 million Internet-based company is available to its customers 24 hours a day, 365 days a year, not being online for 11 days equates to $301,370 in lost revenue. The point to be made here is to have cyber insurance to smooth revenue streams and a business continuity/disaster recovery plan that minimizes downtime.

The University of Washington Medical Center confirmed that a computer hacker broke into its systems and downloaded thousands of private medical records earlier this year. Among the records viewed were the names, addresses, and Social Security numbers of over 4000 cardiology patients, along with each medical procedure the patient underwent.[31] Consumers in the United States and Europe are demanding privacy protection. In the United States, the Healthcare Information Portability and Protection Act (HIPPA) goes so far as to levy severe monetary penalties against healthcare and insurance companies that violate patient privacy through lax information security.

We Do It to Ourselves

Using hardware and software straight out of the box without changing the default security settings, sloppy or no password management, and not using security tools such as firewalls and encryption are in "what we ought not be doing" lists every year. SANS Institute, the CERT Coordination Center at Carnegie Mellon University, and other organizations publish these lists. Other problems are not installing manufacturers' patches to close vulnerabilities or knowing what an application can, and cannot, do. Here are examples of what can go wrong.

Two hackers published a program that breaks the encryption-protecting passwords. They reported that their Web search revealed more than 300 vulnerable sites. IBM confirmed the flaw in a posting to the Bugtraq security mailing list, and asked customers using its older servers to use existing patches to fix the problems.[32]

It is common business practice to attach documents to e-mail. What was in the document that was sent? Word, WordPerfect, and other word processing programs can save up to the last five versions, and recipients can recall these versions. The previous versions of a document are kept because of features such as Word's "Allow fast saves." The feature can be turned on or off by clicking Tools, Options, and then checking or unchecking the box under the Save tab, yet few Word users know this trick exists. Knowing this may prevent proprietary and confidential information from flowing out of the company.

Two thirds of companies are not making sure that confidential information held on their computers is secure because they fail to take into account what happens when outdated PCs leave the premises.[33] The best means of preventing this particular channel for data loss is to remove the hard drives and subject them to physical destruction. A Swedish lab penetrated 17 layers of overwrites on a hard drive and still was able to find information.

Dumpster Diving

Some people will do anything. Witness the dumpster diver. They have been known to put on scuba diving wet suits, and in the early morning hours enter dumpsters in search of treasure. For years, software companies in particular have engaged in industrial espionage. Snooping was characterized as "boys-will-be-boys" exuberance sparked by the hypercompetitive business. Employees of many companies were encouraged to gain any advantage. "People were always going through our trash," said Spencer Leyton, former senior vice president at Borland International.[34] It would be prudent to destroy all paper documents. It will be worth the cost of doubling the number of trash receptacles, with one for garbage and the other for papers to be destroyed.

Electronic Fraud

The third survey by KPMG Forensic Accounting identified that Indian companies have the highest level of electronic fraud, followed by U.K. and German companies. Seventy-two percent said their greatest concern was the risk of reputational damage due to a security breach, yet only 12 percent reported their Web site bears a seal identifying their E-commerce system passed a security audit.[35]

When the bank robber Willie Sutton was asked why he robbed banks, he replied, "Cause that's where the money is." Banks continue to be a popular target. While crimes are difficult to prove because the electronic trail is masterfully covered and there is a code of silence among the perpetrators, ex-government intelligence experts and gifted crackers offer their expertise for hire, and organized crime and other well-funded groups can acquire the capabilities to penetrate bank security. We cite the following bank crimes here because they will result in electronic fraud.

Hackers caused havoc in French banking circles during March 2000 after they posted on the Internet the 96-digit encryption algorithm underlying the Cartes Bancaires system. The release of the encryption code allows fraudsters to create dummy smart-card bank cards that contain account details that match the checksum system applied to the interbank card system.[36]

Credit cards issued by banks, such as Visa and MasterCard, have similar, but much lower security checksum protection. On these cards, the card verification value is a three-digit "extra" number printed on the Visa/MasterCard signature strip.[37] While easily detected and "faked," electronic thieves find it much easier to steal tens of thousands or more credit card numbers from servers where the details are unprotected.

Russian and Ukrainian hackers stole more than one million credit card numbers from 40 American E-businesses, according to the U.S. Federal Bureau of Investigation. The hackers threatened to plaster the purloined customer data all over the Internet unless the targeted companies bought their "security" services. The FBI said it believed the stolen data was sold to organized crime.[38]

Egghead.com said its servers were hacked by network intruders. Credit card industry sources said Egghead might warn up to 3.7 million credit card holders. In 1999, online music seller CD Universe lost more than 300,000 credit cards to a Russian thief; and in 2000, 55,000 numbers were poached from the online credit card clearinghouse Creditcards.com.[39]

Cracking Encryption

In the basement of an old farmhouse deep in the Norwegian countryside, a 15-year-old computer geek named Jon Lech Johansen assembled a 57-kilobyte program over the course of a few weeks on his home-assembled Pentium 600. It enables you to copy movies onto any computer with a DVD drive and send them out unscrambled on the Internet.[40]

Descrambling DVDs just got even easier, thanks to two MIT programmers. Using only seven lines of Perl code, Keith Winstein, 19 years old, and Marc Horowitz have created the shortest-yet method to remove the thin layer of encryption that is designed to prevent people, including Linux users, from watching DVDs without proper authorization.[41]

Misinformation

Using the media to send messages to millions can easily shape people's opinions on a global basis in a very short time. Osama Bin Laden and the Al Qaeda use the media extensively. A well-prepared public relations team must be ready to counter misinformation with the truth, and operate proactively to inoculate employees against misinformation.

Some readers who bought copies of *USA Today* did double-takes after a peace activist group wrapped some papers in a fake front page with the masthead "USA Decay." *USA Today* officials did not know how many newspaper boxes had been targeted or how many newspapers were affected, but the group, Shiftdough.org, claimed to have hit nine cities.[42]

Broadening the Attack Envelope

Attacks on the information environment can come from radios, mobile phones, telephones, and personal digital assistants. To defend against IW attacks, all avenues of approach to the information environment need to be addressed. A little protection creates a false sense of security and is only slightly better than no security. Full-spectrum IW defense protects the entire IE by synchronizing many disciplines: awareness, training, and education; physical, operations, computer, communications, and personnel security; sales; public relations; intelligence; law enforcement; legal; and more. Because one IE connects to others (e.g., banks, suppliers, and customers), their risks become the business' risks. Memoranda of understanding, service level agreements, and contracts are necessary to ensure that the IE will be appropriately defended from IW attack.[43] British Standard 7799, also known as ISO Standard 17799, should be used as a point of departure. Below are examples why these measures are necessary.

Radios

In Australia, Vitek Boden, a "disgruntled" former employee of the company that installed a computerized sewerage system for a district, applied for a job with the district's council but was rejected. He later used radio transmissions and hacking programs to alter pump station operations to release up to one million liters of raw sewage.[44]

In this and other chapters, we have detailed how spies can use special equipment to intercept the radio frequency emissions from computers, monitors, and other electronic devices. In many jurisdictions, possession of this kind of equipment is illegal. Nonetheless, it is a lucrative endeavor in the world of corporate espionage.

A type of "software radio" will enable hackers with a modicum of skill to eavesdrop on computers by tuning in to the radio frequency emissions. A PC circuit board with a plug-in aerial does all the tuning under software control in any waveband. "Equipment to do this [spying] now costs at least £30,000, but in five years it will cost less than £1000, and it's hackers who will be writing the software," predicts Markus Kuhn, a research student who has filed the patent with Cambridge cryptographic expert Ross Anderson.[45]

Mobile Telephones

Experts said it is possible to build a phony base station that jams the signal in the global system for mobile communications (GSM) from the real base station and have the cell phone connect to it instead. The phony base station directs the cell phone how to connect (e.g., don't use encryption).[46] GSM is the most common mobile phone standard. Some businessmen use their phones as much as 100 hours a month, or more. There are even add-on capabilities that permit the phone to be connected to a laptop computer for mobile Internet use. A great deal of proprietary information can be intercepted and compromised.

Telephones

Computer hackers penetrated the phone systems of major firms all over Europe and ran up U.S.$67 million in phone bills, the Austrian trade magazine *Computerwelt* reported. The culprits broke into the systems of more than 100 large firms in Austria in February 2001. Most used the Kapsch phone system, and one had an Ericsson system.[47]

Personal Digital Assistants

@stake, a U.S.-based security consulting firm, wrote software named Notsync to fool a PalmPilot into thinking it is talking to the user's desktop computer rather than a hacker's PDA. The hacker downloads the password via the PDA's infrared port. Infrared ports have a range of 50 to 100 cm, but @stake said amplifying systems increase the range threefold.[48] The best defense is to have the "Beam" function turned off, activating it only when necessary.

A Textbook Example of How Information Warfare Strategies, Operations, and Tactics Can Lead to Maximizing Profit[49]

CBS, a multimedia entertainment company, claimed that over 130 million viewers around the world watched the 2001 Super Bowl, the final game of the American football season. The Super Bowl can command over U.S.$1 million for a 30-second television ad. The game means big money to a lot of people. By one estimate, as many as 200,000 people thought it was less than super because they never had an opportunity to watch the game.

DirecTV is a company that sells television programming via satellite service to about 9.5 million subscribers in the United States. A legitimate satellite TV subscription costs anywhere from $22 to $25 per month, after an initial outlay of about $150 for installation and equipment; the average customer pays about $60 per month. "Legitimate" was the key word in that sentence. It is either the thrill of getting away with stealing (i.e., taking advantage of a pirated signal) or the thrill of cracking the system, but as many as 200,000 people were not paying DirecTV for its service. On Super Bowl Sunday, DirecTV took action and those people were unable to process the signal and watch the game.

DirecTV provides an outstanding example of information warfare (IW) as a means to protect its business rights. A company must actively pursue protecting its business rights and intellectual property (IP) to maintain a defensible legal position. Each time a competitor or cracker infringes upon business practices or IP, and the company does not actively pursue protecting those areas, that company then has a diluted interest. When it comes time for legal action to claim ownership, pursue wrongful infringement, and seek recourse, if there is sufficient dilution and insufficient action by the company, it may well lose its case. It should be standard business practice to determine if IP is being stolen. Routine virtual searches should be conducted to identify external and internal anomalous behavior, and, if there is a business case for it, hire a company to review a broad cross-section of books, newspapers, and magazines.

The tactics of DirecTV's action have been well-documented in the press and on the Internet. To recap, DirecTV sent 63 updates to its TV set-top boxes and to the access cards. The updates themselves were innocuous, although the pirates eventually determined that these were parts of a program. The sixty-fourth update tied the others together, unleashing a program that disabled the illegitimate access cards. Although this was called a permanent action, a remedy was posted on HackHU.com and DishNetHack.com. There is a very small contingent of technically astute signal pirates who use a computer to take the place of the set-top box and access card. They were unaffected. Because it appears they were using it only for themselves and that they are such a small group, DirecTV appears to have ignored them, for now.

The tactics of the action itself were brilliant, culminating over six years of move-countermove between DirecTV and the signal pirates. More compelling is the story about the strategy and operation behind the tactics.

The apparent strategy appears straightforward: allow legitimate customers access for what they have paid, prevent access to those who do not pay for the service, and maximize revenue. The operation, while not disclosed, can readily be inferred.

DirecTV has an Office of Signal Integrity that works very closely with federal and state authorities. DirecTV has worked with the Federal Bureau of Investigation (FBI), won over $55 million in legal actions, and has broken piracy rings.

As the satellite service grew, the access card manipulation techniques became publicly known and shared via the Internet. Access card programming devices were sold through Canadian dealers. Cracker-developed software for the access cards that allowed reception of all programming was posted on the Internet.

Rather than sitting back and taking it on the chin, DirecTV realized it must be proactive in its approach to preventing theft. It actively monitored chat rooms, reviewed Web sites, even watched for newspaper classified ads for sales of pirate cards. These cards can sell for $750 or more. Because the manipulated cards can process the full range of DirecTV programming, they pay for themselves in three months. DirecTV has even worked with Canadian authorities to establish a legal basis for its position. In Canada, sale of the pirate cards is not illegal.

DirecTV also has technicians who comprehend the technology behind the system. Sending the program in 64 parts was technically well done. More impressive was the understanding of what DirecTV's actions meant to the signal pirates and the signal pirates' necessary counteractions. Beyond just throwing technology at a problem, DirecTV went to the trouble of understanding the behavior of those who knew how to reprogram the access cards. Like a brilliantly conceived and executed chess game, the signal pirates were compelled to make moves that led to victory for DirecTV.

Although DirecTV could have taken action at an earlier date, that they chose to disable the illegitimate access cards on Super Bowl Sunday was for psychological effect. The implied message was, "We do not know you individually, but we know how to disable your illegal system and deprive you of viewing popular programming. Let this be a deterrent."

Many facets of IW are evident: overt and covert intelligence, network management, offensive action, law enforcement, legal action, and psychological operations. No different than a well-planned and executed military operation, generally accepted principles of war (expanded on in Chapter 17) were followed:

- *Unity of command:* the only legitimate source
- *Objective:* deny service to signal pirates
- *Initiative:* took action to maintain a good legal position, stop theft, and increase revenue
- *Maneuver:* virtually wheeled forces (software and electronic signals)
- *Mass:* applied tailored software to set-top boxes and access cards

- *Economy of force:* the least amount of force was used to achieve maximum effect
- *Simplicity:* sent 63 portions of a program to appear as normal updates, and then sent the unifying and executing program
- *Surprise:* signal pirates were caught off guard although they were aware a program was on their machines
- *Security:* quietly developed

Posted on the HackHU Web site are counteractions to mitigate the shutting down of the illegitimate H cards. Technically bright and creative people used capabilities such as "unloopers" and software decomposition to restore the ability to steal the signal. Seldom is anything truly permanent, and so the struggle between DirecTV and the signal pirates continues.

There are a variety of measures DirecTV could employ to decrease signal theft. Some of these could involve using hardware and software tamper-resistant technologies, biometric recognition features, or other methods to verify legitimate users before they can receive a signal. The business case, legal implications, and customer acceptance for each will drive their adoption.

Biometrics is the latest information assurance craze. Capturing a person's digital persona and saving it on a computer system can be dangerous. If the "digital picture" is stolen, then that person's true identity may be compromised — much worse than having a social security or credit card number stolen. The cost of protecting and securing the biometric information may not justify a reasonable business case for DirecTV. Another approach DirecTV could use is role-based access for each legitimate user. Based on the "lost revenue" each month from a single signal pirate, breakeven for this approach should be a few months.

Software tamper resistance is appealing because any change is an indicator something has gone awry. This may prevent crackers from writing programs to reprogram the access cards. Hardware tamper resistance is also quite interesting. The technically adept do not mind voiding the warranty, ignoring cautions not to enter the hardware unit, and accepting the risk of ruining their unit. "No user serviceable parts inside" may take on a new dimension when the components of set-top boxes and access cards fall apart, transform into something else, or destroy themselves when unauthorized behavior is detected. Legal implications may offset operational benefits and prevent adoption of hardware tamper resistance.

On balance and in the near-term, software tamper resistance and role-based access seem to be two capabilities DirecTV could financially and legally employ in its ongoing IW campaign.

Another aspect of IW is the legal perspective. DirecTV appears to have done nothing legally inappropriate because they own the set-top boxes and access cards; the users who buy the service lease/rent those devices. The actions DirecTV took to those devices they did to their own equipment. There is little possibility that the signal pirates will join together in a class-action lawsuit against DirecTV. What is the argument? "They reprogrammed their own devices and prevented us from stealing their service?" What DirecTV did was not to bypass the legal system and "duke it out" in cyberspace, but to protect what was legitimately theirs.

Many companies could and perhaps should emulate DirecTV by being proactive in protecting their business rights and IP. This does not imply that businesses should take illegal offensive action. There are many physical and virtual network management, legal, intelligence, and information assurance tools, techniques, and procedures that can be performed individually, as well as in a coherent and synchronized fashion. IW should not be performed strictly at the tactical level, but be part of a business plan and a component of an overall strategy and operation. DirecTV has provided a good example.

Summary

The competition and others are using a vast array of tools and techniques to probe, steal, degrade, deny, or destroy business information environments physical and virtual realms on a daily — even hourly — basis. George Santayana said, "Those who cannot remember the past are condemned to repeat it." Top management must place a priority on offensive, exploitative, and defensive information warfare if they wish to run a profitable business.

Notes

1. Cyber Warfare Incidents to Cost $20 Billion in 1999, mi2g, July 26, 1999.
2. Hackers cost Europe £3 billion, John Geralds and Linda Leung, *Accountancy Age,* February 15, 2001.
3. Management Takes Notice, George V. Hulme, *InformationWeek,* September 3, 2001, pp. 28–34.
4. U.K. Firms Reluctant to Prosecute Attackers, *Security Wire Digest,* November 8, 2001.
5. KPMG E-fraud survey results, KPMG, SecurityWatch, http://www.securitywatch.com/newsforward/default.asp?AID=6667, full text of survey at http://www.kpmg.co.uk/kpmg/uk/direct/forensic/pubs/efraud.pdf, April 30, 2001.
6. Code of Silence: Cyber-Crime Is Rising, but Companies Are Loath to Admit They Have Been Victims, Patricia Richardson, *Chicago Tribune,* http://www.chicagobusiness.com/cgi-bin/news.pl?post_date=2001-09-08&id=3277&feature=1, September 10, 2001.
7. CEO, CIO, Presidents, Vice-Presidents, and Directors Risk Cyber Crime Law Suits, *Security News Portal,* http://www.securitynewsportal.com/article.php?sid=837&mode=thread&order=0, June 18, 2001.
8. The Security Roadblock, Sandra Gittlen, *Network World,* April 12, 2001.
9. What Company Spies Really Want, *Wired News,* Joanna Glasner, December 4, 1999.
10. Post-Industrial Espionage, *The Wall Street Journal,* April 22, 2001.
11. Technology and the Law: The Evolution of Digital Warfare, David Tubbs, Lt. Col. Perry G. Luzwick, USAF, and Lt. Col. Walter Gary Sharp, Sr., USMC (Retired), *Naval War College Blue Book on Computer Network Attack and International Law* (forthcoming December 2001).
12. Competitive Intelligence: The "Blind Spot" of South African Decision Makers, Peet Venter, *South African Association of Competitive Intelligence Professionals (SAACIP) Newsletter,* June 2001, pp. 12–14.

13. Situational Awareness and OODA Loops: Coherent Knowledge-based Operations Applied, Perry Luzwick, "Surviving Information Warfare" column, *Computer Fraud and Security,* an Elsevier Science magazine, April 2000.

14. Competitor Behvioural Analysis: Outwitting, Outperforming, Outmanoeuvering Competitors, George Nel, *SAACIP Newsletter,* June 2001, pp. 1–4.

15. Corporate Espionage or "Intrapreneurship," CI Fortifies Decision Making, Arik Johnson, *KMWorld,* November/December 2001, pp. 8–10.

16. Open Yet Guarded: Protecting the Knowledge Enterprise, Steve Barth, *Knowledge Management,* March 2001, pp. 44–52.

17. Stop the Presses for a Late Breaking Story! Countries and Companies Spy on Each Other!, Perry Luzwick, "Surviving Information Warfare" column, *Computer Fraud and Security,* an Elsevier Science magazine, November 2001.

18. E-mail Wiretapping Used to Spy on Corporate Communications, John Leyden, *The Register,* June 4, 2001.

19. Word Documents Susceptible to "Web Bug" Infestation, Paul Festa and Cecily Barnes, CNET News.com, http://news.cnet.com/news/0-1005-200-2652562.html, August 31, 2000.

20. Computer Genius Took His Revenge, Maurice Weaver, *London Telegraph,* August 16, 1999.

21. Former Employee Steals Internet Radio Stations, Adam Creed, *Newsbytes,* April 18, 2000.

22. Hacking It on Wall Street: A Wall Street Online Trading Company Traces Attacks on its Servers to a Disgruntled Employee, http://www.techtv.com/cybercrime/hackingandsecurity/story/0,9955,3013872,00.html, Tech TV, December 5, 2000.

23. Reverse-Engineering, Mathew Schwartz, *ComputerWorld,* November 12, 2001, p. 62.

24. Mystery of Soviet Bomber Pieced Together, Associated Press, January 26, 2001.

25. Strategy Fills Security Gap for WLANs, *IT Week* Staff, August 28, 2001.

26. Rogue WLANS — The Next Security Battlefield?, John Leyden, *The Register,* September 8, 2001.

27. Type Me Your Password, Robert Blincoe, *The Register,* http://www.theregister.co.uk/content/8/19736.html, June 18, 2001.

28. Financial Security Sites Struggle with Wireless, Rutrell Yasin, November 14, 2000.

29. Often Unguarded Wireless Networks Can Be Eavesdroppers' Gold Mine: It Is Easy to Make a Wireless Network Secure, Lee Gomes, *The Wall Street Journal,* http://www.msnbc.com/news/565275.asp, April 27, 2001.

30. Hacker Virtually Disables Nederland 'Net service, Pippa Jack, Special to *The Denver Post,* October 12, 2000.

31. Hospital Confirms Hack Incident, Bob Sullivan, MSNBC, http://www.msnbc.com/news/499856.asp, December 7, 2000.

32. Hackers Trumpet IBM Software Hole Program Breaks Encryption-Protecting Passwords, Robert Lemos, Ziff Davis News Network, http://www.msnbc.com/news/541378.asp, March 8, 2001.

33. Warning on IT Security, Thisislondon.com, http://63.108.181.201/2001/04/13/eng-newsquest_london/eng-newsquest_london_153550_106_784090632573.html, April 16, 2001.

34. Oracle Isn't Alone in High-Tech Spying, Antone Gonsalves, with contributions by Barbara Darrow, *TechWeb News,* June 30, 2000.

35. KPMG E-fraud survey results, KPMG, SecurityWatch, http://www.securitywatch.com/newsforward/default.asp?AID=6667, Full text of survey at http://www.kpmg.co.uk/kpmg/uk/direct/forensic/pubs/efraud.pdf, April 30, 2001.

36. Hackers Reveal How to Forge a Bank Card, Paul Webster, *The Guardian* (*UK*), http://www.newsunlimited.co.uk/international/story/0,3604,146566,00.html, March 14, 2000.

37. French Banks Hacked, Sylvia Dennis, *Newsbytes,* March 11, 2000.
38. Hackers Steal 1 Million Card Numbers, Maria Bruno, *Bank Technology News,* http://www.electronicbanker.com/btn/articles/btnapr01-5.shtml, April 10, 2001.
39. Hackers Crack Egghead.com, Robert Lemos and Ben Charny, December 22, 2000.
40. Teen-age Norwegian Computer Geek: A Villain to Hollywood, A Hero to Hackers, Doug Mellgren, with contributions by Gary Gentile and Peter Svensson, Associated Press, http://www0.mercurycenter.com/svtech/news/breaking/merc/docs/023590.htm, February 25, 2001.
41. Descramble That DVD in 7 Lines, Declan McCullagh, *Wired News,* March 7, 2001.
42. Group Puts Fake Front Page on USA Today, Brigitte Greenberg, Aponline, January 27, 2000.
43. On the Battlefield and in the Marketplace, Information Warfare is Much More than Attacking Computers, Perry Luzwick, "Surviving Information Warfare" column, *Computer Fraud and Security,* an Elsevier Science magazine, February 2000.
44. 'Sewage' Hacker Jailed, *Herald Sun,* http://www.heraldsun.news.com.au/common/story_page/0,5478,3161206%255E1702,00.html, October 31, 2001.
45. New-Wave Spies, *New Scientist,* June 11, 1999.
46. Cell Phone Flaw Opens Security Hole, Sara Robinson, *ZD Net News,* September 18, 2000.
47. Hackers Run up Million-Dollar Phone Bills in Europe, Deutsche Press-Agentur GmbH, February 21, 2001.
48. Crackers Can Zap Data off Palm Pilots, Ian Lynch, Vnunet.com, http://vnunet.com/News/1116644, January 22, 2001.
49. In Business and War, Information Warfare Strategies, Operations, and Tactics Lead to Successful Outcomes, Perry Luzwick, "Surviving Information Warfare" column, *Computer Fraud and Security,* an Elsevier Science magazine, May 2001.

Chapter 14

It's All about Power: Information Warfare Tactics by Terrorists, Activists, and Miscreants

The terrorists practice a fringe form of Islamic extremism that has been rejected by Muslim scholars and the vast majority of Muslim clerics — a fringe movement that perverts the peaceful teachings of Islam. The terrorists' directive commands them to kill Christians and Jews, to kill all Americans, and make no distinction among military and civilians, including women and children. This group and its leader — Al Qaeda and a person named Osama bin Laden — are linked to many other organizations in different countries, including the Egyptian Islamic Jihad and the Islamic Movement of Uzbekistan. There are thousands of these terrorists in more than 60 countries. They are recruited from their own nations and neighborhoods and brought to camps in places like Afghanistan, where they are trained in the tactics of terror. They are sent back to their homes or sent to hide in countries around the world to plot evil and destruction.

— George W. Bush, President of the United States of America

9/11/01: A Date in Infamy

This book was in the process of its initial editing when the Massacre of September 11, 2001, took place. While it would be wrong to rewrite this chapter or the book in response to that one terrible event, it would be shameful to fail to acknowledge the effects and the losses.

The attacks on the World Trade Center and the Pentagon were extreme but conventional terrorist attacks, but some of the retaliatory action that took place in the following days and weeks occurred in cyberspace. The outcome of these actions must be judged by the results.

This chapter discusses the publicly known terrorist nations, drug cartels, and hacktivists (cyber disobedience) capabilities such as those of animal rights groups, freedom-fighters, and the like. Examples include terrorists like Osama bin Laden using the Internet and encrypted communications to thwart law enforcement; the drug cartels' use of computers to support their drug money laundering operations; and the Zapatista movement in Mexico, outnumbered and outfinanced by the Mexican government, took to the Internet to support its cause. The Zapatistas conducted denial-of-service attacks against the Mexican and U.S. governments.

Information Warfare Tactics by Terrorists

The first group examined are terrorists. The motivation of a terrorist is to undermine the effectiveness of a government by whatever means it chooses. It is worth remembering at this point that a terrorist in one country is a freedom-fighter in another, and as a result, there is no stereotype. When you take into account the differing cultures around the world and the differing political regimes that exist, it is easy to understand that a whole variety of actions may be terrorist actions when carried out for political means, or the actions of a hooligan, or in computer terms a hacker.

Let us first address a term that is in current and widespread use — cyber-terrorism. While it can be accepted that this term can be used to convey a general meaning, it is not possible to accept the current use of the term to be anything more. The definition of terrorism that was adopted by the gateway model in the United Nations in the spring of 1995 is:

> *A TERRORIST is any person who, acting independently of the specific recognition of a country, or as a single person, or as part of a group not recognized as an official part of division of a nation, acts to destroy or to injure civilians or destroy or damage property belonging to civilians or to governments to effect some political goal.*
>
> *TERRORISM is the act of destroying or injuring civilian lives or the act of destroying or damaging civilian or government property without the expressly chartered permission of a specific government, thus, by individuals or groups acting independently or governments on their own accord and belief, in the attempt to effect some political goal.*
>
> *All war crimes will be considered acts of terrorism.*
>
> *Attacks on military installations, bases, and personnel will not be considered acts of terrorism, but instead acts by freedom fighters that are to be considered a declaration of war towards the organized government.[1]*

A very different definition was offered at the Fifth Islamic Summit that was convened to discuss the subject of international terrorism under the auspices of the UN, which is as follows:

> *Terrorism is an act carried out to achieve an inhuman and corrupt (mufsid) objective, and involving threat to security of any kind, and violation of rights acknowledged by religion and mankind.*[2]

It is notable that in the main body of this definition, there is no reference to the nation-state, something that, in the West, would be fundamental to any understanding of terrorism. The author then goes on to make a number of additional points to clarify the definition, the most significant of which are:

- We have used the term 'human' instead of 'international' for the sake of wider consensus, official or otherwise, so as to emphasize the general human character of the statement.
- We have referred to various types of terrorism with the phrase, "security of any kind."
- We have mentioned the two criteria, i.e., religious and human, first to be consistent with our belief and then to generalize the criterion.

This totally different approach to the issue of terrorism is significant and a clear reminder to the nation-states that consider themselves to be "Western" that not all cultures view the issue in the same manner as Anglo/Americans.

Even given these diverse views of the meaning of terrorism, there is an underlying trend of physical destruction and of the actions being of such a magnitude and type as to cause "terror" to the people. This does not fit well within the "cyber" environment because there is no direct physical destruction (other than "0s and 1s") and, without the effect of the bullet, the blast, or carnage of the bomb, the "terrorization" of the people is difficult in our current state of technological advancement. It is more likely that as our cultural values change and we become more highly dependent on technology than we currently are, that the cyber-terrorist in the true sense will come into being. For example, today and more so into the future, as we increase our proliferation and dependence on telemedicine, a terrorist might

- Attack a computer system, shutting off life support to patients
- Change their dosages of medicine, killing them in the process
- Manipulate blood bank information, causing the wrong blood type to be given to patients and thus resulting in numerous deaths

What Do They Want to Achieve?

Let us first look at what a terrorist will want to achieve through the use of the Internet. This may be one or more of a number of things. The terrorist organization may wish to use this medium for the transmission of communications between individuals and groups within the organization. Look at the potential:

- The terrorist has been offered all of the facilities that the Cold War spy always dreamed of. It is possible to be anonymous on the Internet, with pay-for-use mobile phones and free Internet accounts.
- No attempts are made by the service providers to ascertain that the details provided by a customer are real and actually do relate to the user.
- Once the user is online, there are a number of ways that user can further disguise his or her identity.
- There are anonymous re-mailers and browsers that can disguise the identity of the user.
- There is freely available high-grade encryption that law enforcement cannot yet break and civil liberty groups that want to ensure that this situation remains so. The desire of civil liberty organizations to maintain the privacy of messages on the Internet has actually nothing to do with the terrorist — they have the liberty and privacy of the individual at heart, but the terrorist is just one of the winners of the pressure that they seek to exert.

A well-reported example of terrorist use of the Internet in this way is the activity of Osama bin Laden, who is reported to have used steganography (the ability to hide data in other files or the slack space on a disk) to pass messages over the Internet.[3] Steganography has become a weapon of choice because of the difficulty in detecting it. The technique hides secrets in plain sight and is especially important when there is a concern that encrypted communications are targeted.

It was reported that Bin Laden was "hiding maps and photographs of terrorist targets and posting instructions for terrorist activities on sports chat rooms, pornographic bulletin boards, and other Web sites." According to another report, couriers for Bin Laden who have been intercepted have been found to be carrying encrypted floppy disks.[4] Other references to the use of the Internet by bin Laden describe the use of a new form of the Cold War "dead letter box," which was a predetermined place where one agent deposited information to be collected by another agent. A June 2001 report indicated that bin Laden was suspected of using encryption for his messages for at least five years.[5]

According to reporter Jack Kelley,[6] FBI director Louis Freeh stated that, "Uncrackable encryption is allowing terrorists — Hamas, Hezbollah, Al-Qaeda (another name for bin Laden's organization), and others — to communicate about their criminal intentions without fear of outside intrusion." Kelley also reported that according to other unnamed officials, bin Laden's organization uses money from Muslim sympathizers to purchase computers from stores or by mail, after which easy-to-use encryption programs are downloaded from the Internet. As evidence, they cite the case of Wadih El Hage, one of the suspects of the 1998 bombing of two U.S. embassies in Africa, who is reported to have sent encrypted e-mails under a number of aliases, including "Norman" and "Abdus Sabbur" to associates of Al-Qaeda.

Also cited as evidence is the case of Ramzi Yousef, the man convicted of masterminding the World Trade Center bombing in 1993, who is reported to have used encryption to hide details of the plot to destroy 11 U.S. airlines.

The computer was found in his Manila apartment in 1995 and passed to U.S. officials who cracked the encryption and foiled the plot. The same report goes on to say that two of the files took more than a year to crack. This is, in itself, revealing because it gives some indication of the level of effort that government and law enforcement agencies are prepared to invest in their efforts to bring to justice this type of criminal, as well as the level of effort and sophistication that is being used by terrorists.

Osama bin Laden is also skilled in the use of the media to promote the aims and the aura of the organization. This is evident from his use of the press to provide interviews. He is a well-educated and, through his family, a wealthy man. He has a good understanding of the way in which the media can be used to influence public opinion and has used the media to promote his philosophy.

Tactics

Having identified some of the types of effects that terrorists might want to use the Internet to achieve, let us now examine the tactics and tools that they would use to realize their aim. In the case of Osama bin Laden, he is apparently communicating via the Internet using steganography and encryption. Dealing with the two issues separately for the purposes of describing the tactics, this is in no way implying that the two (steganography and encryption) do not go together; in fact, quite the reverse. If you are paranoid and you want to make sure that your messages get through undetected and in a state that is unreadable to anyone that should guess their presence, then the combination of techniques is a powerful one.

Data Hiding

What is steganography? The word "steganography" literally means "covered writing" and is derived from Greek. It includes a vast array of methods of secret communications that conceal the very existence of the message. In real terms, steganography is the technique of taking one piece of information and hiding it within another. Computer files, whether they are images, sound recordings, text and word processing files, or even the medium of the disk itself, all contain unused areas where data can be stored. Steganography takes advantage of these areas, replacing them with the information that you wish to hide. The files can then be exchanged with no indication of the additional information that is stored within. A selected image, perhaps of a pop star, could itself contain another image or a letter or map. A sound recording of a short dialogue could contain the same information. In an almost strange twist in the use of steganography, law enforcement, the entertainment industry, and the software industry have all started to experiment with the use of steganography to place hidden identifiers or trademarks in images, music, and software. This technique is referred to as digital watermarking.

How does it work? Well, the concept is simple. You want to hide one set of data inside another but the way that you achieve this will vary, depending on the type of material in which you are trying to hide your data.

If you are hiding your data in the unused space of a disk,[7] you are not, primarily, constrained by the size of the data because you can break it into a number of sections that can be hidden in the space described below. Storage space on disks is divided into clusters that in Microsoft DOS and Windows file systems are of a fixed-size. When data is stored to the disk, even if the actual data being stored requires less storage than the cluster size, an entire cluster is reserved for the file. The unused space from the end of the file to the end of the cluster is called the slack space. For DOS and older Windows systems that use a 16-bit File Allocation Table (FAT), this results in very large cluster sizes for large partitions. As an example, if the partition on the disk was of a 2-Gb size, then each cluster would be of 32 kb. If the file being stored on the disk only required 8 kb, the entire 32-kb storage space would be allocated, resulting in 24 kb of slack space in the cluster. In later versions of the Microsoft Windows operating system, this problem was resolved (or at least reduced) by the use of a 32-bit FAT that supported cluster sizes of as small as 4 kb, even for very large partitions.

Tools to enable you to do this are available on the Internet for free and examples of this type of tool include:

- *S-Mail.* This is a steganographic program that will run under all versions of DOS and Windows. The system uses strong encryption and compression to hide data in EXE and DLL files. (Yes, it is possible to hide files within full working programs; after all, that is what a virus does.) The software has a pleasant user interface and has functions in place to reduce the probability of its hiding scheme being detected by pattern or ID string scanners (these are tools that can identify the use of steganographic techniques).
- *Camouflage.* This is a Windows-based program that allows you to hide files by scrambling them and then attaching them to the end of the file of your choice. The camouflaged file then appears and behaves like a normal file, and can be stored or e-mailed without attracting attention. The software will work for most file types and has password protection included.
- *Steganography Tools 4.* This software encrypts the data with one of the following: IDEA, MPJ2, DES, 3DES, and NSEA in CBC, ECB, CFB, OFB, and PCBC modes. The data is then hidden inside either graphics (by modifying the least significant bit of BMP files), digital audio (WAV files), or in unused sectors of floppy disks.

If you are attempting to hide data in files, no matter what the type, then you have two options:

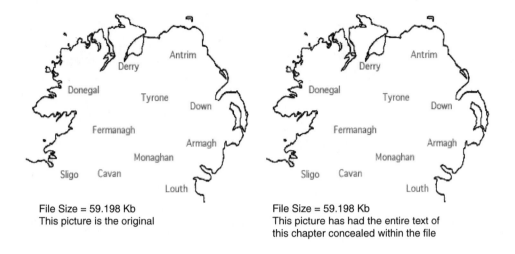

File Size = 59.198 Kb
This picture is the original

File Size = 59.198 Kb
This picture has had the entire text of
this chapter concealed within the file

Exhibit 1.   Steganography

- You can hide your material in the file by adding to the data that is already there and thus increase the size of the file.
- You can replace some of the data that is already in the file with the information that you want to hide and retain the same file length but have a slightly reduced quality in the original representation.

To explain this in more detail, if you are using an image file to hide data, the normal method is to use the least significant bit of each information element as a place to store hidden data. In doing this, the changes to the image are so subtle as to be undetectable to the naked eye. But the changes are significant enough for steganographic software to be able to hide relatively large quantities of information in the image and also for the software to recognize a pattern within the image that it can use to reveal hidden material.

It would not be unrealistic to hide the contents of this chapter in a relatively small image; for example, if you look at the two images that are reproduced in Exhibit 1, they are relatively small and yet it is possible to hide more than 30 pages of text within one of them with no noticeable degradation in the quality of the image.

For the most part, the size of the file and the quality of the image are not significant; after all, if you do not have the before and after copies of the file or image on hand, how can you tell that the file has grown or that the image has been degraded? Even when you look at the two images above side by side, it is not possible to detect any significant difference.

Other methods that can be used to hide data in other types of files include:

- The use of programs such as Snow, which is used to conceal messages in ASCII text by appending white spaces to the end of lines. In a conventional page of text, there are normally 80 columns of information to the page. When we use a text file to save information that we have created on a computer screen, we do not use all 80 columns. If the word at the end of the line falls short of the 80th column, then we get a carriage return character after the last letter. If it is the last line of a paragraph, then there may be a considerable number of unused columns in the row. The Snow program fills in all of these unused spaces and uses the least significant bit of each of the bytes to hold an element of the hidden message.

- Software such as wbStego lets you hide data in bitmaps, text files, HTML, and PDF files. The data is encrypted before it is embedded in the carrier file.

- If you want to hide messages in music and sound files (MP3), then software such as MP3Stego will hide information in these files during the compression process. The data is first compressed, encrypted, and then hidden in the MP3 bit stream. Although MP3Stego was written with steganographic applications in mind, again there is the potential for it to be used for the good of the music and movie industries by allowing them to embed a copyright symbol or watermark into the data stream. If an opponent discovers your message in an MP3 stream and wishes to remove it, they can uncompress the bit stream and recompress it, which will delete the hidden information. The data hiding takes place at the heart of the encoding process, namely in the inner loop. The inner loop determines the quantity of the input data and increases the process step size until the data can be coded with the available number of bits. Another loop checks that the distortions introduced by the process do not exceed the predefined threshold.

- For the Linux enthusiast, there are programs such as StegFS,[8] which is a steganographic file system for Linux. Not only does it encrypt data, it also hides it such that it cannot be proved to be there.

This plethora of choices of software and encoding schema allows the terrorist a wide set of options to suit the chosen method of communication. If the selected method of covering the communications is through a newsgroup that exchanges music, then the use of an MP3 encoder is most sensible. After all, if the other users of the newsgroup have the same taste in music as the sender and recipient of the message, there is no problem; they can download the file, play it, enjoy it, and yet be totally unaware of the hidden content. If the chosen method of communication is one of image sharing, then again, the images can be posted in public, with anyone able to view the images, but only those who are aware of the additional content are likely to use tools to extract it.

On the plus side of this is that, increasingly, it is possible to detect the use of steganography. There is now software becoming available that will identify the use of an increasing range of the steganographic packages in use.

One example of a tool that can detect the use of steganography is the Steganography Detection & Recovery Toolkit (S-DART), which was sponsored by the U.S. Air Force Research Laboratories[9] and commissioned by WetStone Technologies, Inc. The aim of this kit was to develop algorithms and techniques for the detection of steganography, in digital image files, audio files, and in text messages. The aim of the project was to develop a set of statistical tests that could detect the use of steganography and also identify the underlying method that was used to hide the data.

Another tool is Stegdetect, an automated tool for detecting whether there is steganographic content in images. It is capable of detecting a number of different steganographic methods used to embed hidden information in JPEG images. Currently, the methods that can be detected by this software package are jsteg, jphide for UNIX and Windows, invisible secrets, and outguess 01.3b.

While these tools are still limited in the range of data hiding techniques that they can detect, this will increase rapidly. However, as with viruses and most other forms of malicious code on the Internet, the detection tools will always lag somewhat behind the tools that provide the capability.

Cryptography

It makes sense that if you are a terrorist and you want to communicate using the Internet, you are not going to risk your life or your liberty to people not being able to recognize the use of steganography on its own. Because the steganographic software is not interested in the type of material that it is incorporating into the carrier file, it will hide an encrypted message just as happily as it will hide a cleartext message.

An encryption program scrambles information in a controlled manner through the use of a cryptographic key. In the past, you sent a message encrypted with a particular key to someone and they had to be in possession of the same key to decrypt the message. This is known as symmetrical cryptography. This, unfortunately, meant that you had to communicate the key to the person to whom you were sending the message.

This was achievable for governments that have the infrastructure to distribute the cryptographic keys in a secure manner. However, this type of approach is just not realistic for the general public to consider. It is only in recent years that such technology has increasingly been found in the public domain. Perhaps the best known of the publicly available high-grade encryption systems is Pretty Good Privacy (PGP), the system developed by Phil Zimmerman. As a result of the prominence that PGP has achieved, this discussion will concentrate on a description of cryptography on this system.

PGP is a public key encryption software package that was initially intended for the protection of electronic mail. Since PGP was published domestically in the United States as a freeware offering in 1991, it was very quickly taken and adopted all over the world, with the result that it has become the *de facto* worldwide standard for encryption of e-mail.

The author of the PGP software was under investigation for a period of about three years by authorities (the U.S. Customs Service) who were investigating a possible breach in the arms control relating to the export of weapons, which includes high-grade encryption. It is one of the nonsenses of the age of technology that it was considered to be an offense to export the software package that incorporated the encryption algorithm, but there seemed to be no problem with leaving the country with the algorithm printed on a t-shirt.

The investigation into the situation was finally closed, without Zimmerman being indicted, in January 1996.

It is interesting that in at least one interview, Zimmerman stated, as part of the rationale for the development of PGP, that the software was now used all over the world, particularly in Central America, in Burma and by the government in exile from Tibet, and by human rights groups and human rights activists who were documenting the atrocities of death squads and keeping track of human rights abuses. He went on to state that he had been told by these groups that if the governments involved were to gain access to the information that had been encrypted, all of the individuals involved would be tortured and killed. Again, who is the terrorist? Who is the freedom fighter?

Propaganda

Another reason that a terrorist organization might use the Internet is to spread the organization's message and further the cause. For this, the Internet is an outstanding tool. It is the most widely used, uncontrolled medium that has international reach. The number of organizations that have exploited this reach and lack of censorship is huge. Some of the better examples of this are the Provisional Irish Republican Army (PIRA), the Euskadi Ta Askatasuna (ETA), the Mexican Zapatistas, and the Chechen rebels.

The PIRA has a well-founded presence on the Internet through the auspices of its political wing, Sinn Fein, and publications with a strong online presence such as An Phoblact. Web sites that support the aspirations and the "cause" of the PIRA can be found in a number of countries and some good examples are the Sinn Fein home page[10] and Sinn Fein Online.[11] Other informative sites can be found at the Irish Republican Network[12] and the Trinity Sinn Fein Web site.[13] In addition to the large number of sites that provide information on the IRA, other sites provide a different perspective on the conflict in Northern Ireland, some of the sites providing a more balanced view than others, but undoubtedly, that statement in itself demonstrates a prejudice as other people would take a different view of the balance of reporting of the sites. The conflict in Northern Ireland is one of the longest-running "terrorist" actions that has taken place in the English-speaking world and, not surprisingly, attracts a lot of comment and debate and presence on the Web. While the PIRA is the best known of the groups that represent one side of the conflict, there are a large number of other groups that claim to be active in the Province.

The other main groups are:

- Continuity Irish Republican Army
- Combined Loyalist Military Command
- Irish National Liberation Army
- Irish People's Liberation Organization
- Irish Republican Army
- Loyalist Volunteer Force
- Real Irish Republican Army

- Ulster Defence Association
- Ulster Freedom Fighters

The majority of these also have, to a greater or lesser degree, a Web presence, and some of the more notable of these are:

- The Irish People's Liberation Organization[14] which represents another view of the republican perspective
- A loyalist view can be found at the Ulster loyalist Web page[15]
- The Ulster Volunteer Force (UVF) presence with the UVF page of the Loyalist Network[16]

In addition to all of these many partisan views of the situation, there are a number of sites that allegedly attempt to provide a "neutral" view of the situation. Examples of these sites can be found at Rich Geib's Universe[17] or the Irish Republican Army Information Site.[18] Other sites that provide insight into the attitudes of, and toward, the various parties in the province can be found at Vincent Morley's flags Web page[19] and a unionist Mural Art from Belfast page.[20]

An example of a terrorist site from another part of Europe is the case of the Euskadi Ta Askatasuna (ETA). This violent terrorist group, which lays claim to a portion of northern Spain and southern France, has its own Web presence to present the case for its grievances and to explain culture and history and to justify its actions and seek support. As with other similar groups, it has its supporters and detractors, both of which use the Web to try to influence the opinion of the readership.

In the case of supporters of ETA and the Basque state, which they themselves refer to as "Euskal Herria," the primary Web pages are the *Euskal Herria Journal,* which promotes itself as Basque Journal[21] and puts forward the aims and expectations of the group that it represents and the Basque Red Net,[22] which puts forward a very well-developed argument based on the culture and history of the area.

A view of ETA from the Spanish government can be seen at the Ministry of the Interior page that has the title "ETA — Murder as Argument."[23] This Web page is produced in three languages (Spanish, French, and English) to enable the widest reasonable readership of the arguments presented. One French view of the issues can be seen at the Web site of the Mediapaul Project.[24]

In an example from Central America, the Zapatista rebels in the Chiapas region of Mexico have become one of the most successful examples of the use of information systems and communications by a hugely outnumbered and outresourced group of activists. The Zapatistas used the Internet to outmaneuver the Mexican government and to bring world pressure to bear on a situation that was entirely internal to Mexico. The use of the Internet gained the Zapatistas not only support from throughout Mexico, but also from the rest of the world. It will also now be used as a template for actions in other parts of the world, and the implications of the Zapatista rebellion will have an effect on other confrontations with contemporary capitalist economic and political policies.

The surge of support for this, to European and North American eyes, very parochial action in a Central American republic came when a report, written for Chase Emerging Markets clients by Riordan Roett, was apparently leaked to Silverstein and Cockburn's *Counterpunch* newsletter. The report was found to call for the Mexican government to "eliminate" the Zapatistas to demonstrate its command over the internal situation in Mexico. When this news and the report were posted on the Web, there was worldwide reaction against the Mexican government, America, and the American bank that had commissioned the report.

Part of the response to this news was an increase in the hacking of Mexican government Web sites. In addition, the Electronic Disturbance Theater (EDT)[25] released what they referred to as a digital translation of the Zapatista Air Force Action, which they called the Zapatista tribal port scan. This was carried out to commemorate a nonelectronic act that involved, on January 3, 2000, the Zapatista Air Force "bombarding" the Mexican Army federal barracks with hundreds of paper airplanes on each of which was written a message for the soldiers monitoring the border.

Despite the fact that the action in the Chiapas region has effectively been underway since 1994, there was still support and online action such as that by the EDT in 2001.

In the former Soviet Union, the situation with regard to the ongoing conflict in Chechnya is one that the media is now starting to class as an "information war." The Chechen separatists are primarily represented on the Internet by two sites: one from the Chechen Republic of Ichkeria and the other from Kavkaz-Tsentr.[26] The Ichkeria site is seldom updated, but the Kavkaz-Tsentr is reported as an example of a professional approach to information war. This site is kept up-to-date with daily reports on Chechen military successes against Russian forces, as well as more light-hearted items and events that surround Chechnya.

According to numerous reports from organizations, including the BBC, Moscow is applying the same tactics that it observed NATO using in the former Republic of Yugoslavia to try to win the information war in Chechnya. In the previous Chechen war that started in 1994, the then-fledgling commercial station NTV showed graphic pictures from both sides of the conflict; however, now the Russian broadcasters and press are much more selective in their reporting of the fighting.

The Kavkaz-Tsentr site has repeatedly been targeted by hacker attacks since at least 1999. The hackers have repeatedly defaced the Web site with anti-Chechen images and slogans and have redirected traffic intended for the site to a Russian Information Center site; however, the site has normally managed to restore normal operations within 24 hours.

Reaction to the World Trade Center and Pentagon Attacks

This has been inserted here because the case to be highlighted shows the dangers of "vigilantes" and people who, for the best of intentions, take actions for which they have not researched the background information. The action in question was reported by Brian McWilliam of Newsbytes[27] on September 27, 2001, who revealed that members of a coalition of vigilante hackers had mistakenly defaced a Web site of an organization that had had offices in the World Trade Center. The hacker group, called the Dispatchers, attacked the Web site of the Special Risks Terrorism Team, which in fact was owned by the Aon Corporation. The other sites that were attacked by this group were both in Iran, which for the geographically challenged is not in Afghanistan, and is in fact hostile to the Taliban regime and Osama bin Laden. One can understand the anger and frustration and the desire to strike out in the aftermath of the attacks, but this type of action by uninformed and nonrepresentative individuals does much to damage relationships with countries and organizations that have not (at least in recent years) caused any offense and are in fact sympathetic to the cause.

Denial-of-Service

When a terrorist organization cannot achieve its objective by the means that are normally used — the bullet and the bomb — it has the potential to use the Internet and the connectivity of the systems on which we now rely so heavily to gain the desired impact. There are a number of advantages and disadvantages to this approach; but if the normal techniques cannot be used, it allows another vector of attachment to be utilized that has the advantages of being untraceable to the source and nonlethal.

When compared to the average activity of a hacker, who has limited capability in terms of equipment and sustainability, the terrorist will normally have a greater depth of resources and of motivation. An action that is taken in support of a cause that is believed in will have a much higher motivation to succeed than the whim of an idle mind or simple curiosity.

What Is a Denial-of-Service Attack?

A denial-of-service (DoS) attack is characterized by an attempt by an attacker or attackers to prevent legitimate users of a service from using that service. Types of DoS attacks that may be seen include:

- Network flooding, resulting in the prevention of legitimate network traffic
- Attempts to disrupt connections between two machines, resulting in the prevention of access to a service
- Attempts to prevent a particular individual from accessing a service
- Attempts to disrupt service to or from a specific system or person

Not all disruptions to service, even those that result from malicious activity, are necessarily DoS attacks. Other types of attack might include denial-of-service as a component, but the denial-of-service itself may be part of a larger attack.

The unauthorized use of resources may also result in denial-of-service. For example, an intruder might make use of your anonymous ftp area as a location where they can store illegal copies of software, using up disk space and CPU time, and generating network traffic that consumes bandwidth.

The Impact

DoS attacks can disable either the computer or the network. In doing so, this can neutralize the effectiveness of your organization. DoS attacks can be carried out using limited resources against a large, sophisticated or complex site. This type of attack may be an "asymmetric attack." An asymmetric attack is one in which a less capable adversary takes on an enemy with superior resources or capabilities. For example, an attacker using an old PC and a slow modem might be able to attack and overcome a much faster and more sophisticated computer or network.

Types of Attack

DoS attacks can manifest themselves in a number of forms and be targeted at a range of services. There are, primarily, three types of DoS attacks:

- *Destruction or alteration of configuration information for a system or network.* An incorrectly configured computer may not operate in the intended way, or operate at all. An intruder may be able to alter or destroy the configuration information and prevent the user from accessing his computer or network. For example, if an intruder can change information in your routers, the network may not work effectively, or at all. If an intruder is able to change the registry settings on a Windows NT machine, the system may cease to operate or certain functions may be unavailable.
- *Consumption of precious resources.* Computers and networks need certain facilities and resources to operate effectively. This includes network bandwidth, disk space, CPU time, applications, data structures, network connectivity, and environmental resources such as power and air conditioning.
- *Physical destruction or modification of network elements.* The primary problem with this type of attack is that of physical security. To protect against this type of attack, it is necessary to protect against any unauthorized access to the elements of your system — the computers, routers, network elements, power and air conditioning supplies, or any other components that are critical to the network. Physical security is one of the main defenses used in protecting against a number of different types of attacks in addition to denial-of-service.

DoS attacks are normally targeted against network elements. The technique that is normally used in an attack is to prevent the host from communicating across the network. One example of this type of attack is the SYN flood attack. In this type of attack, the attacker initiates the process of establishing a connection to the victim's machine. It does this in a way that prevents the completion of the connection sequence. During this process, the machine that is the target of the attack has reserved one of a limited number of data structures required to complete the impending connection. The result is that legitimate connections cannot be achieved while the victim machine is waiting to complete bogus "half-open" connections.

This type of attack does not depend on the attacker being able to consume your network bandwidth. Using this method, the intruder is engaging and keeping busy the kernel data structures involved in establishing a network connection. The effect of this is that an attacker can execute an effective attack against a system on a very fast network with very limited resources.

According to a report posted on May 23, 2001, the Computer Emergency Response Team/Coordination Center (CERT/CC), one of the most important reporting centers for Internet security problems, was offline for a number of periods during Tuesday and Wednesday as a result of a distributed denial-of-service (DDoS) attack.[28]

The CERT/CC posted a notice on its Web site on Tuesday saying that the site had been under attack since 11:30 a.m. EST that day and, as a result, at frequent intervals it was either unavailable or access to the site was very slow. The CERT/CC is a government-funded computer security research and development (R&D) center that is based at Carnegie Mellon University in the United States. The site monitors Internet security issues such as hacking, vulnerabilities, and viruses, and issues warnings related to such issues and incidents.

According to the report, the organization was still able to conduct its business and had not lost any data. The center issues warnings and sends alerts via e-mail. News of the attack on CERT/CC came on the day after researchers at the University of California at San Diego issued a report stating that over 4000 DoS attacks take place every week.

A DDoS attack, such as the one experienced by the CERT/CC, comes when an attacker has gained control of a number of PCs, referred to as zombies, and uses them to simultaneously attack the victim.

According to an unclassified document[29] published November 10, 2001, by the NIPC, technologies such as Internet Relay Chat (IRC), Web-based bulletin boards, and free e-mail accounts enable extremist groups to adopt a structure that has become known as "leaderless resistance." Some extremist groups have adopted the leaderless resistance model, in part, to "limit damage from penetration by authorities" that are seeking information about impending attacks. According to the report, which was prepared by NIPC cyber-terrorism experts, "An extremist organization whose members get guidance from e-mails or by visiting a secure Web site can operate in a coordinated fashion without its members ever having to meet face to face."

In addition to providing a means of secure communications, the range and diversity of Internet technologies also provide extremists with the means to deliver a "steady stream of propaganda" intended to influence public opinion, and also as a means of recruitment. The increasing technical competency of extremists also enables them to launch more serious attacks on the network infrastructure of a nation-state that go beyond e-mail bombing and Web page defacements, according to the NIPC.

According to a separate article on international terrorism by a professor at Georgetown University, the leaderless resistance strategy is believed to have been originally identified in 1962 by Col. Ulius Amos, an anti-Communist activist and this approach was advocated, in 1992, by a neo-Nazi activist, Louis Beam.

Information Warfare Tactics by Activists

What does an activist seek to achieve by using IW techniques? It is likely that the types of activity that an activist will undertake will be very similar to those of a terrorist group, with the main difference being the scale and the type of target. One of the main aims of an activist is to achieve their goals by exerting pressure through a route other than the government or a corporate process, although they may also use this route. If they can exert this pressure on the targeted organization through denial-of-service or through propaganda, they will do so, but they will also use the Internet to communicate with their colleagues and fellow activists and to gain information or intelligence on their target to identify its weak points.

Activists were, historically, groups of people with a common cause who wanted to bring pressure to bear on the "establishment." The establishment might be a government, an international organization such as the World Trade Organization, or even an industry sector such as the petrochemical industry or the biotech sector.

DoS attacks do not have to be sophisticated to have an impact. In 1995 during the detonation of nuclear tests in the Pacific, a number of groups, including Greenpeace, took online action to put pressure on the French government. The actions ranged in scope and type from those reported by Tony Castanha,[30] who said that the Hawaii Coalition against Nuclear Testing would be conducting its second protest of the summer on Sunday, September 3, 1995, at 8:30 a.m. He reported that the Coalition would be gathering at the Diamond Head end of Ala Moana Park and then march to Kapiolani Park. The Coalition requested readers help to support a nuclear test ban and to voice their concern on French nuclear testing. The online posting also requested that people attending the protest bring signs and banners with them. This was the use of the online resource to inform people of a physical gathering and to keep them informed of the latest local news with regard to their issues.

Another online action that was part of the Greenpeace campaign against the French nuclear tests was an international fax campaign. The campaign was advertised online and details of the fax numbers that were nominated as targets were listed, together with printers that were apparently available. An extract from the material on the Web page is given below.

> *E-Mail the French Embassy in Wellington — Tell Monsieur Chirac what you think mailto:remote-printer.french_embassy/wellington/ NZ@6443845298.iddd.tpc.int*

The Greenpeace postings also advocated that participants should send e-mails to one of the leading French newspapers, *Le Monde* — mailto:lemonde@vtcom.fr. to express their concern. The postings advocated that participants should:

> *inundate these numbers with protest e-mail. Note: Jacques Chirac's e-mail address was closed within one day of posting here so.... If you could send one fax every week to any or every number below, that would be brilliant!*
>
> *THE NUMBERS ARE:*
> *Jacques Chirac, President de la Republic*
> *+33 1 47 42 24 65*
> *+33 1 42 92 00 01 (not working at present)*
> *+33 1 42 92 81 88 (not working at present)*
> *+33 1 42 92 81 00*
> *Fax Number: +33 1 42 92 82 99*
>
> *Charles Millon, Ministere de la Defense (Defence Minister)*
> *+33 1 43 17 60 81 (not working at present)*
>
> *Herve de Charette, Ministere des Affaires Etrangeres*
> *+33 1 45 22 53 03 (not working at present)*

Also given were the fax numbers of a number of leading French individuals and organizations. The individuals included Alain Juppe (Prime Minister) and the organizations included the French Embassy in London, the French Institute in Taipei, the French Nuclear Attaché in Washington, and the Nuclear Information Centre at the French Embassy in Washington. This relatively early example of the use of the Internet by activists to bring pressure to bear, in this case on the French government, showed a range of ways in which the technology could be used. These included e-mail protests to individuals and a newspaper, the dissemination of fax numbers for use by people who could then block these numbers with the volume of calls that were made to them, and the dissemination of information about local actions that could be accessed by a large number of people.

Another example of online activity by pressure groups can be seen in the September 2000 fuel protests that took place across Europe. Not only was the Internet used to post news of the current situation with the fuel protest to keep the people involved informed of the latest situation in each of the countries and regions, but it was also used to mobilize activists to considerable effect.

An example of the effect that was achieved can be seen in the online news posting that was headlined "Berlin stands firm over fuel protest." This was posted on September 20, 2000. The news item reported that Germany's transport minister, Reinhard Klimmt, had said that the government would not hand out any concessions to German haulers, despite the fact that concessions had been handed out elsewhere in Europe, and that any such move would have to be part of a coordinated European Union effort.

This statement was made after German truckers and farmers held up traffic in a series of protests over the high price of fuel on Tuesday, but the government refused to cut taxes and criticized other European governments that had done so, with both France and Italy having offered to cut tax on diesel fuel to appease truckers in those countries.

Another online action by activists targeted the world trade summit. This action was planned by a coalition of cyber-protesters who intended to flood 28 Web sites associated with the free trade negotiations at the Summit of the Americas with e-mail messages and requests for Web pages. The participants hoped to gain enough support to effectively mount a DoS attack. The action was apparently led by a group called the "Electrohippies." This hacktivist action was intended to mirror the summit's schedule, which started on Friday evening and ran through the weekend to Sunday in Quebec City. Leaders from 34 nations were meeting there to discuss the establishment of a single free trade zone that would extend from Canada in the north to Chile in the south.

One of the fastest growing activities on the Web is the defacement of Web pages. The rationale for the defacement and the selection of the target for the attack will be totally dependent on the cause that the attacker is supporting. Examples of this type of attack include:

- The attack on the Kriegsman fur company by the hacker "The Ghost Shirt Factory" on November 12, 1996. The Web site was defaced by the animal rights activists who made clear their dislike of the fur trade.
- An attack on the Web site of the Republic of Indonesia by a hacker known as "TOXYN" on February 11, 1997, this attack was on the Web site of Indonesia's Department of Foreign Affairs and was claimed to be an action taken in protest against Indonesia's occupation of East Timor.
- Another attack on the Republic of Indonesia took place the following year when hackers known as "LithiumError/ChiKo Torremendez" defaced approximately 15 Indonesian domains at the same time. This was claimed to be a part of an anti-President Suharto campaign.
- Another example, this time from France, occurred when the French National Front Web site was defaced by a hacker known as "RaPtoR 666." The attack took place on January 28, 1999, and the hacker defaced the Web site in French, but an English-language version was also made available by a hacker known as the "GrandMeister."

These examples are but a tiny fraction of the thousands of Web site defacements that now take place every day around the world. Archives of hacked Web sites can be found in a number of locations. But some of the more popular sites are the Onething Archive[31] and the 2600 magazine archive.[32]

The use of propaganda by activists is an effective weapon in their armory. Through its distributed nature and the lack of control that exists on the Internet, it is extremely easy to get a message published and with determination and resources, you can put up a very effective presence to support your cause.

It could be said that any Web presence for terrorist or activist Web sites, or the sites of the regimes or topics that they oppose, are placed on the Web for the purposes of propaganda. It is worth remembering that those plain and simple facts that to you or me are indisputable are, to others, propaganda produced by a system that they oppose. There are a number of Web sites that have dealt with this subject in some depth and that largely poke fun at the more obvious cases of propaganda, whether they are from governments or from other organizations. One of these sites, Propaganda & Psychological Warfare Studies,[33] looks at the situation in Africa; and another, the Extremist propaganda Web page,[34] pokes fun primarily at the American culture.

Another group becoming more of a "domestic terrorist" factor in the United States are the "eco-terrorists" who appear to be out to "save the planet from human destruction." Currently, they appear to be happy blowing up buildings and destroying laboratory research equipment, which ironically are in some cases being used to help the environment.

Information Warfare Tactics by Miscreants in General

The catch-all of the miscreant in general is really here because there are all the other people and groups out there that cannot be classed as either terrorist or activist, but who can still create a significant impact on a country, an organization, or an individual. This will include groups such as the drug cartels and other organized crime groups such as the Mafia.

The tactics that they will use will depend on the level of skill they possess, the target of their attention, and the effect they are trying to cause.

One small but significant grouping is that of the anarchists and techno-anarchists. It is surely surprising that the anarchists that are active on the Internet can organize themselves well enough to have an impact. Given that the definition of an anarchist is:

> **An-ar-chist** ***an-er-kist, -ar-** \\ *n 1: one who rebels against any authority, established order, or ruling order 2: one who believes in, advocates, or promotes anarchism or anarchy; esp. one who uses violent means to overthrow the established order.*

Does their joining together in a common cause mean that they are not true anarchists, or does it mean that the definition is wrong?

Typically, the targets for anarchists have been governments and large multinational companies; but in recent years, there has been a significant shift in targeting toward the meetings of the G8 and other institutions perceived to have an effect on the world economy, such as the World Bank. Recent meetings of the heads of governments have increasingly come under violent attack from the anarchists and this has been mirrored in the activity seen on the Internet.

The cause of a denial-of-service attack from this portion of the population will be totally dependent on the relationship between the attacker and the target. The attack may be as the result of a perceived slight on an individual by another individual or an organization, or as part of a concerted attack that is part of a wider event. One set of observed attacks that fall into this group are the well-documented, but totally unexplained, attacks on a site known as GRC.COM, which is described as:

> *On the evening of May 4th, 2001, GRC.COM suddenly dropped off the Internet. I immediately reconfigured our network to capture the packet traffic in real-time and began logging the attack. Dipping a thimble into the flood, I analyzed a tiny sample and saw that huge UDP packets — aimed at the bogus port "666" of grc.com — had been fragmented during their travel across the Internet, resulting in a blizzard of millions of 1500-byte IP packets. We were drowning in a flood of malicious traffic and valid traffic was unable to compete with the torrent. At our end of our T1 trunks, our local router and firewall had no trouble analyzing and discarding the nonsense, so none of our machines were adversely affected. But it was clear that this attack was not attempting to upset our machines, it was a simple brute-force flood, intended to consume all of the bandwidth of our connection to the Internet...and at that it was succeeding all too well. Gibson Research Corporation is connected to the Internet by a pair of T1 trunks. They provide a total of 3.08 megabits of bandwidth in each direction (1.54 megabits each), which is ample for our daily needs.*
>
> *We know what the malicious packets were, and we will soon see (below) exactly how they were generated. But we haven't yet seen where they all came from. During the seventeen hours of the first attack (we were subsequently subjected to several more attacks), we captured 16.1 gigabytes of packet log data. After selecting UDP packets aimed at port 666.... I determined that we had been attacked by 474 Windows PCs. This was a classic "Distributed" Denial of Service (DDoS) attack generated by the coordinated efforts of many hundreds of individual PCs.*

After some investigation, the victim of the attack was contacted by the attacker who posted the following messages to him:

> *hi, its me, wicked, im the one nailing the server with udp and icmp packets, nice sisco router, btw im 13, its a new addition, nothin tracert cant handle, and ur on a t3...so up ur connection foo, we will just keep comin at you, u cant stop us "script kiddies" because we are better than you, plain and simple.*

In this message, the attacker revealed him (her)self to be 13 years old.

> *to speak of the implemented attacks, yeah its me, and the reason me and my 2 other contributers, do this is because in a previous post you call us "script kiddies," at least so i was told, so, I teamed up with them and i knock the hell out of your cicso router*

In this posting, the attacker reveals that he has had the help of a couple of friends, subsequently named as hellfirez and drgreen, but reveals that the denial-of-service attacks (there were six in all) were caused because someone has told him (WkD) that the victim had referred to him as a "script kiddie." If such a perceived (but unconfirmed) insult generates this level of reaction, then the consequences of a real event are impossible to guess.

Some of the easier-to-remember cases of theft on the Internet are cases that originated in Russia, the most notorious being the Citibank theft that was perpetrated by Vladimir Levin. Although the eventual result of this attack was reported to be a loss of $400,000, the exposure of the bank during the attack was reported as $10 million to $12 million. Levin was captured as he passed through London and in 1998 he was sentenced to three years in jail.

Another Russian case was that of "Maximus," a cyber-thief who stole a reputed 300,000 credit card numbers from Internet retailer CD Universe during 1999 and demanded a $100,000 ransom not to release them onto the Internet. When the money was not paid, he posted 25,000 of the credit card numbers onto a Web site. The impact of this was that 25,000 people had their credit details exposed to the world. The only possible outcome of this action would be the replacement of all the affected cards with the respective cost implications.

It is notable that in Russia, according to Anatoly Platonov, a spokesman for the Interior Ministry's "Division R" that handles computer crimes, there had been 200 arrests made in the first three months of the year 2000, which was up from just 80 in all of 1998. He speculated that this rise in the number of arrests may reflect an increased police effectiveness rather than a growth in crimes.

In the United States, an incident that was given the name of Solar Sunrise, which was first reported in 1998 in the "Defense Information and Electronics Report," exposed the Department of Defense's poor state of computer security. The Pentagon initially believed that the attack was very serious and probably originated in Iraq. However, two teenagers in California were eventually arrested for breaking into the military networks. The teenagers were able to breach computer systems at 11 Air Force and Navy bases, causing a series of denial-of-service attacks and forcing defense officials to reassess the security of their networks. The two Californian kids were assisted by an Israeli youth, Ehud Tenenbaum, who was known as "The Analyzer" and were described by Art Money, the acting Assistant Secretary of Defense for Command, Control, Communications, and Intelligence, and the DoD's CIO, at the time as kids "having a hell of good time."[35]

For some of the groups in this category, the online collection of intelligence is currently a major issue. It is now almost irrelevant as to whether you refer to this activity as spying, as open source intelligence collection, or as industrial espionage; the net results are very similar, as are the methods used. In the past, if you were planning an action against an adversary, you would carry out a reconnaissance of the target and gain as much information as possible to enable you to identify the specific targets and to learn as much as possible about their habits, practices, and history.

You would visit the public offices and the libraries and read newspapers to gather background information and you would visit the site to gather more specific information through observation, or through methods such as dumpster diving (yes, it did exist before we had computers; it was just that the information that the dumpster diver was looking for was different). Now, most of the information that exists with regard to a person or an establishment is held in computer text files or databases, so the need for a protagonist to expose themselves to identification by visiting the site or by being seen in local libraries or public offices is greatly reduced.

Another form of attack that this category of attacker might use is *identity theft*. It is now trivially easy to gain all the information you need to assume someone else's identity (identity theft) or draw all of the information needed with regard to an organization or a company. Identity theft is still largely confined to the United States; however, the number of recorded incidents has risen dramatically in recent years. If an individual is the victim of an identity theft, the results can be startling and the restoration of a state that is similar to that which existed before the identity was stolen is extremely difficult and time-consuming. It also has terrorist implications as one can imagine.

If there is a recorded case that exemplifies the damage that can be caused to an organization if details of it are known to hostile activists, it is worth looking at the case of the Huntingdon Life Sciences in the United Kingdom. The organization had resisted intense pressure from animal activists for a considerable time, first experiencing direct action against the organization and its staff and then, more recently, through indirect action which was highlighted by the protesters putting pressure on the banks that were providing finance and banking facilities to the organization. Where did the animal rights activists get the information on where Huntingdon Life Sciences banked? There are actually a number of ways in which they could have obtained this information; but in reality, if you know where to look for it, it is actually freely available online. Once the protesters had this innocuous item of information, they could bring the organization to the brink of disaster by putting intense pressure on the banks and intimidating their staff members.

Since its early days, the Internet has been exploited for espionage. What better medium could the modern information broker, activist, or spy want? They have been provided with a low-risk means of access to a country and a facility or organization, a means of communication that is both anonymous and untraceable, the potential to use cryptography without raising the slightest suspicion, an updated version of the Cold War "dead letter box," and a set of obstacles to overcome to gain access to industrial and government information that, in previous times, would have been considered laughable.

The first case of online espionage was reported when Cliff Stoll documented his actions and discoveries of 1985 in his book *The Cuckoo's Egg*.[36] In this case, the Soviet Committee for State Security (Komitet Gosudarstvennoi Bezopasnosti — KGB) is known to have paid an East German hacker, Markus Hess, to penetrate U.S. defense agency systems. In a present-day case, the heavily reported Moonlight Maze attacks have been occurring for some time, probably since 1997 or before, where hackers from eastern Europe have broken into a large number of systems, including the Pentagon's systems, accessing "sensitive information about essential defense technical research matters." Although the stolen information has not been classified, it is still invaluable to foreign governments, terrorist groups, and private companies because these networks hold information on military logistics, planning, payrolls, purchases, personnel, and routine Pentagon e-mails between departments. The most sophisticated attacks observed to date apparently came from just outside Moscow and were eventually traced to the Russian Academy of Sciences laboratory, the country's leading scientific research body.

The average miscreant in this category will have one of two driving motivators for his activity on the Internet: it will either be for curiosity (the "can I do that" factor) or it will be for financial gain. The following discussion takes a look at some of the techniques that will be used to try to obtain financial gain.

Unusually, there is a report from a country that we consider to be "closed" to us in a number of ways and which, if we believe all the stories we are presented with, is now run by the Mafia and organized crime. According to a report by Ruth Alvey[37] in July 2001, the level of cyber-crime that was recorded in Russia has grown rapidly in recent years. In 2001, there were 1375 crimes registered in the high-technology field, a growth of 18 percent from 1999. The report highlights the fact that this type of expansion is particularly worrying because only approximately 4.5 percent of the Russian population is connected to the Internet, which compares with connectivity rates of approximately 49.1 percent in the United States. The report also gives a conservative estimate of between 250 and 500 hackers operating in Russia today, with 15 to 20 of these hackers available for hire working in the Moscow area and around ten working in the area of St. Petersburg. The reporter also gives further details of hacker activity in Russia, such as the level of sales of hacker magazines (30,000 copies per month) and cites that 1605 Russians participated in a single hacking competition on a Russian Web site (www.hackzone.ru) in the year 2000 suggesting that the actual number of active hackers is much higher.

From the United States there is a report from Florida, in which it was stated[38] that an FBI sting operation resulted in the arrest of Fausto Estrada for allegedly stealing various confidential documents from the credit card company MasterCard International and offering to sell them to MasterCard's competitor, Visa International. A five-count complaint charged Estrada with theft of trade secrets, mail fraud, and interstate transportation of stolen property. According to the complaint, in February 2001, Estrada, using the alias "Cagliostro," mailed a package of information he had stolen from MasterCard to Visa's offices located in California. Estrada allegedly offered to sell to Visa sensitive and proprietary information that he had stolen from MasterCard's headquarters. According to the Complaint, among the items Estrada offered to sell to Visa was a business alliance proposal valued in excess of $1 billion between MasterCard and a large U.S. entertainment corporation.

As part of a sting operation conducted by the FBI's Computer Intrusion and Intellectual Property Squad, an FBI agent posed as a Visa representative and negotiated for the purchase of the MasterCard documents in Estrada's possession. If convicted, Estrada faces a maximum sentence of ten years in prison and a fine of $250,000, or twice the gross gain or loss resulting from the crime on each of the two charges of theft of trade secrets and the two interstate transportation of stolen property charges, and five years in prison and a $250,000 fine or twice the gross gain or loss resulting from the crime on the wire fraud charge. This was a fairly straightforward theft, but hitting at the heart of the electronic trade bedrock — the credit card.

In another report from the United States, a 16-year-old New Jersey teenager, Jonathan G. Lebed, settled a civil fraud lawsuit filed against him by the Securities and Exchange Commission (SEC), which alleged that he had hyped stocks on the Internet before selling them for a total profit of $272,826. He settled the charges brought by the SEC by paying the government $285,000, which included his alleged illegal profits plus interest. The SEC accused Lebed of using the Internet, beginning when he was 14 years old, to tout nine small stocks he owned, driving up their prices. He sold the shares, usually within 24 hours of the promotional e-mail, making as much as $74,000 on a single stock sale, the agency's suit alleged.

This is a classic case of using the power that is provided by the freedom of the Internet, together with the lack of verification that takes place with online publishing, to influence the opinions of people. This is a trivial example of how, when it started, a 14-year-old youth could exert enough influence to affect the price of stocks on the stock exchange. Imagine the potential for influencing people that could be achieved by a well-funded and -trained organization.

The next example is the first of what will inevitably be repeated. In this case, the Italian police arrested 21 people who were accused of involvement in a massive online banking fraud that could have cost the Sicilian regional government more than 1 trillion lire (U.S.$465 million), according to a statement by the Italian authorities in October 2000.

Members of a criminal group with links to the Cosa Nostra allegedly managed to "clone" an online branch of the Banco di Sicilia and were preparing to remove funds from an account belonging to the Sicilian regional government, officials said. The scheme was operated with the assistance of two members of the bank's staff, using stolen computer files, codes, and passwords. With these facilities, the gang managed to gain access to the bank's information systems.

It was alleged that the group was planning to steal 264 billion lire from the bank. According to the Italian news agency AGI, one of the possible destinations of the stolen money was the branch of a Portuguese bank, the Banco Espirito Santo e Comercial of Lisbon, in Lausanne, Switzerland.

Police identified the leader of the gang as Antonio Orlando, 48, described as being close to one of Palermo's leading Mafia families and with previous arrests for fraud, money laundering, and receiving stolen property. According to an official from the Palermo Police, "The operation was certainly authorized by the Mafia, because here in Sicily any operation of economic importance requires the Mafia's permission."

Also within this group is the use of the Internet for the purposes of communication. What is being referred to here is the use by individuals and groups who are engaged in nefarious activities of technologies that will either allow them to remain anonymous or let them send and receive messages that cannot be intercepted and reduced to a meaningful state by either law enforcement of their opposition will always attract them to technology and the Internet.

Let us look at the case for anonymity. In the United Kingdom, because of the way the Internet industry has developed, it is possible to take out a "free" Internet connection through an ISP. While the user is required to provide personal details for the account, because the service provider is not trying to gain any money for the use of the service from the user, there is normally only a cursory check that the details that have been provided are correct. (If you were the ISP and the user was not the direct source of revenue, how much effort and resource would you invest in checking out the details provided?) It is also possible in the United Kingdom to purchase from any High Street store, a pay-for-use mobile phone. These can be purchased for cash and replacement cards or top-up cards can also be purchased for cash from a large number of outlets. The result: anonymous communications and access to the Internet. There are a large number of ways to obtain free telephone calls, most of which are illegal, but the combination of untraceable telephone calls and connectivity over the Internet is a powerful combination.

Having looked at a number of criminal group types, it would be unrealistic not to look at the material available on the Cali drug cartel from Columbia. In a paper written by a Los Angeles policeman,[39] he identifies that not only are criminals using the available technologies to make their illegal activities more profitable, but that they are also using computers, cellular phones, and other sophisticated electronic devices to gather intelligence information on police operations to prevent themselves from being caught. He cites as an example of this that

> *When agents of the United States Drug Enforcement Administration recently conducted a raid at the Cali drug cartel headquarters in Colombia, they discovered two large IBM mainframe computers. The computers were hooked into the national telephone service of Colombia and stored the phone records of millions of Cali residents. These phone records were routinely cross-checked against calls made to the United States Embassy in Colombia and the Colombian Ministry of Defense in an effort to identify Colombians who were cooperating with government drug enforcement efforts.*

In a court case in California,[40] it was quoted that the

> *Cali cartel is reputed to be using sophisticated encryption to conceal their telephone communications and to scramble transmissions from computer modems.*

Also referred to in the same court case was the Italian Mafia downloading copies of Pretty Good Privacy (PGP) from the Internet and the fact that Dutch criminal organizations encrypt their communications and computers with PGP and IDEA.

If the drug cartels and Mafia have this type of capability at their disposal, and there is no reason to doubt that they do — untraceable money will buy you almost anything — the potential is frightening. There is considerable paranoia regarding the capabilities of various "Big Brother" governments to intercept an individual's e-mail (and just because you are paranoid does not mean that they are not out to get you), but governments are at least voted into office and can be removed. Criminals with the same potential powers have no such constraints placed on them.

Historically activists were groups of people with a common cause who wanted to bring pressure to bear on the "establishment." The establishment might be a government, an international organization such as the World Trade Organization, or even an industry sector such as the petrochemical industry or the biotech sector. Another tool in the hands of the activist is the denial-of-service attack. The case below is an illustration of the effect that such an attack can have and the seesaw motion between the capabilities of the hackers and those of the defenders of the systems as they develop countermeasures.

In a report[41] by Rutrell Yasin on February 5, 2001, he stated, "Roughly a year after cyber-terrorists paralyzed some of the Web's most trafficked sites, technology is finally emerging to stop such distributed denial-of-service attacks before they ever reach their target sites. The new tools are designed to thwart attempts to bombard routers with large volumes of bogus requests that overwhelm servers and deny access to Web sites."

The denial-of-service attacks were/are again a major problem for Microsoft, especially after an employee had apparently misconfigured one of the routers on the system. The attackers were able to capitalize on this human error by one person at Microsoft and bombarded the routers with bogus data requests. The defensive measure brought to bear was an intrusion detection system. In this case, Arbor Networks, a relatively new company that has been jointly funded by Intel and Cisco, was about to announce the launch of a managed service that it claims will detect, trace, and block DoS attacks. This type of technology is not unique, and similar services have been produced in the United Kingdom by the Defence Evaluation and Research Agency (DERA) for use by the U.K. Ministry of Defence and have subsequently been used to provide a service for both government and industry. Other commercial organizations such as IBM and SAIC also offer similar services.

The service relies on sensors that are placed at strategic locations within the network to allow the monitoring agent to detect abnormal behavior on the system. The primary type of activity monitored is the system penetration; however, if the sensors are placed in front of the routers, the monitors can collect information about traffic patterns and identify anomalies, such as excessive traffic coming from a given IP address. In some cases, the software is capable of generating a fingerprint that can be used to trace the origins of the attack; however, this type of functionality has proved to have limited success to date (how do you identify the attacker in a DDoS attack that uses thousands of zombies?). Operators at the customer site or Arbor's network operations center can take corrective action, such as blocking excessive traffic.

The defacement of Web sites has been occurring for some time, but has increased in the past two years to the point where the Web site that became famous for its up-to-date reporting of Web sites that had been defaced stopped trying to keep up with the list of sites that had been damaged (www.atrition.org). They ceased activity after more than two years of tracking the defacement of sites.

A German Web site, Alldas.de,[42] now attempts to provide an up-to-date listing of the Web sites that have been hacked each day, together with a considerable amount of useful and related information (see Exhibit 2). This Web site also maintains league tables of which hacker groups have been responsible for which attacks during the period.

An example of this type of information is given in the small extract below, showing the name of the Web site defacer, the number of Web sites that were claimed to be defaced, and the percentage of the overall number of Web site defacements that this represents:

- *A-I-C* defaced four Web sites, which is 0.02 percent of all archived defacements.
- *A-jaX* defaced four Web sites, which is 0.02 percent of all archived defacements.
- *A-Open* defaced one Web site, which is 0 percent of all archived defacements
- *A1L3P5H7A9* defaced one Web site, which is 0 percent of all archived defacements.
- *Abfgnytvp* defaced one Web site, which is 0 percent of all archived defacements.
- *Abu Sayaff* defaced two Web sites, which is 0.01 percent of all archived defacements.
- *Abu Sayaff Boys* defaced one Web site, which is 0 percent of all archived defacements.
- *abuzittin* defaced one Web site, which is 0 percent of all archived defacements.
- *AC* defaced seven Web sites, which is 0.03 percent of all archived defacements.
- *AccessD* defaced eight Web sites, which is 0.04 percent of all archived defacements.
- *ACE* defaced eight Web sites, which is 0.04 percent of all archived defacements.
- *acecww* defaced four Web sites, which is 0.02 percent of all archived defacements.
- *acid* defaced one Web site, which is 0 percent of all archived defacements.
- *Acid Blades* defaced one Web site, which is 0 percent of all archived defacements.
- *aCid fAlz* defaced 13 Web sites, which is 0.06 percent of all archived defacements.
- *acid klown* defaced three Web sites, which is 0.01 percent of all archived defacements.

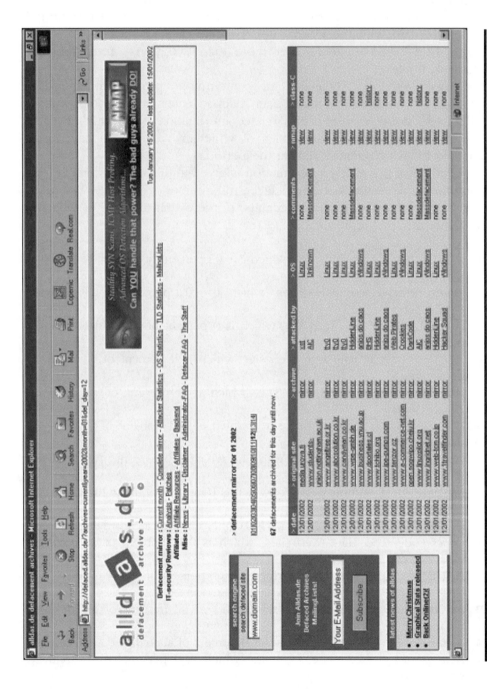

Exhibit 2. Extract of Information from Alldas.de

It is interesting to note that this Web site (Alldas.de) was itself the victim of collateral damage when the service provider on which it depends, Telenor, apparently suffered significant problems at the beginning of July (2001) for more than 40 hours. The site was also the target of a distributed denial-of-service attack during the middle of July 2001 that prevented it from operating for four days.

In Europe during the protest about the cost of fuel and the tax that the governments were levying on fuel, a number of Web sites came into being that provided not only communications within the local environment, but also allowed for the coordination of activity over the wider area. The material that is shown on these pages is from Web pages and newsgroups, all of which are semi-permanent; but in fact, a great deal of the information that was passed during these and other activities is now passed through services such as the Internet Relay Chat (IRC) channels, which can be as public or as private as the participants wish, and for which there is less of a permanent record created.

In the United Kingdom for the fuel protest, sites such as Bogush's Lair[43] provided an excellent example of a Web site that provided communication with the international situation as well as the local events, and provided details of meetings and actions that were kept up-to-date throughout the protest. The Web pages provided a network of related pages that gave a good overall picture of the situation as it developed and which provided a good barometer of public opinion with regard to the situation. It is interesting that governments in the areas affected were slow to realize the potential that was being exploited and did not appear to capitalize on the information that was being made available on the Internet.

Within the United Kingdom, there was an interesting mix of online activists, including concerned citizens who would not normally have been viewed as activists; political parties and groups, such as the West Berkshire Conservative Association;[44] the more expected trade group and industry sites; and truckers' forums.

Electrohippies, a group based in England, used DoS attacks against the World Trade Organization (WTO) in December 1999. The Electrohippies claimed that 452,000 supporters bombarded the WTO's Web site. The Electrohippies are hacktivists (i.e., computer-aided activists who hack) with a conscience. They will not intrude into computer systems and, in fact, abhor physical violence, preferring to send e-mail bombs rather than real ones that can hurt or kill.

iDEFENSE reported that the cyber-activist group RTMark has used eBay to help raise funds to support a variety of cyber-protest campaigns. RTMark utilizes an array of cyber-protest methods to target large companies and organizations. The group also solicits funds for developing hacker tools to be used against its targets.[45]

The Harsher Side of Activism

Urban terrorists from disparate factions across Europe used the Internet and mobile phones to orchestrate the rioting that marred a European summit.

Operating from a back-street bar and neighboring cyber café, under the noses of the 6000-strong security force surrounding Nice's Acropolis conference center, four men dispatched reports.[46]

When the International Monetary Fund and World Bank met in September 2000, the Federation of Random Action and an affiliate, toyZtech, orchestrated thousands of online protesters. Employing a new DDoS tool for people with almost no computer expertise, the attack was to force the Web sites off line.[47] In addition to the inconvenience resulting from this act, the groups also hoped to cause monetary loss.

Activists are usually cash strapped, preventing them from being able to afford the best technology. This creates a capabilities gap, but that is overcome with creativity. Activists adapt and improvise what they have to achieve their goals. This has been the case for thousands of years. Today, activists use that creativity and adaptability to bring to bear the technologies they can acquire.

Summary

In this chapter the different types of techniques and tools that a number of different types of individuals with a cause may use, or be perceived to have used, have been examined. In some cases, the action is intended to be an act of warfare, but the primary issue is that it is now impossible to determine whether an incident on a network or system has been the result of an accident, is an act of warfare, is a criminal activity, or is the action of three curious youths who were experimenting with the tools that they had found on the Internet.

The Solar Sunrise incident clearly demonstrates that what was initially thought to be an action by a hostile nation was eventually traced, some considerable time later, to the activities of three youths (two in California and one in Israel).

Notes

1. Definition of Terrorism Adopted by Gateway Model, United Nations, Spring, 1995, http://www.inlink.com/~civitas/mun/res9596/terror.htm.
2. *Al-Tawhid, A Quarterly Journal of Islamic Thought & Culture* article, Towards a Definition of Terrorism. Ayatullah Muhammad 'Ali Tashkiri.
3. Bin Laden: Steganography Master? Declan McCullagh, February 7, 2001.
4. Bin Laden's Name Raised Again — A Primer on America's Intelligence Archenemies, Robert Windrem, NBC News, http://www.ummah.net.pk/dajjal/articles/ladenagain.html.
5. Terrorist Instructions Hidden On Line, *USA Today,* Jack Kelly, June 19, 2001.
6. Terror Groups Hide Behind Web Encryption, *USA Today,* Jack Kelley, June 19, 2001.
7. Webopedia definition, from http://webopedia.internet.com/TERM/S/slack_space.html
8. StegFS homepage can now be found at http://www.mcdonald.org.uk/StegFS/.
9. Air Force Research Laboratories, http://www.afrl.af.mil/if.html.
10. Sinn Fein Web site, http://www.sinnfein.ie/.

11. Sinn Fein Online, http://www.geocities.com/sinnfeinonline/.

12. http://www.geocities.com/diarmidlogan/.

13. http://www.csc.tcd.ie/~sinnfein/.

14. http://www.irsm.org/irsp/.

15. http://www.ulsterloyalist.co.uk/welcome.htm.

16. http://www.houstonpk.freeserve.co.uk/uvfpg.htm.

17. Rich Geib's Universe, http://www.rjgeib.com/thoughts/terrorist/response1.html.

18. Irish Republican Army Information Site, http://www.geocities.com/CapitolHill/Congress/2435/.

19. Vincent Morley's Flag Web page, http://www.fotw.stm.it/flags/gb-ulste.html.

20. Unionist Murals from Belfast, http://www.geocities.com/Heartland/Meadows/7985/mural.html.

21. *The Basque Journal,* http://free.freespeech.org/ehj/html/freta.html.

22. Basque Red Net, http://www.basque-red.net/cas/enlaces/e-eh/mlnv.htm.

23. Spanish Ministry of the Interior Web page, http://www.mir.es/oris/infoeta/indexin.htm.

24. http://www.ac-versailles.fr/etabliss/plapie/MediaBasque2001.html#ancre45175.

25. Electronic Disturbance Theater Web site, http://www.thing.net/~rdom/ecd/ecd.html.

26. Kavkaz Tsentr Web site, www.kavkaz.org.

27. Hacking Vigilantes Deface WTC Victim's Site, Brian McWilliam, Newsbytes, September 17, 2001.

28. CERT goes down to DoS attacks, By Sam Costello, IDG News Service, 05/23/01.

29. The NIPC publication is available at http://www.nipc.gov/publications/highlights/2001/highlight-01-10.pdf.

30. French Nuclear Protest, Tony Castanha Thu, August 31, 1995.

31. Onething Defaced Web Site Archive, http://www.onething.com/archive/.

32. 2600 hacker magazine defaced Web site archive, http://www.2600.com/hacked_pages/.

33. Propaganda & Psychological Warfare Studies Web site, http://www.africa2000.com/PNDX/pndx.htm.

34. Extremist propaganda Web page, http://scmods.home.mindspring.com/index.html.

35. Defense Information and Electronics Report, Anne Plummer, October 22, 1999, http://www.infowar.com/hacker/99/hack_102599b_j.shtml.

36. *The Cuckoo's Egg,* Clifford Stoll, Doubleday, ISBN 0 370 31433 6, 1989.

37. Russian Hackers for Hire — The Rise of the E-Mercenary, by Ruth Alvey. July 1, 2001, http://www.infowar.com/hacker/01/hack_080301a_j.shtml.

38. FBI Sting Captures New York Man Who Stole Trade Secrets from MasterCard and Offered Them for Sale to Visa, U.S. Department of Justice, March 21, 2001, http://www.usdoj.gov/criminal/cybrcrim/Estrada.htm.

39. Why the Police Don't Care About Computer Crime, by Marc D. Goodman, a sergeant with the Los Angeles Police Department and student in the Public Administration program at Harvard.

40. No. 97-16686. in the United States Court of Appeals for the Ninth Circuit, Daniel J. Bernstein, plaintiff-appellee, v. U.S. Department of Commerce et al., defendants-appellants on appeal from the United States District Court for the Northern District of California.

41. Tools Stunt DoS Attacks, Monitors Dam Packet Floods at ISP Routers, by Rutrell Yasin, Internetweek, February 5, 2001, http://www.internetweek.com/newslead01/lead02051.htm.

42. Alldas Web site, http://www.alldas.de.

43. Bogush's Lair Web site, http://network54.com/Hide/Forum/101883.

44. West Berkshire Conservative Association Web site, http://www.wbca.org.uk/fuel.htm.

45. iDEFENSE Intelligence Service, March 15, 2000, http://www.idefense.com/ or http://www.csmonitor.com/atcsmonitor/cybercoverage/bandwidth/p122899bwice.html.

46. Cyber café is HQ for rioters, http://www.thisislondon.com/dynamic/news/ story.html?in_review_id=342673&in_review_text_id=286292, Colin Adamson, This Is London.com, Associated Newspapers Ltd., December 9, 2000.

47. Hacktivists Chat up the World Bank: 'Pecked to Death by a Duck', http://www. villagevoice.com/issues/0042/ferguson.shtml, Sarah Ferguson, The Village Voice, October 19, 2000.

INFORMATION WARFARE DEFENSES: COUNTERMEASURES, COUNTERATTACK, OR BOTH

This section discusses in more detail than in the previous chapters how to survive in the dynamic global and very competitive information environment when your organization is being attacked from within and from without. This section includes the following five chapters.

Chapter 15: Surviving the Onslaughts: Defenses and Countermeasures

This chapter discusses defensive strategies such as defense-in-depth, counter-measures, and counterstrike as critical elements of any design process intent on negating, thwarting, or delaying the impact of IW strikes from economic espionage agents, techno-terrorists, general miscreants, and military info-warriors.

Chapter 16: Defending against Information Warfare Attacks Begins with Knowledge Management

This chapter discusses a basic approach to defending the business or government agency by establishing a bottom line based on KM.

Chapter 17: What's a CEO to Do? Ya Gotta Have a Plan

This chapter discusses the need for the business or government agency CEO to take an aggressive leadership role if he or she wants to defend the turf. To do so, a bottom-line, no-nonsense plan is required. This chapter discusses the philosophy, strategy, and planning that is necessary.

Chapter 18: Those Who Do Not Accept Change and Adapt Will Be Consumed By It; The Enlightened Will Survive

This chapter discusses survival. Traditional security focuses on the physical world, but today that focus may not always be valid. Controlling the information environment (IE) may be more effective than a physical attack and may be able to prevent it. This leads to the conclusion that controlling the IE to successfully attain corporate objectives is the approach we need to pursue. Using Coherent Knowledge-based Operations (CKO), corporate action would be synchronized and coherent, following consistent themes conducted simultaneously.

Chapter 19: The Twenty-First Century Challenge: Surviving in the New Century

This chapter reveals the writers' impressions of the courses that high technology, businesses, national objectives, and military strategies have charted for us based on current trends. Will our worlds collide? How will we fare if or when they do? What are our future threats, vulnerabilities, and risks brought on by high technology as a weapon of choice in IW?

Chapter 15

Surviving the Onslaughts: Defenses and Countermeasures

Extremism in the defense of liberty is no vice. And moderation in the pursuit of justice is no virtue.

—Barry Goldwater

This chapter introduces defensive strategies such as defense-in-depth, countermeasures, and counterstrike as critical elements of any design process intent on negating, thwarting, or delaying the impact of information warfare (IW) strikes from economic espionage agents, techno-terrorists, general miscreants, and military info-warriors.

Defense-in-Depth

For any defense to be effective, it must be well-planned and based on clearly understood requirements for the protection of the information system and the information that it stores, processes, and transmits. An effective defense will be created from a range of measures that will span the physical, procedural, personnel, and electronic areas (see Exhibit 1). The use of these measures must be balanced to provide the protection required in the most cost-effective manner, consistent with acceptable risks. While not all-encompassing, the information presented forms the very basics required to be implemented. Then one can begin to tailor system defenses and countermeasures based on one's specific environment, threats, vulnerabilities, and risks.

If a system is to be protected, then the starting point of any defensive measures must be an understanding of what is to be protected and the value

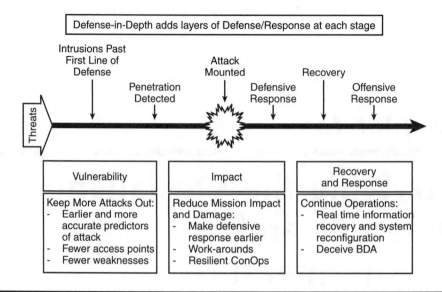

Exhibit 1.   Generic Defense-in-Depth Approach

of the system and the information that is stored, processed, or transmitted by the system.

You might expect to see a heavy emphasis on the electronic environment, but it is important to keep sight of the fact that any security that is implemented must consist of a range of measures that encompass all of the areas (e.g., physical, personnel, and administrative).

The holistic approach to security that is needed to gain cost-effective security is depicted in Exhibit 2. As you can see from this exhibit, security is not just about technology; it is about a mixture of measures that are taken from a range of areas and disciplines. On the outside, you have physical security. These are the fences, walls, security doors, barred windows, and intruder alarms. Next you have personnel security measures. This is most commonly the checking of staff to ensure that you know their background and have some confidence in their morals and relationships. It may include carrying out background checks, criminal record checks, and reference and financial checks. It will also include the training of staff in their duties and the actions to be taken in the event of an incident. In larger organizations, it could also include psychometric testing and lie detector tests. What one is trying to achieve is that one can have confidence that the people you are employing do not have any skeletons in their closets, but also that they can, as far as is testable, be trusted. To ensure that the physical and personnel security measures are working in harmony with the electronic security measures, there is a need for procedures to be initiated.

Risk Assessment and Risk Management

A cost-effective defensive system, complete with countermeasures, must be designed. Its design and architecture must be based on analyses of acceptable levels of risks. This is accomplished by conducting a risk assessment, during

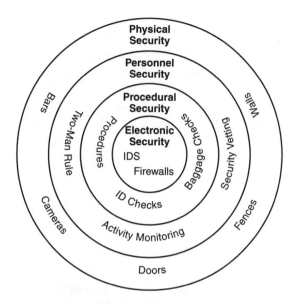

Exhibit 2. The Onion Ring Model

which system assets are identified and valued, and the threats to the systems are determined and understood. The impact of the failure of the security measures that have been applied to the systems is assessed. Once the risk assessment has been carried out, the measures that will have to be taken to protect the system will then be identified and the owners of the system can have a clear understanding of the residual risk that the adoption of the countermeasures will leave. It is now accepted that there is no such thing as perfect security and that the more you impose security, the less useful and usable the system is and the greater the cost of implementing the solution.

Put simply, the risk to a system is a factor of the threat to it, which is measured in terms of the capability of an attacker and his motivation; an exploitable vulnerability; and the opportunity to carry out an attack, together with the impact that an attack on the system would have.

Prior to the morning of September 11, 2001, U.S. airports and airlines believed that their minimum-security measures provided a cost-effective security system based on risks. That was the decision made by numerous executive managers who accepted the risks. Basically, they bet hoping nothing would happen so they would not have to pay for more security and be less profitable. The majority of the security problems can be solved, but not unless executive management in corporations and government agencies decide that the risks are currently not acceptable, and support a defense-in-depth approach. After all, even in information warfare, there are lives at stake.

Physical Security

Physical security is the fundamental building block on which all conventional systems security is based. Without adequate physical security, effective control of access to the system is difficult. It is certainly possible to achieve a high level of security without good physical security, but it will normally mean that the risk to the system from theft, physical damage, or destruction is higher and that the cost of achieving any reasonable level of security will be greatly increased. If you take the laptop as an example of a situation in which the physical security is weak, then the protection that is afforded to the computer must be through the person using it. That person must take care to make sure that it is not removed from that owner, from physical security measures such as wire straps to attach the laptop to a work surface or upright, access control devices, and a software or hardware device to encrypt the storage medium.

Personnel Security

The security aspects that relate to staff are some of the most important aspects to be addressed. Remember that the staff who are employed on a system are uniquely privileged. They have access to the system, they have knowledge of the data that is held on it, the value of the data, and the ways in which information can be imported to or exported from the system. A disaffected staff member can do far more damage to an organization in a far shorter time than any external attacker is likely to be able to achieve. It is also true that the majority of damage that is caused to a system is caused by insiders (e.g., a mole working for an adversary). Some estimates of insider threat put the number as high as 85 percent, but the average estimate is around 65 percent.

Staff Retention

Staff retention is highlighted here under personnel security because it is only when you have happy, stable, reliable, and trustworthy staff that your personnel security will work effectively. The shortfall in skilled IT security staff is growing and will continue to grow for the foreseeable future as all areas of industry and commerce continue to adopt the E-business culture. Staff are an increasingly valuable and precious asset as the investment in their training and the complexity of the subject they must understand increases.

The dot.com companies of the late 1990s and early 2000 made the promise of huge wealth to those who would take the risk of working in a start-up company. For many of the younger people employed in the industry (and let us be honest, this is very much a young person's field), the potential for great wealth in the short to medium term as the share options became valuable was irresistible. Unfortunately, the bubble burst and most of the share values plummeted to near-zero; but for some time period (a lifetime even in IT terms), the focus within the industry was on getting dot.com companies online rather than concentrating on security. The aspirations of individuals had been raised, but in most cases had not been satisfied.

In large organizations over a number of years, there has been the drive to return to core competencies, which had been pushed forward as a way of becoming more efficient and profitable. What this means is that if you are a factory producing boxes, you concentrate your staff's effort on doing just that. If you need transportation to move the product from the factory to the wholesaler, then you employ a haulage firm; you do not have your own trucks. If you need computers to manage your production, finances, and payroll, you outsource this to a company that specializes in providing this type of service to the manufacturing industry. All of this is very rational and sensible. After all, if the company has developed around the manufacturing of boxes, what depth of knowledge does it have in trucking? Also, unless it is a very large manufacturing organization, what economies of scale can it achieve?

This has been a significant factor for more than a decade in the way that organizations have been restructured. Another factor has been the drive to make staff more mobile and likely to move from one job to another to satisfy their career aspirations and to make the organization more streamlined and fitter. The old concept of joining a company as an apprentice and working your way up through the company over a career is now largely history.

All of this is good for the efficiency of the operation; but in recent times, the realization has started to dawn on the organizations that have followed this route to any great degree that while the old ways may carry some baggage in the form of underachievement by a small portion of the staff, it did provide a framework in which the concept of "staff loyalty" was strong. It also provided organizations with their own "in-house" expertise which, while it may have performed in a sub-optimal manner, did belong to the organization and could be rapidly redeployed to cover unforeseen situations. Imagine that it is four o'clock in the morning on a Sunday and the system has just gone down. If you have a well-written service contract, the efficient, outsourced service provider will of course dispatch an engineer from its local depot within the contracted response time. Again, suppose you could only realistically afford a four-hour response, so it will not be unreasonable to expect that you will be back online within six hours or so. Would it not have been so much easier to phone one of your own staff members who lives within easy commuting distance of the site to come in and restore a system that he feels responsible for as a part of the organization?

The point highlighted here is that while it is probably more cost-effective on the balance sheet to keep staff mobile and not offer them long-term commitments, the downside of this is that you do not achieve any form of staff loyalty, a handy thing to have when you are dealing with security, and you cannot generate the level of skill that is required to safely manage your systems.

Security of any organization or system is not about a single measure or set of measures. For example, on its own, a password will not protect a system. The military has for many centuries understood that security, to be effective, must be provided in a number of layers that interoperate. This is known as defense-in-depth and is perhaps best demonstrated by the onion ring model (Exhibit 2). In this model, there are a number of concentric rings that each provide an element of the overall security.

Education and Training

It is arguable whether staff education and training should be in this part of the chapter or the procedural or electronic areas, but it is addressed here because, whatever else it is, it is about personnel. To make the security that is being applied to a system effective, it is essential that the staff who are going to manage, use, and protect the system and its security are adequately skilled through a balance of both suitable training and experience. This is not a one-time purchase, but an ongoing investment in the staff throughout the life of the system.

Unfortunately, the technology field that we are working in is a fast-moving area with numerous reports telling us that the capacity of the hardware doubles every 18 months and that the software refresh rate is around 9 months. When we add into this equation that there are very few schools or universities that can offer effective computer security training courses and add to that the limited number of places that a person can gain experience in this field, it is easy to see why this problem will continue.

Given the complexity of modern systems and the range of security functionality that may need to be applied to a system to make it secure, the maintenance of an effective level of knowledge can be expensive and difficult to achieve. Additionally, the retention of staff will become a significant problem as you invest more in their training, as their market value and attractiveness will increase and their potential to increase their salary by moving will become more attractive. A good management system that cares for its staff will go part of the way in reducing this problem; but in the end, if you make staff precious by training them to a high level, you will inevitably take losses. In large organizations, this may be a tolerable cost, but how can smaller organizations maintain the levels of training and knowledge in their staff in a cost-effective way? The answer is probably that they cannot, and the choice for these companies is that they either accept a lower level of system security or they outsource the problem and employ an organization that specializes in system security to manage their systems in a secure manner.

This might appear to be attractive and give you a good, cost-effective solution; but as with most things in life, it is not quite that easy. If you have gone through the process of identifying the type and level of security that you require to protect your system, can you really believe that an organization that specializes in the delivery of cost-effective security services will be able to tailor the solution that is delivered for your systems to your individual requirements? It is far more likely that, even if you had followed the route of risk assessment and the identification of a suitable solution, the outsourced security provider will actually provide a generic solution. After all, the cost savings of bulk delivery come from being able to generate an easy-to-manage system that fits most of the problems. This does not necessarily mean that the security that you purchase will be to a lesser level than you had identified, but it does mean that the solution that has been provided may not be that which was identified as the most appropriate.

Once you have decided to outsource your security problems to an organization that specializes in providing security solutions, how do you verify that they have the skills and integrity to carry out the task of protecting your information systems? If you are a wealthy company and can afford the big-ticket international names to provide this service for you, then there is no problem. Or is there? How do organizations identify and recruit staff with the appropriate skills and knowledge. We already know that staffs with this type of knowledge and skill are in incredibly short supply; that their motivation in many cases is as much about knowledge as it is about money. We also know that to keep pace with the changes in technologies, computers have to be a way of life and not just a job. Given all of this, together with the fact that a large number of the people who have an interest in this type of technology will have experimented with systems to see how effective their security is, it will not come as a surprise that a good portion of the information security specialists are or have been hackers. The use of the term "hackers" in this context is fairly all-encompassing, and includes both the old-fashioned black-hat and white-hat (ethical) hackers as well as modern crackers. No one who is any good at computer security got that way by reading books; it takes a long time as an apprentice, an innate desire to understand systems, and a lot of common sense.

While this is not a problem for the majority of people who will have experimented within the bounds of systems that they have owned or been given permission to experiment on either in a university, at home, or at work, there remains an issue of those people who are "addicted" and who are hackers because they have a cause or because they just enjoy it and are arrogant enough to think that they will never get caught. If you have given over the responsibility for your security to another organization, how do you convince yourself that they will do a good job for you? You could conduct periodic checks that they are doing the job well, but because you have given them the job because you cannot retain the staff to do it for yourself, where will you get qualified staff to carry out this sort of testing? Of course, you could hire a penetration testing team from an independent supplier, but now you have given access to your information to a third group who really does not need to know it.

Additionally, the users of the system must be educated on the acceptable use of the system and the events that they should be able to identify. The development of the technologies we use today has been rapid, with the result that for the majority of the users, the use of them is not intuitive. Great efforts have been made by software houses such Microsoft and Macintosh to hide from the users the workings of the system. This is done so that they are presented with an understandable representation of the functionality that they need, and not the sometimes-messy reality of what is actually happening, through the use of a graphic user interface (GUI). The GUI provides the user with a point-and-click environment and with little pictures (icons) to represent different programs or functions. Of the two major operating systems, UNIX/Linux and Microsoft Windows, the UNIX/Linux family is considered to be the operating system of choice for the techno-nerd because it is still very much a system that relies on the command-line user interface and requires the user to have a good knowledge of the operating system. Windows is the operating system of choice of the average user because it is designed for nontechnical people.

The security professional must be highly trained and experienced to cope with the information warfare challenges of the Twenty-first Century. One can no longer rely on IT people to be responsible for security; they lack the education and training required. Is it any wonder why our systems are so easily penetrated today? Two reasons: executive management takes unacceptable risks and security people are not up to the task. System administrators are concerned with throughput, and security puts a drag on that. Why should the system administrator be put in a position to have to choose? Even those who are usually are not given the authority equal to their responsibility? This must change and the change must begin, in part, with training these personnel as Twenty-first Century information warriors. It requires advanced degrees in computers, security, criminology and the associated experience of investigations, counterintelligence/counterespionage, anti-terrorism, and information warfare tactics — as a minimum. This type of professional rarely exists today in the private or government sector.

Procedural Security

Procedural security is the set of rules and practices developed and implemented to integrate the physical, personnel, and electronic security measures that have been chosen. An example would be that if you have a 12-foot fence, intruder detection sensors, and cameras to protect the exterior of the building, they are pointless if there is no one manning a monitoring station to take some action if an intrusion occurs.

Only when all of these elements can be used in harmony can any realistic and sustainable level of security be achieved. The concept of onion ring models can be described in many different forms, but almost all of them acknowledge that to be effective, the security of a system must be holistic and take the strengths of each of the individual elements and merge them into effective, multidisciplinary protection.

Cradle-to-Grave Security (Also Known as Life-Cycle Security)

> *Design it securely, build it securely, implement it securely, and maintain it securely.*

The protective measures applied to a system must be enforced and updated throughout its entire life. If the security you are using to protect your system is not developed as part of the overall system architecture, it will not be as cost-effective as would be possible and will probably not operate in the optimal manner. Security must also be considered during the design and development phase, as you will normally neither want your system architecture, configuration, or connectivity to be made public, nor will you want people who do not need to have it to have access to your system, as they may make changes to it before it is brought into use. During the phase in the life of a system when it is in operational use, then the security is most obviously seen as necessary and can be monitored. Finally, when a system is taken out of service and decommissioned for disposal, security must be considered for any parts of the system that may contain sensitive information or technology; for example, storage media for information that may still reside on it or hardware for cryptographic components. This may seem a moot point; but in the past there have been a number of examples of organizations disposing of obsolete systems and it subsequently being discovered that there was sensitive data that was recoverable.

An example of this was reported by the United Kingdom's *Daily Express*[1] under the headline "Secrets of McCartney Bank Cash Are Leaked." In this report it was stated that the Merchant Bankers Morgan Grenfell Asset Management (now renamed Deutsche Asset Management) had failed to make sure that the disk of an old computer that was being disposed of had been erased to an adequate level. The report went on to state that, in all, more than 100 files relating to the private financial transactions of Sir Paul McCartney had been recovered from the computer, together with the details of transactions of a number of other customers.

In many countries today, this type of security lapse will have a more profound effect than the embarrassment that the publicity of the loss will cause. If the people whose data is held on such systems are affected, then companies are likely to start to see litigation taking place for their failure to provide adequate protection for the information they have in their possession. A second line of action may come from the organizations that have been established within governments to ensure that the rights of the individual are cared for and that due care has been taken to protect information about individuals. In the United Kingdom, the legislation that has been enacted is the Data Protection Act, which was updated in the year 2000, and the person who has responsibility for implementing it is the Data Protection Registrar. In addition, from an information warfare perspective, the information in the hands of an adversary could be devastating.

In the case of the information from the Morgan Grenfell Asset Management computer, the Data Protection Registrar had commented that there did appear to have been a breach of security.

Some organizations have recognized the problem and, in producing their computer security policy, have incorporated the instructions for how to safely dispose of unwanted computers and the factors that need to be considered. One of the clearest sets of instructions that can be found comes from an academic institute, Oxford University in the United Kingdom. They have a clear "University Policy for Computer Disposal,"[2] in which, among a number of other issues that must be considered before a computer is disposed, are instructions that highlight the importance of data removal. It states that "An overriding consideration in any move of equipment must be to ensure that any University data on the machine, and any software licensed to the University, is removed. It is, of course, vital to satisfy the requirements of the Data Protection Act, but it must also be understood that any University data that is discovered by a later owner may cause controversy, adverse publicity, etc. Ensuring adequate destruction of data is the responsibility of the unit that owns the equipment, and must not be delegated to any person outside the University without adequate contractual obligations being imposed."

The instructions continue with details of how systems can be disposed of and the technical details of how to delete data. The question must be asked that if academic institutions can lead the way in the awareness of the requirements, not only for the protection of the privacy of the individual but also with the awareness that not to follow the procedures may cause adverse publicity, why cannot government departments and industrial organizations that have considerably more to lose in that area understand or address the problem?

Some of the other tools techniques, methods, and policies that will help in the protection of systems and the identification of attackers are described below.

Graceful Degradation

A consideration that applies to both computer security and business continuity is that of the graceful degradation of systems so that in the event of a loss of availability, the system will fail in such a manner as to ensure that the operating system and data are not corrupted. Another aspect might be to ensure that you continue to operate, in a restricted manner, for as long as possible before the final collapse of the system. One aspect of this is the use of uninterrupted power supplies (UPS) to ensure that, in the event of a power failure, the system can continue to operate for a limited period of time, but for at least as long as is necessary to sequentially shut down the system. Redundancy of capacity in either processing power, storage, or communications bandwidth might be used to extend the resistance of the system to failure.

If a system is under attack from a denial-of-service weapon, then surplus capacity in communications bandwidth can help in dealing with the problem, as can additional processing power. If the attack is a storage resource-consuming attack, then a surplus capacity will delay the final failure of the system. Individually, none of these measures will protect the system or save it from an attack; what they, together with other measures, can achieve is to extend the time from the start of the incident to the eventual failure of the system. This will allow, in the worst case, for the graceful shutdown of the system; or, in a more optimistic view, for those responsible for the system to deflect the attack or reduce the impact and allow the system to continue to operate, or it may even allow for the identification of the source of the attack. It will also, potentially, allow time for the collection of evidence and the gaining of information on the tools and techniques that the attacker is using.

Diversity

In the drive to make the systems that we use easier to manage and to reduce the cost of managing them, there has been a significant reduction in the diversity of equipment and operating systems that we use within a system. After all, why would you want to use two or more different operating systems with all of the overheads of two sets of training for the systems managers, looking after two sets of vulnerability patches and updates, and the problems of cross-platform working? The answer is that in some cases, if the system is critical to your organization, it may be in your best interests to swallow the cost because if you are attacked, it is possible that you will be able to keep on working through the attack because only one of the operating systems will be affected.

One of the simple and yet effective ways in which we can protect our systems and make them more resilient is to ensure that we have diversity within them. This would include diversity in the types of equipment used, so that more than one type of firewall or intrusion detection system (IDS) is used. Different types of anti-virus products and network and workstation operating systems will also help in preventing a complete system failure due to a single type of malicious software. For example, if you have a specific type of firewall in use throughout your organization and a vulnerability is discovered in it, the entire organization is at risk. If you are using a variety of firewall products, it is unlikely that a vulnerability will affect all of them, so you will have reduced the exposure to damage or loss and increased the chance that one of the unaffected firewalls will show abnormal activity. The same is true of anti-virus and IDS software. For the network and workstation operating systems, if a vulnerability is found in a version of UNIX, but you have diversity and are running some of your systems on Microsoft products, then you are likely to retain some functionality and potentially have a platform from which to recover.

As mentioned earlier, one of the problems with this approach is that there is a significant cost associated with this route. Not only is there the cost of purchasing such diverse products, but there will also be a cost to integrate them and to manage this complex network. It is, and must remain, a business decision that is derived from the risk assessment as to whether the cost of implementing this solution is justified and also sustainable.

Use the Delivered Software Features

It is possible to improve the security of the systems by implementing the security features that are delivered as a part of the system. Most operating systems are delivered with a great deal of security built into them but, as mentioned earlier, it is switched off for the benefit of the average user (home user), who neither knows of the features nor cares about them. As long as the Internet remains grossly insecure (and there is no prospect of it becoming more secure in the near future), we will continue to have significant problems. We have to protect our own assets and treat anything that we do not control as being untrusted, the equivalent of the "Wild West."

When a vulnerability in the software we are using is identified and a patch (a piece of software that will negate the vulnerability) is produced, if it is applied in the appropriate way, the system is protected (at least from that problem). Unfortunately, there are a number of complications here. The first is the time delay between the bad guys getting to know about the vulnerability and the producers of the software, or another helpful organization, producing the patch. In the past, this may have taken considerable time as manufacturers did not want to admit that the software that they produced was flawed and preferred to issue the fixes in the form of software updates. Fortunately, this practice is largely in the past, and even the largest corporation, Microsoft, is now very responsive to such problems.

The next issue is that one needs to find out about the problem; but unless you employ someone to monitor all of the potential sources of information for any reference to the software that you are using, how will you find out? Even major corporations find it difficult to justify the cost of employing staff for this type of work. Assuming that you have discovered a vulnerability and have found the patch to fix it, your next problem is that you have to apply the patch, which may mean shutting down your systems — and this will mean more working time lost. Next, you need to consider whether the patch will have any impact on the other software you are running; after all, it has been produced in response to a specific vulnerability and distributed as fast as it can be to stop the vulnerability from being exploited. What if the patch causes a conflict with the other software that you are running? Given the time scales with which we are dealing, it would be unrealistic to expect that the patch will have been tested in all possible system configurations enterprise wide.

The sad reality is that, even when the patches are available, organizations do not apply them to their systems. An example of this is the Solar Sunrise incident referred to previously. The patch to prevent this type of attack had been available for some time but had not been applied to the systems that were affected. If it had been, the attack would have failed.

Redundancy

The use of redundancy as a defensive measure may not seem obvious; however, it has, along with the use of diversity, great potential for improving the survivability of systems. A very simple aspect of having a spare workstation in an office is that if one fails, there is another available to replace it with a minimum of delay and effort. When you scale this to look at the wider aspects of a system, then the potential gains become more important. On a network scale, there should always be sufficient spare processing power and storage medium to allow for the worst-case expected loading that the business will require with, if possible, at least a 100 percent spare capacity. In the past, the cost of this would have been prohibitive, but the cost of processors, memory, and storage media has dropped to such a low point that it is totally achievable.

Other factors that should be considered are things like the power supply: do you have sufficient resilience of supply to meet the demand? We all look at the uninterrupted power supply (UPS) as the answer to the immediate problem, but this will only allow for the limited running of the system and a graceful shutdown. Has an alternative power source been investigated, and is it appropriate? This may take the form of a backup generator, an alternative supplier of mains power, or, in some cases, the organization's own power station.

The path that communications take is yet another area where redundancy should be considered. At the lowest level, this should be "call-off" capacity to allow for the expansion of the bandwidth from the local provider but if the system is of sufficient importance, then the possibility of taking bandwidth from two separate providers or across two or more different paths should be considered. This might mean using two or more providers (as long as you ensure that they bring the access into your organization through different points) or it may mean using two or more different media, such as cable and microwave to link into the public infrastructure.

Finally, on the subject of redundancy, it is worth remembering that the system must be protected against all types of threats, whether they are natural accidents, incompetence, or malicious activity. Do not forget that redundancy will also extend to facilities such as buildings. If a bomb goes off and makes your corporate headquarters unusable for a period of time and you cannot recreate a working environment at another location, then you will not be in business for very long. An example of this was the Canary Wharf bombing in London, which caused an estimated £300 million in damages. Probably more startling was that of the companies affected, it was estimated that 80 percent of those that were not operational within 24 hours of the event went into liquidation.

Remember also that a terrorist attack with a car bomb to take out your computer facility or a facility near you will not only likely cause window damage and loss of power, but may well register as a small earthquake on the Richter scale. Therefore, although you may not be located in an area prone to earthquakes, one should consider the effects of terrorist attacks that have the same force of power. It is then prudent to consider "earthquake-proofing" your systems (e.g., bolt racks and brace equipment). Also remember that any high-pressure water pipes in the ceiling above your systems can burst, thus causing water damage to your system. In addition, if you have those nice, rectangular false ceiling tiles in place and high-pressure water pipes above, the water from the burst pipes can turn those ceiling blocks into 250-pound bricks that can fall and damage equipment and even kill people.

This incident is not unique, as shown by a report from a German publication entitled "Why rebuild a phantom city?"[3] It gave details of other major incidents, along with costs and the actions that were subsequently taken to minimize repetitions:

> *When the level of terrorist attack leads to very heavy insurance claims, "area" measures are taken. This is what happened after the 1993 and 1994 one ton IRA bombs in the City of London, which inflicted £1.5 billion of damage, and the 1996 bomb at Canary Wharf which led to insurance claims of over £300 million. It also happened in the United States after the 1993 World Trade Centre, 1995 Oklahoma, and 1996*

Atlanta bombs. It happened after ETA attacks in Spain, after Tamil attacks in Sri Lanka, Palestinian attacks in Israel and Lebanon, and cult attacks on urban transportation systems in Japan.

It may be realistic for large, distributed organizations to maintain a backup site in a different part of the city or perhaps even farther away, but in most cases this is not a realistic option. The alternative is the type of facilities that are offered by business continuity specialists who maintain facilities that are tailored to the generic needs of a number of types of business sectors, with organizations "booking" space at them in case of disaster. This allows them to move to a working location that has the power, processing, storage, and communications power that they require for their individual businesses in a very short period of time. All that they have to do on arrival, in most cases, is to load the latest copy of their data and continue where they left off.

Security (Defensive IW) Policies and Procedures

Once you have carried out your risk assessment and have decided on the protective measures that need to be applied to the system to make it secure, you will need to develop a system security defensive IW policy to document the system and the security measures that are to be adopted. From a security policy it will be possible to develop operating procedures that will not only identify the roles that will need to be carried out to maintain the security of a system, but also the actions that will need to be taken in the event of an incident. Without a set of operating procedures that are in a form that can be understood by any user, the chances of staff reacting in an appropriate manner to an incident are very low. All of the users need to be able to read and understand the security operating procedures.

Despite years of knowledge and a widespread education effort, it is still true that in many organizations there is no security policy for the information systems. This does not just apply to small organizations, but also to major corporations and government bodies. If the organization does not have an effective and up-to-date security policy, then there is no structure on which to make decisions about the most effective way to integrate the security measures that can be taken. Perhaps the closest analogy to this is having good locks on the doors and windows but forgetting to secure the external access to the cellar because you did not think the problem all the way through. If, in an organization, you do not have the structure and commitment to document the way in which information security will be addressed, you will fail. Good intentions just are not good enough in the defensive IW field.

There are, however, things that need to be done before you start to produce your security policy. It is no good to start documenting how you are going to tackle your defensive IW issues if you have not carried out a risk assessment to identify what risks you are trying to protect yourself against. It may seem surprising, but that is exactly what many organizations actually do. Let me ask you this question. In your organization, has anyone ever attempted to determine the value of the data that is now, in all probability, the lifeblood of your organization?

A security policy or defensive IW policy is the basis for everyone's understanding of the way in which the organization will address the problem. It not only identifies the measures that must be taken, but will also detail what it is that is to be protected. From the security policy, you can derive operating instructions, which will be the document that will be read by staff. This document will tell them who is responsible for what on the system, what their duties are, and, most importantly, how to recognize an incident and what to do if one occurs.

Security or Defensive IW Policy Enforcement

When you have gone to the effort of producing a security or defensive IW policy for your organization, it is worthless unless you enforce it. This is like having road traffic laws but no law enforcement to ensure that they are obeyed. In many organizations, particularly in government, once the policy is produced, it sits on a shelf and gathers dust. The users carry on doing their own thing because nobody is tasked or willing to take on the unpopular task of telling people what they should or should not do on the system.

The Domain-Based Approach

When dealing with large systems or systems that are volatile, in that they constantly need to be reconfigured, there is a problem with approaching defensive IW or security in the conventional way. Under such circumstances, it is difficult to perceive or manage the whole of the system as one entity and it will certainly be extremely difficult to maintain the configuration of the system with any degree of confidence. The most common ways in which this difficulty is highlighted occur when connections to systems are discovered that were not thought to exist (a far more common event than you might believe), or when it is discovered that, to "do their job," a member of staff has negated all of the defensive measures that have been put in place by connecting a modem to the back of his workstation. It is notable that one of the organizations that conducts a large number of penetration tests reports that it has never carried out a test on a large corporation without discovering one of the two problems detailed above.

This problem is not confined to commercial organizations. All indications are that the problem is at least as bad, if not worse, in government departments, where people get fed up with the bureaucracy and delay in having official modifications to the system and make unauthorized connections out of desperation. In the United Kingdom, the Defense Evaluation and Research Agency, now known as QinetiQ, came up with a different way to conceptualize systems and developed the *domain-based approach* to system security.

Using this approach, the system is regarded in the light of "islands" of systems/users with a common need that are connected by causeways that provide the security and control of data flows. This allows the problems of large systems to be dealt with in manageable groupings. The concept was developed for the military to help designers and implementers understand the issues of a highly dynamic environment, but has since been found to have far wider applicability in the mobile and highly dynamic area of business in the modern world.

Management

The management of any system needs to be carried out in an efficient and effective manner. The same is true for the management of the security of a system. It is a little-realized fact that if you manage your system in an efficient and effective manner to meet the business or government agency needs, then the issues related to security will, in many cases, have been addressed in part, if not completely.

When systems that use conventional operating systems are installed, it is normal practice to install the components out-of-the-box and then configure them in a manner that will provide the best security that can be achieved. In most cases, the possibility of removing those elements of the operating system that will provide functionality that is not required in the configuration to be used is not considered. In simple terms, what this means is that a portion of the operating system that provides no benefit to the organization that owns and maintains the system remains in place, and any vulnerabilities that exist in the software could, potentially, be exploited by anyone who might gain access to the system. With a small amount of preplanning, this software could be removed from the installation before it is brought into operation. By the way, always, always get rid of vendor-provided default passwords, a common way for anyone, from an eight-year-old to an information warrior, to successfully attack systems.

Penetration Testing

This has a number of names, depending on the environment in which it is being conducted. In the U.S. military, it is known as Red Teaming or Vulnerability Analysis and Assistance Program (VAAP) testing. In the United Kingdom, for government, it is known as IT Security Health Checking (ITSHC). For the most part, the name by which it is known is irrelevant. What is actually taking place is the physical and logical testing of the configuration of a system to ensure that it is as secure as possible in its working environment.

Penetration testing can be partially automated, but this is not an excuse or a reason for "de-skilling" the process. It is an opportunity for the skilled exponent of the process to cover more tests in the same time period. Penetration testing is about using highly skilled staff to employ their knowledge and skills to make sure that any techniques that an attacker may use will be defeated. The range of tests that can be carried out on a system's configuration are many, and the skill in the task is to identify the tests that would be most appropriate for the environment and to apply a range of tests that will prove the validity of the security measures that have been taken. Penetration testing is a practical and time-efficient method of testing whether the security measures that have been taken are working and of identifying ways in which the measures taken could be improved.

Electronic Security

This is the next layer of security that is identified and this will encompass any of the electronic measures that are implemented on the system. This will include firewalls, sandboxes, virtual private networks (VPNs), public key infrastructure (PKI), intrusion detection systems (IDS), smart cards, biometric devices, cryptography, passwords, and access control devices, to name some of the more common product types. This grouping encompasses anything that is found in the electronic sphere that works toward improving the security of the system.

Routers

A router is an item of hardware or, in some cases, software that does just what its name implies: it routes data from one part of a network to another. Routers also act as traffic regulators, allowing only authorized systems to transmit data into the local network so that private information on the local network can remain secure. In addition, routers also handle errors, keep network usage statistics, and are now smart enough to handle a number of security issues. A router performs its function by creating and maintaining a table of the available routes and their conditions. The router uses this information, along with distance and cost algorithms, to determine the best route for a given packet. Typically, a packet can travel through a number of network points with routers before arriving at its destination.

Firewalls

Most people perceive a firewall as a part of the defense of a system that is designed to prevent unauthorized access to or from a network. In reality, it is the opposite; it is designed to allow authorized transactions to take place between a network and other networks, including the Internet. Firewalls can be either a hardware or a software device, or a combination of both. The different types of firewalls were described previously in this book. A firewall is not the universal panacea to all of a system's security requirements, nor is it a "fit-and-forget" device. It does, however, form an important element in the overall security of a system. By the way, while you use firewalls to secure your systems, do not forget to also secure your firewalls and other security software, firmware, and hardware.

A firewall is a system that, when properly configured and maintained, can prevent unauthorized access to an organization's network. A firewall can be located on the network's server that acts as its gateway to the Internet, or it can be a dedicated system placed between the network and the Internet so that the network is never in direct contact with the Internet. In large organizations, it is not unusual to use firewalls between different parts of the infrastructure of an organization to keep separate its different divisions or functional areas; for example, the marketing, finance, and research and development areas. All messages entering or leaving the protected network pass through the firewall, which examines each message and blocks those that do not meet the specified criteria for passing into or out of the network.

There are a number of different types of firewalls:

- *Packet filtering.* A packet filtering firewall looks at each packet entering or leaving the network and accepts or rejects it based on rules defined for the system. Packet filtering is reasonably effective and is normally transparent to users but can be difficult to configure.
- *Application gateway.* An application gateway applies security mechanisms to specific applications, such as Telnet and FTP. This type of firewall can be highly effective but may cause a degradation in performance.
- *Circuit-level gateway.* A circuit-level gateway applies security mechanisms when either a TCP or UDP connection is established. Once the connection has been established, packets are allowed to flow between the hosts without any further checking.
- *Proxy server.* A proxy server acts as a proxy for the network that it is protecting. It intercepts all messages entering and leaving the network. The proxy server effectively hides the true network addresses.

As with all security measures that you apply, it is sensible to use more than one type of firewall because reliance on any one manufacturer or methodology will leave you vulnerable in the event of a failure.

Personal Firewalls

The personal firewall is a recent innovation but there is already a wide range of software that is available free from the Internet for the home user. This has been a dramatic step forward in Internet security because a large portion of the personal computers in the world are privately owned and are used in home or small office/home office (SOHO) environments and do not enjoy the same protection that commercial systems are given. Examples of some well-known personal firewalls are BlackICE Defender, ConSeal PC Firewall, Internet Firewall 2000, and ZoneAlarm.[4] While the average user does not see a great need to protect the system with a firewall because they do not consider they have any information of value on their systems, this is gradually changing. The way in which people use their systems is developing and they are beginning to understand that any personal information they hold on their computer may be of value to someone who either wants to use their identity (identity theft) or wants to collect information about them.

When you consider that most people currently use their computers to type letters (with their name and address on them), keep their accounts (with their financial details), and make online purchases (with their credit card and account details), these aspects alone make the PC worth protecting. The other aspect that people are becoming more conscious of is that of someone taking over their computer for illicit purposes. The installation of even the most basic personal firewall will make the PC less attractive to the average attacker and they will probably go off and attack someone else's PC.

Audit Logs

Most operating systems are delivered with some type of auditing capability. The quality and usability of such a capability may be debatable, but one major factor that is beyond dispute is that auditing is of no use to anyone if it is switched off. And it is of little use even if it is switched on if it is not regularly monitored. It is of no comfort or benefit to know that a security incident took place two weeks ago!

Patching of Software

Part of the efficient security management of a system is the installation of patches to the operating system and applications. While this is a simple action to achieve, it is probably one of the most overlooked and ignored in the management of information systems. Whether you consider this subject to be a procedural or electronic event, it is addressed here for convenience. The Code Red Bug that attracted so much attention during the summer of 2001 would have been entirely avoidable if the patches that were available had been applied. The same is also true of the previously mentioned Solar Sunrise incident. According to a report from the System Administration, Networking, and Security (SANS) Institute,[5]

> *Most of the systems compromised in the Solar Sunrise Pentagon hacking incident were attacked through a single vulnerability. A related flaw was exploited to break into many of the computers later used in massive distributed denial-of-service attacks.*

While there are a large number of software vulnerabilities that are discovered every year, only a portion of them affect security; and of that portion that does affect security, very few are actually exploitable or exploited by information warriors or miscreants.

Forgotten Hardware, Firmware, and Software

When developing information warfare defenses for your information, networks, and other systems, do not forget to "harden" your other hardware and software systems that are often forgotten. These are the networks and systems that operate the physical security alarms, control physical access to facilities, and the fire control systems. Especially in today's environment, the threats of terrorists (e.g., eco-terrorists, religious fanatics, and theft of hardware and information through physical access) are also part of information warfare defenses.

The other "forgotten" software includes software used for maintenance and also the information systems security software. These products are crucial to information warfare defensive systems and must not be overlooked when validating and verifying that no malicious codes have been embedded in these products.

As with any other vendor-provided software, hardware, or firmware product, information warfare defenders must work with their contract office personnel to ensure that contracts for these products include a clause stating that the product is free of malicious code and viruses. And furthermore, that the vendor is liable for any damage caused by such embedded viruses and malicious codes. Although you will undoubtedly find that vendors will fight such clauses in purchasing agreements, this liability issue must be continually pursued. One wonders about the product being purchased and the vendor's quality control processes if they refuse to agree to such a clause.

Other options include the following:

- Examine the process used by vendors in their product development.
- Where is the code being written?
- What quality control processes, such as validation and verification, are in place?
- What security checks are made on the programmers?
- What security checks are made on the vendors?
- How is access controlled to their facilities?
- How is access controlled to their product development processes?

Sandboxes

It is becoming increasingly common practice these days to constrain the activities of applications to a "sandbox" to contain any unexpected outcomes from their operation to a virtual environment. The term "sandbox" is used similar to the way it was initially conceived — the children's sandbox, a place where they could play in relative safety. In the computer world, this concept has been retained and a sandbox is normally considered to be somewhere that you can play with code that you do not have confidence would not cause damage to your system if it were allowed to interact with it, or somewhere that you might ship code that you believed might be malicious.

Virtual Private Networks (VPNs)

The discussion of what constitutes a VPN is now a long-standing debate with one group, largely from the vendors, maintaining that any network service provided over a shared infrastructure while at the same time appearing to offer the users exclusivity is a VPN. The other main argument equates privacy with security, and defines a VPN as an encrypted pathway (tunnel) that has access control and host or user based authentication. Depending on the size of the organization, use of a VPN will either be from the organization's own facilities or will be purchased from a service provider. The way in which an organization will use a VPN will depend on its business needs. In some cases, a VPN will allow organizations that are spread over a wide geographical area to use the public infrastructure to connect elements of their internal corporate networks. In others, it will allow individuals who need remote access to the corporate systems to connect to the system while maintaining the integrity of the system boundary.

One of the downsides of using this type of technology is that, while it will, in most cases, prevent anyone from gaining access to your information, it does so by encrypting the traffic. In most cases, the firewall and software such as the anti-virus scanner and the intrusion detection software will be unable to read the encrypted traffic. This will expose the organization to a new set of problems, such as viruses propagating through the infrastructure and the loss of the ability to filter and monitor the information that is traveling across the VPN. One way in which these problems can be overcome is if rather than from the desktop to the desktop, the VPN exists only from the firewall of one system to the firewall of the other (this is effectively "point-to-point encryption" versus the "end-to-end encryption" as discussed in this chapter). Other solutions involve carrying out the checking that has traditionally been carried out at the firewall; at some other point in the network, the loading that will take place on processors and communications paths within the network would need to be carefully monitored to ensure that the loadings were balanced.

Public Key Infrastructure (PKI)

The widespread use of PKI systems is a goal that is being sought by a large number of organizations but, at the moment, still mainly by those that are trying to sell the concept. The potential power of a PKI system to provide a high level of confidence that a message that is sent from one person to another can be proved to have been sent, to have arrived, and to not have been altered or viewed in transit is highly desirable. Unfortunately, for any large-scale PKI to work, there is a requirement that at some level there will be trust (someone must fulfill the role of the certification authority), cooperation (the different vendors and user organizations must agree), and implementation standards.

If commerce is to be encouraged to adopt PKI systems, then government must enable their use, not hinder it by insisting on national schemes that allow them to retain "control." Multinational and transnational companies have the ability to move to areas that provide a beneficial environment for them to conduct their business. If governments try to use cyber-crime and cyber-terrorism as a reason for these companies to hand over access to their company's information to government agencies through any type of imposed key escrow or similar schemes, then it is likely that those companies will perceive the environment to be less conducive to them conducting their legitimate business and move elsewhere. Both in the United States with the "Clipper Chip" fiasco and in the United Kingdom with the E-commerce "Key Escrow" debacle, exactly these arguments have been used. In both cases they have failed, partially because of pressure from the commercial sector and partially because the initial arguments were weak. Criminals who use cryptography to disguise their activities are not likely to supply a copy of their key to a third party that has been nominated and authorized by government. One of the closest comparisons that can be drawn is perhaps in the United Kingdom where, as a result of a number of shooting incidents, guns have effectively been removed from private ownership. The net result: the law-abiding user is penalized while the criminals, who were not well-known for registering their weapons, have not been significantly inconvenienced.

Bruce Schneier and Carl Ellison raised some of the issues relating to PKI in an article entitled "Ten Risks of PKI: What You're not Being Told about Public Key Infrastructure,"[6] which highlights a number of concerns that have yet to be addressed to the satisfaction of the community. There is no doubt that within an enclosed group that falls under the control of one overarching authority, if the level of assurance that can be provided by a PKI system is required, then it can be managed at a realistic cost. In this environment, the organization can be confident that the system will provide "what it says on the box." It is only when the use of the system crosses the boundaries of organizations or trust groups that the difficulties start to occur. If there is more than one certification authority, how do you, in your organization, know that the other Certification Authority (CA) is applying the same stringent tests on the validation of certificates that your CA is applying? How can this be tested and proved? Does a value that is implied on your certificate mean the same on the certificate issued by the other CA? Who will accept liability for any losses that result from failures in the system?

Encryption

Encryption can be used in two ways that are similar but achieve very different effects. The encryption algorithms that are used may be very similar, but the way in which they are implemented is the primary difference. What is achieved is either "end-to-end" encryption, where the communications path is encrypted from the sender of the message to the recipient of the message or "point to point encryption," where the traffic is decrypted at the intervening nodes. The advantage of the former method is that only the sender and receiver of the message have access to the plaintext message. The disadvantage of this method is that the users (both the senders and the recipients) must each be responsible for their own cryptographic keys.

With the introduction of an asynchronous cryptographic system such as PGP, this has become less of a burden; but in reality, it is still beyond the average user's capability or ambition. The other option of point-to-point encryption means that the message is encrypted in transit between systems, but will not normally be encrypted within the system. The disadvantage of this is that anyone who has access to the system can potentially gain access to the message, but the advantage is that the cryptographic key management is restricted to the system staff who will manage the keys for all of the users of their system.

When these techniques are combined, by using both systems together, an incredibly strong encryption system is generated.

Anti-Virus Measures

We hear a lot about viruses and the devastating effect they can have. You only need to look at the Love Bug, the Anna Kournikova,[7] and homepage viruses, and the speed with which they spread, to understand the potential damage that can be caused. The cost of cleaning up these virus infestations is immense. An example of the costs of virus infections can be found in an article in Computer Economics[8] that reported that the financial impact of virus attacks on information systems around the world amounted to $12.1 billion in 1999, $17.1 billion in 2000 and $13.2 billion in 2001. Another example can be found in a report from the insurers Lloyds of London,[9] who estimated that the cost of the Love Bug virus could run into tens of millions of pounds in the United Kingdom alone. Yet another example is the FunLove virus,[10] that infected Windows 95, 98, and NT systems. This virus spread across the globe within two weeks of initial discovery. According to an article in *The Irish Times* magazine, work at Dell's factory in Ireland had to be suspended for a period of at least two working days and approximately 12,000 computers had to be recalled for checking. The estimated cost of the disruption to Dell could be as much as U.S.$22 million. In addition to the direct losses, some of the PCs that had been infected by the virus had reached customers before the virus was detected, causing further damage.

It would almost be easy to not include this type of measure because we have gotten, over the past 12 years or so, very used to the ever-present threat of viruses. The anti-virus measures that are available are effective and are largely keeping pace with the problem. It is debatable whether viruses and anti-virus measures should be discussed in the present environment, or whether we should in fact talk about the wider issue of malicious code and the measures that we take against them. A long time ago, there were well-defined groups of viruses, worms, Trojan horses, and other malicious code, and each could be and was dealt with pretty much in isolation.

It is not possible to do so today, as what we refer to and deal with as a "virus" is actually a mixture of a virus and malicious code. For many, these are the information warfare weapons that are launched against government and private computer systems. For example, the Melissa virus, which was first seen in early 1999, was a "macro virus" that propagated through the medium of e-mail attachments. When a user opened an infected Microsoft Word 97 or

Word 2000 document, the macro virus was immediately activated. When activated, the virus lowers the macro security settings to permit all macros to run in the future when documents are opened. As a result of this, the user is not notified when the virus is executed at some future time.

The next action that the virus takes is to check the registry of the system to see if the key "HKEY_Current_User\Software\Microsoft\Office\Melissa?" has a value of "...by Kwyj"bo." If the registry key does not exist or does not have that value, the virus attempts to propagate itself by sending an e-mail message to the first 50 entries in every Microsoft Outlook address book that is readable to the user executing the macro. To make matters worse, if any of the addresses is a mailing list, the number of people that the message will be sent to will increase significantly. However, if the system does not have Microsoft Outlook installed, the system will still become infected but the virus will not be able to propagate itself. The virus next infects the Normal.dot template file. By default, all Word documents will use this template and, as a result, any new Word document will also be infected.

The point here is that while the macro virus is achieving its aim, which is to replicate, it uses techniques to achieve this that are sophisticated and complex and that go far beyond what is normally considered to be a virus.

The anti-virus industry is mature and there are a number of software manufacturers that provide an excellent service in maintaining software that will deal with all of the currently known viruses and worms. This is achieved by the distribution of regular updates of libraries of signatures for the malicious code. As new viruses occur and their signatures are isolated by the anti-virus companies, the signatures are distributed to their client base and the incidence of viruses is controlled. There will probably never be a point when viruses will be eradicated because the intellectual challenge of developing a new strain is high.

Another way of "inoculating" one's systems against a particular virus strain is to modify it and send it back into cyberspace. For example, one can take the Code Red Bug, look at its source code to see how it works, and then change the code so that when it gets into a system or network it actually inoculates the system against the actual Code Red Bug. An obvious note of caution: be sure you know what you are doing.

Such an approach was taken and the information warriors who had developed that inoculation contacted members of several U.S. government "three-letter" agencies in an effort to work with them to do just that. All of them were interested but either said that it was another agency's responsibility or declined to do it — but would not mind if the developers did it themselves. Any wonder why U.S. systems are still being successfully attacked when you have bureaucrats in charge of systems' protection instead of information warriors?

How do we improve matters? We can start by buying and using the anti-virus software that is available. The general advice for a corporation is that you should use at least two different products from different producers; in this way you stand the best chance of intercepting a known virus. For the home user, there are plenty of free anti-virus downloads that will give you some protection or, if you can afford it and your information is valuable enough, you could actually buy one. The second part of this is educating the users. In the working environment, who is actually vain enough to open a message with "I Love You" in the subject line? Who thinks that it is appropriate and sensible to receive and open an e-mail during work hours that says it has a photo of Anna Kournikova? If you educate users as to what is acceptable usage and give them some awareness of the potential problems, you will do a great deal to reduce the impact of such an attack.

Biometrics

Biometric devices have started to play an increasingly important role in the armory of tools that can be used to defend systems. The term "biometric" is largely defined as a measurable, robust, distinctive, physical characteristic, or personal trait of an individual, that can be used to identify or verify the claimed identity of that individual. The accepted definition of biometric has further been developed so as to assume that the biometric process is an automated one. A biometric device has two main functions:

- *To prove that the user is who he says he is.* This is achieved by relating the person that is presenting himself to the system with an identity previously registered on the system.
- *To prove that a user is not who he says that he is not.* This is the most common use of biometric systems. The purpose of this type of biometric system is to prevent a single person from using multiple identities. If it fails to find a match between the presented sample and all of the enrolled templates, then it will allow the user to operate on the system. A match between the sample and one of the enrolled templates results in a "rejection."

The range of biometric devices and types is growing rapidly. There are two fundamentally different types of devices:

- Physiological methods, which include fingerprint, hand geometry, iris scanning, retinal scanning, facial recognition, and voice verification
- Vein pattern scanning and behavioral methods, which include signature verification and keystroke dynamics

Again, on their own, they are of limited value; but when incorporated with other devices and technologies such as smart cards, their potential as security enhancing devices has yet to be fully realized.

Information Warfare Software and Hardware Development

Not talked about by government suppliers or government agencies involved in information warfare activities is the development of specific defensive (IW-D) and offensive (IW-O) information warfare software, hardware, and firmware. Although COTS makes up the majority of the IW-D programs at this time, others may be in development under national security classified contracts. It is hoped that if so, such products are also free of malicious codes and that they will eventually be made available for commercial use by corporations and individuals.

One does wonder, however, if the government agency customer is verifying and validating that such products are free of malicious code? If not, such products will not be much good in time of an information warfare attack by an adversary. Are foreign companies who are the low bidder for the contract developing such programs or subsets? Is any customer representative checking to find out?

As part of IW-D, there are many government agency suppliers working under contract to develop IW-D visual aids. Also, some are using artificial intelligence and neural networks to analyze attacks to predict when they may next occur, by whom, and from where. Such developments will eventually lead to products that find their way into the commercial market. However, it is expected that the trend will continue — IW-D will always be behind IW-O.

IW-O hardware, firmware, and software development is indeed classified at the highest levels of national security (e.g., Top Secret and special compartmented information). These weapons are treated much in the same manner as access to classified nuclear-related information. This is logical because, for an information-based, information-dependent, and information-driven nation-state, access to such information warfare weapons by an adversary can have the same effect as the release of a nuclear weapon on a modern nation-state.

Areas of Technology that Must Be Addressed

While we are already using the technologies and concepts that are detailed above to help in the defense of systems, it is possible that in the future there may be no option but to continue to use them or other technologies. The following paragraphs look at just a few of the issues that currently need to be addressed if the infrastructure is to be protected in a coherent manner.

Wireless Local Area Networks (LANs)

Wireless LANs are a fairly recent innovation in the general marketplace, but are already being widely adopted for a number of reasons. The primary reason for their deployment is the low cost of adoption. If you are extending a network or refurbishing an office block, the difference in cost in rewiring compared to the installation of a wireless infrastructure is huge. Another reason is flexibility. In the military, the wireless LAN is an attractive option because it allows a system that is highly dynamic and that, in tactical situations, needs to be dismantled and rebuilt every few hours to be easily moved. There are, however, issues that need to be addressed if this technology is to be used beneficially as part of the overall defensive means adopted for the system.

Because the use of wireless LANs has become relatively common, it has become very noticeable that you can get in a car and drive through any industrial area and acquire a large number of wireless networks that are deployed. This is because the owners of the system are not taking even the most basic of countermeasures. They are not tying access to the system to known machines or network cards, which can be uniquely identified; they are not encrypting the network traffic; and they are not monitoring their system to see what systems are active as a part of the system.

This technology has the potential to provide good access control and to give flexibility and mobility in a safe and secure manner — if it is properly implemented.

Digital Subscriber Line (DSL)

DSL technology uses the existing copper pair wiring that runs to almost every home and office. To maximize the capability of the twisted copper pair, special hardware is attached to both the user and switch ends of line that allows data transmission over the wires at far greater speed than standard phone wiring. DSL equipment gives a potential for 2 Mbps from the switch to the user and 56 Kbps from the user to the switch. In addition, it is a constant and permanent connection, so you have access to the Internet and to e-mail 24 hours a day, seven days a week.

There is no need to dial in to your ISP each time you want to get online. This unfortunately also has a downside. Because people are leaving their systems permanently switched on and because they are, for the most part, not secured effectively, they provide the perfect playground for attackers. In simple terms, there are now thousands of systems online that are not securely configured, with no one attending them; thus, attackers can easily take over the system and then use it as a base to launch attacks on other systems. In particular, the personal systems connected via DSL can be used to launch distributed denial-of-service attacks. Such systems provide a huge unprotected base that can be used by an individual or small group to swamp a targeted site. If the environment in which we need to operate is to be made more secure, then this issue will need to be addressed.

Electromagnetic Pulse (EMP) and High-Energy Radio Frequency (HERF)

Protection against attacks with weapons that utilize electromagnetic pulse (EMP) technologies have not thus far been given much attention outside of military. An initial flurry of concern was brought about by articles in the press in 1996[11] that stated, "Several financial institutions, such as banks, brokerage firms, and other large corporations have paid extortion money to sophisticated international 'cyber-terrorists.' Huge sums of money have reportedly been paid to these criminals who have threatened to destroy computer systems and... have proven that they can do it."

The *London Sunday Times,* in a report on June 2, 1996, ran an article under the headline of "City Surrenders to £400m gangs." In this article it was reported that financial institutions in the City of London had paid huge sums to gangs of cyber-terrorists who had, supposedly, gained in the region of £400m from financial institutions around the world by threatening to destroy their information or even the computer systems themselves. The article went on to report that financial institutions in America had also acceded to the blackmail attempts in an effort to avoid the denial-of-service that this would cause and the subsequent loss of confidence from their customers and other institutions.

A second report in the *London Sunday Times:*[12]

> *In 1994, a consultant working for a company which undertakes com-
> puter risk assessments for City institutions compiled a table of 46 attacks
> on banks and finance houses in New York, London, and other centres,
> starting in January 1993.*

This *Sunday Times* report also indicated that the methods that the attackers had supposedly used included "logic bombs" and electromagnetic pulse (EMP) and high-energy radio frequency (HERF) guns. The report went on to quote a spokesman for the Bank of England who confirmed that his bank had come under attack by the cyber-terrorists. At the time, it was speculated that the attacks were the result of HERF weapons being fired to black out trading positions in city finance houses. The weapon disables a computer by firing electromagnetic radiation at it.

In a separate report from around the same time period, entitled "The Cyber Terrorists…,"[11] the author stated that the American National Security Agency was taking a "very hard look at" four cases, which were detailed as:

- *January 6, 1993.* A computer crash halted trading at a British brokerage house. A £10 million pound ransom was paid into a Zurich, Switzerland, bank account.
- *January 14, 1993.* A British bank paid a ransom of £12.5 million pounds.
- *January 29, 1993.* £10 million pounds was paid by a British brokerage house in ransom after threats were made.
- *March 17, 1995.* A British defense firm paid a ransom of £10 million pounds.

In each of these cases, it was reported that senior executives in the organizations had been threatened by the blackmailers. They had received a demonstration that their systems could be affected; and in every case, the company is reported to have paid the blackmailers within hours of the demand being made. A separate report indicated that the executives had received messages on their systems from the blackmailers indicating that they could gain control of the systems.

While this activity in June of 1996 caused a great deal of consternation, and caused Winn Schwartau, one of the first people to publicize this type of weapon to answer criticism for the "hype," there has been no substantiated evidence of the presence of this type of capability outside of the military and there has been no evidence of the use of this type of weapon in the commercial environment. However, it should also be noted that Sweden appears to be very active in researching these potential information warfare weapons and especially their impact on systems and how to defend systems against them.

Countermeasures

Having identified the issues above, what countermeasures can be adopted to defend the systems? There are a wide range of potential actions that could be taken but their use will be dictated by the level of commitment and funding that is available and the laws that exist in the country in which the system is located. Detailed below are some of the countermeasures that can be employed, although this will change with time and the situation.

Intrusion Detection Systems (IDSs)

Intrusion detection systems (IDSs) are the new generation of system protection that has been developed. In reality, there is nothing particularly new about the concepts of logging activity on a system to identify what is occurring.

In a perfect world, an IDS would be able to detect an intrusion in real-time; that is to say, as it occurs. This would enable the person monitoring the system to react and set in motion such processes as are required.

It is around this point that one of the major disadvantages of the way in which the technology has developed to date becomes apparent. By relying on an operator to intervene in a process, there is delay induced in two ways:

- The time it takes an operator to notice that an event is occurring
- The time it takes the operator to initiate some form of remedial action

To identify an "in-progress" event and to have any chance of identifying the perpetrator of the event will depend on many factors, including the attacker's skill, the logging mechanisms that are in place and the events that they have been set up to capture, the placement of the sensors on the system, the skill of the operator, the location of the attacker in relation to the victim (are they in the same country?), and several other variables. This does not make it impossible to catch the attacker, but it does make it increasingly difficult.

We have in the past relied, for the most part, on the post-event detection of an incident. This may have been the review of the security log the following morning or a periodic review of the configuration of the system that detects that changes have been made. This will not be good enough in the future. It is doubtful whether even the intervention of an operator will be acceptable in future systems if we are to have any chance of remaining secure and attackers are to be caught.

At the moment, there are two primary types of IDS: network-based and host-based sensors. Both have benefits and disadvantages, but a combination of the two, well located on the network, can give a very good view of what activity is taking place on the system. Most of the IDS systems in use work by using a set of identified attack signatures in a manner similar to anti-virus software. This means that for them to detect an attack, it must have been seen before and some unique characteristics isolated. Another family of IDSs that is developing is the intelligent knowledge-based system (IKBS), which looks at the network over a period of time and identifies what is normal behavior. Once it has learned this, it can identify behavior that is anomalous. While this group of IDSs has huge potential, they have not been deployed effectively on any large scale to date.

What it is important to remember is that an IDS will do what it says — it will detect intruders or, in the case of the anomaly detection system, it will tell you that something that it does not understand is happening on the system. What it will not currently do is react to the event. Once an event has been identified, then the human operator must, with the technology that is currently available, decide on what actions to take. This may include changing the configuration of the system to minimize or prevent damage, initiating action to identify or track an attacker, or gathering forensically sound evidence. This leaves a window of opportunity for any automated attack that is large enough to allow an attack that could cause significant damage to a system. In the future, the systems will have to be more highly integrated so that once the system identifies that it is under an attack, it will cause the system to reconfigure to a state in which the attack is ineffective.

This has a great deal of potential, but will, in itself, have issues that need to be resolved. How many times can a system be reconfigured without human intervention? With all that has been previously discussed in this chapter about risk assessments, creating policy, and operating procedures, how can a self-modifying system ever be adequately controlled to give system owners confidence that the system has remained in a safe state. How would this be provable? How long would it be before a determined attacker managed to find a way to subvert such a system? While all these questions need to be answered, there is no doubt that any reduction in the time between the identification of an attack and the reconfiguration of the system to negate it must be welcomed.

Computer Forensics

The continued development of computer forensic techniques that can be used to investigate activity on a network or workstation, or to capture evidence after an attack has occurred, will be essential in keeping up with the developments in technology and the creation of new techniques. The problems that must be addressed by computer forensic techniques range from the capture and preservation of evidence on a stand-alone PC, through to the capture of information from a wide area network (WAN) that spans continents. The PC is normally the simple end of the problem, with the preservation of evidence on WANs still being almost impossible with the current state of the technologies.

In the past, computer evidence meant the print-out of logs or files from the computer and, even today, the majority of the evidence that is presented to the court is just that. One of the most important issues in the collection of forensic evidence is that it must be provable that the evidence presented in court has been handled properly at all times and that any changes that have been made to the system have been accounted for. It is normal to make "images" of the original evidence so that one of the images can be retained and another can be examined, and apart from the initial imaging, the original evidence is not touched.

Computer evidence is like any other evidence that is to be presented in court; it must be:

- Complete
- Accurate
- Admissible
- Authentic

Unfortunately, the collection of evidence from computers has some specific problems that must be addressed. Unlike most evidence that is placed before a judge or a jury, computer evidence cannot be seen and can only be observed when a computer has processed it. Computer evidence is transient and can change from nanosecond to nanosecond, and the very act of collecting computer evidence may change the very evidence itself. Finally, most people do not have a deep enough knowledge of the way computers work to be able to make a judgment as to the significance of the evidence being presented to them.

This last comment is also, unfortunately, true for the majority of the judiciary, so getting computer evidence admitted to and understood by a court remains difficult, although the situation is improving.

Honey Pots and Honey Nets

The honey pot is a relatively new measure that has been brought into use to enhance the security of systems and to deflect an attack from a system. A *honey pot* is a system that is intended to provide a more attractive target than the real system and is designed in such a way that once an attacker has entered the honey pot, he is not liable to move on to the real system. A *honey net* is a larger-scale implementation of the honey pot and is part of a complete network. There has been a lot written about the potential for honey pots during the past two or three years but there are factors that must be taken into consideration when considering their use, including:

- *What makes a honey pot so attractive to an attacker?* If you make access to the system too easy, then any competent attacker will smell a rat and become very difficult to convince.
- *What do you put into this site that will be convincing to an attacker, but which will not be obvious to them?* The material must be close enough to

the real thing to make the attacker believe that he is in the correct site, but at the same time, must be far enough from the reality that you do not give away more information than intended.

While there is no sympathy for any person who is attacking a system, there are legal and moral issues that must be considered when using a honey pot. Is it, or might it, constitute entrapment? Is it right to take the "law" into your own hands and trap the aggressor? Another issue that should be considered is that in creating a honey pot, you may well attract attackers to your environment. What will be the reaction of an aggressor when he discovers that he has been misled and that you have been wasting his time and feeding him less-than-accurate information?

The SANS Institute describes honey pots as "decoy servers or systems set up to gather information regarding an attacker or intruder into your system."

In one publicized case of the results of a honey pot,[14] a group of suspected Pakistani hackers thought that they had broken into a U.S.-based computer system in June 2000 and thought they had found a vulnerable network to use to mount Web attacks on Indian sites. What they had actually done was to enter a honey pot and all of the actions they took were recorded and examined. The aim of the honey pot was not only to protect real systems from damage, but also to learn the hackers' techniques and identities.

Honey pots are not a replacement for other, more traditional Internet security systems, but they do provide another avenue for protecting the system. A honey pot can be created anywhere in the system. This can mean that it is created inside, outside, or in the demilitarized zone (DMZ) of a firewall design.

Counterstrike

The whole concept of counterstrike is still in its infancy as governments and the military try to determine what can be considered a balanced response to a cyber-attack. There is currently no explicit national or international law to deal with the situation, and a number of nations have expressed disparate views as to how they would regard an attack via computers. Walter Gary Sharp, Jr., has written a seminal work on this subject entitled "Cyber Space and the Use of Force." Russia has made clear statements on the significance it would attach to such an attack, which is probably best reflected in the statement from Vladimir Markonenko, First Deputy General-Director, Federal Agency for Government Communications & Information (*Federal'naya Agenstvo Pravitel'stvennoy Svayazi i Informatsii* (FAPSI)), who stated that, "...the danger of an information war breaking out is coming to the fore, and information warfare will soon rank second only to thermo-nuclear war by its consequences."

A second but very similar comment from Russia was made by Professor Tsymbal,[15] who stated: "From a military point of view, the use of information warfare means against Russia or its armed forces will categorically not be considered a non-military phase of a conflict whether there were casualties or not." He also went on to explain that, "...Russia retains the right to use nuclear weapons first against the means and forces of information warfare, and then against the aggressor state itself." This is a very powerful statement and reflects the concern within the Russian military as to the possible effects of a cyber-attack.

In addition, there has been to date no case law that could be used to determine an acceptable response to an attack. If an individual living in a country that is at war with your country launches a denial-of-service attack against a significant element of your critical national infrastructure, does this represent an act of war? Was the individual acting on his own or was he acting under orders from an organization in the hostile country? Does it make any difference?

What constitutes an adequate response? In physical terms, if an individual from a country that you were at war with was caught trying to damage or destroy part of your infrastructure, you would consider this an enemy action and would take whatever steps were required to capture the individual and prevent him from repeating this type of action, including killing him. Is the same legitimate in cyberspace? If it is possible to identify the cyber-attacker, is it acceptable to kill him? What constitutes reasonable action to prevent him from repeating the attacks? Does the cyber-attacker have any rights under the Geneva or other conventions? In answering these questions, it is easy to be cavalier and dismiss any possible defense that the attacker might have, but it is also worth remembering that, to date, more American citizens have been caught in this type of activity than the people of any other country.

It will be a considerable time before this subject is fully tested in action and in any subsequent court or war crimes tribunal. Why war crimes tribunals? Well, if, during a conflict, the enemy information warrior has infiltrated a medical facility and changed the records of an individual or for blood supplies, this would be a war crime in the conventional arena and thus would probably also be considered to be the same in cyberspace.

Law Enforcement

The role of law enforcement agencies as part of a counterstrike would have been a laughable concept even five years ago; but today and in the future, there is no doubt that they will increasingly have a significant role to play. A large number of countries now have dedicated computer crime organizations and, through a variety of organizations and forums, have started to become effective internationally. It is only with the assistance and collaboration of the law enforcement community in all the countries that have been used by the attacker that any possibility of an arrest and conviction is going to be possible.

Some good examples of this type of collaboration that have already occurred include the arrest and trial of Datastream Cowboy, one of two British hackers who attacked a large number of American military systems in 1994. It was only because of collaboration between American law enforcement and the British police that this arrest was possible. Another good example is that of the creator of the I Love You virus, Onel de Guzman, in which the technical skill and assistance of the U.S. FBI were, apparently, instrumental in identifying the culprit.

Emergency Response Teams

Computer Emergency Response Teams (CERTs) are the fire brigade of the IT world. When an incident has occurred, members of a CERT will assist an organization in determining what happened and what action needs to be taken to remedy the situation. CERT team members will normally be highly knowledgeable in all aspects of software and will be able to understand what has caused a problem on a system. They will provide expert advice to all parties that have become involved in an incident. A good example of a CERT is the Carnegie Mellon University CERT Coordination Center (CERT/CC),[16] which describes itself as an organization that studies "Internet security vulnerabilities, handle[s] computer security incidents, publish[es] security alerts, research[es] long-term changes in networked systems, and develop[s] information and training to help you improve security at your site."

Another good example of a CERT is the Australian Computer Emergency Response Team (AusCERT).[17] The AusCERT claims to provide a single trusted point of contact in Australia for the Internet community to deal with on issues relating to computer security incidents and their prevention. AusCERT's stated aims are to reduce the probability of successful attack, to reduce the direct costs of security to organizations, and lower the risk of consequential damage.

CERTs are normally government or academia sponsored and work together, both nationally and internationally, to improve the computer security situation. In a number of cases they can achieve far more in the short term than law enforcement agencies because they are not acting with the same motives or with the same authority and constraints as law enforcement.

Cyber-Vigilantes

Increasingly there is a call, particularly from the United States, to justify the actions of and to allow cyber-vigilantes to retaliate to incidents on the Internet. There is already some evidence of this type of activity taking place in response to perceived, but usually unvalidated, events. In 1999, Winn Schwartau published a set of "Guidelines for would-be corporate vigilantes,"[18] in which he provides a very logical set of steps to take in the event of an incident, until he advocates "strike back if you choose, but only with adequate legal counsel. There is a range of actions you can take — some more offensive than others." It may be difficult to believe that any legal counsel would advocate a strike-back — the potential for being able to gather adequate supporting evidence to support the strike back that would be a defense against future litigation is almost zero.

The potential danger associated with cyber-vigilantes is huge. A look at recent high-profile cases in the United States is enough to sound a cautionary bell against any thought of adopting this course of action. The first case that is worth looking at is that of Richard Pryce, the British teenager who, along with another British hacker, Matthew Bevan, hacked into a large number of systems around the world. The time taken and the level of resources from around the world that were required to track down these miscreants were immense. Along the way in the investigation were indications that this attack was emanating from Colombia (Bogota) and from Latvia. A cyber-vigilante might have felt that this was sufficient information for a strike-back against those sites and the consequences could have been disastrous.

A second example is that of Solar Sunrise, the attack on U.S. Air Force systems during a military buildup related to continued Iraqi infractions. Initially, given the timing and the target of the attacks, it might have been reasonable to assume that the attack was from Iraq; but as the investigation progressed, there were indications that it might be coming from China, until it was found to be two kids in California and one in Israel. Imagine the consequences of a vigilante strike on either Iraq or China!

There has already been one case of a response that was taken by the U.S. military against a group that was conducting a denial-of-service attack against them. In this attack, the Pentagon came under attack from the Electronic Disturbance Theater (EDT) on September 9, 1998. The EDT launched Floodnet, a denial-of-service program, against a Pentagon Web site. The attack was made to draw attention to the Zapatista rebels of Mexico and was launched against the Pentagon because the United States supported the Mexican government in opposing the Zapatistas. A Pentagon source revealed that the potential attack was known about prior to its launch and, as a result, the Pentagon was able to prepare for it. The response, which was managed by the Defense Information Systems Agency (DISA), was to mirror the denial-of-service attack back at the attackers. The result was that the systems used by the individuals launching the attack on the Pentagon froze and required re-booting before they could be used again. It is interesting that the people involved in the attack on the Pentagon now claim to be considering taking legal action against the U.S. government.

The procedure employed by the Pentagon does raise a number of questions regarding whether the U.S. government and the military have the moral or legal right to launch cyber-attacks within the United States. An 1878 law known as Posse Comitatus bans the use of the military for domestic law enforcement. It is one of the areas in which the applicability of the law to events in cyberspace will continue to be the subject of debate. In this particular case, it is debatable whether this was a cyber-attack, or a straightforward defensive maneuver and it was taken against a source outside the United States. The only systems that were adversely affected were those that were actively attacking the Pentagon's systems.

Summary

This chapter has looked at, on a very superficial level, a wide range of security considerations, defenses, and countermeasures that can be employed. When used sensibly and in conjunction with each other, they can be used to reduce the likelihood of a successful attack against you. In the event of a successful attack, they can minimize the damage caused and provide information with regard to who was conducting the attack and the tools and techniques that were used. In this chapter, the resources that can be called on to respond to an attack were reviewed; and while it is accepted that unless the attack that is occurring is a nationally significant event, the possibility of successfully having the attacker brought to justice is low if the attacker is not operating from within the same country as the target.

As law enforcement and the computer community in general become more computer literate, and as the impact of the damage that can be caused becomes better understood by governments, the situation is improving. Within the next few years, we should see the introduction of up-to-date international laws and treaties that will improve the cross-border situation, while at the same time protecting the rights of the individual.

A key point to keep in mind: the defenses, countermeasures, and counterstrikes should be incorporated into a holistic security program and not implemented in a piecemeal fashion. Furthermore, such a program must always be maintained with the latest security upgrades and patches. The proper implementation of a defense-in-depth program will minimize the risk of successful attacks by nation-states' information warriors, terrorists, hackers, and other miscreants.

Notes

1. Secrets of McCartney Bank Cash Are Leaked, by Jonathan Calvert and Peter Warren, *Daily Express,* February 9, 2000.
2. Oxford University Policy for Computer Disposal, Alan Gay, Deputy Director, OUCS, last updated November 27, 2000.
3. http://www.heise.de/tp/english/inhalt/co/2209/2.html.
4. http://www.firewallguide.com/.
5. SANS Institute. How To Eliminate The Ten Most Critical Internet Security Threats. The Experts' Consensus, Version 1.33, June 25, 2001.
6. Ten Risks of PKI: What You're not Being Told about Public Key Infrastructure, Carl Ellison and Bruce Schneier.
7. http://www.antivirus.about.com/.
8. 2001 Economic Impact of Malicious Code Attacks, Updated: January 2, 2002; http://www.computereconomics.com/cei/press/pr92101.html.
9. Article titled "Love Bug Bites U.K. hard," BBC News Online, May 4, 2000.
10. Data Fellows Corporation.
11. "The Cyber Terrorists...," excerpted from EmergencyNet NEWS Service, Tuesday, June 4, 1996, Vol. 2-156, Steve Macko, ENN Editor.
12. *London Sunday Times,* item titled "Secret DTI Inquiry into Cyber Terror," dated June 9, 2001.

13. The Complete, Unofficial TEMPEST Information Page, http://www.eskimo.com/~joelm/tempest.html.

14. Hackers caught in security honeypot, by Keith Johnson, December 19, 2000, http://www.zdnet.com/zdnn/stories/news/0,4586,2666273,00.html.

15. Professor V. I Tsymbal, paper titled "Concept of Information War." 1995.

16. Carnegie Mellon University CERT/CC, www.cert.org/.

17. Australian Computer Emergency Response Team, http://www.auscert.org.au/.

18. Guidelines for Would-Be Corporate Vigilantes, Netword Fusion News. Jan 1, 1999. http://www.nwfusion.com/news/0111vigitips.html.

Chapter 16

Defending against Information Warfare Attacks Begins with Knowledge Management

---

Some day on the corporate balance sheet there will be an entry which reads "Information;" for in most cases the information is more valuable than the hardware which processes it.

—RADM Grace Murray Hopper, USN

This chapter discusses a basic approach to defending the business or government agency by establishing a bottom line based on knowledge management.

Introduction to Knowledge Management and the Value of Information

Now that you have a good understanding of the source and scope of information warfare (IW) attacks and some basic IW defensive concepts, this chapter offers approaches and concepts for businesses and government agencies to use knowledge management (KM) to protect and defend their information environments (IEs). Central to this are quantitative methods for managers to assist in the rational allocation of constrained resources. These quantitative approaches identify which information should have higher levels of protection. The methods will also assist managers in making winning arguments for additional resources. Although information technology and information assurance return on investment (ROI) is very difficult to qualitatively (much less quantitatively) determine, managers will be able to determine which objectives and campaigns will deliver optimal returns.

KM is a multidisciplinary subject that combines the development and integration of technologies, processes, and cultural changes as it provides a means for well-informed, rapid decision making via collaborative information and knowledge sharing by varied and dispersed organizations and individuals. KM is a complex, holistic, and symbiotic subject.

KM draws heavily from the business, behavioral science, and engineering disciplines, integrating both hard and soft sciences. These are woven throughout the four pillars of KM:

- Leadership
- Organization
- Learning
- Technology

Tenets include:

- Support for organizational processes
- Tailored content delivery
- Information sharing and reuse
- Capturing tacit knowledge as part of the work process
- Situational awareness of information
- Knowledge assets valuation

KM enables an organization to be more agile, flexible, and proactive. The approach is ideal for information warfare. One example is integrating intelligence (e.g., economic and open source) and security (e.g., physical, personnel, and operations). Organizations can use KM to both gain competitive advantage and protect their information environment (IE).

KM is not a "pop management fad," although reading trade periodicals one might get that impression based on the many proprietary, narrowly focused solutions being offered. When correctly applied, KM is the smart integration of technology and people to achieve objectives. KM supports rapid, well-informed decision making for sound use of resources.

Use KM to assess and apply organizational capabilities. Sun Tzu wrote that, "If you know the enemy and know yourself, you need not fear the result of a hundred battles. If you know yourself but not the enemy, for every victory gained you will also suffer a defeat. If you know neither the enemy nor yourself, you will succumb in every battle."

What makes information "superior"? Is "inferior" the opposite? Perhaps unrealized potential is more accurate. Information is neutral, awaiting an actor to realize its value. An inferior position is the result of unrealized potential. Except for the national security capabilities of intelligence agencies and law enforcement, all actors have a reasonably equal opportunity to acquire information. It could be at a government printing office, a patent and trademark office, a securities and exchange commission (e.g., corporation-required filings), on the Internet or the Web (there are probably two orders of magnitude more information on the Internet than on the Web), or on other forms of media. Sophisticated and focused search capabilities, data mining, etc. support the decision-making process.

How can an organization achieve superiority or dominance relative to other actors in the arena? Corporations need to acquire information and act sooner than their competitors. Superiority is binary: either you have it or you do not have it. Information quality (e.g., content, context, timeliness, and delivery to decision makers and executors) are core constructs of superiority. Lags within and between the cyber and physical realms (i.e., speed of thought, speed of electrons and photons, speed of physical world) affect planning.

Before one can have an appreciation for the value of information, there is a need to understand the constructs and challenges of the IE, both within the organization and across the enterprise. As pointed out in Chapter 1, the IE consists of several interrelated areas; it contains all of the organization's data, information, and knowledge (ubiquitously called information), plus information across the enterprise. Formal constructs such as intellectual property are included, as well as the tacit knowledge of individuals. To recap Chapter 1 concepts, information moves across information infrastructures to support information-based processes. Information infrastructure is the media within which information is stored, processed, and transmitted. Examples are people, computers, fiber-optic cable, lasers, telephones, and satellites. Information-based processes are the established ways to obtain and exchange information. This includes people to people (e.g., telephone conversations and office meetings), electronic commerce, command and control, customer relations, public perception, data mining, and surfing the Web. IEs are physically and virtually interconnected. The reason IEs exist is to develop, field, maintain, and withdraw products and services to attain and maintain a competitive advantage. Whatever comprises a specific IE, the important fact remains that if its elements are not defended, consequences can range from irritants to catastrophes.[1]

Rather than banter about definitions and which is more important, the issue is how to value the intangible of information, be it data, information, or knowledge. At any given time, one could be of greater value than the others. The tacit knowledge of employees is the most difficult resource to quantify and value.

The value of information will be the cornerstone of electronic commerce and mobile commerce in the Knowledge Age. Proper use of information is central to profitable business and successful military operations. Information as used here means data, information, and knowledge. Value, or perceived value, drives resource allocation. This leads to the need to use proper protection features because not all information has the same value. Why are there not accounting standards for information? What information is critical enough to require the use of deadly force by law enforcement and the military? Quantitative approaches elude us. The value of information is the underpinning of Coherent Knowledge-based Operations (CKO) the concept introduced in Chapter 1.

Now let us delve into the perceived and quantitative values of information, and describe practical applications using a model to rationally allocate resources.

Information Has Value and Is the Underpinning of Knowledge Management

On a flight to Los Angeles, a fellow passenger and I (Perry Luzwick) discussed our jobs and had other light conversation. When we deplaned, he said, "Thanks for keeping us safe." He paused and thought about what he had said. His next comment was, "Well, at least for keeping us ahead." This second comment characterized the situation as a race. The very real stakes in this race are national sovereignty, economic stability, corporate survival, and individual freedom. My fellow passenger's comment about "keeping us safe" is of increasing concern to industry and government. *InfoWorld* has published a list of the top-ten challenges keeping information technology executives awake at night, including:

- Number 10 was security. Trusted insiders gone bad, crackers breaking in, and viruses were considered problem areas.
- Number 3 was measuring the business value of information technology. This is the Holy Grail.
- Number 2 was getting the most from corporate data.
- And the number 1 issue? Enabling business initiatives.

The value of information is directly related to the above four issues, and three of them topped the list. If you do not know what information is worth, how can you wring the most out of it to attain and maintain a competitive advantage? How can you be sure that the proper protection and security mechanisms are being used? How can you insure it? In disaster recovery and continuity of business operations scenarios, how do you know which information to recover and restore first? The value of information is central to answering these questions.

Headlines tell us that we are in the Information Age, ride the information highway, and are part of a knowledge-based economy. Information must be important. The number-one issue facing businesses and governments is how to value information beyond subjective estimates, converting it from an intangible to a tangible asset. Some countries, companies, and people are entering the Fourth Wave — the Knowledge Age — while others are just entering the Information Age. The need for standards and metrics for the value of information is essential because information will be the cornerstone of E-commerce and the Knowledge Age. Standards and metrics are necessary for broad acceptance of information as a tangible asset, especially if change is desired within a generation.

We do not do intelligence operations for the sake of the intelligence community. We do not undertake the great effort of a new advertising campaign or establishing an E-business for the sake of doing hard work. And we certainly do not engage in KM for the sake of doing KM. We do these actions because we expect them to help us attain and maintain a competitive advantage — and that translates into hard dollars at the bottom line. The need for a tangible value of information should lead directly to a rigorously based rational approach for protecting and securing information. It beckons a new approach for the protection of information. Until national and international standards are established, information will have simultaneous multiple intangible values, and the protection and security of information will continue to be secondary because valuing it is "too hard" a problem.

The speed of thought drives everything — creativity and innovation, accomplishment of normal processes, and survival. The speed of electrons and photons through the IE is man's attempt through the use of technology to approach the speed of thought. The speed of physical forces lags the virtual domain, and the virtual domain in turn lags the mental realm. These lags must be factored into plans and KM applications.

Information can be maneuvered in cyberspace as materiel is in the physical realm. For IT and information assurance (IA), maneuver in cyberspace is done by wheeling information (e.g., software, data mining results, and tacit knowledge). This virtual maneuver enables organizations to effectively and efficiently service many customers. Initiatives, drives, and thrusts can be rapidly shifted, thereby maximizing opportunities through the swift reallocation of resources.

Your Information Environment Is Under Siege

Attacks against and defense of the IE, from mundane physical to sophisticated virtual, is information warfare (IW). Using IW to attack (i.e., deny, alter, or destroy) one or more of the IE's components can result in the loss of tens of millions of dollars in corporate profit and degraded national security. The competition and others are using a vast array of tools and techniques to probe, steal, degrade, deny, or destroy your IE's physical and virtual assets on a daily — even hourly — basis.

The electronic genie is out of the bottle and cannot be put back in. Malicious attacks can be externally or internally mounted to steal, alter, or destroy information, disrupt or destroy information infrastructure, and to disrupt information-based processes. Your competition wants your information; so do not believe that "Gentlemen don't read other gentlemen's mail." While conducting business, your competitors attack your physical and virtual domains every hour of every day. They gather intelligence on you via their sales force while on your premises, at trade shows, searching your Web site, and via other means either legally, in the gray area, or illegally. It does not matter if what they gather is physical, electronic, or optical; they have something of yours.

Mark Rasch, a vice president with Global Integrity, revealed during testimony before the U.S. Senate Appropriations Subcommittee hearing on Internet security that the lure of big, fast-money scores in virtual commerce is making it common for skilled hackers to attack competitors in search of free intellectual property. He said, "The present era of dot.com millionaires and initial product offering (IPO) frenzies and the ease of starting your own business on the Web is creating a tremendous amount of competition to acquire intellectual property by any means at hand. We see sophisticated attacks against computer systems to steal intellectual property which can be used in competition with other companies."

In a research report entitled "Titanic 2020," Dr. Richard Lysakowski notes that within ten years the total number of electronic records produced on the planet could double every 60 minutes. At the same time, this report claims that the integrity of these records as legal evidence will be frequently violated until new systems are designed to meet the needs for long-term archiving of electronic data and legal records. Lysakowski concludes that the world will soon begin to experience trillion-dollar losses of critical data and legal records due to inadequate software infrastructure.

Who is behind the IE attacks? The cyber-activist group RTMark (pronounced art-mark) is using eBay to raise funds to support a variety of cyber-protest campaigns. Although computer network attacks make the news, receiving less attention are the phreaked telephone systems, damaged or destroyed facilities, misinformation in the media, and people being subverted.

These IE threats pose nontraditional challenges. Detecting, assessing, and attributing capabilities and attacks is very difficult in the virtual world. We have a broad spectrum of capabilities to identify missiles, tanks, and submarines. Unlike air defense systems using identification friend or foe/strategic identification feature (IFF/SIF), it is extremely difficult to do the same with electrons and photons.

Kids today have more computing power than many countries had in the 1980s. The sharing of techniques and tools in the cracker and phreaker underworld would in itself make a great KM study. Internet chat rooms, crackerzines, conventions, e-mail, and other media and fora are used to share knowledge. Neophytes can buy (for $29.95) two CD-ROMs with exploit scripts for UNIX (name the flavor), NT, Windows, Solaris, Java, and Active-X, plus a war dialer, password cracker, and more. Similar to the free sharing of information within the scientific community, this knowledge is shared to raise the general level of understanding and is open to peer review.

Vulnerabilities stem from security not being designed into products and services. The worldwide connectivity of infrastructures increases the risk of attack due to interdependencies and shared vulnerabilities. A risk accepted by one is imposed on all. System interdependencies make attack consequences more difficult to predict and more severe due to cascading effects.

If You Do Not Know What a Pound of Your Information Is Worth, Then You Cannot Manage What You Cannot Measure

Are we effective at what we are doing? We cannot know without rigorous, quantitative metrics and measures of effectiveness (MoE). The lack of accounting standards for the value of information hinders E-commerce, M-commerce, and protection of the IE.

It is difficult to believe that we are this far into the Information Age — even entering the Knowledge Age — and still do not have standards for the value of information. Here are some ideas that hopefully will advance the study of the value of information and foster significant change.

The value of information is an area few wish to be involved with because it is a complex, multiplanar problem. What is the value of information in a database? How about after it has been profitably data mined? Converting information from an intangible to a tangible asset is essential for information to be the currency of the Knowledge Age. But the value fluctuates, as do national currencies. Perhaps an accounting category such as "Transient Virtual Assets" would be appropriate. Several countries, the Financial Accounting Standards Board, universities, businesses, and consultancies have efforts underway. At the present rate, it is unlikely that any will have their approach adopted as an international standard within the next five years.

It is important to identify where and when knowledge occurs. In a link and node process flow diagram, explicit information can be identified at formal decision points. At these times, the summation of intellectual capital is captured in documents. A far greater amount of information, especially tacit knowledge, exists during the links. The understanding of why one approach was used over another, new relationships between apparently disparate information, why certain information was used for calculations, and the sharing of information between colleagues can be as valuable as the decision itself. This informal information has reuse value, especially in shortening other groups' decision cycle times.

Exhibit 1 depicts value, as can be found in both the links and nodes of any process. From a process perspective, value resides in both the links and the nodes. The nodes are decision points, and here the work of a team or committee is captured and recorded. The knowledge becomes explicit knowledge in the form of a report, briefing, directive, or other administrative medium. This explicit knowledge has value. But what about the knowledge used in the link? Some of the knowledge made it to the node. Other knowledge was rejected. That knowledge also has value. Perhaps not appropriate in that situation, but it may be able to be reused in another. The knowledge can also be used as how not to do something. There is also the tacit knowledge of those on the team or committee. This informal knowledge probably was not captured, but everyone brought important "how-to" to the team and learned from the experience.

We have identified formal, or explicit, and informal, or tacit, knowledge. Both are important to an organization's processes. Knowing they exist is the beginning of valuing them.

Different groups have dissimilar interests, so they place emphasis on different elements within the value of information. Individuals are concerned with confidentiality and privacy, banks depend on integrity, and the legal community wants non-repudiation. One way to account for this is to use a multidimensional approach to determine a perceived value of information.

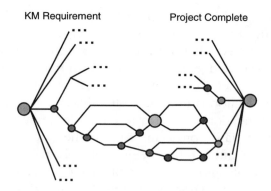

KM Requirement   Project Complete

Explicit/Tacit knowledge is at the nodes.
New knowledge is created in the links.

Legend
⊙ Process Start and End Points
● Tacit Knowledge Flow Points
◉ Explicit Knowledge Flow Points
◯ Core Competencies
— Link (formal and informal path of information flow

Exhibit 1.   Value Is in the Links and the Nodes

A proposed concept until a formal method is available is to use a multidimensional approach to determine a perceived value of information. Exhibit 2 provides a multidimensional approach for determining the perceived value of information.

From a contextual perspective, the information is of either tactical, operational, or strategic nature. From a time perspective, the information is either routine, important, or critical. Keeping the categories to a small number is essential; otherwise, subjectivity will creep in and result in a rating that is either under- or over-inflated. At any given time, selecting an information element, its contextual perspective and its time perspective will result in the perceived value of that information element. The way to differentiate between identical ratings is to add a weighting to the information elements. That unique information in time and context will then be rated relative to other information elements. Does this produce a tangible dollar figure? No. Does it help value intangibles? Absolutely. Can there be more than one perceived value at the same time? Possibly; when two or more people view the contextual and time perspectives differently. A policy can be written to achieve common understanding.

Aspects of information are content, context, and timeliness. Some content is of a lasting nature and thus has durability. Content can be maneuvered in cyberspace, wheeling software, data mining results, and tacit knowledge. This virtual maneuver allows swift resource reallocation and enables effective and

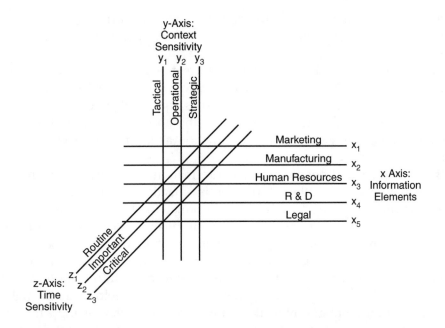

Exhibit 2. Multidimensional Approach for Determining the Perceived Value of Information

efficient service of many customers. Another important aspect is tailored format. Graphics are okay if you are at a desktop with a T1 connection; but if you are in the field with a 28.8 modem, graphics are a hindrance. A subset of context is perishability. Information at the tactical level may be meaningful for a few minutes, while at the strategic level it would have no value. An important subset of timeliness is capturing explicit information and tacit knowledge, organizing it, and delivering it via either smart-push or pull-on-demand.

Here is an example of perceived value. The U.S. Department of Defense sends a roll-on/roll-off (RORO) ship with 100 M1A1 Abrams main battle tanks to South Korea to support an exercise. The ship encounters bad weather in the North Pacific, suffers damage from mechanical problems and cargo that became unsecured, takes on water, and sinks. The nation bemoans the loss of life, Military Sea Lift Command calculates sealift shortfall workarounds, and Materiel Command orders more tanks. The value of the tanks, ship, and loss of life can be accurately calculated by traditional accounting methods. Change the scenario. North Korean actions indicate probable conflict. The United States wishes to show its resolve and support for an ally, so it sends an RORO with 100 M1A1s to meet activated Army and Marine Corps Reservists airlifted to South Korea. The ship sinks. What is the value of the tanks? The perceived value is definitely higher than the accounting value. What is the value of the information to the North Koreans that the ship sunk?[2]

Although time and contextual perspectives can frequently shift, thereby altering perceived value, not having hard figures results in an inability to attain and maintain a competitive advantage, inefficient resource allocation, inadequate IE protection, and a less than optimal bottom line. With values derived through rigorous methods, information can then be carried on the books as an asset. This will spur E- and M-commerce, accelerating us into the Knowledge Age.

How does a corporation know what information to acquire, retain, maintain, defend, and dispose? Laws and practices cover some areas, product line and consumer base others. What information is more valuable? It depends on content, context, and time sensitivities. In the absence of national and international accounting standards for information as a tangible asset, qualitative approaches are necessary.

Business units and people produce information elements, or content. From a contextual perspective, the information is of tactical, operational, or strategic nature. From a time perspective, the information is routine, important, or critical. At any given time, selecting an information element, its contextual perspective, and its time perspective will result in the perceived value. If needed, weights can be used to improve granularity. That unique information in time and context can then be rated relative to other information elements. Does this produce a tangible dollar figure? No. Does it help value intangibles? Absolutely. Can there be more than one perceived value at the same time? Probably. Standards need to be developed for common approaches.

Proactive measures must be taken to protect operations and the bottom line. Businesses could self-insure — in other words, eat a loss. Such a decision needs to be made in the presence of hard facts, not a gut feel. Does the corporation and its individual business units know how much profit they make per year? Per quarter? Per month? Per day? Per hour? Per minute? Is the information protected? How do you know the information has not been stolen or altered? If transactions at Citibank cannot be accomplished, tens of millions of dollars of business can be lost, and that does not count the ill-will of customers and permanent loss of future business to competitors.

Quantitative Methods to Use Knowledge in Information Warfare

A valuation of information would help prosecutors in computer crime cases. The jury must be convinced there was a loss. What is information in a database worth? A simplistic approach would be to take some number of people times their compensation plus a pro-rata portion of IT assets to acquire, process, store, and maintain the information. More complex, is there a competitive or national security loss? What is the cost to replace the information; and what is the cost of lost business/profits or national security?

The lack of rigorous, quantitative metrics, measures of effectiveness (MoE), and accounting standards for the value of information hinders E-commerce and IE protection. What is the value of information in a database? How about after it has been profitably data mined? Converting information from an intangible to a tangible asset is essential for information to be the currency of the Knowledge Age. Because the value fluctuates, an accounting category such as "Transient Virtual Assets" would be appropriate.

Precision is required to derive the perceived value of information because business decisions focus finite resources on products and services and need to be based on more than perceptions. How granular is the information? How much will it cost to acquire more information? What are the costs of an information-based process? Have performance measures of effectiveness (MoEs) for information been developed, such as developing leading indicators (e.g., expected sales or reduced development time); incremental change and rate of change toward those goals; and comparing against lagging indicators (e.g., the corporation's and industry's last quarter's and year's sales, market share, and profits)? What are information development and reuse costs? What is the perishability of the information? What is the individual, district, region, and enterprisewide effect of the information? Have unique MoEs been developed, such as return on time? From which budget will the money come? What are the trade-offs? Could better return be achieved elsewhere? What extra business will be generated? Whether or not information is a tangible or intangible asset, the fact remains that the Internet, intranets, extranets, virtual private networks, and electronic commerce have a common denominator: moving information to support a process. Those corporations that leverage information most effectively will lead their industries. Wal-Mart is a prime example.

Wal-Mart has spent hundreds of millions of dollars on its IT infrastructure to store, process, and transmit information. Its information processing capability is the best in the industry. Wal-Mart has developed its own algorithms for data mining, and can tell you down to a thousandth of a cent what its costs are at any stage of a process. Because of this, Wal-Mart is able to articulate a significant number of MoEs. Wal-Mart also understands that much of the information it needs resides with its suppliers. Wal-Mart works closely with its suppliers to ensure inventories are kept to a minimum. To do this, specific information is exchanged on a proprietary and secure basis.

Should all information be equally protected and secured? Some information is more valuable than other information. Protection and security measures need to be added as the value of the information increases. How much should a corporation spend to protect its information with firewalls, radio-frequency shielding, intrusion detection devices, personnel checks, motion sensors, encryption, training, anti-virus software, etc.? A more rigorous approach than using a perceived value for managers to allocate resources to protect and secure the information is clearly necessary.

The formula in Exhibit 3 is proposed as a point of departure for further study. Information has a cost to acquire, store, maintain, and dispose; thus, one can identify a minimum, simplistic value. The formula requires the attributes for a process, product, or service to be identified. Each attribute is adjusted by a number of weights. We have already discussed the three information aspects of content, context, and timeliness; and to these can be added the five IA aspects of confidentiality, integrity, authentication, availability, and non-repudiation. Some weights are important, while others can be deleterious; thus, a weight can range from negative to positive two (−2 to +2). An example is that the financial community considers integrity the most

$$\sqrt{\{[(wt_1)(wt_2)(wt_3)...](attribute_1)^2\}+\{[(wt_1)(wt_2)(wt_3)...](attribute_2)^2\}+\{[(wt_1)(wt_2)(wt_3)...](attribute_3)^2\}...}$$

$$\sqrt{n_{attributes}}$$

Attributes based on business process or product/service.

Weight based on:
- Information factors: content, context, timeliness
- Security factors: confidentiality, integrity, authentication and identification, availability, and non-repudiation

Range from -2 to +2.

Exhibit 3. Proposed Quantitative Point of Departure for a Multidimensional Approach for Determining the Value of Information

important feature, so it can be assigned a factor of two. If needed, more weights can be added for greater precision. This will derive a unique number that can be used as a basis for "what-if" analyses. Similarly, if a delivery company cites timeliness as its most important feature, but customers are complaining of late deliveries, then negative two (–2) would be assigned. As needed, more factors can be added for greater granularity. This will derive a unique number that can be used as a basis for "what-if" analyses.[3]

Examples of information that can be captured, organized, and measured are decisions from meetings, e-mail, telephone conversations, stored information, and the tacit knowledge of employees. Given the difficulty in having employees devote even one minute each hour to security, they are not likely to take time from their busy day to input their unique tacit knowledge. The process needs to be a transparent normal course of doing business. Privacy advocates no doubt will rail against such a system; so if they have a better conceptual and technical approach or wish to recommend safeguards, they can propose a solution amenable to all concerned.

Theft or disruption of information will cause loss of revenue and profit. What is the intellectual property, infrastructure, or process that you must have, or else sales will decline? What would the effect on sales be if the competition knew what you had in R&D or knew the contents of the contracts with your customers? Change the questions to ask what if you knew this information about your competition. Would your competitive advantage go up or down? Of course, you would like to know as much as possible about the competition's essential elements of information.

Building on the proposed perceived and explicit mathematical approaches for the valuation of information, we can examine two practical applications.

- The "To What Extent" model proposes a method for a rational lay-down of protection and security mechanisms.
- Coherent Knowledge-based Operations (CKO) provides a means of controlling an IE — both the physical and virtual aspects — through the use of KM, IW, and network-centric business (NCB).

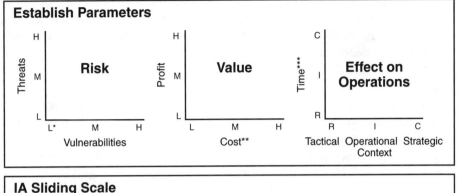

Establish Parameters

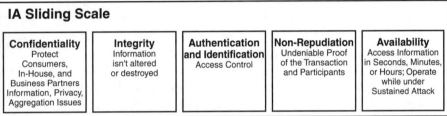

IA Sliding Scale

Confidentiality	Integrity	Authentication and Identification	Non-Repudiation	Availability
Protect Consumers, In-House, and Business Partners Information, Privacy, Aggregation Issues	Information isn't altered or destroyed	Access Control	Undeniable Proof of the Transaction and Participants	Access Information in Seconds, Minutes, or Hours; Operate while under Sustained Attack

Rational Laydown of IA Mechanisms

* Low/Medium/High
** To obtain and replace information and information infrastructure
*** Routine/Important/Critical

Exhibit 4. The "To What Extent" Model Can Be Used by Managers for Resource Allocation and Budget Preparation

Configuration management records are normally not accurate, so there is no complete, reliable baseline of fielded IE products and services. There is usually no real-time auto-discovery/mapping of the IE to keep track of dynamic changes in hardware, software, information, and physical assets, thus preventing just-in-time (JIT) IT and IA insertion and abandonment strategies. There is no identification of friend or foe on the network (e.g., IFF/SIF [identification friend or foe/strategic identification feature] that the military uses for battle management exist in computer network traffic). Finally, no accounting standards for the value of information and no rigorous, quantitative cost-benefit analyses (CBA)/trade-off analyses and MOE result in subjective conclusions. People say, "Well, that is an 80 percent solution." Eighty percent of what? Values based on sound philosophical understanding and intellectual insight are difficult to argue against.

A rigorous approach needs to be used to determine a rational lay-down of IW products, services, and processes that will be tailored for each corporation. Exhibit 4 depicts a process to determine a rational lay-down of IA mechanisms.

The "To What Extent" model is an approach to determine a rational lay-down of security products and services. The more valuable an IE's components are to achieving goals and gaining a competitive advantage, the more necessary a full-spectrum defense of the IE. Competitors either overtly or covertly acquire products, then reverse-engineer hardware and decompose software to understand how they work and to identify vulnerabilities. Tools are then developed to exploit the vulnerabilities with the purpose of narrowing or widening the competitive edge. Without protection and security to defend, and the use of deception to mislead competitors, they will continue to search and exploit, if for no other reason than their own survival is at stake. The model can easily embrace more complexity.[4]

Leading off the "To What Extent" model is validated requirements. A streamlined process to request, analyze, prioritize, and fund them should be in place. The two-dimensional graphs are based on the perceived value model. The model is scalable, so more graphs can be added as needed. For the "Risk" graph, there is no a one-for-one relationship between threat and vulnerability. System administrators should know the vulnerabilities of their networks, systems, and applications. Threats can be identified through the releases of the Computer Emergency Response Team Coordination Center (CERT/CC) at Carnegie Mellon University, the FBI's National Infrastructure Protection Center (NIPC), IA companies, and others. The intersection of the two is risk; and assuming consequence management indicates the need, those risks are what need to be defended against. Risk management and consequence management are important because there is not enough time or money to eliminate risk.

Another graph is "Value." The cost to obtain and replace information can be calculated. The corporation's leading and lagging indicators and measures of effectiveness can be used to determine profits. Knowing profits and costs can lead to questions such as, "What is the value of not having the information?" The third graph uses the context and time perspectives to determine the "Effect on Operations."

The third graph uses the contextual and time perspectives of the perceived value of information to determine the effect on operations.

In this example, the IA sliding scale is applied to all the graphs. Confidentiality, integrity, authentication and identification, non-repudiation, and availability are applied to risk, value, and effect on operations. The summation of these 15 calculations is used to determine the rational lay-down of IA mechanisms. "What-if" analyses can be performed on the graphs and the calculations then rerun to fine-tune the solution. Other parameters can be used in the sliding scale, as appropriate.

The "To What Exent" model can be used for disaster recovery and contingency operating plans. IW attacks make disaster recovery and business continuity plans essential. For every minute that information systems are not up and fully running, revenue, profit, and shareholder value are lost. The last thing a general counsel needs is a lawsuit from unhappy shareholders who are suing for millions because standards and best practices to protect information, the lifeblood of any organization, were not followed.

Putting Knowledge Management to Practical Use

Coherent Knowledge-based Operations (CKO) combines the three powerful concepts of KM, information warfare, and network-centric business (NCB) to

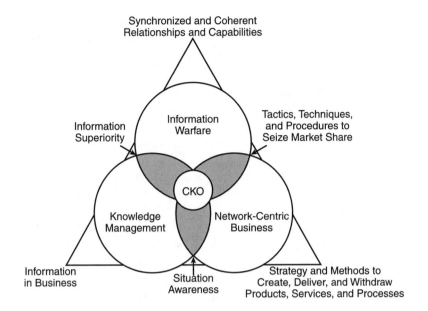

Synchronized and Coherent
Relationships and Capabilities

Information
Superiority

Information
Warfare

Tactics, Techniques,
and Procedures to
Seize Market Share

CKO

Knowledge
Management

Network-Centric
Business

Information
in Business

Situation
Awareness

Strategy and Methods to
Create, Deliver, and Withdraw
Products, Services, and Processes

Use CKO to bring resources to bear and
to seize and maintain a competitive advantage.

Exhibit 5.   Coherent Knowledge-Based Operations (CKO) Model

control IE. This control enables an organization to attain and maintain a competitive advantage while at the same time preventing, or at least complicating, the competition from reducing your advantage and increasing their own.

CKO brings together finite resources — people, money, and time — to achieve an organization's goals and objectives. Using KM, IW, and NCB in a coherent and synchronized fashion is far more powerful than the individual components, creating a decisive means to survive IW attacks, as well as to attain and maintain a competitive advantage. Islands of knowledge and stove-piped functional areas are the wrong construct in the Knowledge Age. Exhibit 5 depicts the Coherent Knowledge-based Operations model.

IW is the ability to control your own IE while at the same time disrupting your competitors. The goal is to influence others' perceptions through the use of physical and virtual capabilities to attain and maintain goals and objectives. IW is about synchronized and coherent relationships and capabilities from many disciplines such as operations, intelligence, legal, law enforcement, public affairs, computer network attack and defense, and counterintelligence that are brought to bear as appropriate to control the IE.

The purpose of undertaking any action is to achieve a goal and an objective. In a nation-state's defense ministry, IW occurs across the spectrum of conflict: peace (are we, and our networks, ever really at peace?), operations other than war (e.g., peacekeeping and humanitarian aid), war, and returning to peace. The goal of offensive IW is to influence the perceptions of decision makers or groups so they take actions toward the goals and objectives you want to attain. Milosevic's use of the press, radio, and television to influence public opinion and decision makers was virtual IW. NATO bombing Serbian television transmitters to limit Milosevic "getting out the word" was physical IW. The commercial sector cannot normally destroy manufacturing sites, monitor e-mail, and steal intellectual property. To avoid criminal and civil penalties, scale offensive IW to the realm of the possible. One approach is to use open source information to identify a competitor preparing to launch a new product and take appropriate actions to prevent or delay the launch, limit its success, or turn it into your advantage.

Sharing information is a necessary, powerful means to conduct effective operations. Withholding information, especially tacit knowledge that is so vital in fast-paced businesses and on the battlefield, has a negative effect by creating information islands. These islands are not in the best interest of attaining and maintaining a competitive advantage. KM brings to bear explicit information from disparate corporate and other locations, along with employee tacit knowledge for leaders to make informed, rapid decisions. KM can be used to organize and deliver tailored information, identify issues and trends, and lead to decisions to modify or develop new goals. KM cannot be bought off-the-shelf. To make KM work, employees must understand goals and be encouraged and rewarded for collaborating to eliminate information islands. KM is as much about culture as it is about technology.

NCB enables an organization to enhance its strategies and tactics by using the IE to create and deliver products and services faster and better than the competition. NCB is recognition that government, industry, and the public are becoming reliant on computers, evidenced by computer-based features as central to our standard of living. Shunting railroad cars, load balancing power grids, navigation warfare, and even refrigerators that contact the manufacturer via the Web if a problem arises are examples of our acceptance of, and in many cases dependency on, computers. We conduct E-commerce, allow employees to telecommute and have remote access, and spend millions of dollars on Web sites to attract customers to sell products and services. Computers and robots are in manufacturing plants, personnel and medical records are automated, and many of us participate in automated deposits and bill payments. If the computers stopped, not enough trained and skilled people could take over the functions, and many businesses and governments would quickly come to a halt.

There is no faster, more effective, or more efficient means to beat the competition than using NCB. It allows an organization to take maximum advantage of its business processes, such as taking and placing orders, using the supply chain, conducting R&D, just-in-time production, and using distribution channels to field products and services better and faster than the competition. NCB leverages not only all the resources within an organization, but also includes customers and business partners. They are all part of the solution set that drives the bottom line. The resources within the organization — people, money, and time — are finite but can be effectively and efficiently allocated to provide optimal support to customers and to maximize the bottom line.

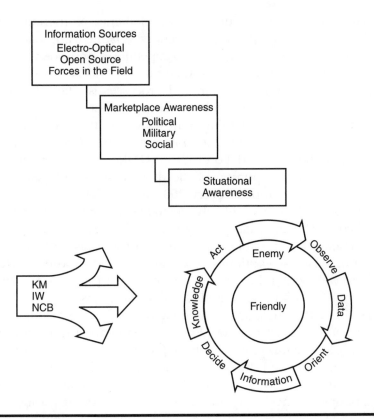

Exhibit 6. CKO in Action

Knowing about your competition is good, but it is not enough. Situational awareness of the "big picture" is essential. Understanding not just the "local" market, but also the political, social, economic, and military factors affecting the business environment is vital to making sound decisions. Situational awareness must be conducted in both the micro and macro sense. Examples of micro awareness are monthly and quarterly sales metrics, and comparison with the competition's and your sales for last quarter and the same fiscal period last year. Macro awareness considers leading and lagging indicators for the sector, industry, and country; government policies, laws, and regulations; cultural shifts; and national and international economic variances.

Exhibit 6 depicts a CKO in action.

KM enables situational awareness to be woven synergistically throughout an organization. KM, the fuel for the Observe-Orient-Decide-Act (OODA) furnace, fully enables IW and NCB. Businesses that execute decisions sooner than competitors will have the edge in influencing customers' perceptions and gaining market share. Rapid decision cycles enable a quick-to-market offering of modified or new goods and services.

This situational awareness permits you to understand what is influencing your organization and what influence you can wield. It also helps bring into focus what information you need to go after, as well as the IE (i.e., information, information infrastructure, and information-based processes) you need to defend. Concentrate your resources on the competition's essential elements of information (EEIs) and your essential elements of friendly information (EEFIs). This traditional operational security (OPSEC) approach requires a thorough understanding of both your competition's and your own organization's IE, especially the value of information and critical processes.

Being able to stratify information and processes is vital. The World Trade Center and Pentagon terrorist incidents on September 11, 2001, are examples that "They will never cross that line" and "Odds are that will not happen" are not reality. Worst-case scenario planning is essential. One example is when the Red River flooded Grand Forks, North Dakota, and concurrently a fire wiped out the remaining portion of the downtown. A 100-year flood is not called that because it happens every 100 years. It has to do with the flood stage. Although statistically unlikely, a 100-year flood could happen three years in a row. The information, knowledgeable people, and hardware to keep the organization a viable entity need to be available to meet a worst-case scenario. Business continuity, contingency, and disaster recovery plans need to be developed and exercised. Exercises are performed to prepare the people as well as to find flaws and improve the plans.

When satisfied that the plans are in good order, define a "Minimum Essential Emergency Knowledge System" (MEEKS). This system is the absolute minimum necessary for the organization to deliver products, services, and processes. It is not a thin-line system. If triple redundant, fault tolerant networks, alternate hot-site operating locations, a line of credit, transportation resources, and additional people are necessary, then that is a cost of doing business. Exhibit 7 shows the Minimum Emergency Essential Knowledge System (MEEKS).

Recommendations for Further Study and Research

The value of information is an amorphous and extremely complex subject. Even the most acknowledged experts (e.g., Karl Erik Sveiby from Sweden) have stated that placing hard figures on intellectual assets is not meaningful. Although time and contextual perspectives can shift frequently, thereby altering perceived value, not having hard figures results in:

- The inability to attain and maintain a competitive advantage
- Inefficient resource allocation
- Inadequate information environment defense
- A less-than-optimal bottom line

Traditional accounting methods are insufficient, as shown by Paul Strassmann. Aggregated information also needs to be considered; standard data labeling and automated reclassification of information would be useful.

Organizations need to be proactive in protecting their information-based assets because the risk of information loss is real. Quantitative approaches embracing both the "hard" and "soft" sciences for valuing information need to be developed. A crack team of accounting, operations research, mathematics, industrial psychology, anthropology, sociology, other subject matter

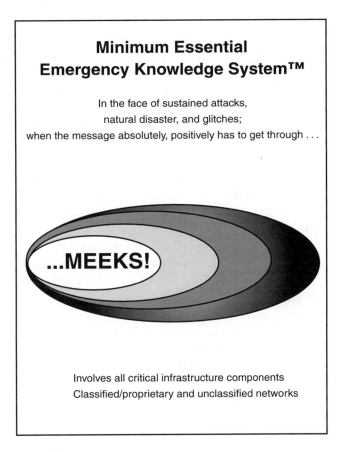

Exhibit 7. Minimum Emergency Essential Knowledge System (MEEKS)

experts from the business, legal, insurance, and academic communities should develop methods, formulas, measures of effectiveness, and standards. With values derived through rigorous methods, information can then be carried on the books as a tangible asset. This will spur E- and M-commerce, and accelerate us into the Knowledge Age.

Each arena of conflict has its own metrics and means for valuing items in the arena (i.e., methods to determine what is valuable and worth fighting for and what is not). The capability to valuate items in the arena has directly led to principles and guidelines for engaging in conflict within the arena. One example is the Principles of War, such as mass, economy of force, offensive actions, and security. These principles are valuable because commanders are always resource constrained and the principles, coupled with the valuation, determine where resources should be focused for optimum effect. In the IA, IW, and critical infrastructure protection (CIP) arena of conflict, information and the associated infrastructure make up the arena and are the equivalent of terrain. As is well known, IA/IW/CIP have no principles or theories of conflict as have been developed for other arenas of conflict. This state has persisted because practitioners in the field have no reliable means for valuating information. The lack of a rigorous means for valuation, and hence principles for conflict, have led to the current defend everything, ad hoc, expensive approach to information defense. As in all other arenas of conflict, this approach also fails in the IA/IW/CIP field because resources are limited. As a result, defense is weak everywhere, intrusions continue to occur, and defense becomes ever more expensive and ineffective. Defense in the IA/IW/CIP arena cannot succeed, as has been proven in other fields of conflict, until we know what to defend (i.e., where to concentrate resources) by making at least relative statements concerning the value of information. Rigorous means for information valuation will lead directly to the ability to make rational, informed choices concerning where to employ scarce resources.[5]

The lack of rigorous means for information valuation, quantitative metrics, MoE, and accounting standards for the value of information hinders E-commerce as well as IA/IW/CIP. Questions such as, "What is the value of information in a database?" and "What is the value of information after it has been profitably data mined?" cannot be answered. Data, information, and knowledge are intangible assets. As such, they are, at best, derivatively valued from qualitative assessments, perhaps subjectively valued ("gut feel"); but more likely than not, they are not valued, monetarily or in any other manner, at all. Lacking a rigorous, scientific-based approach to the valuation of information (i.e., embracing data, information, and knowledge), the best IA/IW/CIP solutions will never be achieved and some solutions may even be contrary to the best defense.

The ability to determine the value of information will put teeth into IA, IW, and CIP. Definitively knowing what information is worth will directly contribute to determining the best national security and business competitive advantage courses of action. Information valuation helps ensure that the proper security mechanisms are used, resources are conserved, essential information is fairly insured, and, in disaster recovery and continuity of operations situations, the best sequence to recover and restore information. With a means for information valuation in hand, managers will be able to rationally allocate resources — people, time, and money — and be able to defend their decisions as well as their information.

Established value of information models and formulas will enhance disciplines such as artificial intelligence (AI) and modeling and simulation (M&S) because AI and M&S are the basis of risk and consequence management. Additionally, by knowing the value of information and applying reasoned protection measures, the military's and industry's ability to operate effectively under a sustained IW attack will be improved.

Valuing information is akin to nailing Jell-O to the wall, as there can be simultaneous correct valuations of information. The problem is exacerbated by the lack of accounting standards. There are content (e.g., durability), context (e.g., information at the tactical level may be meaningful for a few minutes, while at the strategic level it would have no value), and time sensitivity (e.g., perishability) issues. A further confounding issue is that there are many disciplines, such as accounting, finance, economics, statistics, mathematics, and psychology that must be brought together to provide an integrated, rigorous, and coherent solution. The difficulty is on a level similar to Shannon's mathematically interpreting information theory.

The problem is hard because there are many dynamic components of information value that must all be addressed and their influence determined. A few such components include leading and lagging indicators, performance MoEs, and unique MoEs such as return on time. An analysis of multiple variables must be used to determine a rational deployment of IA/IW/CIP mechanisms.

Study and research need to develop a rigorous, mathematically based set of methodologies for determining the value of information in a variety of venues. Research should include assessing the problem from accounting, command and control, IA, and psychological and sociological perspectives; developing hypotheses for information evaluation; development of mathematical formulas; testing, simulation, and evaluation of the formulas in a number of venues; and submission of the results to military and civilian standards organizations.

The value of information will be the cornerstone of successfully conducting conflict and E- and M-commerce in the Knowledge Age. Although time, content, and contextual perspectives can frequently shift and thereby alter value, an inability to valuate information results in the inability to attain and maintain a competitive advantage, inefficient resource allocation, and inadequate IA/IW/CIP policies, strategies, tactics, and operations. With information values derived through rigorous methods, information can then be carried on the books as an asset. The results of this research will spur E- and M-commerce, guide defense, and accelerate us into the Knowledge Age.

Summary

The value of information has been the key missing factor in risk management and consequence management. Not knowing the value of the intangible assets they are protecting has been the bane of IT and IA managers. Without these valuations, they are hard-pressed to win their fair share of the budget battle, and this is one reason why security is almost always underfunded. To assist resource allocators, models to calculate the perceived value of information, calculate a "hard" value of information, and allocate resources were presented. Because IW is meant to enhance the national security of nation-states and competitive advantage of businesses, it was shown in relationship to knowledge management and network-centric business. These three concepts, when combined and executed in a synchronized and coherent fashion, will improve national security and competitive advantage.

Notes

1. Perry Luzwick, If You Don't Know What a Pound of Your Information is Worth, Then You Can't Manage What You Can't Measure, May 2000.
2. Perry Luzwick, Information Warfare Attacks Can Be against Trade Secrets and Intellectual Property. How Much is a Kilo of Your Information Worth?, "Surviving Information Warfare" column, *Computer Fraud and Security,* Elsevier Science, July 2000.
3. Perry Luzwick, InfoWarCon 2000, Washington, D.C.
4. Perry Luzwick, InfoWarCon 2001, Washington, D.C.
5. Dr. Martin Stytz and Perry Luzwick, White Paper on Rigorous Means for Information Valuation, May 1, 2001.

Chapter 17

What's a CEO to Do?
Ya Gotta Have A Plan

Fortune favors a prepared mind.

—Louis Pasteur

This chapter discusses the need for the business or government agency CEO to take an aggressive leadership role if he or she wants to defend the turf. To do so, a bottom-line, no-nonsense plan is required. This chapter discusses the philosophy, strategy, and planning that is necessary.

Introduction

The "prepared mind" to which Pasteur refers can be extrapolated to be a prepared organization. Pasteur spent much time preparing — studying, lecturing, consulting, and seeking counsel — so that as he methodically conducted his research, if anything out of the ordinary should occur, he would be ready to take advantage of it. Call it fortune, luck, chance, or fate; if you are not prepared to take advantage of an opportunity, you will not be able to attain and maintain a competitive advantage. No mater how meticulous your planning may be, opportunities will arise in using information warfare (IW) to attack a competitor's and defend your own information environment (IE).

Execution is a result of preparation; and the better the preparation, the better the execution. General Tommy Powers, Commander-in-Chief, Strategic Air Command, said, "If it wasn't for communications, all I would command is my desk, and that's not a very lethal weapon." Preparation and execution cannot happen in a vacuum. Communications is the key to rapidly bring together all the key players to develop as good a plan as possible.

Following our convention of "from the battlefield to the boardroom," "lethal weapon" can be restated as "a prepared organization." If you are "running with the wolves" or "swimming with the sharks," you view your organization as lethal: well prepared and devastating in execution. Rather than market dominance, the only thing that will happen in the absence of a plan is getting shot in the foot. The number-one reason for business failure is a poorly conceived plan or a poorly executed plan. It does not matter if your organization is small and simple or large and complex; the market is unforgiving. By example, profits do count, as shareholders and venture capitalists reminded the dot.coms. A plan built on strategies and principles is considerably more likely to succeed.

Business CEOs and government organization leaders need to take an aggressive leadership role to defend market share and national security. To do so, bottom-line, no-nonsense planning is required. Underpinning the plan must be a strong commitment to strategies, and these in turn must be based on sound principles. Strategies and principles enhance agility, adaptability, effectiveness, efficiency, and innovation. Without strategies and principles, an organization is like a ship without a rudder. Strategies and principles serve as guides in good times and bad, providing a unifying theme and common thread for strategic, operational, and tactical decisions.

The strategy that is ecumenical in nature is Coherent Knowledge-based Operations (CKO). Principles that have been proven over the centuries are the Principles of War (PoW). While CKO and the PoW may sound militaristic, it is their underlying themes and concepts that need to be understood and embraced to attain and maintain a competitive advantage. When the Japanese invaded Okinawa several centuries ago, they outlawed weapons. The people developed a martial art, Ishin Ryu, based not only on unarmed combat, but also on weapons — adapted from farm implements.

Exhibit 1 depicts Coherent Knowledge-based Operations, and how IW relates to knowledge management (KM) and network-centric business.

Warfare is not confined to the military. For example, there is social warfare, economic warfare, and business warfare. While conducting business, your competitors lay siege to you every hour of every day. They gather intelligence on you via their salesmen while on your premises, at trade shows, searching your Web site, and via other means either legally, in the gray area (legal but not ethical), or illegally. It does not matter if what they gather is physical, electronic, or optical, they have something of yours that you did not want them to have.

Enlightened people mine gold nuggets from many virtual hills to achieve their goals. So it is with CKO and the PoW. Rather than exclude or belittle good work, or to propose a proprietary closed approach, CKO and the PoW allow planners to use almost anything that will enable them to achieve their goals.

Using Coherent Knowledge-Based Operations as a Planning Framework

Be it Secretary, Under Secretary, and Assistant Secretary in a government organization or Chief Executive Officer (CEO), Chief Operations Officer (COO),

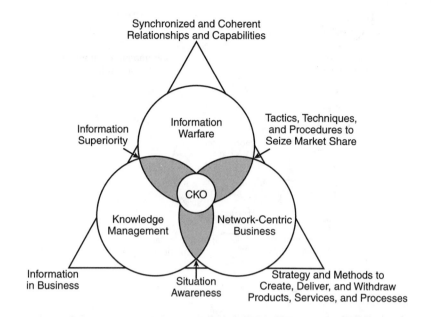

Use CKO to bring resources to bear and
to seize and maintain a competitive advantage.

Exhibit 1. Coherent Knowledge-Based Operations

Chief Information Officer (CIO), Chief Finance Officer (CFO), and Chief Knowledge Officer (CKO) in business, they need to be of like philosophical approach when it comes to using the organizational, technical, and cultural aspects of the organization to attain and maintain a competitive edge. Their executive commitment and involvement is necessary for all areas of an organization to be positively affected.

Technology brings speed, but not always clarity. Artificial intelligence and expert, learning, and brilliant systems have not stopped us from drowning in information. In some cases, decision cycles have been lengthened. Throwing technology at a problem is insufficient. A plan, to include training and organizational change, is needed because the people who use the capabilities need support. Yet this is still not enough. An embracing strategy is needed to bring to bear all the organizational capabilities in a coherent and synchronized manner. Coherent Knowledge-based Operations (CKO) does just that. CKO fuses knowledge management (KM), IW, and the increasingly network-centric business (NCB) processes of an organization. Networks, both internal and external, are now inextricably linked for the purpose of attaining and maintaining a competitive advantage.

The CKO strategy we previously alluded to in this book is the framework under which to do planning. CKO has many benefits. First and foremost, CKO can be performed on the strategic, operational, and tactical levels. This better enables a government organization to do those functions necessary to deliver good and services. For a commercial business, increased revenue and the bottom line are the compelling reasons to adopt it. CKO promotes structure

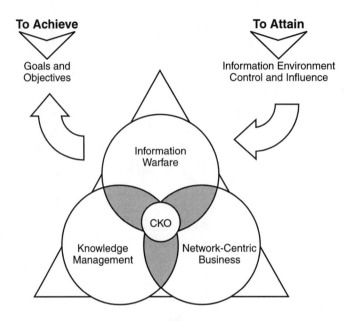

Exhibit 2.   How CKO Is Used for Attaining Goals and Objectives

and common purpose because it combines technology, operations, and culture. It also leverages technology expertise throughout the company as well as government organizations by embracing many strategies and disciplines. To turn and burn faster, cross-team information sharing encourages knowledge reuse and reduces reinvention, thus shortening time to market. By being ecumenical in nature, CKO avoids overemphasis on organizational boundaries, working at cross-purposes, and misfocused or missing groups. Because IW is an umbrella concept that achieves its best results by appropriately tailoring to a situation the capabilities of many disciplines and functions, CKO is the ideal vehicle to ensure that this happens in a synchronized and coherent fashion.

The planning process is improved through the use of CKO. Planners can draw on subject matter expertise of the many areas in IW, tailoring what functions are used as the situation dictates. A basic KM system enables planners to tap not just the explicit knowledge of the organization, but also the tacit knowledge. A well-designed system takes planning to a higher level, tapping the data, information, and knowledge from across the enterprise. Recall that the enterprise includes customers, employees, financial stakeholders, teaming and strategic partners, suppliers, and public perception.

Exhibit 2 shows how CKO is used for attaining goals and objectives. CKO blends IW and KM in a coherent and synchronized manner with network-centric business (NCB). NCB includes those strategies and methods an organization uses to create, field, and withdraw products, services, and processes.

The intersection of NCB and KM is situation awareness. Combining KM and IW results in information superiority. The intersection of IW and NCB comprises the tactics, techniques, and procedures to seize market share. Together, these will result in an organization that is "lethal." Strategy alone is insufficient. Central to any plan, and supportive of the strategy, are principles that, although best used judiciously as guides, need to be considered. Experience and judgment will determine on a case-by-case basis which principles are used.

Gaining Competitive Advantage and Protecting the Information Environment by Using the Principles of War[1]

Applying the Principles of War (PoW) to business is premised on the notion that business is war. Phrases abound: "economic warfare," "competitive intelligence," "casualties on the information highway," and "the marketplace is a battleground where companies fight for domination." Direct competition and other threats relentlessly attack — there is no better word — 24 hours a day, seven days a week, 365 days a year. There is no peace. Every company fights for survival. It does not get more primal than that.

The Principles of War (PoW) are battle-proven guides for conducting military operations across the spectrum of conflict. The PoW can be readily adapted from the battlefield to the boardroom for gaining competitive advantage in the information environment (IE), the interrelated set of information, information infrastructure, and information-based processes. Information is handled in information infrastructures that store, process, and support information-based processes. Being attacked (i.e., denial, alteration, or destruction) in one or more of the IE's components can result in the loss of billions of dollars in profit and degraded national security. Change the paradigm, and attacking (i.e., denying, altering, or destroying) one or more of the IE's components can result in the gain of billions of dollars in profit and enhanced national security.

Because "competition" is analogous to "enemy," other business–military analogies can be made with profit, shareholder value, competitive edge, and industry rank to achieve brand recognition, customer loyalty, and market share. A business leader, similar to a military leader, must plan to train and equip forces; gather intelligence; assemble, deploy, and employ forces at decisive places and times; sustain them; form coalitions with other businesses and nations; and be successful. There are many physical and virtual world parallels, as can be seen in the following headline: "Cisco to use SNA as weapon against competition.... Cisco believes its experience in melding SNA and IP internetworks can be used as a weapon in the company's battle with Lucent and Nortel for leadership in converging voice, video, and data over IP networks."[2]

The Principles of War (PoW)

Based on analysis of military actions for several thousand years across the spectrum of conflict, the U.S. Army after World War I enshrined principles that guided successful warfare at the strategic, operational, and tactical levels. Sun Tzu, Jomini, Clausewitz, Mahan, and experts in many countries have developed versions appropriate for their cultures and the way they conduct war (see Exhibit 3).[3]

Exhibit 3. Principles that Guided Successful Warfare Planning at the Strategic, Operational, and Tactical Levels

United States	Great Britain/Australia	Former Soviet Union	People's Republic of China	France
Objective	Selection and maintenance of aim	Initiative	Selection and maintenance of aim	Concentration of effort
Offensive	Offensive action	Simultaneous attack on all levels	Offensive action	Liberty of action
Mass	Concentration of force	Massing and correlation of force	Concentration of force	Surprise
Economy of force	Economy of force	Economy and sufficiency of force	Security	
Maneuver	Flexibility	Mobility and tempo	Mobility	
Unity of command	Cooperation	Interworking and coordination	Coordination	
Security	Security	Surprise	Surprise	
Surprise	Surprise	Preservation of combat effectiveness	Initiative and flexibility	
Simplicity	Maintenance of morale		Morale	
			Freedom of action	
			Political mobilization	

Although the mutually dependent and intertwined PoW have stood the test of time, they must be adapted and judiciously applied because every situation is unique. The PoW are useful for shaping Coherent Knowledge-based Operations (CKO) strategy and for planning. A good plan requires a clear vision of an end state before resources are expended. This is reasonable because resources (i.e., time, money, and people, especially their vital brain power) are finite.

Use the PoW and CKO to Make the Organization More Competitive and to Protect the Information Environment

Combining CKO with the PoW provides organizations with a powerful approach to seize the IE high ground. Doing this results in competitive advantage. Obtaining and maintaining this advantage is difficult because the competition and IE are dynamic and unpredictable. Here are descriptions and implementations of each principle that should be considered and incorporated into the planning process.

Objective

> *Every operation ought to be directed towards a clearly defined, decisive, and attainable objective.*[4]

Every activity a business, government, or military organization undertakes is based on information and driven by a bottom line, be it profit, social good, or victory in battle. It follows that IE investments exist not for themselves, but must support *Objectives* that provide a competitive advantage. A lot of money can be spent on building and protecting an IE, but not result in an improved bottom line.

Whether attacking or defending, the purpose of the *Objective* is to focus efforts. Indirect routes may be required to achieve the *Objective*; being obvious is never a requirement of achieving an *Objective*. There should be one strategic plan everybody knows and understands, a limited number of supporting operational plans (related to *Mass, Economy of Force, Maneuver,* and *Simplicity*), and tactical plans as necessary to achieve the *Objective*. These plans cover the long, intermediate, and near terms. A clear vision of the *Objective* focuses on planning. Vision, mission statements, annual goals, and similar documents must follow a consistent theme and be linked. As situations change, posing new opportunities and dangers, agility and timeliness to modify or create new *Objectives* is essential. Williams Corporation, reinventing itself as a long-distance communications provider from an oil industry company, and Security First Technologies, refocusing itself as an online banking software and services provider from a virtual bank, are examples.

There is never a perfect strategy. Frederick the Great said, "No plan survives contact with the enemy." A major cause of defeat is lack of or insufficient planning. By devoting more time to planning, *Objectives* are more likely to be achieved with the desired results. In battle, the *Strategic Objective* is defeat of enemy combat forces, by direct destruction, elimination of their capability to conduct operations, or other goals as directed by the political leadership. In business, it is doing better than, and even eliminating, the competition. *Tactical and Operational Objectives* are developed to achieve the *Strategic Objective*. A *Tactical Objective* may involve teaming with businesses. *Operational Objectives,* as in chess, analyze several moves ahead. How will the landscape change? Which competitors will be stronger or weaker? What challenges will customer needs bring? Where and how should the organization position itself? *Objectives* must be reconsidered because the competition is dynamic; partnering with other businesses may or may not work out, and IT can quickly level the playing field.

Sharing information is a powerful means to achieve *Objectives*. Withholding information by practicing "information is power" (especially tacit knowledge, so vital in fast-paced businesses and on the battlefield) has a negative effect, causing information islands to emerge. These islands are not in the best interest of achieving the *Objective*. KM brings to bear information from disparate corporate and Internet locations along with employee tacit knowledge for leaders to make informed, rapid decisions. KM can be used to organize and deliver tailored information, identify issues and trends, and lead to decisions to modify or develop new *Objectives*. KM cannot be bought off-the-shelf. To make KM work, employees must understand the *Objectives* and be encouraged and rewarded for collaborating to eliminate information islands.

Offensive (also known as Initiative): Seize, Retain, and Exploit the Initiative[4]

> *Being proactive, sustaining drives and thrusts, and taking advantage of opportunities are focused actions to achieve Objectives.*

Knowing what customers want by conducting market research, doing research and development to support current and future operations, and having exit strategies (e.g., IT and IA abandonment strategies) are examples.

Napoleon said, "Seize the initiative and keep it at all costs" because objectives cannot otherwise be realized. Maintain and reinforce, as necessary, successful drives. Initiatives may have one or multiple drives in parallel at the strategic, operational, and tactical levels, and their timing can be staggered. Ensure there is sufficient mass to be effective. As the competition counters, rapid decision cycles enabled by KM should be used to shift the main drive (i.e., if it falters, then press on with a secondary drive) to maintain the *Initiative*. Sun Tzu wrote, "All war is deception."[5] Deceive the enemy as to which is the main *Objective* and *Initiative* with determined proactive efforts and *Security*.

Initiative means waking up and asking, "How can I beat Corporation X today?" If organizational units are not moving toward *Tactical, Operational,* and *Strategic Objectives, Initiative* is lost. During a lull in activity, organizations should do something beneficial, such as supporting a business partner or building good will in the community. Defeat or elimination of the competition can be accomplished without a Pyrrhic victory. In the military, decisive battles (e.g., Dien Bien Phu) can result in heavy attrition, a notion anathema to business. Sun Tzu emphasized indirect and circular approaches, today known as asymmetric warfare. The great Chinese general said, "Therefore, the skillful leader subdues the enemy's troops without any fighting; he captures their cities without laying siege to them; he overthrows their kingdom without lengthy operations in the field."[6] Perception management, accomplished via psychological operations, misinformation campaigns, and advertising campaigns, may enable conflict to be avoided. If it cannot be avoided, then goals may be able to be achieved with minimum conflict, thus conserving resources.

A company may consider itself an aerospace business with a focus on manufacturing aircraft or a personal products company with a focus on personal care. Doing one thing well may not be enough; so to remain competitive, it makes thrusts and sub-thrusts into related areas. A drive (e.g., air superiority aircraft, Gillette Mach 3 razor shaving system) can have thrusts (e.g., suppression of enemy air defenses (SEAD), shaving gel), and sub-thrusts (e.g., anti-radiation missiles, after-shave). Within each drive and thrust, opportunities to specialize or expand may emerge.

There are times when assuming a defensive posture is appropriate. During these times, increase communications with customers and business partners because they find comfort in knowing they will continue to be serviced. The defensive should be dynamic and only temporary: always fight to regain the *Initiative*. You never know how badly off opponents are or when they will collapse (e.g., the Confederate breach of the Union lines at Gettysburg and the German unexploited first use of chlorine gas in WWI). Make competitors lose their *Initiative* via your attack. If there is no competition, seize as great a market share as possible. Internet stocks are examples, turning traditional approaches upside down by having years of financial losses just to be first to market and seize a dominant share. Doing this significantly raises the bar to entry by competitors.

What are examples of attacks and self-imposed losses in the private sector? Grocery store shelf space, especially the correct aisle and level, is a fierce battleground. All cake mix products were in the same type of vertical box. One company redesigned its box to be horizontal, with no change to the contents. The result was to take shelf space and customers' eyes away from competitors, two inches at a time. Similar battles are waged in the automotive sector. Ford ran a commercial for its pickup trucks. A father let his son use the truck for a date. Upon return, the father gave the son and truck a once-over and asked, "Is that a scratch?" The son paled, ran his hand over the spot, and, with a sigh of relief, wiped away some dirt. Within a few weeks, Chevrolet ran a commercial with two cowboys using a pickup truck. One remarked to the other about a scratch. His friend replied it was okay. "Scratches build character. Besides, girls like them." TechWeb contributor Jeffrey R. Harrow wrote about being unable to purchase items in a computer store in "What, No Failover System? I'll Shop Elsewhere!" "'The computer is down' equated directly to 'the store is down.'" If there is no contingency plan to make sales and manage inventory when the computer is down, customers will vote with their wallets by shopping at a competitor's store. This is an example of many elements across the IW spectrum (e.g., financial and technology leadership, network management, business continuity, disaster recovery) not acting in a synchronized and coherent fashion supporting the *Objective* and *Initiative* to attain and maintain a competitive advantage.

Perception and expectation management are strategies going back to Sun Tzu. Sun Tzu and T.E. Lawrence (better known as Lawrence of Arabia) were masters who controlled the battle space purely by shaping their enemies' perception of the conflict. Griffith wrote that in war we shape the enemy by "moving him where we want him to be," and in business we shape the customer and market place.[7] Shaping is an essential factor in business and military operations, and is accomplished by executing decision cycles faster than the competition or enemy. Avoid paralysis of analysis because a good decision can be more profitable than the perfect decision for which there is never sufficient time, money, or information. What is the difference? Billions of dollars in lost profit or compromised national security. Businesses that execute decisions sooner than their competitors will have the edge in influencing customer perceptions and gaining market share. Rapid decision cycles enable quick-to-market offerings of modified or new goods and services; and if the new drive is successful, it could become a primary *Initiative*. Knowledge management (KM) is the fuel for the Observe-Orient-Decide-Act (OODA) loop furnace.

The competitor that moves through this OODA loop cycle the fastest gains an inestimable advantage by disrupting his enemy's ability to respond effectively. Boyd, who was the first to describe the OODA loop, based his theory of conflict on an updated and elaborated reinterpretation of Sun Tzu's classic *Art of War,* written around 450 B.C. All of Boyd's work is consistent with an effort to use the ideas of Sun Tzu to overcome what he considered to be the central flaw in the Clausewitzian paradigm: an overemphasis on overcoming friction (i.e., those factors in war that impede vigorous activity, like bad weather, broken equipment, and uncertainty) to seek decisive battle, and an underemphasis on strategic maneuver. Summarizing the difference between Sun Tzu and Clausewitz, Boyd concluded that Sun Tzu aimed to pump up his adversary's friction, while Clausewitz aimed to reduce his own friction.[8] In *Patterns of Conflict,* Boyd concluded that operating inside an opponent's OODA loop generates "uncertainty, doubt, mistrust, confusion, disorder, fear, panic, and chaos."

Exhibit 4 depicts CKO in action.

Mass

Concentrate combat power at the decisive place and time.[4]

The purpose of *Mass* is to mobilize superior power at crucial points for decisive success. The business analogy of the military phrase "Put iron on target" is "Put products and services at the right place and at the right time." Effectively concentrate available resources — money, time, information, and the right people with the right skills. *Mass* does not mean "more." Superiority can be

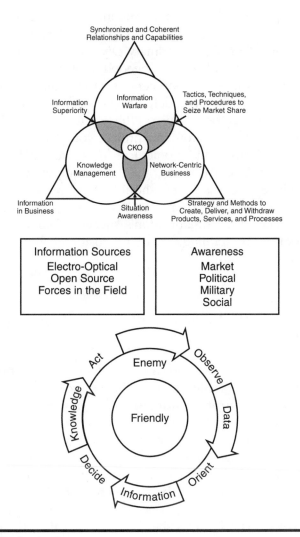

Exhibit 4.   CKO in Action

achieved against larger competition through qualitative superiority in IT, leadership, morale, and training. *Mass* is usually accomplished by *Maneuver* and is closely related to *Economy of Force*. Establishing partnerships with academia and other businesses is a means of building *Mass*. Success may not depend on a lot of resources if the competition is unfocused or has overextended itself by engaging in too many *Initiatives*. Use KM to assess organizational surge capabilities. Sun Tzu wrote, "If you know the enemy and know yourself, you need not fear the result of a hundred battles. If you know yourself but not the enemy, for every victory gained you will also suffer a defeat. If you know neither the enemy nor yourself, you will succumb in every battle."[5]

Economy of Force

Allocate minimum essential combat power to secondary efforts.[4]

This does not imply "do the job with minimum combat power." Only the fewest necessary resources are committed to secondary *Objectives* and *Initiatives,* thus achieving *Mass* for the primary *Objective* and *Initiative.* Secondary efforts are those products and services needed to attract and retain customers, but are not the main profit generation centers. Do not leave markets, products, or services open to attack because defending consumes resources better used for other purposes. It may be resource effective to have another business or academia take on a task to reinforce or protect you. Although the knowledge is not developed in-house, this would free resources to be used for other *Objectives* and *Initiatives.* By doing this, the competition's flexibility and *Initiative* are reduced because they should divert resources for a counter-move. Loss of subject matter experts (SMEs) goes against *Economy of Force.* Retain SMEs essential to the organization's success: compensation is cheaper than loss of productivity, training costs, and customer ill-will.

Maneuver

> *Place the enemy at a disadvantage through the flexible application of combat principles.*[4]

Maneuver itself cannot achieve decisive results; but when properly used, it makes decisive results possible through the application of *Offensive, Mass, Economy of Force,* and *Surprise. Maneuver* is a powerful means to defeat a larger, better financed competitor. For IT and IA, *Maneuver* is done predominantly in cyberspace by wheeling information (e.g., software, data mining results, and tacit knowledge). This virtual *Maneuver* enables organizations to effectively and efficiently service many customers. Additional examples are just-in-time IT and IA insertion and abandonment. Initiatives, drives, and thrusts can be rapidly shifted, maximizing opportunities through swift reallocation of resources. The virtual world moves significantly faster than the physical world, so the lag associated with physical resources must be factored into planning.[8] Another use of *Maneuver* is to attack where the enemy does not anticipate it, which results in *Surprise.*

Unity of Command

> *For every objective, there should be unity of effort under one responsible commander.*[4]

This is the most straightforward Principle of War, yet its proper application is frequently wanting. Committees, matrix management, outsourcing, and other similar constructs diffuse authority and responsibility. Leaders bring focus and direction. Responsibility, delegation of authority, accountability, and followership stem from leadership. A lack of *Unity of Command* results in fragmented IT investment, insufficient IE protection, failure to achieve *Objectives,* and thus a loss of competitive advantage.

Security

Never permit an enemy to acquire an unexpected advantage.[4]

This Principle of War refers to not only the need to prevent competitors from achieving *Surprise,* but also to prevent a disadvantage developing over time. This implies guarding against incremental attack and also maintaining the *Initiative.* Full-spectrum IE protection integrates and synchronizes many disciplines: comprehensive, enforceable security policy; awareness, training, and education; physical, operations, computer, communications, and personnel security; law enforcement; legal; and others. To facilitate IE attack response (i.e., recovery, reconstitution, redesign, retaliation), maintain backup and shadow systems. These also facilitate operating while under sustained attack. Perfect IE *Security* is not possible, so have and exercise risk management, consequence management, disaster recovery, and business continuity plans.

Risk is the intersection of threat and vulnerability. Risk reduction is the mitigating actions (e.g., awareness, training, and education; anti-virus/malicious code software; encryption; hardened operating systems; firewalls; and intrusion detection devices) used to reduce risk, leaving residual risk. Risk management is a term truly erroneously used. Too often, risk management is synonymous with risk reduction. Management is the effective and efficient use of resources, and associated with that are quantitative, mathematical, rigorous approaches. Risk management should therefore link threats and vulnerabilities to establish risk; apply appropriate risk reduction actions across the IA spectrum; and establish a level of residual risk. Consequence management is used to determine if residual risks can be accepted. To improve both risk and consequence management, modeling and simulation (M&S) of IE attacks are needed. Without M&S, leaders may be able to make a decision, but it probably will not be the best decision. What is the difference? Potentially tens of millions of dollars in profit or degraded national security.

The more valuable the IE's components are to achieving *Objectives* and gaining competitive advantage, the more necessary a full-spectrum IE protection program. Competitors either overtly or covertly acquire products and reverse-engineer them. They exploit the vulnerabilities found to narrow/widen the competitive edge. Without protection to defend and deception to mislead them, they will continue to search and exploit, if for no other reason than their own survival is at stake.

An *Intelligence* capability is needed to protect the IE. *Intelligence,* such as economic and open source, needs to be in strategic, operational, and tactical plans and be synchronized with organizational functions such as sales and public relations. KM supports information gathering (also known as competitive intelligence, business intelligence, and corporate espionage) by facilitating timely information sharing to achieve situational awareness. If leaders are not cognizant of the myriad of activities and their meaning within the battle space — the arena in which products and services are sold — *Surprise* will happen frequently.

It does not matter if it is called competitive intelligence, business intelligence, market intelligence, or corporate espionage — it happens. Two Taiwanese were convicted of economic espionage for stealing trade secrets from Avery Dennison.[11] Hacking, cracking, social engineering, and related malicious activity can lead to significant losses. Nokia estimated that its damages from hacker Kevin Mitnick included development costs, $7.5 million for testing, and $120 million in lost revenue due to new developments delayed in reaching the market.[12] Organizations are recognizing the need to make *Security* part of the regular business process. This approach is essential to achieve *Security* and avoid *Surprise.*

Surprise

> *Strike at a time and/or place and in a manner for which the enemy is unprepared.*[4]

Before the enemy can effectively respond, the purpose of conducting the *Surprise* must have achieved most, if not all, of its *Objectives. Tactical* or *Strategic Surprise* does not mean catching the competition unaware. The competition may be overwhelmed by an attack it has seen coming if the attack is too fast or powerful for it to counter. Just-in-time IT insertion enables providing innovative products and services sooner and better than the competition. Always assume that one third of the competition is smarter and more motivated, one third is the same, and one third are less. Therefore, estimate that two thirds of the competition is expecting attempts to *Surprise* them. KM, by integrating the tacit knowledge of employees with open source information, can provide important insight into the where, what, when, and how to achieve *Surprise.*

Prevent *Surprise* as well as avoid a disadvantage from developing over time. The more valuable the IE's components are to a competitive advantage, the greater the need for full-spectrum IW defense. Perfect IW defense is not possible. To facilitate operating while under IW attack, maintain backups and shadow systems, and use risk and consequence management, disaster recovery, and business continuity plans.

Simplicity

> *Provide clear, uncomplicated plans and clear, concise orders to ensure thorough understanding and execution.*[4]

Relatively simple plans are usually best. The market changes and so must your plans. Simple plans are easier to change and to explain to employees and business partners. Everybody in the organization should know the *Objectives* and generally what needs to be done to achieve them. Vision and mission statements, annual goals, operating procedures, and other documents must provide the basis of guidance.

One IT corollary of *Simplicity* is "design for modest skill levels." A consistent approach and implementation supported with clear communications and training ease transitions to new technologies. *Simplicity* also means to design in robustness and resilience for and ease of reconstitution in a disaster recovery situation. "Simplicity appeals to the mass of new online users, to techies chagrin."[13] "We've got to remove complexity. We've got to add relevance," said Microsoft president Steve Ballmer on the need for suppliers to offer computer users a more manageable platform.[14] Set simple policies and use measures invisible to end users; obsess about ease of management.[15] To bring about KM cultural changes and implement the right technologies, obtain constant employee feedback. This feedback will make change less contentious, improve productivity, and ultimately result in competitive advantage.

Conclusions

Competition, either in the marketplace or on the battlefield, has been part of human nature for 10,000 years. The Principles of War learned from successfully executed warfare can be adapted to the business realm. Information is fundamental to both business and the military; therefore, enhancing and protecting the information environment are vital to gaining a competitive advantage.

The Principles of War and intellectual works by Clausewitz, Sun Tzu, Boyd, and others are not academic constructs to be taught in the classroom and forgotten. The study of successful warfare is a serious undertaking. For example, to grasp Sun Tzu, an understanding of the historical context (i.e., social and political), cultural considerations (e.g., Taoism), and martial arts (e.g., beyond physical combat to the meta-physics of nature) is essential. The Principles of War are guides — not absolutes — to provide structure for planning and operations. The judicious use of the Principles of War will more likely than not result in achieving *Objectives*.

KM is not a pop management fad, although by reading trade periodicals one might get that impression based on how proprietary, narrowly focused solutions are offered. When correctly applied, KM is the smart integration of technology and people to achieve *Objectives*. KM supports rapid, well-informed decision making for sound use of resources.

Together, the Principles of War and Coherent Knowledge-based Operations (CKO) offer a means to enhance and protect the IE. The combination of these will increase the bottom line — be that either profit or national security — and result in competitive advantage.

Martial Arts Are Directly Applicable to Information Warfare Planning[16]

On the surface it might appear a stretch to relate martial arts to information warfare, but the similarities are striking (no pun intended). There are physical, spiritual, and metaphysical aspects to martial arts that are directly and indirectly analogous to the physical, virtual, and perception management aspects of IW.

The perfect example to portray fundamental basics are immutable is when Lloyd Bridges in the film *Airplane 2* proudly and contentedly exclaimed, "Ahhh! Some things never change!" Martial arts are more than a sport and many in Japan, both in government and in business, read Miyamoto Musashi's *A Book of Five Rings* written in 1645, interpret it for modern times, and adapt it to politics and business. Enlightened strategy supported by enduring principles is timeless.

Most techniques in martial arts are based on the five Ds:

- Discern
- Defend
- Disrupt
- Deliver
- Discern

These may sound like simple concepts, but mastering them until they are as natural as breathing is a lifelong pursuit. Proficiency of the basic skills in one 900-year-old Japanese art, Ninjitsu, can be attained in eight to twelve years. The one test to permit progression to the intermediate level is pass–fail. Striking, grappling, rolling, and other physical skills are not part of the test. These skills must have been passed at exams years earlier. The student kneels facing away from the grandmaster, who is standing behind the student with a raised sword and attempting without warning to kill the student. (Today, a training sword made of bamboo is used instead of a steel sword, and the student is allowed two chances.) A passing score is given if the student senses intention and moves out of the way without being struck. Analogies can be made to a cracker breaking into your personal or corporate system, planting destructive Trojan horses and other malicious logic, stealing personal and proprietary information, erasing information, destroying years of work and customer goodwill, and disappearing into the computing ether. The organization's defenses should be sufficiently aware to sense attacks and sufficiently agile to respond before damage occurs.

Discern

The first technique is to discern, to be aware of what is going on, to have situation awareness. On a top level, intrusion detection devices provide tactical warning when someone using a pattern not permitted is knocking on your door. Detailed analysis of intrusion detection logs may uncover an attack, either forthcoming or underway. Audit logs from network management, operating systems, databases, storage networks, and other areas, when properly analyzed, may disclose improper behavior that has already occurred. That is in the cyber-domain. Do you have people monitoring chat rooms, associating with the cracker underworld, and using other less-common means of uncovering information? Do you belong to a group that shares information? Just what is your situation awareness? At the other end of the spectrum is the ability to have strategic indications and warnings (I&W) that malicious activity will occur. These capabilities require technical and human collection means, mostly in the realm of law enforcement and government organizations using sources and methods not available to businesses. To protect national interests, law enforcement and government organizations may distribute warnings.

Full-spectrum IW goes beyond computer network attack and defense. The purpose of IW is perception management using ways and means that include information manipulation, physical destruction, and other operations to influence decisions to support the desired outcome. Discern, therefore, covers the entire spectrum of situation awareness — social, political, military, and information. Ideally, situation awareness, when supported by respected offensive and defensive capabilities, will deter attacks. If deterrence is insufficient, you move on to the next step.

Defend

After you have discerned that there is a problem, you need to defend. If a punch were to be thrown at you, there are three things you can do: avoid it, block it, or absorb it. Absorbing punches is not a good thing. Several unchecked IW attacks on a business or country could bring down those entities. Blocking punches has merit. The good news is you did not eat the punch; the bad news is you are still in a position to be punched again. Are you fast enough to block all the punches? You cannot seize the initiative when you are on the receiving end. Good offense and defense involve movement, also known as maneuver, one of the Principles of War. Get off the line of attack. You cannot be punched if you are not there. In the cyber-domain, at nodes where storage and processing occur, strong access controls based on multiple rotating identification and authentication (e.g., passwords and biometrics) and encryption ought to be employed. Frequency hopping and spread-spectrum communications techniques will be used in the future, in addition to biometrics and encryption, to strengthen links. In the perception management area, public relations, advertising, publicity, and community relations may all be necessary for damage control.

Disrupt

Smart movement places you in position to avoid strikes and to begin the process of conducting your own attack. Once you have defended, you can then disrupt the attacker and begin to seize control of the situation. Disruption is done by either actively or passively unbalancing the assailant(s). Shifting the focus of attention by a new product announcement, legal action, or strategic hiring are examples of active unbalancing. Some passive unbalancing means can be done through advertising and publicity to influence competitor and customer perceptions, and conducting visible and unusual market research. Disruption is essential because most attacks fail when the enemy is not unbalanced.

Deliver

At this point you can, if appropriate, deliver the attack. The disruption was a counterattack. This is a new attack. Do not be associated with lawlessness; therefore, do not physically or virtually do things that will result in civil or criminal penalties. Law enforcement and government organizations have the capability and legal authority to conduct offensive physical and cyber-actions. That does not preclude you from aggressively taking the initiative. You can buy or sell companies, field products and services that one-up the competition, control customers' eyes in stores using appealing displays, and be a force in cyberspace with attractive and user-friendly Web sites that promote E-commerce.

Discern

After you have done what is necessary to control the situation, return to discern. Masters discern during all the Ds so as not to be surprised and to effectively deal with multiple attackers. The five Ds are simple to explain but complex to execute. They require constant practice to maintain proficiency. Agility, creativity, and learning are important for adaptation to change.

Discern, defend, disrupt, deliver, and discern. A strategy, business plan, and information warfare defense and offense can be built based on these five Ds. The five Ds have survived the test of time because they have been a successful basis of conduct on the battlefield and have been used in other venues with equally positive results.

Do Not Let Your Plans Be Deterred by Naysayers[17]

"Oh, it was just an idea" is arguably the saddest phrase. How many good ideas and plans have been lost because the naysayers drowned them out? Consider these examples of what successful people have been up against, and learn from them and plan accordingly.

> *"It would appear that we have reached the limits of what it is possible to achieve with computer technology, although one should be careful with such statements, as they tend to sound pretty silly in five years."*
>
> —John Von Neumann, 1949

> *I think there is a world market for maybe five computers.*
>
> —Thomas J. Watson
> Chairman of IBM, 1943

> *I have traveled the length and breadth of this country and talked with the best people, and I can assure you that data processing is a fad that won't last out the year.*
>
> —Editor in charge of business books for Prentice-Hall, 1957

Everything that can be invented has been invented.

—Charles H. Duell
Commissioner, U.S. Office of Patents, 1899

There is no reason anyone would want a computer in their home.

—Ken Olson
President, Chairman, and Founder
Digital Equipment Corporation
1977

The wireless music box has no imaginable commercial value. Who would pay for a message sent to nobody in particular?

—David Sarnoff's associates
in response to his urgings
for investment in the radio in the 1920s

Plan for Cyber-Insurance

What if, after thorough planning and plan execution, you realize there are insufficient resources? Turn to cyber-insurance.[18]

There are some events that were not expected. Hannibal crossed the Alps. Clay defeated Liston for the heavyweight boxing title. CD Universe did not think crackers would break into its systems. Buy.com did not expect a distributed denial-of-service attack. It is going to happen: one day your information environment defenses are going to be beaten. When they go down, your revenues and profits will also go down.

Even if you have taken all the appropriate measures to protect and secure your physical and virtual assets, much falls outside your span of control: protected and secured power and communications infrastructures, security-rich and bug-free commercial off-the-shelf software, and the creativity of crackers and phreakers to find new vulnerabilities in technology to exploit. Also, you probably cannot control your customers', financial stakeholders', and suppliers' information environments that are connected to yours. If you are an Internet-based company, then electronic and mobile commerce account for the majority of your revenue. Any disruption and your customers will go to your competitors. If you are a traditional brick-and-mortar company expanding into the Internet to enhance your customers' ability to do business with you, business interruptions and disclosure of customer data will taint your reputation and credibility. Business interruption can be costly on many levels.

The Internet Age has once again proven the adage that "Time is money." Suppose a company has U.S.$1 billion in electronic and mobile commerce revenue. That is, $2,739,726 per day; $114,155 per hour; $1903 per minute; and $32 per second. According to *Information Week*, hacker attacks cost the world economy about U.S.$1.6 trillion in 2000. That is, $4,383,561,644 per day; or $182,648,402 per hour. How long can your business afford to be offline?

You are up against a lot of smart people, from hostile intelligence sources with deep pockets to industrial espionage to script kiddies. Competitive intelligence, economic warfare, and hacking are being taught. Idaho State University has courses on spying, officially called competitive intelligence (CI). At least a dozen universities in the United States offer students a course on CI, including business schools such as Harvard, the Wharton School at the University of Pennsylvania, and UCLA. Iowa State University has an information warfare course; in Paris is a business school named the "School of Economic Warfare"; MI5 in England teaches businesses how to fight foreign spies; and some of the best hacking schools are in St. Petersburg, Russia.

Fred Cohen, a principal member of Sandia National Laboratory's technical staff, is credited with inventing in 1983 the computer virus while a graduate student. He directs a cyber-security research staff as well as Sandia's Cyber Defenders Program. He demonstrated to clients how attacks against them could cause factories to be shut down, cause environmental damage and kill people through release of chemicals or biologicals, or drain millions of dollars. His research staff has modeled more than 400 attacks, and about 2000 more are planned.

Are you sure your information environment defense is up to the challenge? Can you withstand multiple simultaneous physical and virtual attacks?

Suppose malicious crackers manipulate or destroy your databases, post your customers' contract information on the Web, or steal your intellectual property (IP). To avert a public relations disaster, you work with marketing, legal, public relations, CIO, and systems administration, and perhaps even the stock analysts. What about law enforcement? Now there is an interesting dilemma.

In the United States and many other countries, computer crimes default to criminal events and are not considered national security incidents unless proven otherwise. Law enforcement seeks evidence during the investigation so they can hand off for prosecution. It is a cold statement, but law enforcement usually does not care about your business. They will want to seize your equipment (or, at a minimum, shut down your site to investigate), maintain a chain of custody, and perform forensic analyses. If you do not copy your data and mirror your site so that legally acceptable copies can be presented to law enforcement, your highly reliable, but not robust, system is not doing you any good when it is not available due to seizure. If you cannot afford backup, it is your call: do not report the crime and put on a public face that all is right, and potentially lose business in the mid-term because almost everything is eventually disclosed, or tell law enforcement and perhaps lose business in the near-term.

Can your business afford loss of revenue from any of these situations? There is a way out of this predicament: cyber-insurance. You have to decide on the business case for these products. Cyber-insurance is more than financial remuneration if there is a loss. It can be used as a forcing mechanism to improve your information environment defense. Because there is little actuarial history to fall back on, insurance companies selling "cyber" policies limit their risk by requiring a stringent assessment of your security and protection capabilities. These assessments fall into three additive tiers, and your premium will depend on the one you elect: questionnaires (some are 250 questions and require documentation to be attached), remote testing, and on-site visits. These structured and in-depth assessments will identify the strengths and weaknesses of your information environment defense.

A usual precondition of being awarded a cyber-insurance policy is to undergo a rigorous security assessment by a third-party information technology (IT) security firm. The cost ran into the tens of thousands of dollars for start-up dot.coms with no security track record. You want an independent review of your INFOSEC to point out weaknesses before they are pointed out for you. Get reviewed by an IT company specializing in full-spectrum information environment defense. They may not have some of the arcane security functions in-house, but as a systems integrator they have subcontractors who do. They will tailor a team based on the level of risk you are willing to take on. For example, if you do not want personnel security or physical security evaluated, they should advise you as to risks you will be incurring.

A brief Web search disclosed 25 insurance companies, from small independents to large corporations, selling cyber-insurance. There are several other companies that do not sell such policies but have expanded their property and liability coverage.

In August 1999, comprehensive cyber-insurance sold for $45,000 to $50,000 per $1-million coverage (4.5 to 5 percent of the amount insured). In August 2000, the range came down to $10,000 to $25,000 per $1-million coverage (1 to 2.5 percent of the amount insured). Coverage limits can be as high as $200 million; deductibles can be in the millions of dollars. Companies with revenues of U.S.$1 billion or less can expect to pay premiums of between $25,000 and $125,000 for $25 million in coverage. Some applicants have received premium differences of 500 percent from different companies. It is certainly worth your time to comparison shop.

Types of insurance cover generally include:

- First-party coverage (property): computer network security, business interruption, public relations coverage, IP coverage (loss of IP, information, and information systems resulting in lost revenue)
- Third-party coverage (liability): media liability, interruption of service liability
- Misuse of trademarks, domain names, plagiarism, copyright infringement, defamation, libel, privacy violation, and cyber-extortion

Lloyd's of London said a survey of U.S. insurance workers at a San Francisco conference found that 70 percent of the delegates identified E-commerce as the major emerging risk of the new century. Andrew Tanner-Smith, an industry analyst at Frost & Sullivan, said, "It's very difficult for firms to put a value on the confidential information which is needed to establish the extent of insurance coverage." One problem of traditional insurance policies is that some insurance carriers do not protect against data loss or the consequences because data is considered intangible. To get around this accounting problem, "Some commercial property policies treat data as tangible property," said Matt Camp-bell, Assistant Vice President and Associate Counsel at Chubb Corporation.

Cyber-insurance has quantitative and qualitative management return on investment:

- Higher earnings by reducing improper IP disclosure and fraud
- Reduced improper behavior (e.g., viewing pornography at work), resulting in lower overhead costs (e.g., available bandwidth should increase, delaying the need to purchase a larger network)
- Information environment defense assessments will reduce cyber-insurance premiums
- A staff on the lookout for electronic crime and a well-tended information environment defense program promotes business continuity
- Support legal due diligence
- Smooth earnings when you sustain damage in your information environment, making shareholders glad you had foresight and were enlightened

Cyber-insurance is not a substitute for establishing and maintaining a sound information environment defense. On balance, cyber-insurance may be what you need to avoid or reduce the effect of attacks and provide compensation in the event of a loss.

For business continuity, disaster preparation, and disaster recovery to be successful, there must be:

- Management commitment
- Sufficient resources (i.e., people, money, and time), including reserves
- Underlying principles
- Synchronized and coherent strategy
- Contingency considerations detailed in disaster recovery and business continuity plans
- Insurance to fill in those unexpected gaps and to smooth revenue streams

Practical Planning

Abraham Lincoln said, "Things may come to those who wait, but only the things left by those who hustle." What are your plans for dealing with a distributed denial-of-service or other virtual attack, natural disasters, a trusted employee who sells trade secrets, a flood of new viruses, physical destruction of key facilities, or a misinformation campaign against your company? What if these happened simultaneously? Be prepared for the worst case and the mundane will be easy.

Lead the Planning Efforts and Expect the Unexpected

CEOs must not just rely on their staff to do the right thing. They must plan, lead by example, and be active participants in the information warfare battleground.

They must ask the right questions and not be satisfied with general answers. Then the information must be incorporated into plans. For example:

- How well are we protected?
- How do we know that?
- Will we know if we are attacked?
- How will we know it?
- When will we know it (e.g., immediately, four hours after the attack, when?)?
- What responses to the attacks have we established?
- Where are we most vulnerable?
- How do we know that?
- What has been done to mitigate the vulnerabilities?
- If attacked, how fast can we recover?
- How do we know that?

Another question to ask along the way is, "How much will it cost in terms of time, people, and equipment?"

Risk management planning may be done, but that is not enough. Consequence management planning must follow each proposed risk management decision with its own decision tree of "what-ifs." The CEO should also know:

- How much does it cost to produce?
- How much does it cost to replace?
- What would happen if I no longer had that information?
- What would happen if my closest competitor had that information?
- Is protection of the information required by law? If so, what would happen if I did not protect it?

Have a Perception Management Plan in Place

Be prepared for disasters, from nature, man-made external to the organization, and self-inflicted. One company wanted to release information about its products and services. Lacking central focus, each organization within the company "did its own thing." Public relations released one version of information while the organization's Web site had different information, and sales representatives and legal had yet different versions. This confused customers. Worse still, the competition used the poor implementation against the company, stating this was a way to deceive customers (which it was not) by masking problems within the company and its products. "Ya gotta have a plan."

Summary

Good planning is the key to almost anything — examinations, sports events, business, and war. The more time you spend planning and in preparation, the less time you will need to spend in execution and, as a benefit, the better the outcome should be. Coherent Knowledge-based Operations, the framework to conduct KM, IW, and network-centric business in a synchronized and coherent fashion, should be considered in planning your IW defenses and offensive operations. Because there are many similarities between the battlefield and the marketplace, the Principles of War can be applied by businesses and other organizations. Combining CKO with the Principles of War for preparation and execution should enable a nation-state or business to be dominant in almost any endeavor. There are many other models that can be used for planning and execution; the martial arts offer several. Choose those that are congruent with the culture of your organization.

Bad things happen to good businesses, and to be on the receiving end of an IW attack is not unlikely. While not a substitute for a good defense, planning and cyber-insurance can fill in revenue gaps during and after an IW attack.

Most of all, leadership is required for the best preparation. Anything less than the best will result in less than the best execution. In today's world of real-time action, in which IW is used to enhance operations and deceive competitors, anything less than the best plans can result in loss on the battlefield (or loss of national sovereignty) and removal from the marketplace.

Notes

1. Gaining Competitive Advantage and Protecting the Information Environment by Using the Principles of War with Knowledge Management, Perry Luzwick, *Information Security Bulletin,* Chi Publishing, July 1999.
2. *Network World,* August 10, 1998.
3. http://www.au.af.mil/au/cpd/cpdgate/prinwar.htm, adapted from Armed Forces Staff College Pub 1, 1997; Joint Publication 1; Army Field Manual 100-1; Naval Doctrine Publication 1; Air Force Manual 1-1; Fleet Marine Force Manual 6-4; Military Review (May 1955 and September 1981); *U.S. Naval Institute Proceedings,* November 1986.
4. Principles of War, http://www.wpi.edu/Academics/IMS/Depts/MilSci/BTSI/prinwar.html.
5. Sun Tzu, *The Art of War* (Lionel Giles, trans.), http://www.all.net/books/tzu/tzu.html.
6. Surprise and Anticipation: The Principles of War as Applied to Business, Dr. Chet Richards; http://www.belisarius.com/modern_business_strategy/richards/chi_and_cheng/cheng_and_chi.html; draft of November 28, 1998.
7. Franklin C. Spinney, *Proceeding of the U.S. Naval Institute,* May 1998, p. 27.
8. Applying the Principles of War in Information Operations, Major Bradford K. Nelson, U.S. Army, http://www-cgsc.army.mil/MILREV/English/SepNov98/nelson.htm.
9. U.S. Companies Lost $250 Billion Last Year from Trade Secret Theft, Loss of Information in the Workplace Creates Growing Concern for Corporate America, New York — *Business Wire* via NewsEdge Corporation, March 31, 1999.
10. *Information Security,* May 1998, p. 12.
11. *Baltimore Sun,* April 29, 1999, Business section, p. 2.
12. How Much Damage Did Mitnick Do?, Douglas Thomas, May 13, 1999, http://www.wired.com/news/news/politics/story/19488.html.
13. *ComputerWorld,* March 22, 1999, p. 40.
14. http://www.techweb.com/wire/story/TWB19990409S0022?ls=twb_text.

15. IT Embracing Security Policies, *PC Week,* May 10, 1999, p. 24.
16. The Five Ds in Martial Arts Are Directly Applicable to Information Warfare, Perry Luzwick, "Surviving Information Warfare" column, *Computer Fraud and Security,* Reed Elsevier, Elsevier Science, September 2000.
17. (In)Famous Predictions, Perry Luzwick, "Surviving Information Warfare" column, *Computer Fraud and Security,* Reed Elsevier, Elsevier Science, May 2000.
18. If Most of Your Revenue is from E-Commerce, Then Cyber Insurance Makes Sense, Perry Luzwick, "Surviving Information Warfare" column, *Computer Fraud and Security,* Reed Elsevier, Elsevier Science, March 2001.
19. Security? Who's Got Time for Security? I'm Trying to Get My Job Done, Perry Luzwick, "Surviving Information Warfare" column, *Computer Fraud and Security,* Reed Elsevier, Elsevier Science, January 2001.

Chapter 18

Those Who Do Not Accept Change and Adapt Will Be Consumed By It; The Enlightened Will Survive

---

If you consciously try to thwart opponents, you are already late.

—Miyamoto Musashi
Japanese Philosopher and Samurai, 1645

Whereas Chapter 1 summarized what we were going to discuss, this chapter summarizes what we discussed. This chapter discusses survival in an information warfare environment. The reader may find some of the topics presented herein covered in other chapters throughout this book. However, it was felt that the information presented in this chapter provides a good "wrap-up" of the state of the global information warfare environment, addressing and reinforcing key issues that every security officer, information warrior, and executive manager in government and business should understand.

Introduction to Survival

Traditional security has always focused on the physical world, but today that focus may not always be valid. Controlling the information environment (IE) may be more effective than physical attack and may be used to prevent it. This leads to the conclusion that controlling the IE to successfully attain corporate or nation-state objectives is the approach that should be pursued to gain the maximum benefit. Using Coherent Knowledge-based Operations (CKO), which was described in detail in a previous chapter, corporate or government action would be synchronized and coherent, following consistent themes conducted simultaneously.

Accepting this approach requires breaking with traditional strategy. Senior corporate and government leadership support is necessary to develop the appropriate planning guidance, strategy, skilled workforce, and plant and equipment. Corporations and nation-states need to boldly accept the new reality lest they wish to lose and not be able to re-attain the competitive edge. Bureaucracy has no place in an information warfare environment with nanosecond attack weapons requiring nanosecond responses.

Senior leadership is essential for security to be meaningful to the bottom line or national security of nation-states. Corporate espionage is becoming as big a threat as government espionage — maybe more so. Netspionage has become a valuable tactic in support of a corporation or government agency's overall espionage strategy.

According to an FBI report, U.S. companies are under economic attack from 23 countries trying to steal trade secrets and other intellectual property. A recent survey conducted by the American Society for Industrial Security (ASIS) reports that 56 percent of the 172 companies surveyed discovered at least one attempted theft of intellectual property last year. Sixty two percent of surveyed companies have no procedures for reporting information loss; 40 percent do not have a formal program for safeguarding proprietary information. The Computer Security Institute (CSI) and the FBI reported in a 2001 survey that 35 percent of 538 organizations interviewed reported financial losses due to computer security breaches, from financial fraud, and theft of proprietary information to sabotage.

There is no silver bullet, no one-time expenditure of money to "fix the problem," and no means to put the genie back in the bottle. Enlightened and dedicated leadership willing to stay the course is necessary to guide governments and businesses.

The Need for a New Approach to Security — Defensive IW

Throughout the previous chapters, the general situation, the threat, and a range of countermeasures were discussed. One of the most important aspects that will affect the ability of governments and organizations to create, and to operate, systems that can sustain during the future has yet to be discussed in detail. That aspect is the approach that responsible organizations and infrastructures must take to ensure that we have the correct environment to endure.

This will require a significant change in the attitude and approach that is taken at all levels of governance and management. What will be required in order for the structures that we understand to survive is a large-scale adjustment in the attitudes taken on the whole subject; a realization that the threats are real; and the adversaries are serious about IW. Do we need a "pearlharbor.com" as Winn Schwartan puts it? We have already seen it in the physical world. Can the virtual world's Pearl Harbor be far behind?

The Changing Environment

To the present day, we have a long history of understanding the issues that are related to security that is imposed by physical, procedural, or personnel means. We are also beginning to understand the IW offensive and defensive worlds. For as long as we have had groupings of individuals who were, are, or at some time in the future may be, in conflict, there has been the need for security. Places, people, and things have all needed to be protected and we have learned, over an extended period of time (thousands of years) and through trial and error, what works and what does not. The things that we were, in the past, trying to protect could be seen and touched and were comprehensible (even an enciphered piece of information is comprehensible — you can see it and understand what it is, you just cannot understand what it says). The methods that were used to protect the information were well tried and tested; and while all, potentially, are flawed, they could be tested by the user who would gain confidence in the defensive measures from an understanding of metrics such as the size of the guard or the thickness of the walls.

Unfortunately, in the very recent past — that is, in terms of years and decades rather than centuries and millennia — this has all changed and we now have a situation in which the things we are trying to protect are not tangible. All of the conventional barriers that we had grown to know and have confidence in are now, individually, pointless and ineffectual. In a very short space of time this new concept that is not tangible, (you cannot see the information) and it has no comparable characteristics to that of which we have experience.

As an example, take a look at the physical volume that information occupies. A box of paper contains 2500 sheets. A 3.5-inch floppy disk, which weighs a few ounces and occupies the same space as around four sheets of paper would store in the region of 550 pages (just over a ream of paper) of text on a floppy disk. A huge saving, but this is already *passe,* with both hard and removable storage media holding orders of magnitude more material. Perhaps the comparator now is the LS120 disk (approximately 47,000 pages of text or nearly 20 boxes of paper) or the Zip 250 disk (approximately 98,000 pages of text or close to 40 boxes of paper), both of which are of similar size to the old 3.5-inch floppy disk.

The physical barriers, those that we have previously relied on, such as walls and fences and even national borders can now be easily circumvented. To function properly and effectively, we introduce the computer and the communications that are required to allow us to gain the maximum benefit from it. We have, for the first time in history, allowed technologies to outstrip our understanding of how to use them in an effective and controlled way. The only option that we have if we want to be totally sure that we are not at risk of information leakage or contamination is not to use the technology. However, one can go back to almost as easily stealing information by copying documents or stealing them.

In the present-day environment, we rely heavily on gaining security from methods and mechanisms that are well tried and tested. Since groups of people first got together, there has been a requirement for security measures. The procedures that are used have been developed and tested over centuries and millennia. The triumvirate of physical, procedural, and personnel security has always been required and none of them, individually, are totally effective. If you have good physical security (high walls) but do not have good personnel security (you have not checked who is inside the walls), then your security is worthless. If you have good physical security and good personnel security but do not have good procedures (you do not patrol the wall), then again your security is pointless. Finally, if you have good personnel security and good procedural security but do not have good physical security, then all of your efforts will be in vain. How can you defend an undefined boundary?

In our new electronic environment, the whole concept needs to be revised. In this first departure from what is known and understood, the expertise in security does not reside with the people who have developed and gained experience over the whole range of the subject. Security "experts" traditionally serve a long apprenticeship in the military or one of the three-letter intelligence or law enforcement agencies, where they receive training, gain experience, and learn from their more experienced colleagues. Suddenly in the world of information security, these people are largely redundant. They do not have the domain knowledge of IT and information systems. As a result, security has suddenly moved from the domain of the mature, experienced professionals to smart kids with or without degrees who have never had to get their hands dirty in the field.

This poses a number of dilemmas. The first is that the seasoned professional has a track record and experience. He is normally more mature and is known by the "security" community. In reality, he has a pedigree that can be judged. With the new breed of information security "gurus" that are being produced by the universities, institutes, and learning by doing, all of this traditional background is missing. Organizations are asked to take, at face value, the skill and knowledge of a person who may or may not have a relevant degree but who claims to be an "information security" specialist.

To make matters worse, by employing information security specialists, we are making the provision of comprehensive security much more difficult to achieve because we are creating a specialization and treating the subject in isolation rather than as a part of the holistic security environment. This can be managed and incorporated with good management but there are significant difficulties as a result of the immense cultural rift between the two types of expert. One security professional who wished to remain anonymous put it as shown in Exhibit 1.[1]

Exhibit 1.   It Is the Twenty-First Century and about Time for Doing the INFOSEC Job Right

We must change the way we do the job of protecting information, but not just that information that is being processed, stored and transmitted by our massive computer networks. But where to begin? In thinking about that, we thought about two major areas on which to focus and they are (1) the flow of information throughout a corporation or government agency and (2) the INFOSEC officer as a profession. If we are to make the changes that are undoubtedly required to adequately protect sensitive information then those are the two areas where we should start.

Why? Simply put, it is because the information flowing throughout a corporation or government agency, by whatever means, provides a holistic look required to adequately protect it. Also because the INFOSEC officer is the one that must take the lead in the protection efforts.

We are spending a great deal of time and money, in fact on a global basis, trillions of pounds annually to protect information stored, processed, and transmitted by computer systems. However, that is really only a piece of this valuable, information asset, protection process. One can have the very best information systems security mechanisms in place, but still not adequately protect the information. We are losing sight of the goal of information protection. We have been putting the cart before the horse so to speak. We have been concentrating so much on information systems security we have forgotten the protection of that same information in other environments...

The segregation of the corporate security profession and the INFOSEC profession has led to a segregation of true information protection. Today's' information protection processes are like a piece of Swiss cheese. It is so full of holes, no wonder information can easily be stolen or compromised.

We must begin to establish a holistic approach to information protection no matter where it is within the corporation or government agency. To do that, we must look at information from cradle to grave. For example, right now there are basically no adequate controls on information in hardcopy, nor are there any real controls on what is sent out via e-mails or carried out in notebook computers, removable media, and the like.

So, one may have the latest firewalls, intrusion detection systems, massive monitoring, and aggressive actions where INFOSEC policy violations occur, but what about the user who prints out a copy of the sensitive information and walks out the door with it? Some corporations or government agencies may have guards that check for sensitive documents going out the door. Nice try, but no way does that work. The guards do not know what is sensitive and what is not. Even if they did, they would have to read the entire document and make judgments based on what they read. A ridiculously impossible and unrealistic task. So, they look for caveats on the document. No problem. As a user, even if the document is caveated coming off the printer, one can tape over it and make a copy of the document or even cut off the caveated borders.

As for removable media such as CDs or floppies, one can just hide them in their pockets. No corporation or even government agency will be doing body searches.

So what is one to do? Let's start where the information starts, at the desk of the creator of that information. If it is determined to meet the criteria for placing a caveat on it as corporate sensitive, then that is where the information protection process should begin. It may be on a PDA, desktop computer, a notepad, or any other information-holding device. The flow of that information should then be tracked and protection mechanisms put in place to protect

Exhibit 1. It Is the Twenty-First Century and about Time for Doing the INFOSEC Job Right (Continued)

that information. This is a needed process to start looking at the information in a holistic manner. It is not perfect since we do not live in a perfect world and we are not perfect human beings. However, it is the right approach to begin to protect sensitive information.

The information systems security officer has been developing into a formally recognized profession for a decade or more. However, just as we have piecemealed together how we try to protect information, as noted above, we have done the same with the profession.

If you recall, before the profession of the INFOSEC officer was established, physical security guards were used to protect the computer systems and the information that was stored and processed. In the beginning, little transmissions of information took place. As systems grew in power, et al., the information technology folks gradually took on the role of INFOSEC officers. In previous issues, we've discussed that issue so there's no need to repeat it here.

This approach has led to information technology professionals doing INFOSEC, auditors looking for computer fraud, and security professionals and retired or ex-cops conducting high-technology investigations. Again, a totally separate approach to the entire problem.

It is time to begin looking at the entire INFOSEC officer profession and establishing a mandatory skills (education and experience) base. The Certified Information Systems Security Professional (CISSP) is a good start. There are others over the years who have tried to jump on the bandwagon coming up with different INFOSEC certifications, as well as still others trying to be the certifying agency for high-technology crime investigators. Then there are also the Certified Information Systems Auditor (CISA), the Association of Certified Fraud Examiners Certified Fraud Examiners (CFE) credentials, as well as the American Society for Industrial Security's Certified Protection Professional (CPP) credentials. All worthwhile credentials as far as they go, but again they only look at a piece of the problem. They do not address the profession in a holistic manner.

What is needed is professional standards and a certification program that encompasses all the disciplines required to protect information.

The ideal INFOSEC officer — no let's not use that term. Actually what is needed is a corporate information protection officer (CIPO). This person should report to the highest level of management possible. In some corporations, it may report to the corporate information officer (CIO). However, let me caveat that by saying that should only happen if the CIO is responsible for information protection and not the IT vice president with a protection vision limited to sensitive information processed, stored, or transmitted by information systems.

You notice I use the term "protection" and not "security." That term is used because that is what the information security program is all about; however, the term security has often had a reputation of "guards, badges, and guns." In addition, the need to concentrate on privacy issues and liability issues should also be encompassed in the professional CIPO position.

The credentials of a CIPO should include as a minimum a multitude of education in global business and marketplace; business security; information systems; INFOSEC; social science; human relations; marketing; general investigations; and psychology. The CIPO must know the global business environment; so undergraduate degrees and certifications in such fields are necessary, as well as an MBA in international business. The CIPO should also be trained in project and time management, and other management subjects such as teambuilding, creative problem solving, resource allocation, and budgeting.

Exhibit 1.   It Is the Twenty-First Century and about Time for Doing the INFOSEC Job Right (Continued)

The CIPO should be experienced in all aspects of investigations to include interview and interrogations, evidence collections, as well as systems development and architecture, networks, Internet, auditing, business security, and a minimum of ten years of experience in management.

This of course is a difficult task for someone to accomplish but we must set some high standards and begin developing professionals to fill such positions. INFOSEC managers are already beginning to reap the financial rewards (vice president of INFOSEC in New York City may command an annual salary and benefits of over U.S.$300,000) for their experience and education. However, most also lack the true skills necessary to do the job in a holistic manner.

It is about time we take a hard look at protecting information as it flows through various environments throughout the corporation or government agency — not just through computer systems. We must also begin to re-look at the INFOSEC profession and develop specific standards that require the true INFOSEC professional to be able to provide a holistic information protection program.

Control and the Information Environment

The information environment has assumed a hugely increased importance with the advent of the computer and the supporting communications capability. In the Gulf War, we saw what came to be referred to as the CNN syndrome when, for the time, the decisions of the military commander in the field were affected by information straight from the media that had gone through no validation or assessment. Up to this point, the intelligence staffs had prepared the intelligence picture for the military commander from all of the information sources that were available to them.

This intelligence was processed, integrated, evaluated, and delivered as a single coherent picture. Unfortunately, this process of collection and assessment and integration is time-consuming, with the result that this intelligence is not available for some time after an event. In the Gulf War, the cameras of the media were broadcasting in near-to-real-time from both sides of the conflict, with the result that there was reporting of the results of an action almost as soon as it had occurred. This was available to the commanders and, of course, they wanted to see it for themselves. Unfortunately, for the intelligence communities, what this meant was that the first reporting of events that the commanders saw was from the media. For the first time in history, when the intelligence staffs presented the fused intelligence picture to the commander, they were now at the disadvantage and were trying to change his perception of what had occurred, because his first impression had been obtained from the media.

When you consider that in the competitive world of the media, it is the most graphic and exciting presentation that wins the ratings wars, it is not too difficult to see that the presentation of events may not be even and, of course, there is not the same level of validation of the source of the information that is presented.

If we can allow ourselves to be caught in this unintentional trap, how vulnerable are we to an adversary who understands these issues and could manipulate the media to present a view of events that would affect our decision makers?

The Russians clearly understand the power of information and their attitude is reflected in the academic text[2] from as long ago as the mid-1990s that states "Information Warfare is a way of resolving a conflict between opposing sides. The goal is for one side to gain and hold an information advantage over the other. This is achieved by exerting a specific information/psychological and information/technical influence on a nation's decision making system, on the nation's populace and on its information resources structures, as well as by defeating the enemy's control system and his information resource structures with the help of additional means..."

There was another aspect that has been very little discussed in the military, but this was probably the first occasion when a commander's decisions were dissected as they became apparent, in front of a world audience as the outcome of those decisions were realized. This, without a doubt, has put pressure on military commanders to consider the outcome options of a course of action and to look at the impact of the various options. It has also concentrated the minds of a large number of politicians, who tend to be sensitive to world opinion when it is vocal and against the issues that they have supported. When you know that a decision that you make will be reviewed for its impact, will you make a choice that, as a direct result causes loss of life, if there is an alternative that can be taken that does not?

One advantage of information warfare is that it is very often covert, out of the view of the media. Therefore, the commanders can make decisions focused on achieving their objectives without being second-guessed by the news media and politicians. Furthermore, there is less loss of life and physical destruction of other than electronic assets. Consequently, if or when politicians or media discuss the matter, there is less of a negative impact and the public would also tend to be more supportive.

What if one of the options for a course of action will cost a small number of lives now, but will shorten the conflict and save considerably more lives in the overall conflict. What if another option causes no loss of life now, but will lead to considerable loss of life at some future date? As the commanders are now judged on their every decision, almost as they make them, will they and can they still make the best decisions? If they make a sound decision that appears to be expensive in the short term, will they still be in command to make the subsequent decisions.

This is a very political/military example of the changing role and importance of information but it does not take a huge stretch of the imagination to picture a similar scenario in the commercial sector. The role of information and the way it is utilized must be clearly understood if it is to be used effectively in the future to ensure that we retain control of the information environment.

Information superiority is a term that the military has understood for some considerable time. In military terms, it is viewed as a force multiplier in that by gaining "information superiority" in the given environment, the effectiveness of deployed resources can be significantly incremented. This requirement for information superiority is recognized by armed forces around the world, and this is shown by a comment from the Siemens Company in Germany, who supply the German military. It stated, on their English language homepage,[7] that "Command efficiency depends directly on information dominance. The German armed forces will only be able to acquire and maintain information dominance and, hence, continue to meet the requirements placed on them by making full use of progress and innovation in technology."

The Siemens' site also demonstrates that this organization recognizes the concept of force enhancement through the use of information and identified that "information dominance is a force multiplier." The defense contractor recognizes that information dominance represents a potential advantage that is based on information technology.

Another expression of the way in which information is considered is the Observe, Orient, Decide, and Act (OODA) loop. This term is used primarily in the military to describe the decision cycle. What the OODA loop is in fact describing is the decision cycle that applies to all rational decisions that are made. It is expressed in terms that are common to the area in which it was first defined and used, but the concept behind it holds good in all areas. In the conventional decision cycle:

- First you Observe what is happening.
- You then Orient your thinking to understand what you are observing in the context of everything else that is happening that is relevant.
- When you understand the contextualized input, you can make a Decision on the appropriate course of action that needs to be taken.
- Finally, you can take that Action.

There is no advantage in understanding this cycle unless you can use the knowledge to speed up your decision-making cycle and make decisions in less time than your opponent. This is known as getting inside your opponent's OODA loop (see Exhibit 2). If you can complete the decision cycle faster than your opponent, you can act more rapidly when both sides have received the same information inputs, whether you are fighting a war or are engaged in business competition, and you will have increased your chances of winning.

The use of the appropriate information at the correct time has been part of military thinking for a long time, and terms such as information flooding and information overload are now well-recognized. In the past, there was always the problem of getting sufficient information to the decision makers for them to make the most effective choice of action at the appropriate time. Now, in many ways, we have the reverse of the problem, where we flood them with so much information that they cannot elicit the items of information that are relevant to the issue at hand.

If we allow our decision makers to be in this position, we have, potentially, handed an advantage to our adversary because we have caused the making of a decision to be more difficult and possibly outside the OODA loop of the

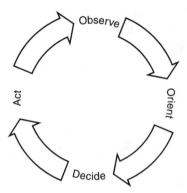

Exhibit 2.   The OODA Loop

adversary. If we can force this situation on our adversary, while at the same time maintaining control of our own information flows, we have gained a potential advantage. The concept of information superiority is not new but the potential to achieve it has developed enormously with digital communications and computers.

The absence, at the moment it is required, of a critical element of information can spell the difference between success and failure in the modern political, military, or commercial arena. The side with the tightest OODA loop operates at a higher tempo, forcing the opposing side to react to its moves. Through a successful campaign of subversion (marketing), deception (product development protection), and psychological operations (advertising), friendly forces can increase the size of an opponent's OODA loop without affecting the size of their own. If the information warfare (business intelligence) campaign is fully implemented, the enemy (competition) may ultimately be compelled to work toward targeted objectives and lead its force to ultimate destruction.

Based on the premise that information is a strategic asset, a portion of IW doctrine seeks to disrupt or deny access to information to seize the initiative from an adversary. The other half of IW doctrine seeks to maintain the integrity of our information gathering and distribution infrastructure.

In a paper written by Captain G.A. Crawford,[4] he identifies that, "In the past decade we have witnessed phenomenal growth in the capabilities of information management systems. National security implications of these capabilities are only now beginning to be understood by national leadership." It is noteworthy that with all of the material that has been written on the subject of IW, this is one of the few texts that recognizes the relevance of systems that manage the information and the potential power that can be developed from them.

The Need for Enlightened and Dedicated Leadership

There will, if an environment in which organizations can feel safe to operate in is to be achieved, need to be significant changes in the attitudes of both government and management at all levels of organizations. There will need to be infrastructures put into place that will allow, at an international level, for collaboration between governments and law enforcement agencies. Perhaps one of these will be forums in which incidents can be reported in a suitable manner by individuals, companies, and governments and where best advice can be gained. While these exist in some countries and communities, they must be ubiquitous and easy to access. If attacks are taking place at Internet speeds over structures that do not recognize national borders, then any impediment that the current structures and organizations impose will encourage the perpetrator.

In government, in most of the democratic nations, an individual who will champion the cause of creating the correct environment for the protection of information systems is a conundrum. It would require a political nominee that was willing to put the cause that they were supporting not only above their own ambitions (information systems security is not an area that has a track record of producing new party or national leaders) but above party loyalty. They would need to have seniority within their own party, cross-party support, and tenure in the post for a period of more than one term of office if they were to have any significant effect.

This is, in some ways, summed up by a statement to the U.S. House Judiciary Subcommittee on Crime, by the president of the Information Technology Association of America (ITAA), Harris Miller, who, when referring to the sharing of information in the fight against cyber-crime, stated, "Obviously, leadership in information sharing and strategic analysis has not been, for want of a better term, very stable. You can't run a railroad with leadership changing every year."

There was light at the end of the tunnel, at least in the United States. The U.S. government is taking the matter seriously and, according to a press release by the ITAA on the same subject, in a statement to the U.S. House Judiciary Subcommittee on Crime, Harris Miller also identified that "government leadership in a meaningful partnership with industrial leadership is essential and that all of the involved parties need to share information to reduce vulnerabilities and improve network security."

In his testimony, Miller went on to state that, "Industry and government must share the view that, given the nation's extensive dependence on information systems, information security means economic security," adding that information sharing and analysis centers (ISACs) were one of a number of mechanisms that were being adopted by industry, financed by the private sector, to combat cyber-threats and respond to attacks.

Miller said otherwise highly competitive information technology companies are working to overcome their natural reluctance to share information. He added that the Partnership for Critical Infrastructure Security is providing an important venue for cross-sector industry exchange.

In a separate, realistic article on the same subject, Ronald Dick, a career FBI agent commented in an interview[5] on the maturing scope and capabilities of the National Infrastructure Protection Center (NIPC). He identified some of the outstanding work that remains to be carried out with regard to improving and integrating the efforts of the security community to improve the flow of information between government and commerce. He stated that, "some of my infosecurity colleagues have been frustrated when they've tried to work with the NIPC. They find that the FBI culture and the more informal worlds of information security are often in conflict." This insight gives an indication of some of the cultural issues that must be overcome if collaborative working is to be achieved.

The cooperation and joint effort of government, industry, and academia is increasingly being recognized as essential, yet we have much to learn if we are to progress with the speed that is required. It is noteworthy that governments such as those of the United States and the United Kingdom have shown an innate ability not to learn from the past.

In the context of information security, we first saw, in the United States, the concept of the Clipper Chip. This was much discussed and explained as way to make it "safe" to do business on the Internet. Why was the government so surprised when the commercial sector and the private individual rebelled and refused to accept it? Well, the answer to that is that the arguments that were put forward were totally unconvincing.

In the first case, can the government be so naïve as to imagine that international and multinational companies are willingly going to hand over to the government access to all of their corporate information? Given the number of times that the United States has been caught passing on the results of information that it has gained through spying to create an advantage for U.S.-based companies, did they really think that companies based overseas were ever going to allow access to their systems? And for what — to prevent criminals from hiding their activity through the use of encryption? It does not actually take a national security expert to see the flaws in these arguments. In the first case, criminals already use high-grade encryption.

Organizations such as the Colombian Cali drug cartel can afford the best in all aspects of equipment and technology. Second, they do not tend to be the sort of organization that is likely to hand over a means of access to its information to the government. Take a look at the potential leverage that the United States could apply to the criminals. Oh yes, we will not let you trade in the United States unless you comply. Not totally convincing. Unbelievably, in the United Kingdom in 1998, the government there tried to introduce a key escrow clause into the E-commerce bill that was published for discussion. Not only had they failed to learn from the lessons of the government of the United States, but they did not even try to re-package it in any sort of convincing way.

This is one of the areas that will require considerable attention and effort if there is to be any kind of environment in which the individual or the organization can have confidence. Unfortunately, to achieve that status, national governments are going to have to accept a lesser role in the management of the infrastructure. This flies straight in the face of national security, so a dichotomy exists. However, if there is to be cooperation at a realistic pace across national borders and it is to work in such a way that businesses from the multinational to the medium and small enterprises will be content to take part, government must accept the change.

Appropriate Management

To be competitive in today's global markets, organizations must develop and implement effective network strategies. The design and implementation of a security strategy requires input from the disciplines of management, research, and analysis and must be fully integrated with the business strategy. If this is to take place, it will be imperative that the decisions are made and supported at the highest level in an organization.

Information security is critical to the welfare of the organization, whether it is a government agency, business, or association. Security awareness is vital to the success of the information security program. What most organizations are poor at is putting that knowledge into practice; and unless we can improve this aspect of our staff education, we are destined for failure. If you talk to any security person, they too find the awareness training dull, "a waste of their time." If they feel that way, how can one expect the non-security people to react with more understanding and enthusiasm? The entire security awareness process must be overhauled. Until then, as the saying goes, "You keep doing what you are doing, you'll keep getting what you are getting."

As both industry and business communications renew themselves to take advantage of the business opportunities presented by the Internet and technologies such as virtual private networks (VPNs) and public key infrastructures (PKIs), software and hardware products are evolving to meet these needs.

Enterprise networks must, increasingly, learn how to defend themselves against the increasing number of internal and external security threats. Some of the types of attacks they are exposed to include Internet Protocol (IP) spoofing, Trojan horses, and denial-of-service (DoS) attacks. Effective prevention strategies will need to be developed to combat them.

Ultimately, network security is a business issue, and information systems managers are responsible for developing the appropriate security solutions for their organizations and must understand all aspects of network security.

Corporate Objectives and the Control of the Information Environment

To control an information environment (IE), it is necessary to have some understanding of what that IE comprises. To fail to understand the IE will mean that the corporate objectives that we aspire to achieve will either not be met or will cost more than they should have. This is not good business, no matter what the organization is involved in.

To maximize the benefit that can be gained from the IE, it is necessary to ensure that publicly available information, which is becoming more commonly known as open source information (OSI), that is of benefit corporation and to the corporate objectives is collected and exploited. At the same time, it is necessary to explore the IE to ensure that corporate information that would be of benefit to a competitor is minimized or removed from the environment.

In an article by Robert Steele,[6] he puts forward one U.S. official definition of open source intelligence (OSINT) used by the U.S. Intelligence Community: "publicly available information appearing in print or electronic form. Open source information may be transmitted through radio, television, and newspapers, or it may be distributed by commercial databases, electronic mail networks, or portable electronic media such as CD- ROMs." This definition also indicates that OSI can be disseminated by a number of means to the broad public, or to a more select audience, through literature, which might include conference proceedings, company shareholder reports, and local telephone directories. Open source information is, by definition, unclassified and it is not subject to any proprietary constraints other than copyright.

One of the upcoming tools of the intelligent organization is the concept of Coherent Knowledge-based Operations (CKO), as discussed previously in the book. The attributes of CKO that will separate the successful from the unsuccessful are described as decisive, proactive, adaptable, and agile. The physical world will always lag behind the virtual domain that, in turn, will lag behind the realm of the human mind. This must be considered when new products and services are introduced. Situational awareness, the OODA model, and CKO will significantly increase an organization's chance of being successful.

If the CKO approach is to be accepted, then traditional strategies will have to be abandoned or, at the very least, severely modified. The role of the system security officer will have to change and there will have to be a role for information or KM. The defensive and offensive aspects of information warfare will have to be dealt with as part of a coherent whole rather than as two totally separate functions, and all will have to accept and adapt to the fact that they will need to operate in partnership to be effective. The choice of where to invest resources will be (hopefully) better informed and will be spread across the three main areas.

If there is to be any progress in gaining support for the adoption of this philosophy, then an element of what will need to be implemented will have to be metrics that will let the decision makers and the accountants be informed as to the cost benefit of the investments and also provide the operations staff with indications of the effectiveness of the measures they are using.

Adaptation

We have seen the rise and fall of the first wave of dot.com companies and there are a number of lessons that can and must be learned from this phenomenon if we are not to be doomed to repeat the past. The unsupportable valuations of the companies based on the hope of future earnings was driven by a failure of the investors to understand that there was no substantive value in the companies in the terms that have been commonly understood and used to value companies of the type that have existed until now. In conventional markets, the value of a company is made up of a number of elements. Part of this value is the bricks-and-mortar assets, part the track record, part the brand-name value, part the intellectual property, and part the earnings and future earning potential. In the dot.com companies, there was no history (track record), there was no bricks-and-mortar value, there was no brand awareness (they were mostly new starters), and there was little likelihood of profit in the early years. Yet people still invested. Why? Because they felt that the dot.com route was the future of commerce and that they had to be a part of it or get left behind in the rapidly moving electronic world.

The early adopters of the dot.com philosophy were driven by the opportunity and need to get to market and become a known name. Security awareness and the need for it to enable them to succeed, even if the product they were selling was viable, was not high in the minds of the entrepreneurs who led the way. Given that the dot.com boom was based on network technologies, it was essential that senior leadership understood the requirement for security that would ensue and the impact that it would have on the bottom line of the accounts of any venture.

For any security that is applied to a system to be effective, it is imperative that any actions taken have the support and backing of management. This starts with senior members of government who need to understand the issues and provide support for high-level policy, and then to senior management within organizations, without whose support there is no possibility that any security within an organization will be implemented or maintained.

The Threat of Corporate Espionage[7]

It is becoming increasingly clear that the threat to profitability and survivability of a venture is real and one of the vectors through which the threat to the venture is manifested is corporate espionage. The concept of industrial or business espionage is not new. The act has been taking place since the very early days of trade and has long been accepted as something that, while unsavory and distasteful, is a necessary evil if competitive advantage is to be maintained. A number of cases that illustrate how this activity has translated into the information age are given below.

In 1999, Will Knight reported that Egyptian police were investigating[8] an extremely serious, if low-tech, computer crime involving the theft of floppy disks from laboratories at Cairo University. The disks were reported to contain classified data giving details of Egypt's oil, gas, and uranium reserves in addition to the location of gold and copper deposits. While no value was placed on the information, it is easy to understand that the value would, potentially, have been huge.

In July 2001, industrial espionage apparently hit the high-tech world of Formula One motor racing when the Benetton team revealed that its season's world championship hopes had been damaged after computer hackers "stole" engine design data. The news was broke that it had been discovered that hackers had broken into the company's computer system in 2000 when the team was working on the final designs for the 2001 season's Benetton car. In 2001, the team has only managed to score one point in the first 11 races of the season.

One of the more infamous recent industrial espionage cases was that in which the former vice president for worldwide purchasing from General Motors, Jose Ignacio Lopez de Arriortua, was reported to be involved. He was indicted on charges of defrauding the company by taking boxes of its confidential documents with him when he took up a top post at Volkswagen in 1993. Lopez was indicted on six counts of wire fraud and interstate transportation of stolen property. It was reported at the time that the Justice Department had asserted that Lopez had copied a range of GM computer files on the cost of auto parts and of future car models and took them to VW when he moved to the German car maker.

In the United States in 1999, it was reported[9] that an Internet bookseller called Alibris, based in California, agreed to pay $250,000 to settle federal charges that its corporate predecessor had intercepted e-mails sent by its business rival Amazon.com and also that it had in its possession unauthorized password files. The company was charged with ten separate counts of unlawful interception of e-mail messages and a separate count of unauthorized possession of passwords with intent to defraud. The charges related to the company's predecessor, Interloc Inc., which had been an online bookselling organization that had also, through a business called Valinet, provided Internet service in the Greenfield area.

The information presented alleged that in the period between January and June 1998, Alibris/Interloc had intercepted e-mail messages from Amazon.com to clients of Alibris/Interloc that used Interloc e-mail addresses. The prosecution argued that the interception of the e-mails was intended, if possible, to gain a competitive advantage for Alibris' online book-selling business.

Prosecutors also alleged that Interloc kept unauthorized copies of the confidential password files and customer lists of its competitor Internet service providers. In defense of the actions of Interloc, a lawyer for Alibris, Ethan Schulman, stated that while Alibris accepted that Interloc had improperly intercepted the e-mail traffic, the intention was not to spy on Amazon.com, but to trace the source of a problem with a computer system.

In another case, it was reported[10] that the Italian police had arrested 21 people who were accused of taking part in a huge online banking fraud. If successful, the fraud could have cost the Sicilian regional government more than one trillion lire (U.S.$465 million) after accessing an account that contained European Union structural funds for regional development.

According to Italian officials, members of a criminal group, with reported links to the Cosa Nostra, allegedly succeeded in creating a bogus online branch of the Banco di Sicilia and were thought to be preparing to remove funds from an account that belonged to the Sicilian regional government. According to a senior member of the Palermo police department, the organization was reported to have used two bank employees who had been subverted to gain access to and utilize stolen computer files, codes, and passwords to enter the bank's information systems. If they had been successful, the funds would have been transferred to a bank in the Bologna area, and from there to accounts that were located overseas.

In a report in the *New York Times*,[11] it was revealed that computer companies affiliated with the Japanese Aum Shinrikyo doomsday sect, which had been responsible five years before for the release of nerve gas in the Tokyo subway that killed 12 people, had developed software for at least ten separate government agencies, including the Defense Ministry and law enforcement. According to police, who had carried out surprise raids on a number of the group's sites, they had also worked for more than 80 major Japanese companies in recent years.

There was considerable concern that affiliates of the Aum Shinrikyo sect should have had access to the systems of these government agencies and companies, and a number of organizations immediately ceased to use any software that had been created by these companies.

According to another report from William Malik, an analyst at Gartner Group, there were two cases of electronic espionage during the past couple of years that cost the companies involved over $500 million. In one of the cases, two companies involved in heavy engineering were bidding against each other for a $900 million contract. In the end, the winning company managed to outbid the other and won the contract by a fraction of a percent. According to Malik, this was not just due to bad luck. The company that had lost the competition had been testing network monitoring software during the bidding process and had later discovered that an unknown person had broken into the company's computer network and accessed files that contained bidding strategy information. If not for the monitoring software, they would never have known that anything was wrong.

In another case reported by Bill Hancock, the security chief of Exodus Communications (a French defense contractor) became aware that its designs were being given away. In the ensuing investigation, a computer criminal was identified who had taken a job inside the French company. Once there, he had laboriously embedded the organization's trade secrets inside Web site images that he then posted on the company's public Web site. A colleague then downloaded the secrets from the company's own home page. The method was only discovered as a result of Hancock noticing slight variations in the size of the image file.

While the problem is not new and the motives have not changed, the difference now is the level of access that it is possible due to networked computer systems and the speed with which they can be used against their owners.

The Dilemma-Security and ROI

Many chief executive officers (CEOs) and chief information officers (CIOs) have been slow to understand and invest in information security because they do not know and have great difficulty in establishing any method for measuring their return on investment (ROI). This has been the age-old problem with information security. Intangible assets such as information do not tend to be valued and, to date, no real effort has been made to calculate the cost of acquiring and maintaining the information.

Given that the "thing" that is to be protected does not have an allocated value to the organization, how can the decision makers make rational decisions as to how much they should spend on keeping the information safe and available? The other issue is actually, in part, that we have people like CIOs making decisions such as these. Do we involve the production manager in the decision making for the security of the operation? No, we use security professionals for this and they consider all aspects of the environment in forming their decision. In the case of information technology, we tend to regard it in isolation and this can be a very expensive and inefficient way to tackle the problem.

In an article by Dr. Anita D'Amico,[12] a very bright and articulate woman who is the director of Secure Decisions, a Division of Applied Visions, Inc.[13] from the Northport, New York area, she outlines some of the considerations that should be made when trying to measure the cost of providing security. In the article, she identifies that the costs of an information security breach can be both tangible and intangible. She then goes on to describe that the tangible costs can be derived from estimates of the:

- Lost business, due to unavailability of the breached information resources
- Lost business, which can be traced directly to accounts fleeing to a "safer" environment
- Lost productivity of the non-IT staff, who have to work in degraded mode, or not work at all, while the IT staff tries to contain and repair the breach
- Labor and material costs associated with the IT staff's detection, containment, repair, and reconstitution of the breached resources
- Labor costs of the IT staff and legal costs associated with the collection of forensic evidence and the prosecution of an attacker
- Public relations consulting costs, to prepare statements for the press, and answer customer questions
- Increases in insurance premiums
- Costs of defending the company in any liability suits resulting from the breached company's failure to deliver assured information and services.

Not all of these tangible costs will occur with each breach; some will only occur with major, well-publicized breaches.

The intangible costs refer to costs that are difficult to calculate because they are not directly measurable — but are nevertheless very important to the business. Many of these intangibles are related to a "loss of competitive advantage" that results from the breach. For example, a breach can affect an organization's competitive edge through:

- Customers' loss of trust in the organization
- Failure to win new accounts due to bad press associated with the breach
- Competitor's access to confidential or proprietary information

Dr. D'Amico wrote this article with a focus on the commercial world, but noted that the military environment has similar cost issues, even if the costs are measured differently. In the military, the most obvious tangible costs are measured in human lives, the cost of replacement of equipment, and prolonged military operations. The intangible costs would include the loss of a tactical advantage, the loss of international prestige or credibility, and a weakened negotiating position.

D'Amico also introduced a number of hypothetical examples of the cost impact of security breaches and quoted a report in which Forrester Research estimated the tangible and intangible costs of computer security breaches in three hypothetical situations. Their analysis indicated that, if thieves were to illegally wire $1 million from an online bank, the cost impact to the bank would be $106 million. They also estimated that, in the hypothetical situation that cyber-techniques are used to divert a week's worth of tires from an auto manufacturer, the auto manufacturer would sustain losses of $21 million. Finally, they estimated that if a law firm were to lose significant confidential information, the impact would be almost $35 million.

D'Amico comments that Forrester Research used both tangible costs and intangible costs in its estimates and included the loss of confidential information and reputation and gives some real-world examples of cost impacts which are based largely on the tangible costs.

> *In December, 1998, Ingram Micro, a PC wholesaler, had to shut down its main data center in Tucson, Arizona due to an electrical short.*
>
> *While the reason for the shutdown was not a security breach, the loss of Ingram's Internet business and electronic transactions from 8:00 AM to 4:00 PM mimicked what could happen with a Distributed Denial of Service (DDOS) attack or a major intrusion. As a result of its one day of lost sales and system repairs, Ingram estimates that it lost a staggering $3.2 million. This figure is comparable to Forrester's projection of a $21 million loss for an auto manufacturer who is unable to get tires for a week.*

D'Amico then went on to discuss the cost impacts across industries and gave some examples including: Some research and consulting firms such as Computer Economics (www.computereconomics.com) measure the impact of computer breaches across several companies or industries.

Open Source Intelligence

The concept of OSI and OSINT was mentioned previously in this chapter, but is a significant area that requires further discussion.

OSI is being put across as a "new" concept that has been made possible by the advent of the Internet. Nothing could be further from the truth. OSINT has been available and utilized for as long as we have maintained records of our activities. The factor that has changed is the ease of access and the sheer volume of information that is now available to anyone that wants to see it. Individuals and organizations give out vast amounts of information about themselves, some of it intentional, but for the most part in total ignorance. Governments spend vast amounts of money creating and maintaining agencies to gather and maintain information on anything that could be considered a potential threat to their security.

One of the cants of hackers in justifying their actions is that "information" should be free. A number of them, from members of the "Legion of Doom" in the early days of hacking, have been tried, convicted, and sent to prison because they individually and collectively tried to access proprietary information.

It would be a reasonable expectation that if information is a strategic asset, organizations will have made an assessment of its value based on the cost of acquiring and maintaining it and the potential gain or loss that may occur as a result of the availability of the right element of this information at the right time. Unfortunately, this is not the case. Information is, in most organizations, valueless because the organizations do not view it as an asset. As a result, the protection that could be offered to the asset is not applied.

Even within the information security community, there is widespread ignorance of the volume and type of information regarding an organization that can be collected from "open sources." This is not surprising because it has, historically, been a business intelligence function to gather information on competitors and the publication of information to the world has not been a security issue. The bad news is that if organizations do not take this problem a lot more seriously than they have in the past, they will do so at a high cost.

Robert Steele[6] defines OSINT as "intelligence that is derived from public information. This will provide tailored intelligence which is based on information that can be obtained from public sources that is acquired in a manner that is both legal and ethical." The use of OSINT can act as a force multiplier (make your manpower more effective, unit for unit, than that of the opposition), a resource multiplier (make your equipment more effective, unit for unit, than that of the opposition), or both. The use of OSINT can provide a practical political, military, or commercial advantage that complements the advantage that can be provided by effective and timely traditional intelligence. Unlike the traditional intelligence sources, OSINT is available at low cost and is increasingly ubiquitous.

The conventional definition of OSINT is confined to the standard commercial sources of traditional information and this fails to take into account the importance of unpublished materials. These may include information that is in an electronic form and human knowledge, as long as it can be accessed in any way that is both legal and ethical.

In both the military and government, there is still a very dated dislike for the collection of intelligence, with it being considered as not being a gentlemanly thing to do. This has not inhibited commerce, but the skills in intelligence gathering and analysis still lie, primarily, with the military.

Across most communities, there seems to be reluctance by individuals to assume responsibility for the collection and processing of OSINT. It does not have the same credentials as the more conventional forms of intelligence but does have a number of advantages, the main ones being the speed of access to the information and the ubiquity of the information.

Experienced intelligence professionals within the government and military communities have found that OSINT is not a substitute for traditional intelligence disciplines, such as human intelligence (HUMINT), imagery intelligence (IMINT), electronic intelligence (ELINT), and signals intelligence (SIGINT). However, OSINT does offer a number of advantages for the military when it is planning and conducting operations.

In the changing world of conflict that exists today, symmetrical conflict between two well-armed and balanced forces is increasingly uncommon. The more common event these days is a peace-keeping or peace-enforcement role by a coalition force in the Third World. The result of this is that there will probably not have been the depth of conventional intelligence collection performed on the forces' factions that are involved in the confrontation. The use of OSINT will provide a basis that can rapidly be acquired and from which a more complete intelligence picture can be developed in due course.

OSINT is also a potential means of achieving significant savings, in that many of the essential elements of information required by the commander and his staff can be acquired from commercial sources at a lower cost and in less time than from the conventional resources. OSINT, because of its origins and the range of resources involved in its collection, is often more up-to-date and involves no political risk in its acquisition.

An additional advantage of the collection of OSINT is that, whether it precedes or follows traditional intelligence collection, if appropriate, it can be used to protect the classified source of the information. The quantity and quality of OSINT will depend on, and will vary according to, the area in which the operations are due to take place.

The correlation between this type of use by government and the military and industry is not difficult to visualize. All of the disciplines mentioned above and the types of use can easily be of benefit in the commercial world.

A number of organizations have now come into existence that will provide assistance to an organization or perform the entire task of open source intelligence collection for an organization. Some examples of this include:

- The Online Intelligence Project describes itself as being oriented toward individuals and professionals who have an interest in international news, commerce, and references. It uses a government-type intelligence service model to provide the appropriate resources into departments or regional desks as appropriate
- The Open Source Marketplace describes itself as offering busy international professionals a single source through which they can gain the best-available open source intelligence, software, and services. OSS Inc. claims that it will provide its members with a selection of the best sources of information, including commercial imagery, useful and appropriate software for open source intelligence processing and analysis needs and additional open source information services in niche areas. They also offer their clients a proxy service to manage their open intelligence sources.

■ The Intelligence Network offers a service described as being dedicated to providing individuals and groups with news and documents from around the world that have been aggregated into a suitable and convenient format. This organization describes its offering as a fusion of software and human experts that traverse a network of more than 500 Web sites to deliver up-to-date news, documents, photographs, videos, and other content. It also claims to provide original content based on reports received from their own sources on the ground.

OSINT is, potentially, both an asset and a liability. There is a huge amount of information available on almost any subject if the searcher has the ingenuity to think in a lateral manner. A lot of this information is still in written form but increasing amounts are available electronically, and that means easily. To give an example, an individual posted messages to a newsgroup from an e-mail address at an organization. He was an engineer who had been contracted by the organization to install firewalls. Because he was inexperienced on some of the issues related to installing a firewall in the environment in which he was working, he had posted to the newsgroup asking for advice on how to configure the equipment.

Over a period of time, he asked for and received a range of advice in response to a number of questions. No great damage you might think. However, in his chats on the newsgroup, he had revealed quite a large amount of information about the system he was working on, including the platform, the network operating system, and a number of the applications. The situation was now that an individual, who was associated with the organization, had revealed a significant amount of information about the network topology. He had also used information from "experts" on the Internet, whose credentials he did not know, to configure the organization's firewall. To the criminal, this is as close to perfect as it gets; he can now, with minimal effort, by pressuring a man who has exposed too much information on an organization for whom he is a contractor, gain access to that organization's networks. When this story was used to illustrate the subject at an industry forum, the security manager for another organization in the same industry suddenly went white and started to make phone calls — his organization had used the same contractor! To put the situation in perspective, this was the banking industry.

Summary

In our increasingly interconnected environment — in which citizens need to be able to communicate with government and industry, and industry needs to be able to communicate and share information with government and other areas of industry — it is clear that doing more of what we have already done will give us more of what we already have, which is a situation where the cost of operation is rising to a level that will make it prohibitive.

If we are to succeed, we must operate in a very different way. Governments must make themselves trustable by industry and the citizen, a fundamental and nontrivial task. Both government and industry must start to use information effectively and efficiently and this will require sharing, collaboration, and joint effort on a level basis.

Within organizations there will need to be a massive change in the way in which information is perceived, handled, and valued. If we cannot better manage the information and understand its relevance to the organization's effectiveness, then the cost of ownership may prove to be too high.

There are already efforts underway to tackle a number of the major issues being led, not surprisingly by the United States. This is also a potential problem because, based on past performance, the government will try to retain control; and if it does so, it will cause the inevitable failure of any such initiatives. Internationally, there is likely to be minimal rapid progress, so any actions that are taken in the short term that have any hope of success will have to be led by industry. It is industry alone that has the resources and interest to make this work.

Notes

1. Excerpts from an article in Reed Elsevier's *Computer Fraud & Security* magazine, reprinted with permission.
2. Grau and Thomas, *The Journal of Slavic Military Studies,* September 1996.
3. Siemens Corporation home page, http://www.siemens.com/.
4. Information Warfare: New Roles for Information Systems in Military Operations, Captain George A. Crawford.
5. Center of Attention, by Richard Thieme, August 2001.
6. Open Source Intelligence: What Is It? Why Is It Important to the Military?, Robert D. Steele.
7. See also *Netspionage: The Global Threat to Information,* published by Butterworth-Heinemann, September 2000.
8. Monday, November 15, 1999, 12:03:24 GMT, Will Knight.
9. Internet Service Provider Charged with Intercepting E-mail, *Boston Herald,* November 23, 1999, http://www.businesstoday.com/techpages/isp11231999.htm.
10. Mafia Caught Attempting Online Bank Fraud, by Philip Willan, *IDG News,* April 10, 2000, http://www.idg.net/.
11. Japan Software Suppliers Linked to Sect, by Calvin Sims, *The New York Times,* February 3, 2000.
12. What Does a Computer Security Breach Really Cost?, Anita D'Amico, September 7, 2000.
13. Secure Decisions, a Division of Applied Visions, Inc., www.securedecisions.com.

Chapter 19

The Twenty-First Century Challenge: Surviving in the New Century

The future is disorder. A door like this has cracked open five or six times since we got up on our hind legs. It is the best possible time to be alive, when almost everything you thought you knew is wrong.[1]

—Tom Stoppard, Arcadia

This chapter discusses the global courses in our future world — societies, nation-states, high technology, businesses, associations, individuals, and their effect on information warfare. What will the future world look like? What are our future threats, vulnerabilities, risks, and conflicts in which information warfare will play a vital role? This chapter attempts to answer those and similar questions based on current trends.

Global Trends Study and Comments

In December 2000, the United States' National Intelligence Council (NIC) published a document entitled "Global Trends 2015"[2] (GT2015). The report was developed over a 15-month period by the NIC "in close collaboration with the U.S. government specialists and a wide range of experts outside the government, has worked to identify major drivers and trends that will shape the world of 2015." Although the report was sponsored by the U.S. government and is therefore probably somewhat self-centered, it does offer some expert insight into the future. Having the study limited to no farther in the future than the year 2015 makes sense because the world is changing so rapidly;

anything longer than that is impractical and would soon be outdated. In fact, even that period may be too long a time to attempt to divine the future.

One of the interesting sections of GT2015 is the identification of the "key drivers" of future trends. They are:

- "Demographics
- Natural resources and environment
- Science and technology
- The global economy and globalization
- National and international governance
- Future conflict
- The role of the United States"

Throughout this book we have presented the global environment in which information warfare has been fought, is being fought — and in this chapter, will be fought. We have come to many of the same conclusions as stated in GT2015, to include these drivers. Concerning these drivers, GT2015 also went on to say that:

> *"In examining these drivers, several points should be kept in mind:*
> - No single driver or trend will dominate the global future in 2015.
> - Each driver will have varying impacts in different regions and countries.
> - The drivers are not necessarily mutually reinforcing; in some cases, they will work at cross-purposes.
> - Taken together, these drivers and trends intersect to create an integrated picture of the world of 2015, about which we can make projections with varying degrees of confidence and identify some troubling uncertainties of strategic importance to the United States."

We disagree with the statement that "No single driver or trend will dominate the global future in 2015." It is quite obvious that people around the world are becoming increasingly more dependent on high technology. This driver is a dominant force and will continue to be so into the future. How can it not be when everything in today's information-based, information-dependent societies is dominated by high technology? One has to go back to the Stone Age, or Afghanistan under the Taliban, to escape the dependencies on high technology. For that matter, even the Taliban are dependent on it for communication.

Let us discuss the drivers and trends as identified and stated by GT2015. Although we have discussed information warfare on a global basis and tried not to be biased based on the U.S. view, we have included the role of the United States. This is necessary because we see that the United States will continue to have a great deal of influence in the world for the foreseeable future. Like it or not, that fact cannot be discounted and must be included in any look into the future. To do otherwise would be to hide one's head in the sand and that is a dangerous position to be in.

Demographics

"The world population in 2015 will be 7.2 billion (up from 6.1 billion in the year 2000) and in most countries, people will live longer. Ninety-five percent of the increase will be in developing countries, nearly all in rapidly expanding urban areas. Where political systems are brittle, the combination of population growth and urbanization will foster instability. Increasing lifespans will have significantly divergent impacts.

- In the advanced economies — and a growing number of emerging market countries — declining birthrates and aging will combine to increase health-care and pension costs while reducing the relative size of the working population, straining the social contract, and leaving significant shortfalls in the size and capacity of the workforce.
- In some developing countries, these same trends will combine to expand the size of the working population and reduce the youth bulge — thus increasing the potential for economic growth and political stability."

As the burden of increased populations takes place in developing countries, and based on the developing nation-states' inability to adequately support that growing population, conflicts will increase, exacerbated by already unstable governments. Thus, there will be increased risks of internal terrorist activities, especially in urban areas. These unstable governments may strike out against their neighboring nation-states, blaming them for their internal problems.

The information warfare tactics that will be available to these developing nation-states will be primarily in its current form of propaganda attacks, coupled with limited offensive information warfare capabilities such as denial-of-service and virus attacks. This will be coupled with the industrial-period warfare tactics of bombs, rockets, and infantry — as we have seen over the years (e.g., Israel versus the Arab nation-states, Balkan conflicts, African conflicts, terrorists versus anyone they do not like).

As these developing nation-states increase their high-technology dependencies and resources, they will become more vulnerable to more sophisticated forms of information warfare attacks, as well as use these tools as their offensive weapons. For the most part, the use of these information warfare tactics will be limited in developing nation-states and the violence of the Industrial Age weapons will continue to be the norm. It should also be borne in mind that those nation-states and regional groupings that have recently adopted high technology will also be those that are most able to revert and continue to function without it.

In developing nation-states with decreasing populations and increasing longevity, the private-sector corporations will take on an increased social role to support their employees as the nation-state's ability to provide adequate social safety nets fail.

As a nation-state's population declines and its stability increases, the population will begin to focus more on other issues such as that of the society, their place in the world, and natural environment of the nation-state and the world. Thus, there will be a growing shift in the reasons for information warfare. At the same time, these developing nations can draw in the more advanced nation-states, as we have seen in NATO versus Serbia. If that occurs, and that appears likely as nation-states continue to vie for allies, global influence, and natural resources, the vulnerable information infrastructures of these more technologically advanced nation-states will come under attack. These attacks by developing countries require little more than a microcomputer and Internet access to unleash a barrage of malicious software against vulnerable government and corporate information infrastructure targets.

Natural Resources and Environment

"Overall food production will be adequate to feed the world's growing population but poor infrastructure and distribution, political instability, and chronic poverty will lead to malnourishment in parts of sub-Saharan Africa. The potential for famine will persist in countries with repressive government policies or internal conflicts. Despite a 50 percent increase in global energy demand, energy resources will be sufficient to meet demand; the latest estimates suggest that 80 percent of the world's available oil and 95 percent of its gas remain underground.

- Although the Persian Gulf region will remain the world's largest single source of oil, the global energy market is likely to encompass two relatively distinct patterns of regional distribution: one serving consumers (including the United States) from Atlantic Basin reserves, and the other meeting the needs of primarily Asian customers (increasingly China and India) from Persian Gulf supplies and, to a lesser extent, the Caspian region and Central Asia.
- In contrast to food and energy, water scarcities and allocation will pose significant challenges to governments in the Middle East, sub-Saharan Africa, South Asia, and northern China. Regional tensions over water will be heightened by 2015."

There will continue to be increasingly information warfare-based conflicts between the "haves" and the "have-nots." Those that have been controlling the natural resources will want to continue to do so and use all means available to ensure that such controls still remain in place (e.g., OPEC). As we come to understand more about the world we live in, there are those who will see the destruction of the Earth's natural resources as a threat, regardless of which nation-states control those resources. Those nation-states still relying on the Industrial Age-based processes that pollute the atmosphere (e.g., oil refineries, steel producers, and chemical manufacturers) will come under increased attacks by the groups that seek to protect the environment around the world. These groups are already communicating with those of similar environmental views around the world. We have already seen the use of what the United States calls "eco-terrorist" attacks against government and private institutions that are acting in what these activists consider to be detrimental to the global environment and mankind.

However, such conflicts will not be limited to global environmental issues. The issue of water rights within nation-states will also be evolving into conflicts in which information warfare weapons can be used. In the northwestern part of the United States, for example, there are conflicts between farmers' water needs for their crops in time of limited water supplies and those who want to limit that water supply to keep it in the rivers for the salmon and other aquatic species. Water issues between the United States and Mexico may grow as the United States drains more of the water headed down the Colorado River and the Rio Grande.

In the United States, we have also seen recent conflicts between the sellers of resources (e.g., electricity and individual states within the United States, like California). Furthermore, conflicts are arising between individual states in the United States, such as the states of Washington and California where power is being diverted to California from Washington. Both states are modern, information-dependent states and have vulnerable information infrastructures. If such states cannot receive what they believe to be fair treatment by the federal government, conflicts may arise in the future that pit one state against another for the valuable resources of water, electricity, gasoline, and heating oil. Are such conflicts likely in the future? We do not know; but we do know that it is quite possible as these resources become more scarce and the political careers of state governors, senators, and other politicians are left hanging in the balance. An IW civil war may take place between warring states where each (e.g., California and Washington) attacks and defends the computer-driven and controlled electrical grids.

Science and Technology

"Fifteen years ago, few predicted the profound impact of the revolution in information technology. Looking ahead another 15 years, the world will encounter more quantum leaps, both in information technology (IT) and in other areas of science and technology. The continuing diffusion of information technology and new applications of biotechnology will be at the crest of the wave. IT will be the major building block for international commerce and for empowering non-state actors. Most experts agree that the IT revolution represents the most significant global transformation since the Industrial Revolution beginning in the mid-Eighteenth Century.

- "The integration — or fusion — of continuing revolutions in information technology, biotechnology, materials science, and nanotechnology will generate a dramatic increase in investment in technology, which will further stimulate innovation within the more advanced countries.
- Older technologies will continue "sidewise development" into new markets and applications through 2015, benefiting U.S. allies and adversaries around the world that are interested in acquiring early generation ballistic missile and weapons of mass destruction (WMD) technologies.
- Biotechnology will drive medical breakthroughs that will enable the world's wealthiest people to dramatically improve their health and increase their longevity. At the same time, genetically modified crops will offer the potential to improve nutrition among the world's one billion malnourished people.

■ Breakthroughs in materials technology will generate widely available products that are multifunctional, environmentally safe, longer lasting, and easily adapted to particular consumer requirements.

■ Disaffected states, terrorists, proliferators, narco-traffickers, and organized criminals will take advantage of the new, high-speed information environment and other advances in technology to integrate their illegal activities and compound their threat to stability and security around the world."

Tomorrow's high technology will continue as the trends suggest: devices will be smaller, cheaper, more powerful, and offer more mobility. In addition, the integration of information devices started in the Twentieth Century will continue and grow in this, our Twenty-first Century.

There are as many views as to what technologies will be developed and their implication to mankind as there are futurists making there prognostications. For example, Dan Miller (IDG) posted an article online on May 15, 2001, in which he wrote that "Analysts see five technologies as key to the future." These five technologies are discussed below.

Extreme Ultraviolet Lithography (EUL)

As discussed in Chapter 2, Moore's law may be reaching a wall. It is thought that extreme ultraviolet lithography (EUL) might be a solution, and it appears scientists are making progress. They may be able to shrink circuits to a width of about ten nanometers (one nanometer = one-billionth of a meter). Thus, microprocessors, according to the article, may be 100 times more powerful than today's microprocessors. In fact, if all goes well, EUV-based chips might be available by 2005.

Imagine the consequences of such a chip, not only on everything not using microprocessors but also from the viewpoint of information warfare. The power, size, and cost of today's modern information warfare weapons systems will be the size of yesteryear's mainframe computer as compared with today's PDA. Information warfare weapons of various kinds (e.g., HERF guns, EMP weapons, other vastly more powerful offensive and defensive computer systems, and surveillance systems) may be available for soldiers in their backpacks. A number of countries, including the United States and the United Kingdom, already have programs in place to achieve the "digital soldier" within the next ten years.

What is also significant is that as the cost, power, and size are driven further down, they will be more readily available to a greater number of nation-states, groups, and individuals than ever before. Thus, the opportunity for more adversaries to have such information warfare technology at their command will also make information-based systems of the modern nation-states more vulnerable to more adversaries than ever before.

Organic Light-Emitting Device (OLED)

OLEDs are based on electroluminescence. They are simpler in design, cheaper to make, brighter, have better color saturation, and provide a wider viewing angle than LCDs. The LCDs of today will be as outdated as carbon paper is to duplicate copying. Thus, the combination of the EUL technology and the OLED will be integrated into more powerful, miniaturized, and cheap information warfare-related weapons. However, as Miller continues in his article, that is not all that is needed and coming in the future. He cites:

Optical Switching

Today's switches delay our information flow because the information must be converted from today's fiber-optic light signals to the electronic signals of the switches. This is needed to advise the switch of the network addressing and then the information is converted back into light for further transmission to its destination or delayed again and again through other electronic switches.

Optical switches will eliminate these delays. Thus, there will be increases in information transmission speeds — and information transmission speed is a key factor in information warfare. Remember, whoever can obtain the information first and then act on that information sooner (e.g., within the OODA loop of the adversary) will have an advantage in warfare of any kind.

Presence Technology

Basically, presence technology will be able to "sense" your presence online. For example, Microsoft's "HailStorm" is projected to detect your location and your accessibility; for example, where you are located and what you may be doing (working at your desk, in a meeting, in transit from where to where).

This has some interesting information warfare connotations. It will be much easier to contact people rapidly when needed and possibly be able to target the location of adversaries (e.g., terrorists and the like who may eventually use such a system to keep in touch with their cells or target their next victims).

Business Intelligence (BI)

Currently, according to Miller, "data mining" and "knowledge management" are the tools for extracting and using the "golden nuggets" of information. Business intelligence goes further. Through the use of artificial intelligence, using advanced data visualization tools, the massive amounts of raw data and volumes of information are extracted, examined, and presented in more useful human-oriented ways, thus allowing an individual to immediately be able to take action based on the information already presented in a totally developed manner. In fact, the deployed artificial intelligence devices will look for interfaces and patterns that may be overlooked by people — not to mention with more speed than literally humanly possible.

Business intelligence (BI) will provide the decision makers with the information they need faster and in a manner that will allow them to make decisions faster. This will result in a further compression of the OODA loop cycle. Also, from an information warfare perspective, the penetration of such a system to obtain the same information in real-time that the adversary is getting and integrating that into one's BI system will increase the advantage. Once accomplished, seeding the adversaries systems with mis-information and false information will increase that advantage. That information seeded into the adversaries system will cause certain information and options to be presented to the adversary. Based on that information and knowing the type of information warfare training that the adversary has received (e.g., what action they are trained to take based on certain events), one would know what action the adversary will take. That action will be exactly what one would want the adversary to take. And at the same time, the adversary's ability to similarly exploit one's own systems must be avoided at all costs.

Other changes we can see in the future can be broken down into:

- Robotics
- Nanotechnology
- Integration of devices
- Quantum computing

Robotics

Robotics will play an ever-increasing role in our future lives, and especially in warfare and information warfare. They are increasingly being used to manufacture weapons, and even other robots. Cyborgs are already being developed in experiments by integrating life forms with robotic parts. The "Borg" of Star Trek fame may not be that far into the future. Robots are already taking the place of humans in hazardous duties such as deployed as a part of bomb squads. So, why not as tomorrow's infantry as well as other duties, such as spies?

One can easily see the use of intelligent robots in warfare to replace human soldiers. It is not a long stretch from the robotic type of armor suits being developed for the military to complete robotic soldiers integrated with sensors, cameras, and weapons of every type. In fact, do we not have that now? It is not an exaggeration to view today's unmanned air vehicles as primitive, human-controlled robots. In fact, in cyberspace can we not consider our offensive information warfare weapons of viruses and worms as "robot soldiers"? Why not? Scientists are already experimenting with "robonauts" for doing space station repairs; "humanoids" to care for the aged; and even all-robot soccer teams that are projected to defeat a human team by the year 2025.[3] With the finances available by the nation-states' military that have led the high-technology developments, can robot soldiers be far off? If so, will it make it easier to want to go to war since the human death toll will be considered "minimal" for those deploying robo-warriors?

We have already been exposed to the term "smart weapon" for some time; but to date, this term has mostly been used to describe the guidance systems of kinetic weapons (e.g., "smart bombs" that can be preprogrammed with their target and laser-guided bombs that will follow a laser beam onto the target). In the future, we can expect to see increasingly sophisticated weapon systems being deployed that will, perhaps, make their own decisions as to target selection from a predefined list, depending on the weather conditions, resistance encountered, and a number of other variable factors.

Nanotechnology

Nanotechnology is an exciting field and one that will have as much impact on our future as the first transistors had in the past and our "modern" microprocessor has today. What is nanotechnology? Simply speaking, it manipulates things on a molecular scale. That is, we will be able to build very, tiny things — atom by atom. We are talking about nanometers.

One meter divided by 1000   = One millimeter[4]
One millimeter divided by 1000   = One micron (a human cannot see a micron)
One micron divided by 1000   = One nanometer

When talking about nanotechnology and the future, we are talking within the next 10 to 20 years if we are to continue the progress in microprocessors and computer hardware, according to Ralph C. Merkle, Principal Fellow of Molecular Technology at Zyvex and a recognized expert in nanotechnology.[5] One can easily envision nano-robots for uses from medicine to spies — and everything in between. Imagine that fly buzzing around your head as you work. It may be a nanobot viewing everything shown on your computer screen and transmitting it in real-time to your adversary. In fact, why not have them infiltrate the data lines themselves and just read and modify the bits as they pass by? They can be used to infiltrate and destroy communication centers, no matter how secure. They can fly in through the air vents, migrate in through communication lines, or maybe even electrical circuits getting powered up as they ride the circuits. Their uses are endless; and the defenses? Maybe fine air filtering systems such as electrically charged walls that are then insulated from the systems that would be sensitive to such electrical charges. The "bug-zapper" may be used for killing more than mosquitoes.

According to some,[6] there are five major implication of nanotechnology:

- Faster computers (e.g., using nano-size structures to replace grooves etched in today's silicon)
- Medicine (e.g., sending nanobots into the human system to do repairs)
- Environment (e.g., sent airborne to repair ozone layer)
- Nano-manufacturing (e.g., using nanobots to build anything atom by atom)
- Extinction of the human race (e.g., no human labor as nanobots could make everything, create improved nanobots, and then seeing no need for humans, create a "de-assembler" to destroy humans)

One can see the application to warfare and information warfare through the above-cited implications, the last one being our ultimate in information and high technology warfare.

Integration of Devices and Other High Technology

The integration of other devices and the advancement of technology will continue in the future — faster than ever before as high technology begets high technology. Some examples include:

- *3G.* 3G is the third generation of mobile phones; however, not mobile phones as we know them today. These 3G devices "will become a combination of phone, personal organizer, remote computer link, portable radio, television — you name it and your phone will probably be able to do it, both faster and clearer."[7] From an information warfare viewpoint, imagine the possibilities once one has broken into such a system! The information that could be gathered, as well as the mis-information that could be transmitted, would be devastating in warfare (e.g., compromise generals' or politicians' devices). The more information is consolidated and relied on by an adversary, the more devastating an information warfare attack can be mounted. Dependency is a vulnerability.

- *IBM Unveils World's Fastest Silicon Transistor.*[8] This transistor is claimed to yield microchips that within two years will run five times faster than current models. This is just a rest stop on the way to the future. However, it does indicate how fast high technology is advancing.

- *Professor to Wire Computer Chip into His Nervous System.*[9] Kevin Warwick, head of the Cybernetics Department of University of Reading in England, plans to implant a silicon chip that communicates with his brain. This, of course, is not a new idea and has been thought of by scientists and science fiction writers for years. Suppose that a nano-processor containing the history of mankind and all the languages and warfighting tools known to man can be integrated and accessed by the human brain. Combined with the cyborg body, this would make a powerful info-warrior. However, can one rely on that chip's programming being done by someone in China, India, or even in one's home nation-state? Is it possible to bypass the moral and ethical code "built" into the human mind and replace it with mis-information? Imagine the consequences of totally dedicated cyborgs and robo-warriors acting without fear, morality, or ethics or any kind? It is a general's dream and humanity's nightmare. In 1998, Warwick had a silicon chip transponder surgically implanted in his arm to enable him to interact with an experimental environment. In 2001, he had the implant, and the tests continue.

- *Breakthrough Promises Crack-Proof Code.*[10] "This breakthrough in quantum physics promises to encrypt national secrets on a particle of light not even the best hacker could crack." Great — except that Peter Pearson, senior cryptographer for Cryptography Research, a consulting firm, says that it will not be in our lifetime. This is because the photon sent must be the same one received and not converted (e.g., photon to electron and back [remember the above discussion of the optical switch?]). However, with dramatic advances in high technology, this may happen sooner than Pearson expects. One fact that we have learned over the years is that nothing is completely protected. With man's ingenuity and enough money, nothing is impossible. Therefore, one should never look for the "golden nuggets," the "Holy Grail" of security, that one panacea to our information protection ills. We would undoubtedly do so at our peril.

- *HP Focuses on Molecular Computers.*[11] Hewlett-Packard is working to create microprocessors equivalent in power to today's chips, but 1,000 times smaller and considerably cheaper. "Scientists at Hewlett-Packard and other labs have proved that parts of some molecules can be made to swing open or shut, making the molecules themselves switches. Such molecular-scale switches can be connected with chemical "wires" that are just 6 to 10 atoms wide, according to Hewlett-Packard researchers Stan Williams and Phil Kuekes and scientists at UCLA. They also have found a way to make the molecular systems run, even with the imperfections found in nature." The impact on information warfare, as with other high-technology breakthroughs, will be ability to carry out more devastating attacks at less cost while making the job of defensive information warfare practically impossible.

- *Interplanetary Internet in the Works.*[12] One cannot look into the future of information warfare and high technology without considering not cyber-space, but real space. Some see a future there "and the future commercialization of space, according to a joint government and industry group that is developing the InterPlaNetary (IPN) Internet.

> *"Starting this year, with NASA funding, the IPN will roll out in pieces over the next several decades to support communications among space-ships, robots, and manned and unmanned outposts in the solar system. "It's conceivable that the IPN could go like its terrestrial counterpart, starting out as a network supporting scientific research and eventually evolving into something of commercial interest," says Vinton Cerf, senior vice president of Internet architecture and technology at WorldCom, Inc."* Of course space has been a warfare environment for decades, although no nation-state is willing to admit it. It then follows that as the Internet goes into space to support space commercialization efforts coupled with global business and economic competition and warfare, it naturally follows that information warfare will be used in space to support the terrestrial information warfare efforts.

As if the above are not interesting enough, some side benefits may include free RAM and hard disks of almost any size as it will be so cheap as to be free. Not likely? One just has to look as the costs of today's ink-jet printers. They are being given away with systems. Look at the cost of replacement cartridges for them. It is becoming cheaper to throw out the old printer and buy a new one than replace the black and color cartridges.

From an information warfare viewpoint, the continued trends in high technology will provide more people with more powerful and sophisticated systems, and information warfare tools at lower and lower costs, thus increasing the number of potential adversaries and attacks on a global basis.

Quantum Computing

Quantum computing has the potential to unleash computing power that is orders of magnitude greater than we can currently conceive. If achievable, quantum computers will allow for the miniaturization of computers and allow them to overcome the barrier that currently exists in which wires cannot be made that are less than one atom in diameter. One group has already stated that it believes it will be possible to create quantum computers that are based on the molecules of a liquid.

Global Economy and Globalization

"The networked global economy will be driven by rapid and largely unrestricted flows of information, ideas, cultural values, capital, goods and services, and people; that is, globalization. This globalized economy will be a net contributor to increased political stability in the world in 2015, although its reach and benefits will not be universal. In contrast to the Industrial Revolution, the process of globalization is more compressed. Its evolution will be rocky, marked by chronic financial volatility and a widening economic divide.

- "The global economy, overall, will return to the high levels of growth reached in the 1960s and early 1970s. Economic growth will be driven by political pressures for higher living standards, improved economic policies, rising foreign trade and investment, the diffusion of information technologies, and an increasingly dynamic private sector. Potential brakes on the global economy — such as a sustained financial crisis or prolonged disruption of energy supplies — could undo this optimistic projection.
- Regions, countries, and groups feeling left behind will face deepening economic stagnation, political instability, and cultural alienation. They will foster political, ethnic, ideological, and religious extremism, along with the violence that often accompanies it. They will force the United States and other developed countries to remain focused on "old-world" challenges while concentrating on the implications of "new-world" technologies at the same time."

GT2015 stated that "its evolution will be rocky, marked by chronic financial volatility and a widening economic divide." Such things as "volatility" and "economic divide" cause conflicts. These conflicts can erupt into national and international conflicts. The sophistication of high technology of the nation-state or states in question will drive their information warfare tactics and how much those tactics will be used in lieu of the industrial-period methods of waging wars, such as bombing and the use of human soldiers to do the killing.

What we are saying here is nothing new to any of us who have lived through at least the last half of the Twentieth Century. Mankind will continue to wage wars and use whatever means they have available to win those wars. This, too, is nothing new. What is new is that the use of high technology may make warfare "cleaner" in that less human life may be wasted in information warfare with the same results being accomplished as in any other form of warfare.

As stated earlier, does this mean that mankind will be more apt to go to war? If our past is any indicator, the answer is "Yes!" because we cannot seem to resist the urge and are not that far removed from the time of cave dwellers when it comes to our emotions. Although economic warfare is not new, economic warfare on a global scale is a fairly recent phenomenon brought about by the globalization of businesses and the business interests of a corporation viewed on a global scale. The economic future of nation-states, corporations, and individuals may be more precarious in the future as the divide between the "haves" and "have-nots" grows. The "have-nots" want their share and if they cannot get it, they will be sure that others do not either; and information warfare is a cheap way to make sure that happens.

Another growing battle that is sure to increase and move more into the information warfare battlefield is the battles between the World Trade Organization (WTO) and individuals, such as anarchists and anti-globalists. Currently, most of these wars are being fought in the streets of Seattle, Genoa, and Washington, D.C. To help "support" the hostile "have-nots" are the anarchists that we have seen at every location in the world in which the WTO meets. Although they are really harming others around the world that can be helped by globalization, that is not their point. These miscreants are nothing but young, spoiled brats with nothing else to do, no clue as to how others have sacrificed for them, and who take advantage of the freedom provided them in our modern nation-states. Maybe a WTO meeting in Beijing or another Third World country would be worthwhile.

They are not contributing to the betterment of the human race. Their only goal is to create havoc and have fun. If they ever won (and they will not, of course), they would not long survive because corporations and government agencies will not be there to clothe them, feed them, and be sure that they had their cappuccino after a hectic day of destruction. However, make no mistake about it, they are dangerous and their use of information warfare tactics could be devastating to a corporation or government that they target. Remember that they are made up of people from many nation-states, and they are also moving into cyberspace. In the future, more of these battles will be fought there. Therefore, they may become a powerful global force to be reckoned with in the future.

The trend is toward more economic-based versions of information warfare conducted by global, informal groups, and international businesses against each other, some with and some without the help of the nation-states that have an interest in the success of specific international corporations. Today, many corporations have offices throughout the world; and to be successful, they need a stable environment — unless of course they are in the business of making money from destabilized governments and conflicts, like arms dealers. Many of these corporations have more money and power than many of the nation-states in which they have offices. Therefore, they can be a great stabilizing force and have significant influence on nation-states' governments.

Because international corporations often have the same quest for power, control leveraged through a dominant position in a nation-state or globally by market share, they may use information warfare tactics against other businesses and the nation-states that do not support them — regardless of the nation-state. Just as, in many cases, modern-day nation-states' warfare has not been officially declared (e.g., NATO versus Yugoslavia, Desert Shield/Desert Storm, Vietnam, Korea), global corporations waging a war for global market share domination will not declare a state of war on their competitors or some nation-state. However, they are in fact today using information warfare tactics, albeit mostly covertly, to gain the advantage. The words may be different, such as business competitor instead of enemy or adversary, but many of the same principles of war will continue to be used — and in fact, possibly increase as the global economic and business competition intensifies.

Nation-states and global corporations already realize that the global power of a nation-state is no longer based on military strength, but on economic strength. They are increasing their support of corporations headquartered in their country. Even the U.S. government, which has a history of attempting to divide major, global corporations (e.g., IBM, AT&T, Microsoft), is allegedly using the power of its spy agencies to obtain information concerning foreign corporations and is allegedly intensifying the collection of economic information of nation-states — information that can be used not only for the benefit of the United States, but also its U.S.-headquartered corporations. The use of ECHELON and its successors may be directed more toward economic and business intelligence collection than military intelligence collection of their potential adversaries.

This trend is expected to continue and to intensify. Thus, we see a growing shift from nation-states' military warfare to economic and global business warfare while at the same time these global corporations seek to ensure that the nation-states where they have business interests are stable. A perfect example is to watch the stock markets in the United States, Asia, and Europe. They are linked, and what affects one usually has an impact on the other two. If a crisis or conflicts breaks out anywhere in the world, the markets react. Thus, the targeting and use of information warfare tactics against global financial processes and systems by adversaries (e.g., terrorists, environmentalists, and anarchists) is expected to intensify.

An interesting phenomenon that might occur is the covert use of information warfare tactics by global corporations against a nation-state by adversely affecting the economy of the targeted nation-state. This is quite possible and likely as powerful global corporations use such tactics to topple a government and support one more favorable to their business. Why not? Nation-states have being doing so for many years. The downside may be minimal while the upside can be phenomenal. After all, business people are accustomed to weighing the risk and making decisions accordingly. If the risk-takers see the risk as low after weighing the advantages and disadvantages, do you think they would act accordingly? Have some already done so? Is this legal? No. Is it ethical? No. However, one should remember the old saying, "All is fair in love and war." There may be more truth in at least part of that saying in the future than ever before.

Another possible future relationship between global corporations and nation-states can be seen in China. If you want to do business in China, you play by China's rules. Their rules may require that you share your proprietary information and train their citizens to produce your products. What is the alternative? Turn down a lucrative, multi-billion dollar contract? So, some aspects of information warfare may be more subtle and hidden in legitimate contracts (e.g., an anti-virus firm vying for business in China but, as a prerequisite, giving up valuable information that can help China develop its own offensive and defensive information warfare weapons).

How does one continue to compete? To be successful, one must make newer and better products faster and cheaper than the competitors — operating within the OODA loop of your competitors. That in the future will be a business' only advantage. Thus, the need for more information about competitors will be greater than ever spawning new and more sophisticated business intelligence collection methods — netspionage. To be successful in the future, global corporations will think and act on a "war footing" — and information warfare tools and tactics will be incorporated into their future marketing strategies on how to gain the competitive advantage.

As the competition for market share intensifies, so will the need for more and more information — not only about competitors and nation-state governments, but also individuals (the consumers). This ever-increasing need will grow in intensity and will continue to be in direct conflict with the individuals' desire for privacy. We may look back in fondness years from now and wish for the "good ol' days" when cookies and spam e-mails on our systems were the least of our privacy concerns.

An example of the sophistication of such tactics is the ability of a corporation to download, without your knowledge, personal information. If you have a stock portfolio and want to monitor it through some online (Internet) financial corporation, how do you know that they have not downloaded your stock portfolio, is maintaining it in their database, and is selling the information, all without your permission? Is this being done without your knowledge and approval now? Yes, it is; and that is just the beginning as the information warfare battles between privacy-rights advocates and corporations grow.

National and International Governance

"States will continue to be the dominant players on the world stage, but governments will have less and less control over flows of information, technology, diseases, migrants, arms, and financial transactions, whether licit or illicit, across their borders. Non-state actors, ranging from business firms to nonprofit organizations, will play increasingly larger roles in both national and international affairs. The quality of governance, both nationally and internationally, will substantially determine how well states and societies cope with these global forces.

- "Nation-states with competent governance, including the United States, will adapt government structures to a dramatically changed global environment, making them better able to engage with a more interconnected world. The responsibilities of once "semiautonomous" government agencies will increasingly intersect because of the transnational nature of national security priorities and because of the clear requirement for interdisciplinary policy responses. Shaping the complex, fast-moving world of 2015 will require reshaping traditional government structures.

- Effective governance will increasingly be determined by the ability and agility to form partnerships to exploit increased information flows, new technologies, migration, and the influence of non-state actors. Most but not all countries that succeed will be representative democracies.

- States with ineffective and incompetent governance will not only fail to benefit from globalization, but in some instances will spawn conflicts at home and abroad, ensuring an even wider gap between regional winners and losers than exists today.

- Globalization will increase the transparency of government decision making, complicating the ability of authoritarian regimes to maintain control, but also complicating the traditional deliberative processes of democracies. Increasing migration will create influential diasporas, affecting policies, politics, and even national identity in many countries. Globalization will also create increasing demands for international cooperation on transnational issues, but the response of both nation-states and international organizations will fall short in 2015."

The statement about the cooperation between a nation-state's government agencies is somewhat optimistic. Even military services fight each other over budgets, weapons systems funding, and leadership roles. Can then civilian agencies not act in a similar manner to preserve their little "fiefdoms"?

Nation-states that do not take the opportunity to utilize high technology or attempt to control the information of its citizens by denying them access to the Internet will potentially cause increased internal and external conflicts. The lack of high technology and the lack of development of an internal information infrastructure by these nation-states, usually run by despots, will make them less vulnerable to information warfare tactics. However, they will still be able to increase their ability to use unsophisticated information warfare tactics (e.g., microcomputers attached to the Internet sending malicious code to global businesses and nation-states in which they have conflicts).

These nation-states are developing some limited information infrastructures as well as information warfare weapons. These weapons have been primarily relegated to Web-based propaganda sites and simple denial-of-service and virus attacks. However, as time goes on, they will continue to increase their information warfare expertise by buying that expertise from more advanced nations such as the United States, United Kingdom, Russia, Germany, France, and China. Those advanced nations that do not develop internally and sell to developing nation-states will have those information warfare weapons stolen from them. History has shown that such activities are already taking place with the industrial-based weapons of bombs and rockets, and it is only logical to continue that process to obtain more state-of-the-art information warfare-based weapons.

What information warfare-based weapons nation-states cannot or are not willing to sell will be sold by arms dealers who will increase their portfolios to include such weapons. After all, if there is a market, there will be those that will find a way to sell such weapons. There will continue to be an increase in such activities.

The problem for those nation-states run by fascists, Communists, and other despotic regimes is that, with the probable exception of the current Taliban in Afghanistan or the Khmer Rouge in Cambodia, both of which effectively shunned the Twentieth Century, they must use high technology and Internet access to compete for power and respect in the world. However, they cannot do so while at the same time trying to control information and limit the access of their citizens to information. This has been tried, is being tried, and will continue to be tried. However, they will not be succeed unless they relegate their citizens into the Dark Ages and are willing to isolate themselves from the world community of nation-states. Even so, it is only a matter of time before they go the way of the dinosaur. Human beings are resilient and the inherited human rights of each human being will eventually overcome any obstacles. That trend can be seen throughout history and that struggle will intensify in the future as such people are supported on a global scale.

Future Conflict

"The United States will maintain a strong technological edge in IT-driven 'battlefield awareness' and in precision-guided weaponry in 2015. The United States will face three types of threats:

- "*Asymmetric threats,* in which state and non-state adversaries avoid direct engagements with the U.S. military but devise strategies, tactics, and weapons — some improved by "sidewise" technology — to minimize U.S. strengths and exploit perceived weaknesses
- *Strategic WMD threats,* including nuclear missile threats, in which (barring significant political or economic changes) Russia, China, most likely North Korea, probably Iran, and possibly Iraq have the capability to strike the United States, and the potential for unconventional delivery of WMD by both states or non-state actors also will grow
- *Regional military threats,* in which a few countries maintain large military forces with a mix of Cold War and post-Cold War concepts and technologies

"The risk of war among developed countries will be low. However, the international community will continue to face conflicts around the world, ranging from relatively frequent small-scale internal upheavals to less frequent regional interstate wars. The potential for conflict will arise from rivalries in Asia, ranging from India–Pakistan to China–Taiwan, as well as among the antagonists in the Middle East. Their potential lethality will grow, driven by the availability of WMD, longer-range missile delivery systems, and other technologies.

Internal conflicts stemming from religious, ethnic, economic, or political disputes will remain at current levels or even increase in number. The United Nations and regional organizations will be called on to manage such conflicts because major states — stressed by domestic concerns, perceived risk of failure, lack of political will, or tight resources — will minimize their direct involvement.

Export control regimes and sanctions will be less effective because of the diffusion of technology, porous borders, defense industry consolidations, and reliance on foreign markets to maintain profitability. Arms and weapons technology transfers will be more difficult to control.

Prospects will grow that more sophisticated weaponry, including weapons of mass destruction — indigenously produced or externally acquired — will get into the hands of state and non-state belligerents, some hostile to the United States. The likelihood will increase over this period that WMD will be used either against the United States or its forces, facilities, and interests overseas."

That the risk of war among developed countries will be low is probably correct, but only if one looks at it from a military viewpoint. As often stated, business and economic warfare are seen as intensifying between not only corporations, but also among nation-states well into the future.

Although this section of GT2015 is obviously slanted more toward the United States, it does address some basic issues of possible future conflicts and weapons. Because this book is about global information warfare, it has not dwelled on the Industrial Age forms of warfare — except where it helped in the discussion of information warfare. However, one must never forget that information warfare tactics and weapons will be used more and more in the future, but as with any other weapon, where it can be the most effective. Those without information warfare weapons will use any weapons at their disposal; while at the same time, some nation-states (or groups) will prefer the use of the non-information warfare weapons of mass destruction (WMD) to have more sensational and abhorrent effects on their adversary just as the Middle East terrorists continue — and will continue — to do.

A Look at a Future War

So, what would a future information war look like? There are many "war games" being played out by the government agencies of the modern nation-states. Take a look at a summary view of the future information war. The future high-technology environment will cause changes, including:

- Mass armies to smaller armies
- More firepower employed from greater distances
- Ground forces only used to identify targets and assess damages
- Blurring of air, sea, and land warfare
- Electronic mail and other long-range "smart" information systems weapons
- Smaller and stealthier ships
- Pilot-less drones replacing piloted aircraft
- Less logistical support required
- More targeting intelligence
- Information relayed directly from sensor to shooter

- Satellite transmissions directly to the soldier, pilot, or weapon
- Military middle-management staff eliminated
- Field commanders access information directly from drones, satellites, or headquarters on the other side of the world
- Immediate recognition of friend or foe
- Military command decisions made under the scrutiny of media cameras

Technology, Menu-Driven Warfare

Information warfare technology will be standard and will be a menu-driven system, with databases, to allow the IW commanders and warriors to "point and click" to attack the enemy; that is:

- Select a nation.
- Identify objectives.
- Identify technology targets.
- Identify communications systems.
- Identify weapons.
- Implement.

The weapons can be categorized as:

- Attack
- Protect
- Exploit
- Support weapons systems

The following examples are provided.

- *IW-network analyses (exploit):* defined as the ability to covertly analyze networks of the adversaries to prepare for their penetration to steal their information and then shut them down
- *Crypto (exploit and protect):* defined as the encrypting of U.S. and ally information so it is not readable by those who do not have a need-to-know; and the decrypting of the information of adversaries to be exploited for the prosecution of information warfare
- *Sensor signal parasite (attack):* defined as the ability to attach malicious code (e.g., virus, worm, etc.) and transmit that signal to the adversary to damage destroy, exploit, or deceive the adversary
- *Internet-based hunter killers (attack):* defined as a software product that will search the Internet, identify adversaries' nodes, deny them the use of those nodes, and inject dis-information, worms, viruses, or other malicious codes
- *IW support services (services):* defined as those support services that support the above or that provide any other applicable services, including consultations with customers to support their information warfare needs: services such as modeling, simulations, training, testing, and evaluations:

Some current and future IW tactics include:

- Initiate virus attacks on enemy systems
- Intercept telecommunications transmissions; implant code to dump enemy database
- Attach worm to enemy radar signal to destroy their computer network
- Intercept television and radio signals and modify their content
- Mis-direct radar and content
- Dis-information: bushes look like tanks, trees like soldiers
- Information overload enemy computers
- Penetrate the enemy's GII/NII/DII nodes to steal or manipulate information
- Modify maintenance systems information
- Modify logistics systems information

Role of the United States

"The United States will continue to be a major force in the world community. U.S. global economic, technological, military, and diplomatic influence will be unparalleled among nations, as well as among regional and international organizations in 2015. This power will ensure not only the preeminence of the United States, but will also cast the United States as a key driver of the international system. The United States will continue to be identified throughout the world as the leading proponent and beneficiary of globalization. U.S. economic actions, even when pursued for such domestic goals as adjusting interest rates, will have a major global impact because of the tighter integration of global markets by 2015. The United States will remain in the vanguard of the technological revolution from information to biotechnology and beyond. Both allies and adversaries will factor continued U.S. military preeminence in their calculations of national security interests and ambitions.

Some states — both adversaries and allies — will try at times to check what they see as U.S. "hegemony." Although this posture will not translate into strategic, broad-based, and enduring anti-U.S. coalitions, it will lead to tactical alignments on specific policies and demands for a greater role in international political and economic institutions.

Diplomacy will be more complicated. Washington (D.C.) will have greater difficulty harnessing its power to achieve specific foreign policy goals. The U.S. government will exercise a smaller and less powerful role in the overall economic and cultural influence of the United States abroad.

In the absence of a clear and overriding national security threat, the United States will have difficulty drawing on its economic prowess to advance its foreign policy agenda. The top priority of the U.S. private sector, which will be central to maintaining its economic and technological lead, will be financial profitability — not foreign policy objectives. The United States will also have greater difficulty building coalitions to support its policy goals, although the international community will often turn to Washington, even if reluctantly, to lead multilateral efforts in real and potential conflicts.

There will be increasing numbers of important actors on the world stage to challenge and check — as well as to reinforce — U.S. leadership: countries such as China, Russia, India, Mexico, and Brazil; regional organizations such as the European Union; and a vast array of increasingly powerful multinational corporations and nonprofit organizations with their own interests to defend in the world."

It is not our intent to dwell on the future role of the United States in global information warfare. However, there is no doubt that the United States is, and will continue to be for the foreseeable future, one of the leading nation-states in researching, developing, and implementing operational offensive and defensive information warfare weapons and tactics. However, as every 14-year-old hacker knows, it is also the most vulnerable nation-state and one that has a dismal record of defending its systems.

It is ironic that one of the world leaders in the development and use of high technology with the strongest, state-of-the-art military cannot even adequately defend its Department of Defense computer systems. In the future, it is expected that improvements will be made. However, the question to be answered is: Will they ever be able to proactively stay one step ahead of the adversaries? Probably not, and the results may one day be CATASTROPHIC!

Individuals and Groups

Current trends indicate that the future will include:

- Rapid spreading of computer literacy throughout the world
- Increasing dependency on information-based systems and high technology by more and more nation-states, businesses, groups, and individuals
- The miniaturization and exponentially increasing power and exponentially decreasing costs

These trends lead one to believe that increasingly more information warfare tactics will be used on a smaller scale by individuals and groups. Some of these information warfare reasons and tactics were discussed above and in previous chapters. This is a very important aspect of future information warfare. The primary battles of the future will decrease between nation-states and increase between individuals and groups against nation-states and global corporations. Currently, groups such as those concerned with human rights, animal rights, and the environment are not well-organized; however, their organizations are becoming more formalized thanks to the massive communications and information distribution center — the Internet. It is expected that information warfare will grow in volume and intensity on a global scale. One just has to look at recent attacks, such as against the World Trade Organization, by various interest groups and anarchists.

One of the primary information warfare battles that will increase in volume and intensity is the right to individual privacy by the world's citizens. As the power of the nation-states decrease, they will require more information on more individuals who will be fighting them for more personal freedoms and the power of the "sovereign individual" over the nation-state.[13] In addition to those conflicts, there will be those by groups and individuals against the corporation, both national and global, that pollute the environment and that collect massive amounts of information on the individual to target them for marketing purposes. The right to privacy, human rights, and personal privacy will be a battle cry of the future. However, these battles will be fought by "a few, for the many." One can see this as people are continually willing to give up their freedoms for security. However, it is hoped that at some point the citizens of the world will say, in the words of the American patriot Patrick Henry, "Give me liberty or give me death!"

In the future, will the security professional be able to resolve the growing conflicts of protecting the nation-state against all adversaries, even its own citizens, or protecting the massive corporations and their databases that violate individuals' rights to privacy? If history is any indication, the security professionals will put "God, country, and corporation" first.

In Defense of the Privacy of Individuals and Groups — The INFOSEC Officer?

Today, because of computer networks and particularly the Internet, many people are talking about privacy — the right to keep your personal information private. However, like INFOSEC, everyone is talking about it but it is still a low priority for businesses and government agencies. In fact, for some, obtaining private information is a lucrative business and often a necessity in the competitive bid for customers.*

When one thinks of the role of the INFOSEC officer, one thinks of the people in charge of protecting computers, networks, Web sites, and the company or government information that systems process, store, and transmit information. However, what about that personal information of the employees, customers, and others that is also being stored, processed, and transmitted by the massive systems of governments and businesses? Who is looking out for their interests, from a privacy point of view? The answer is that there is, in reality, probably no one.

It may be assumed that privacy is being protected by overall protection of the systems. However, is that really the case? For example, much of the personal, privacy-related information would probably reside in the databases of the Human Resources department. However, does everyone in that department have access to that information? Does everyone in that department need that information? Remember that statistics indicate that the biggest threat to information comes from insiders, those with authorized access to information. It is not that uncommon for someone inside the company to sell personal information of employees to other companies (e.g., selling employees' personal information to insurance companies so they can call or mail insurance information

---

* One may ask what this has to do with information warfare? Besides the obvious of identity theft and its uses by info-warriors, such information about individuals helps identify them as IW targets: targets by terrorists, corporations, and governments for a multitude of uses — IW uses.

to them to get them to be customers). In the "good old days," this was being done by printing out the lists and giving it to some sales representative. Now it is just an e-mail attachment away.

Much has been said and written about privacy issues relating to information stored on computers, networks, and their massive databases. It appears that the entire issue of online privacy is following the trend of information security (INFOSEC); that is, we are losing the battle.

The "bad guys" are winning the battles against INFOSEC and they are also winning the battles against ensuring the privacy of everyone's personal information. Much of the privacy is being lost in places where one least expects it. There are databases of customer information being kept by almost every business in our information-dependent nations. These include:

- Groceries stores know what you buy and when you buy.
- Video stores know what videos you like to watch.
- Credit card companies know what you purchase, when, and from whom.
- Cable television knows what channels you watch.
- E-businesses know what you purchase and what Web sites you visit.
- Telecommunications companies know whom you call, when, and for how long.

There are many more examples too numerous to mention here, but you get the idea. Some of the above businesses may even sell their information to others — and many do. Thus, massive databases are being compiled that are building profiles of who you are, what you like, and what you do not like. Your entire lifestyle is on more than one of these massive databases. This information is then, in turn, sold to others.

With all that said, take a quick look at privacy issues and the role of the INFOSEC officer. Begin with the basics; the basics begin with definitions. This is always useful because it gives a clearer understanding of what this entire subject of privacy is all about. After all, INFOSEC officers, high-tech crime investigators, info-warriors, and actually everyone must understand what it is before they can figure out their responsibilities as they relate to its protection — at least they should.

What do we mean by privacy? Private, according to *Black's Law Dictionary,* means, "Affecting or belonging to private individuals, as distinct from the public generally; not official..." So, as it relates to private information, we are talking about information that belongs to us as individuals. Do corporations have private information? Yes they do. We call that private, sensitive, trade secret, or some other name to designate the information as that information that the owner does not wish to be made public.

Now we know what it is, but what gives us all the right to privacy? Is it something that we human beings all inherited at birth? Well, in a perfect world, we should all agree that yes, it is our birthright as human beings. However, because we live in a world far from perfect, we human beings have come together over the centuries in clans, villages, and recently nation-states to physically protect ourselves from our fellow human beings that we consider different and often our adversaries. Then we developed laws, regulations, treaties, and customs to regulate the conduct between ourselves and our fellow citizens, as well as between ourselves and others in other nations.

So it would seem that anything related to privacy has come from laws enacted by our individual nation-states and local governments. Based on that assumption, we citizens, through our governments' legislative bodies or governments' legislative bodies without citizen input (depending on the type of government in power) have enacted policies, rules, laws, regulations, agreements, and treaties that determine how individual and business privacy will be protected. Thus, based on these official, legal documents, we individuals have a right to privacy, and thus a right to have our private information protected.

Again, going back to *Black's Law Dictionary*, the right to privacy is the "right to be left alone, the right to be free from unwarranted publicity.... The right of an individual (or corporation) to withhold himself and his property from public scrutiny, if he so chooses. It is said to exist only so far as its assertion is consistent with law or public policy, and in a proper case equity will interfere, if there is no remedy at law, to prevent an injury threatened by the invasion of, or infringement on, this right from motives of curiosity, gain, or malice..."[14]

Thus, each government has its own definition of what is private, how that privacy is enforced, and what is the punishment for violations of such acts. Businesses also do — or should — have specific guidelines for protecting the privacy of its business and individuals, whether they are employees, customers, potential customers, or "others."

As an INFOSEC officer or manager in a corporation or government agency, do you know what laws have been enacted, what policies have been established to protect the privacy of individuals in whatever categories they are placed? Very few INFOSEC officers know the privacy laws of their nation, although they hopefully know their own government agency's or corporation's privacy policies, rules, or whatever term you choose to call them. In fact, do you know that as an INFOSEC officer or manager, you may be held "personally liable" for failure to adequately protect private information? Yes, that does mean that they can take your house, car, and savings.

What is your role as an INFOSEC officer, or what is the role of your INFOSEC officer in protecting that information deemed private and requiring protection by law, or policy? Is it written in your INFOSEC officer job description? If not, why not? If not you, then who? If not you, then maybe the legal department? The Human Relations department? If one of them or others, and the information is stored in automated databases or other digital forms, then what is your role as an INFOSEC officer? If none, then is anyone really ensuring the privacy of information?

Some may say, "I don't care what information is collected on me." These are probably many of the same people who, when it comes to government surveillance, say, "If you don't do anything wrong, why worry about it?" Think of it this way: most of us have nothing to hide but we still do not want others to know about our personal lives. If someone were to take all the information that is known about you and stored in today's databases and places it in a newspaper or on a specific Web site for all to see, and you do not care, then that is your right. Most of us, however, would not want that information to be made public, just on general principles.

There are many sources to go to and become more familiar with privacy issues. A search of the Internet using the search engines noted disclosed the following: InfoSeek listed 416,958 hits; AltaVista listed 3,282,800 hits; and Yahoo listed 12 categories of privacy leading to thousands of sites.

Think about it. What are you as an INFOSEC officer or manager of a government agency or business doing to protect the privacy of your employees, customers, and others? Please do not say that it is not important or that it is not a high priority. History is replete with examples of private information in the wrong hands. It can be devastating, not only to our freedom but also may actually contribute to death. In Winn Schwartau's book, *Cybershock: Surviving Hackers, Phreakers, Identity Thieves, Internet Terrorists, and Weapons of Mass Destruction,* he writes: "The story had to do with identity theft, which can only be successful if your personal information is stolen: .... The incredible stress sent Bill to the hospital.... Six weeks later, Bill was dead. A massive stroke and a heart attack caused by stress killed him..." Enough said. As an INFOSEC professional, do your job to protect private information; someone's life may depend on it. Some reading this may have thought that, yes, it is important but it has been overdone a bit — life and death and all that stuff. Well, like it or not, it is true. Once more, look at history and what compiled lists of personal information in the wrong hands led to in the past. Just ask any of the Jewish people who lived in Europe in the 1930s and 1940s.

If you, as an INFOSEC officer or manager with INFOSEC responsibility, do not take it very, very seriously, then who will? The right to privacy in our information-based societies is of paramount importance if you believe in human rights and freedom. If you are one who believes that the nation-state (or corporation) has priority over individual rights, then God help those who rely on you to protect their privacy.

Take a look at one scenario of a government, "with all good intentions," violating our privacy rights. A nation-state legislature passes a law outlawing gambling on the Internet. This is a stupid idea because no nation-state can pass laws and have them enforced in other nation-states. No matter, it is being tried by at least one nation-state.[15]

A person operates an Internet Web site for casino gambling. That operator resides in Monte Carlo and takes all bets from anyone in the world. The gambler lives somewhere in another nation-state and, using a credit card or whatever payment system the Web site casino accepts as appropriate, places a bet. It may be against the laws of a nation-state, but the casino is not in that nation-state. The gambler uses an Internet service provider (ISP) to access the Internet and send in the bet. Perhaps the ISP is even in a third country.

Who will know that the gambler broke a nation-state's law? Then the problem arises: who and how will police know that this person violated the nation-state's Internet anti-gambling law?

Will the nation-states' police agency through an ISP, or by other means, like ECHELON, read all our mail for indications of such violations? How many of us would want that? Not even the ISPs want to be involved in that. It is almost like having the person who built the public road also being required to enforce the speed laws. In addition, with literally millions (if not billions) of daily e-mail messages circulating along the Internet, who is capable of reading and screening the content of each of these? Some governments are trying to do that — and beginning to succeed. By the way, what else are they reading as they scan the messages looking for gamblers?

Should government agents be allowed total access to ISPs and then, with a "keyword search program," search for the word "bet," scan all e-mail messages of the ISPs customers to look for such keywords, and then finding a violation of law, initiate an investigation?

This may sound unrealistic but indications are that something similar is already being done in some countries and being proposed in others. In the case of a nation-state or groupings of nation-states, the European Community and other nations have stated that they are concerned about protecting the rights and privacy of Internet users. Sure they are, except when it is not in their best interest to be concerned, which is most of the time. Remember that government employees' first responsibility is the protection of their nation-state, and not the individual — power and job security.

The following provides at least the public view on the subject:

> *U.S. administration officials said they are aware of, but have no imme-diate response to a European Parliament committee's opinion that U.S. online data protection policies don't make the grade for EU-U.S. data privacy collaboration. The sudden hint of turbulence in gaining Euro-pean Union support for the data protection arrangement throws into potential jeopardy the Safe Harbor plan that would allow data transfers to continue between the EU and U.S.*[16]

As time goes on, this issue will become more serious and will be more relevant to those responsible for Internet security. For example, as tele-medicine becomes more closely integrated into how medical practices are performed, the issue of patient records being transmitted over the Internet from one doctor, hospital, and others will undoubtedly be compromised, based on today's Internet security and the lack of the use of encryption. Thus, the potential for public disclosure of records will occur, undoubtedly followed by lawsuits.

If such problems and "invasions of privacy" have occurred in the past on non-Internet-connected computers, the chances of it happening on Internet-connected systems are pretty much 100 percent. For example, according to *InformationWeek* magazine as far back as the May 17, 1993 issue, the Reso-lution Trust Corporation saw nothing wrong with copying the files of a whistleblower. This was done, according to the report, with the permission of the U.S. Justice Department and the corporation's legal counsel.

A U.S. newspaper[17] printed an article discussing a U.S. Federal Trade Commission survey that determined that "hundreds of companies are collecting personal information about consumers, and without telling them, selling the data.... The U.S. Federal Trade Commission also recommends that Congress put into law restrictions from and about children who use the Internet... People who don't use the Internet frequently cite privacy concerns as their reason for staying offline..." In other words, there are people who are afraid, for reasons of privacy concerns, to get on the Internet. Therefore, they are missing the many benefits of such Internet access, probably to their own detriment in many instances.

A U.S. Federal Trade Commission (FTC) survey looked at 1400 Web sites selected at random and found that 85 percent collected some personal information; 14 percent offered notice of that collection; and less than two percent had a "comprehensive privacy policy."

Are additional privacy laws necessary? Another newspaper article[18] states that "...A group that says it represents small Internet business has threatened to make public the e-mail addresses of 5 million America Online members if AOL continues to bar their businesses from pitching products to its (AOL) subscribers..." (in other words, spamming AOL users). AOL said, "We would avail ourselves of any legal remedies we need to protect our members...from this threat.... We see this threat as some sort of cyber-terrorism.... AOL members have made it clear 'they do not want junk e-mail...'"

Based on the above, a case can be made for some type of legislation so that ISPs such as AOL can protect their customers' privacy. The U.S. FTC has developed a model for Internet privacy legislation that includes the following four elements:[19] Web sites would be required to advise those that access their site how they collect information and what information they collect. The Web site owners would offer their visitors' choices as to how the information gathered about them can be used. The Web sites would allow access to the private information collected from their visitors and the opportunity for the visitors to correct inaccurate information. The Web site would be required to provide "reasonable" security to protect the integrity and security of the visitors' information.

If a nation-state's legislative body were to pass additional Internet privacy legislation, the issue then, as with all other Internet related laws becomes: Who will enforce the laws? Who will monitor the Internet to ensure the law is obeyed? In addition, how will the laws be enforced when the corporation collecting the personal information resides outside the borders of the nation-state?

Will the nation-state implement trade sanctions against the country where that corporation resides? Bar the corporations of the nation-state from any business relations with that corporation? All things in life are possible, but some are not likely. This is one that falls into the "not likely" category.

In the United States, the Electronic Communications Privacy Act (ECPA) prohibits unauthorized eavesdropping and unauthorized access to messages and information stored or transmitted by computers. The law was primarily enacted to stop alleged abuses after the time of Watergate but has since been updated to include the newer issues of computers, Internet, and wireless, especially as they relate to privacy.

However, privacy issues of the nation-state may be getting some outside support. According to *Wired* magazine's May 1998 issue, page 135:[20]

> *Beginning October 25, 1998, a group of Brussels bureaucrats…will*
> *oversee implementation of a new privacy policy throughout Europe.*
> *Under this regime, known as the European Data Protection Directive,*
> *any country that trades personal information with the U.K., France,*
> *Germany, Spain, and Italy to any of the other 10 EU states will be*
> *required to embrace Europe's strict standards for privacy. No privacy*
> *no trade. It's that simple…. The new rules will oblige every country within*
> *the European Union to a common set of standards that bind governments*
> *and corporations to a rigorous system of privacy protection…*

This is a very interesting issue,[21] not only because of its requirements, but also because it is indicative of something very far-reaching that is caused by the Internet. Because of the privacy concerns, coupled with today's globally networked computers and electronic commerce on the Internet, a nation or group of nations can pass a law or issue a formal directive that basically dictates how another nation will deal with an issue, such as implementing some mandatory Internet security measures. That is, if you want to connect to my networks, follow my rules.

So, if for example, you are an INFOSEC person working for a nation-state's international corporation with clients and business relationships with the European Community and you use the Internet to conduct business, you may want to get a copy of this new directive because you may be required to provide the required protection based on the European security requirements.

But wait! Not so fast! Supposedly the Organization for Economic Cooperation and Development meeting in Paris, France, approved a plan to allow law enforcement agencies to "eavesdrop on the Internet." This organization consists of 29 countries. However, according to the report, the guidelines included "The fundamental rights of individuals to privacy, including secrecy of communications and protection of personal data, should be respected in national cryptography (coding) policies." "Should" is *not* the same as "will" or "must." Based on the above, how is the INFOSEC officer to interpret such guidelines?

It has also been reported that only three of the fifteen European Union countries have enacted the privacy plan into law. The report also stated that the "White House has been scrambling to ensure that U.S. Web sites aren't hindered by a strict European Union electronic privacy directive set to hit this October…" Hindered? Is that an indication how much the United States currently supports privacy?

In addition, the former Director of the U.S. FBI, Louis Freeh, told a meeting of the International Association of Chiefs of Police, according to a report, that "software vendors should be required by law to offer a security feature allowing law enforcement to decrypt encrypted communications."[22]

If you are required as part of your responsibilities to ensure the privacy of information, do not take the matter lightly as many a lawsuit for millions of dollars can be the result and your name may be on them. Additionally, it is your professional INFOSEC responsibility, so deal with it!

Balancing Security, Human Rights, and Privacy

No common-sense individual wants the total elimination of controls or so much privacy that it is impossible to identify criminals. What the future requires is that we find a way to balance protection, privacy, and personal freedoms — a difficult if not impossible task. It requires mature, educated people who accept responsibility for their actions and are accountable for their actions. However, it appears that every time people are asked about these matters, they contradict themselves; to coin an old cliché, they want to have their cake and eat it too. For example, a CNN.com poll on April 7, 2001 asked: "Do you fear an erosion of your privacy in this digital age?" Some 81 percent (5485 respondents) said they did, while 19 percent (1271 respondents) said they did not. Although this was an informal online poll, the results were what would be expected.

In another CNN.com poll taken on June 3, 2001, the question asked was: "Are you bothered by security cameras that compare photos with a database of mug shots?" Some 49 percent (19,806 respondents) said yes, while 51 percent (21,030 respondents) said they did not mind if it deters crime. So, almost evenly split.

On July 12, 2001, another CNN.com poll asked: "Should human rights concerns prevent Beijing from hosting the Olympics?" In this poll, 84 percent (64,746 respondents) said yes, while 16 percent (12,228) said no. Yes, this was an informal poll but the overwhelming majority were concerned with human rights and the violations of those human rights by nation-states such as China.

So, these polls — albeit informal — do give some indication that there are mounting concerns throughout the world about human rights and privacy. It is expected that these issues will take center stage in the future and increase information warfare attacks against nation-states such as China and others that fail to adhere to basic human rights.

The problem is who will decide what is the proper balance? Can we leave it up to the nation-state agencies? Could they be more objective in the future than they are now? The answer is no — not with corporations pouring massive amounts into the coffers of elected officials and others to stop legislation that would provide more privacy. So, what is the proper course of action? One alternative may be the maturation of the global hacking community to provide a service for mankind and stop being juvenile delinquents.

Hackers: Info-Warriors and Freedom-Fighters of the Twenty-First Century — A Commentary

The elite are not those who destroy or cause havoc in cyberspace, but rather [those who work] to protect the Net.

—Kevin Manson, Senior Instructor
U.S. Federal Law Enforcement
Training Center's Financial Fraud Institute[23]

Most security and law enforcement people look on hackers as the archenemy of the Twentieth and Twenty-first Centuries. This view is also supported by the news media. In their interest to sell, sell, sell, the news media have made the name "hacker" synonymous with anything from juvenile delinquents to mass murderers, and even linked them to the Information Age equivalent of the nuclear bomb in their ability to destroy the information-based world as we know it. New laws have been written in countries around the world to support the investigation, apprehension, prosecution, and incarceration of these "vicious and violent" threats to societies, and even mankind itself!

The initial definition of the term "hacker" and the goal of the early-day hacker, basically a computer enthusiast whose goal was to learn all he or she could about computers and make them the most efficient machines possible to support mankind and his search of knowledge, has sadly been manipulated, modified, destroyed, erased, or deleted. Let us separate hackers from the basic criminals who defraud users, damage, and otherwise destroy systems and information for "fun" and without a meaningful purpose, as the drive-by shooters of large U.S. cities.

Take a look at what is happening today and the trends that forecast tomorrow. There are miscreants, juvenile delinquents, and "wannabes" out there who are causing us some grief; the people who are changing Web sites are like the graffiti writers of the walls and highways of our cities. The only difference is that the people with the paint cannot afford a computer; they can only afford the spray paint. Others are destroying computer-based information, modifying it, and denying access to the systems that store, process, and transmit that information. This is wrong. However, one can put this in perspective. Do they deserve five to ten years in prison for this, and in the company of murderers, rapists, and child pornographers? If you are in law enforcement or security, you may say, "Yes, if they can't do the time, they shouldn't do the crime." That would not only be sad, but unprofessional as well.

Maybe the people we should be putting in jail and charging them with malfeasance, failure to apply due diligence, and apathy are the managers responsible for that information and the information systems. As stockholders in these companies, we want our assets protected. Many of these people are not even adding the basic protection most of us have been recommending for decades. Maybe they deserve to be attacked just for that.

We also have the "civilized, information-based nations" using these threats as an excuse to gain more power, more control over our lives. The U.S. Secretary of Defense was allegedly quoted a year or so ago to have said that if Americans want more security they will have to give up some of their freedoms. Wait! It gets worse. Look at Intel's Pentium III chip that can allegedly provide government agents and others the chip serial numbers of a user's computer without their knowledge, and thus allow monitoring, tracing, and trapping of user's communications.

Also take a look at similar activities by Microsoft, Novell's "balanced" approach, and do not forget that this does not include the alleged classified contracts and programs that government agencies have with vendors for providing covert backdoors. And we will not even discuss the encryption fiasco with the FBI trying to protect us from — who was that? The FBI Director wanting ISPs to give them information about users at their request. Oh, but you say a search warrant is required, probable cause, etc. Hey, we have been there, done that. Some judges are more liberal and pro-law enforcement than others. In other nations, it is even easier.

All this in the name of helping provide all of us with convenience, efficiency, better products, and a more secure life. Products are being developed and used that require more and more of our personal identification — no, more of who we are, what we do, and how we think: a fingerprint required before cashing a check, genetic records of, so far, only convicts, and the beat goes on. The Oklahoma City bombing occurred and the U.S. government tried to use the opportunity to ram through a new anti-terrorist law that, if it had passed in its original form, meant less freedom for all of us, and still does. Why? Could the bombers not be charged with murder? Why do we need new, freedom-limiting laws to prosecute murderers? Does it matter whether they get the death penalty or life in prison for murder or anti-terrorist conduct?

Now the excuse being used by the federal agencies for more power and control is the protection of the information infrastructure of a nation. Wait a minute! Is not most, if not all of that, privately owned? Look at the Russian's SORM and the Chinese attacks against a U.S. Web site devoted to the outlawed Falun Gong sect. If so, then have the Chinese attacked the United States? Also look at other attempts by Asian nations at spying on Internet users, monitoring and controlling the Internet and their citizen's use of it, and again the beat goes on.

In the United States, the Clinton administration planned to create a government-wide security network of "electronic obstacles complete with monitors and analyzers to watch for potentially suspicious activity on federal computer systems." Who will define "potentially suspicious activity"? Add to that Janet Reno, the former U.S. Attorney General, pressuring other nations to restrict, curb, or otherwise control encryption products. The reason is obvious: to read other people's mail.

Further add to this the fact that in the United States, according to the FBI, major crimes are down and the trend is less major crimes in the future; while at the same time, the federal government's discretionary budget is decreasing. That means that federal agencies such as the FBI, NSA, CIA, DIA, and others must all fight for fewer budget dollars. Thus, they are all out there looking for new missions, new money to support their bureaucracies. Hey, you cannot blame them; they do not want to be "downsized" either. The new mission? Hype the hacker threat and the FBI gets $30+ million to go after the teenage hackers — at a time when it appears that the Chinese have stolen and continue to steal our nuclear secrets; at a time when the Russian bear is coming out of hibernation; at a time when real terrorists are gaining new weapons and attacking the interests of the free world in the old fashion way...by blowing it up. Talk about misallocation of available resources!

Nations, and especially the military, are gearing up for information warfare in the Twenty-first Century. They are employing people with "clean" backgrounds and putting them in a bureaucracy with policies and procedures. These are the "good hackers," also known as the information warriors of the future. Actually, they are employed to protect the status quo and the power of nations, even if it means depriving citizens of their rights and freedoms, so sacred to all of us as human beings — a God-given right.

These info-warriors are the ones who, when they were children, had coloring books and were praised by their parents for not coloring outside the lines. They grew up still coloring and being praised for coloring within the lines of security policies, procedures, and doing things by the numbers. Sorry, these are not the people who one should rely on to protect one's freedom as a "sovereign individual."[24] Talk to some of these people and the security and law enforcement people. For many, their perspective of the hacker threat is somehow lost in their emotions, although they are supposed to be objective. They are there as the line of defense to protect the nation. No offense, but so were some of the people in Germany in World War II.

U.S. President Ronald Reagan once called the Nicaraguan rebels "freedom fighters," while their government called them criminals and terrorists. The same holds true throughout the world. Those in power want to keep it and make the rules. Those that are left out of the governmental process want their voices to be heard. Thus, the conflict. As the nation-state begins to have less importance, its employees will fight to keep their power and conflicts will rise at the expense of personal liberties. The global hacker community, although often misguided as to worthwhile objectives in their attacks, are at least beginning to establish better communication lines among themselves.

The hackers of the world are using the Internet to communicate and at the same time to attack systems on a global scale. Many of the attacks are aimed at totalitarian governments, government agencies, political parties, and against the slaughter of animals for their fur, all of which can be considered politically motivated attacks. These are the worst kind and the most feared by nation-states. The attacks mounted by global hackers, based on a "call to arms" by hackers, against the government of Indonesia's Web sites is an example of what they can do and, more importantly, what is yet to come.

These hackers, when growing up, were not given a coloring book and praised for coloring within the lines. The really good ones were given a set of crayons and a blank sheet of paper by their parents and told to draw something using their imagination. These are people who, when attacking systems, write code on-the-fly, using new and imaginative ways to penetrate systems.

The global hackers must rise above the juvenile delinquency level and begin to go back to the true hacker mentality. These are the ones who are the true freedom-fighters fighting totalitarian governments and any governments, institutions, or corporations that want to restrict the rights of individuals and limit their freedom as stated in the U.S. Bill of Rights or other national and international laws such as the European Human Rights legislation.

These hacker freedom-fighters are the ones — based on current trends of governments to control the Web and thus the information flow to its people, the governments that want more and more laws to control us, the governments that want to prosecute their citizens if those citizens criticize that government, and those governments that want to deprive us of the pure freedom that we all inherit as a God-given right when we were born human beings — we will need in the Twenty-first Century to protect our freedoms on a global scale. We will not be able to rely on our governments or the military's info-warriors because they will be protecting the nation-state and the status quo — and it will not be the first time that has happened.

As a final thought: we should have a "Global Hacker Appreciation Day" dedicated to all the hackers, phreakers, crackers, nuts, weirdos, and associated other human beings who surf, spam, use, misuse, and abuse the global information infrastructure. Because of their crazy personalities, criminal conduct, and all-around blatant disregard for rules, laws, and government controls, they have kept millions of people employed, made all our lives more interesting, our work more challenging, and our information security market — and world economies — growing.[25]

What Does the Future Hold?

Another security professional, info-warrior, and futurist put it this way: During the next few decades we will continue to witness the convergence of wide range of technologies. We will also see new technologies that we currently have no perception of come into reality and be adopted and integrated. We may also see a number of the technologies that we consider to be fundamental to the GII become obsolete. A good, current example of this would perhaps be the Wireless Application Protocol (WAP) technology that was much heralded in the recent past as the great hope for mobile Internet connectivity. The 3G technologies and, potentially the Iphone concept that has been adopted in Japan are to likely replace this, in a very short period of time.

What we will undoubtedly see is an increase in the effort placed on developing technologies that will provide sustainable products and renewable energy. These developments will fundamentally affect the types and uses of the GII. We will undoubtedly continue to see demand for resources on the GII outstripping the ability of the infrastructure to supply it. Processors will continue to get faster, although there is a theoretical limit as to how much faster they can operate using currently available technologies. There is already considerable research into alternatives to the silicon chip, with concepts such as optical computing already being tested in the laboratory. Quantum computing has arrived, but as yet has not had the opportunity to demonstrate its potential. We will see our infrastructures become more robust and reliable as the problems of trust and interconnectivity become more widely understood. We will begin to see a more flexible use of the infrastructures as they continue to converge, and we will see users being connected rather than terminals.

Some of the other technologies that we can anticipate are solid-state storage devices and organic memory that may replace the conventional magnetic and optical storage media in use today. New display technologies that will allow us to better interact with the electronic environment and implants that will allow the individual to merge with the computerized environment in which we live and work (this is already being tried in the United Kingdom) will also become a reality.

It is likely that with the advance of technology, coupled with the changing demographics of an aging population in the developed world, the nature of future conflicts will change and there will be an increased likelihood that future wars will be fought, at least in part, in cyberspace. The display of high-tech weapons that the West has demonstrated since the Gulf War is a clear indication of this, but it also gave the first conflict that was judged live on television, with the media transmitting from both sides of the conflict to a global audience. It is likely that the globalization of economies in the form of multinational corporations and the interdependence of nations on each other will reduce the risk of major conflicts.

As the economy becomes more globally based, national governments are less likely to hold the levels of power that they currently enjoy, and the emphasis will shift to centers of gravity such as regional and functional leadership where there is a common cause and benefit.

While the GII has already demonstrated that it has tremendous potential for benefiting all areas of society, it has also been demonstrated that there is potential for harm to be caused by miscreants who will abuse the facilities that it offers.

Summary

As we have seen in the past and as we will continue to see in the future, our world, our global society is shrinking. Yes, physically, we have the same old Earth; however, due to high technology, the world seems so much smaller. When we think back to the past and how long it took for news to travel from one end of the world to the other and then look at the world of today, we find that we have the ability to see and hear what is happening almost anyplace in the world as it happens.

What happens in another part of the world now greatly impacts our "private" world. We can no longer escape it. The current trend will continue, one in which the United Nations will play a greater role in not only world politics but also world economics, social issues, global pollution, and the like. Many modern information-based nations do not like the thought of their declining global power and becoming more "equal" with Third World nations; however, that trend will continue. Thus, there will be new conflicts over the loss of power from the world's larger nations to more equality with the smaller nations of the world.

We see such changes today, from human rights, to global pollution issues, to regional conflicts that draw in the larger nations, to the global AIDS problem — and everything in between.

On a global scale, the race for economic and technological superiority will continue to cause more aggressive and competitive attitudes among the participants. The use of covert information warfare tactics will be used against adversaries for a variety of reasons.

The trend of the people of the world becoming more educated, computer-literate, and aware of the world around them will continue to rapidly increase. It is expected that such issues as human rights, personal freedoms, respect for the environment, and other species in addition to humans will continue to cause conflict among the various sides on the issue; for example:

- Environmental activists versus environmental polluters
- Animal rights activists versus researchers using animals for experiments
- Religious zealots versus religious freedom activists
- Human rights versus despots
- Freedom versus security

It is expected that these conflicts will provide for increased opportunities and use of information warfare weapons and tactics by nation-states, corporations, groups, and individuals.

Let the games continue!

Notes

1. Quote from *The Sovereign Individual, Mastering the Transition to the Information Age,* by James Dale Davidson and Lord William Rees-Mogg, Touchtone, 1999.
2. See http://www.cia.gov/cia/publications/globaltrends2015/ for the detailed report. Much of the quoted material in this chapter is from this report.
3. Robot Technology Marches Forward, by Jim Krane, AP Technology Writer, June 21, 2001.
4. TechTV/Fres Gear, May 9, 2001, broadcast.
5. See http://www.techtv.com/print/story/0,23102,3325337,00.html, Nanotechnology by Fresh Gear Staff, May 2, 2001.
6. See http://www.techtv.com/print/story/0,23102,3326336,00.html.
7. See http://www.cnn.com/2001/TECH/05/14/3g.george/index.html.
8. See http://www.cnn.com/2001/TECH/ptech/06/25/ibm.transistor.reut/index.html.
9. CNN.com article posted December 7, 2000, Reading, England.
10. See http://www.techtv.com/print/story/0,23102,3302633,00.html.
11. http://www.cnn.com/2001/TECH/ptech/07/18/molecular.computing.ap/index.html.
12. http://www.cnn.com/2001/TECH/internet/07/18/nasa.tech.advances.idg/index.html.
13. See *The Sovereign Individual: Mastering the Transition to the Information Age*, by James Dale Davidson and Lord William Rees-Mogg, published by Touchstone, New York, 1997.
14. For those, especially in the United States, concerned with "politically correct" (often a term used in the United States to stifle free speech), I am quoting the dictionary, albeit probably an old version. Thus, the terms in this context imply people of all genders.
15. Some information paraphrased and reprinted by permission of publisher, Butterworth-Heinemann from the book *I-Way Robbery: Crime on the Internet,* by Dr. Gerald L. Kovacich and William C. Boni, 1999.
16. Parliament Pauses on EU-US Privacy Plan — Update, by Robert MacMillan, Newsbytes, Washington, D.C., U.S.A., June 30, 2000, 10:38 AM CST.
17. US newspaper, *Orange County Register,* Business Section, June 4, 1998, Web-Site Privacy Lacking, FTC Finds.
18. *Orange County Register,* Nation/World Section, January 1, 1998, Group Says It'll Post AOL Members.
19. *InternetWeek,* July 27, 1998.

20. http://www.infowar.com/class_1/Class1_051498b_j.html-ssi.

21. While preparing this article, I went through some old research material. I use it here to make a point: although the citations are several years old, nothing has really changed since then — except maybe to get worse!

22. *Information Security* magazine, December 1997.

23. http://www.computerworld.com/cwi/story/0,1199,NAV47_STO62181,00.html.

24. Please read *Sovereign Individual,* by James Dale Davidson and Lord William Rees-Mogg, where they postulate the coming conflicts and demise of the nation-state due to the information age and the rise of the "sovereign individual." An excellent example to support the coming battle for human rights and individual freedom between governments and its citizens.

25. From the column by ShockwaveWriter, published in Reed Elsevier's *Computer Fraud and Society* magazine, Sept. 1988. Reprinted with their permission.

Appendix A

Recommended Reading and References

Recommended Reading and References

1. High-tech karma. *U.S. News & World Report,* August 21, 1995, Business, p. 45.
2. Home Page Speeds Ordering. *Information Week,* August 21, 1995, p. 64.
3. Alberts, David S., Gartska, John J., and Stein, Frederick, P. DoD C4ISR Cooperative Research Program, 1998.
4. Aburdene, Patricia and Naisbitt, John. *Megatrends 2000.* New York: Avon Books, 1990.
5. Anonymous. *A Hacker's Guide to Protecting Your Internet Site and Network,* Maximum Security. Indianapolis, IN: Sams.net Publishing, 1997.
6. Banks, Michael A. *How to Protect Yourself in Cyberspace: Web Psychos, Stalkers and Pranksters.* Scottsdale, AZ: The Coriolis Group, Inc., 1997.
7. Bellovin, Steven M. and Cheswick, William R. *Firewalls and Internet Security.* New York: Addison-Wesley, 1994.
8. Bintliff, R.L. *Complete Manual of White Collar Crime Detection and Prevention.* Englewood Cliffs, NJ: Prentice-Hall, 1993.
9. Boni, William C. and Kovacich, Gerald L. *I-Way Robbery Crime on the Internet.* Boston: Butterworth-Heinemann, 1999.
10. Boni, W.C. and Kovacich, G.L. *Netspionage: The Global Threat to Information.* Boston: Butterworth-Heinemann, 2000.
11. Carroll, John M. *Computer Security.* London: Butterworth Publishers, 1995.
12. Cheswick, William R. and Bellovin, Steven M. *Firewalls and Internet Security: Repelling the Wily Hacker.* Reading, MA: Addison-Wesley, 1994.
13. Clark, Franklin and Diliberto, Ken. *Investigating Computer Crime.* Boca Raton, FL: CRC Press, 1996.
14. Computer Emergency Recovery Team (CERT). *Internet Security,* Information Systems Security Association Conference, Toronto, Canada, 1994.
15. *Computers at Risk: Safe Computing in the Information Age,* by National Research Council, National Academy Press, 1991.

16. Conversations with E. Cummings, Prosecuting Attorney, State of Maryland, 1994–1995.

17. DeMaio, Harry B. *Information Protection and Other Unnatural Acts: Every Manager's Guide to Keeping Vital Computer Data Safe and Sound*. New York: Amacom, 1992.

18. Desman, Mark B. *Building an Information Security Awareness Program*. New York: Auerbach Publications, 2002.

19. Edwards, Mark J. *Internet Security with Windows NT*. Loveland: Duke Press, 1998.

20. Felten, Edward W. and McGraw, Gary. *Java Security*. New York: John Wiley & Sons, 1997.

21. Fialka, John J. *War by Other Means: Economic Espionage in America*. New York: Permissions, W.W. Norton & Company, Inc., 1997.

22. Fites, Philip and Kratz, Martin P.J. *Information Systems Security: A Practitioner's Reference*. New York: Van Nostrand Reinhold, 1993.

23. Futrell, M. and Roberson, C. *An Introduction to Criminal Justice Research*. Springfield, IL: Charles C Thomas, 1998.

24. Garfinkel, Simson, and Spafford, Gene. *Practical UNIX Security*. Sebastopol, CA: O'Reilly & Associates, 1991.

25. Gurbani, Vijay K. and Pabrai, Uday O. *Internet & TCP/IP Network Security*. New York: McGraw-Hill, 1996.

26. Hafner, Katie and Markoff, John. *Cyberpunk: Outlaws and Hackers on the Computer Frontier*. New York: Touchstone, 1992.

27. Held, Jonathan and Bowers, John. *Security E-Business Applications and Communications*. New York: Auerbach Publications, 2001.

28. Herold, Rebecca. *The Privacy Papers: Managing Technology, Consumer, and Employee Actions*. New York: Auerbach Publications, 2002.

29. Herrmann, Debra S. *Practical Guide to Security Engineering and Information Assurance,* New York: Auerbach Publications, 2002.

30. Icove, David, et al. *Computer Crime: A Crimefighter's Handbook*. Sebastopol, CA: O'Reilly & Associates, 1995.

31. Jailhouse Takes Away Prisoners' Cash: IBM Runs Disk Head Dash. *Computerworld,* August 14, 1995, p. 116.

32. Kabay, Michel E. *The NCSA Guide to Enterprise Security: Protecting Information Assets*. New York: McGraw-Hill, 1996.

33. Kenney, John P. and More, Harry W. *Principles of Investigation*, 2nd ed., St. Paul, MN: West Publishing Co., 1994.

34. Knightmare, The. *Secrets of a Super Hacker*. Port Townsend, WA.: Loompanics Unlimited, 1994.

35. Kovacich, G.L. Hackers: From Curiosity to Criminal. The White Paper. ACFE, 1994.

36. Kovacich, Gerald L. *The Information Systems Security Officer's Guide*. Boston: Butterworth-Heinemann, 1998.

37. Kovacich, G.L. and Boni, W.C. *High-Technology Crime Investigator's Handbook: Working in the Global Information Environment*. Boston: Butterworth-Heinemann, 2000.

38. Kovacich, G.L. and Blyth, A. *Information Assurance: Surviving in the Information Environment*. London: Springer-Verlag, 2001.

39. Levy, Steven. *Hackers Heroes of the Computer Revolution*. New York: Anchor Press, 1994.

40. Li, X. and Crane, N.B. *Electronic Style, A Guide to Citing Electronic Information*. Westport, CT: Meckler Publishing, 1993.

41. Marcella, Albert J. *Cyber Forensics: A Field Manual for Collecting, Examining, and Preserving Evidence*. New York: Auerbach Publications, 2002.

42. McBride, Patrick et al. *Secure Internet Practices*. Cornelius, NC: METASeS, 2002.

43. McClain, Gary. *21st Century Dictionary of Computer Terms.* New York: Dell Publishing, 1994.

44. McGraw, Gary and Felten, Edward W. *Java Security: Hostile Applets, Holes, and Antidotes. What Every Netscape and Internet Explorer User Needs to Know.* New York: John Wiley & Sons, 1997.

45. Middleton, Bruce. *Cyber Crime Investigator's Field Guide.* New York: Auerbach Publications, 2002.

46. Middleton, Bruce. *Investigating Network Intrusions.* New York: Auerbach Publications, 2002.

47. Naisbitt, John. *Megatrends Asia.* New York: Simon and Schuster, 1996.

48. Naisbitt, John. *Megatrends.* New York: Warner Books, 1982.

49. National Research Council. *Computers at Risk.* Washington, D.C.: National Academy Press, 1991.

50. Ohmae, Kenichi. *The Mind of the Strategist.* Middlesex, England: Penguin Books, Ltd., 1982.

51. Peltier, Thomas R. *Information Security Policies, Procedures, and Standards.* New York: Auerbach Publications, 2002.

52. Peltier, Thomas R. *Information Security Risk Analysis.* New York: Auerbach Publications, 2001.

53. Pettinari, D. *Using Internet to Communicate with the Public on LE Issues.* Available: FTP: polcomp.ilj.org/pub/polcomp. File: 720.12A.txt., July 20, 1995.

54. Platt, Charles. *Anarchy Online Net Crime, Net Sex.* New York: Harper Paperbacks, 1996.

55. Rose, Lance. *Netlaw: Your Rights in the Online World.* Berkeley, CA: Osborne McGraw-Hill, 1995.

56. Schwartau, Winn. *Time Based Security.* Florida: Interpact Press. 1998.

57. Schwartau, Winn. *Information Warfare: Chaos on the Electronic Superhighway.* New York: Thunder's Mouth Press, 1994.

58. Schwartau, Winn. *Information Warfare: Cyberterrorism: Protecting Your Personal Security in the Electronic Age,* 2nd ed. New York: Thunder's Mouth Press, 1996.

59. Schweizer, Peter. *Friendly Spies: How America's Allies are Using Economic Espionage to Steal Our Secrets.* New York: The Atlantic Monthly Press, 1993.

60. Shaffer, Steven L. and Simon, Alan R. *Network Security.* Cambridge, MA: AP Professional, 1994.

61. Siegel, Pascale Combelles. DoD C4ISR Cooperative Research Program, Institute for National Strategic Studies, 1998.

62. Spencer, Donald D. *The Illustrated Computer Dictionary.* Columbus, OH: Charles E. Merrill Publishing, 1980.

63. Stang, David J. and Moon, Sylvia. *Network Security Secrets.* San Mateo, CA: IDG Books Worldwide, Inc., 1993.

64. Steele, Guy L. Jr., Woods, Donald R., Finkel, Raphael A., Crispin, Mark R., Stallman, Richard M., and Goodfellow, Geoffrey S. *The Hacker's Dictionary.* New York: Harper & Row, 1993.

65. Sterling, Bruce. *The Hacker Crackdown: Law and Disorder on the Electronic Frontier.* New York: Bantam Books, 1992.

66. Stoll, Clifford. *The Cuckoo's Egg.* New York: Doubleday, 1989.

67. Tafoya, W. A Delphi Forecast of the Future of Law Enforcement. Dissertation submitted to Faculty, Graduate School, University of Maryland, 1986.

68. Tipton, Harold F. and Krause, Micki. *Information Security Management Handbook,* Volumes 1–3. New York: Auerbach Publications, 2002.

69. Toffler, Alvin and Toffler, Heidi. *Creating a New World Civilization.* Atlanta: Turner Publishing, Inc., 1994.

70. Toffler, Alvin and Toffler, Heidi. *War and Anti-War.* Boston: Little, Brown and Company, 1993.

71. Toffler, Alvin. *Future Shock.* New York: Bantam Book, 1971.

72. Toffler, Alvin. *Powershift.* New York: Bantam Books, 1990.

73. Toffler, Alvin. *The Third Wave.* New York: Bantam Books, 1980.

74. Tudor, Jan Kilmeyer. *Information Security Architecture.* New York: Auerbach Publications, 2000.

75. U.S. Department of Justice, Bureau of Justice Statistics. *Computer Crime, Electronic Fund Transfer Systems and Crime.* Washington, D.C.: U.S. Government Printing Office, 1982.

76. U.S. Department of Justice, Bureau of Justice Statistics. *Computer Crime, Electronic Fund Transfer Systems Fraud.* Washington, D.C.: U.S. Government Printing Office, 1986.

77. U.S. Department of Justice, Office of Justice Programs, Bureau of Justice Statistics. *Directory of Automated Criminal Justice Information Systems 1993, Volume II: Corrections, Courts, Probation/Parole, Prosecution.* Washington, D.C.: U.S. Government Printing Office, 1993.

78. U.S. Department of Justice, Office of Justice Programs, Bureau of Justice Statistics. *Dedicated Computer Crime Units.* Washington, D.C.: U.S. Government Printing Office, 1989.

79. U.S. Department of Justice, Office of Justice Programs, Bureau of Justice Statistics. *Organizing for Computer Crime Investigation and Prosecution.* Washington, D.C.: U.S. Government Printing Office, 1989.

80. Vold, G.B. and Bernard, T.J. *Theoretical Criminology,* 3rd ed., New York: Oxford University Press, 1986.

81. Walker, Bruce J. and Blake, Ian F. *Computer Security and Protection Structures.* Stroudsburg, PA: Dowden, Hutchinson & Ross, Inc., 1977.

82. Wentz, Larry (contributing editor). Lessons from Bosnia: The IFOR Experience, DoD C4ISR Cooperative Research Program, Institute for National Strategic Studies, 1997.

83. Wood, Charles Cresson. *Information Security Policies Made Easy.* Sausalito, CA.: Self-published, 1994.

Appendix B

Glossary of Terms

ADSL	Advanced Digital Subscriber Line
AEA	American Electronics Association
ARPA	Advanced Research Projects Agency
ASEAN	Association of South East Asian Nations
ASIS	American Society for Industrial Security
BLS	Bureau of Labor Statistics
BO	Back Orifice
BO2000	Back Orifice 2000
CEO	Chief Executive Officer
CERN	Conseil Europeene pour la Recherche Nucleaire
CERT	Computer Emergency Response Team
CESG	Communications Electronic Security Group
CFE	Certified Fraud Examiner
CIA	Central Intelligence Agency
CIO	Corporate Information Officer
CIP	Critical Infrastructure Protection
CIPO	Corporate Information Protection Officer
CISA	Certified Information Security Auditor
CISSP	Certified Information Systems Security Professional
CKO	Coherent Knowledge-based Operations
CMA	Computer Misuse Act
CNA	Computer Network Attack
CND	Computer Network Defense
CNI	Critical national infrastructure
COMSEC	Communications security
CONOP	Concept of Operations
COTS	Commercial off-the-shelf
CPP	Certified Protection Professional
CSCE	Conference for Security and Cooperation in Europe

CSI	Computer Security Institute
CSPP	Computer Security Policy Project
C2W	Command and control warfare
C2IS	Command, control and intelligence systems
C³CM	Command, control, communications, and countermeasures
DARPA	Defense Advanced Research Projects Agency
DDOS	Distributed denial-of-service
DERA	Defence Evaluation and Research Agency
DIA	Defense Intelligence Agency
DIO	Defensive Information Operations
DISA	Defense Information Systems Agency
DIW	Defensive information warfare
DoD	Department of Defense (United States)
DSL	Digital Subscriber Line
DTI	Department of Trade and Industry (United Kingdom)
DVD	Digital video disk
EC	Electronic commerce (E-commerce)
ECM	Electronic countermeasures
EDI	Electronic data interchange
EDT	Electronic Disturbance Theater
EEFI	Essential element of friendly information
EEI	Essential element of information
EMP	Electromagnetic pulse
ETA	Euskadi Ta Askatasuna
EU	European Union
EW	Electronic warfare
FBI	Federal Bureau of Investigation (U.S.)
FTP	File Transfer Protocol
GDP	Gross domestic product
GII	Global information infrastructure
GPS	Global Positioning System
GUI	Graphical user interface
HIRF	High-intensity radio frequency
HPM	High-power microwave
HUMINT	Human intelligence
IA	Information assurance
IAAC	Information Assurance Advisory Council
IDS	Intrusion detection system
IE	Information environment
IFF	Identification friend or foe
III	Integrated information infrastructure
IIU	Israeli Internet Underground
IM	Instant message
IMINT	Imagery intelligence
INFOSEC,	
InfoSec	Information security
IO	Information operations

IP	Internet Protocol
IPO	Initial product offering
IPR	Intellectual property rights
IRA	Irish Republican Army
IRC	Internet Relay Chat
ISDN	Integrated Services Digital Network
ISO	International Standards Organization
ISP	Internet service provider
IT	Information technology
IVDA	Individual Virtual Direct Action
IW	Information warfare
IW-D	Information warfare-defensive
IW-O	Information warfare-offensive
JCS	Joint Chiefs of Staff
KM	Knowledge management
LAN	Local area network
MASINT	Measurement and signature intelligence
MDB	Multilateral development banks
MEII	Minimum essential information infrastructure
MEDII	Minimum essential defensive information infrastructure
MP	Member of Parliament (United Kingdom, Canada, etc.)
MoE	Measure of effectiveness
MVDA	Mass virtual direct action
NATO	North Atlantic Treaty Organization
NCB	Network-centric business
NIC	National Intelligence Council
NII	National information infrastructure
NIPC	National Infrastructure Protection Center
NISCC	National Infrastructure Security Coordination Center
NIST	National Institute of Science and Technology
NSA	National Security Agency
NSC	National Security Council
OIO	Offensive information operations
OIW	Offensive information warfare
OMENS	Other Middle East nation-states
OODA	Observe, orient, decide, act
OOTW	Operations other than war
OPEC	Oil and Petroleum Exporting Countries
OPLAN	Operations plan
OPSEC	Operational security
OSINT	Open source intelligence
OSW	Other signature warfare
PBX	Private branch exchange
PDA	Personal digital assistant
PGP	Pretty Good Privacy
PKI	Public key infrastructure
PSN	Process serial number

PSYOPS	Psychological operations
R&D	Research and development
RF	Radio frequency
RFA	Regional Financial Associates
RII	Regional information infrastructure
RMA	Revolution in Military Affairs
ROI	Return on investment
RORI	Roll on-roll off
RSA	Revolution in Security Affairs
SEAD	Suppression of Enemy Air Defense
SIC	Standard industrial classifications
SIF	Strategic identification feature
SIGINT	Signals intelligence
SIMM	Single in-line memory module
SIW	Strategic information warfare
SME	Small to medium enterprise; subject matter expert
SOHO	Small office/home office
TECHINT	Technical intelligence
TLD	Top-level domain
TMENS	The Middle East nation-state
TOR	Terms of reference
UK	United Kingdom
UN	United Nations
UPS	Uninterruptable power supply
USAF	United States Air Force
USC	United States Congress
UUCP	UNIX to UNIX copy
UWB	Ultra wide band
VPN	Virtual private network
WAP	Wireless Application Protocol
WEU	Western European Union
WTO	World Trade Organization
WWW	World Wide Web (Web)
Y2K	Year 2000

Appendix C

A Short History of Technology and Its Relationship to Warfare

Technology from the Chinese

Some of the earliest known traces of humans, if not the earliest, are found in China. The "Yuan Mou Man" is said to date back to 1.6 million years B.C.[1] This was followed by Lan Tian Man and Peking Man in about 700,000 to 500,000 B.C.; where there were indications that the humans used stone tools and fire. Other technological wonders some believe started in China and were later reinvented or part of a technology transfer to Europe and the rest of the world. Many can be related, directly or indirectly, to prosecuting wars.

There may be some who say that several inventions attributed to the Chinese were in fact invented in Europe. However, those inventions (e.g., the flame-thrower) were "invented" in Europe hundreds of years later. Therefore, it is more likely that technology transfer had taken place between the Chinese and the European merchants, or the European inventors did not know that the technology had already been invented when they "invented" it.

Abacus

From the Greek word *abax,* meaning "calculating board" or "calculating table," the abacus was invented by the Chinese. The first record of the abacus (see Exhibit 1) was from a sketch in a book from the Yuan Dynasty (Fourteenth Century). Its Mandarin name is *Suan Pan*, which means "caculating plate." Its inventor is unknown but the abacus is often referred to as the "first computer" because it was used as a mathematical model for early electronic computers.[2] Some believe that it may have been invented as far back as 3000 B.C.

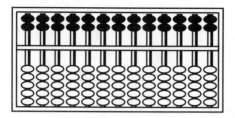

Exhibit 1.   An Abacus

Exhibit 2.   Chinese Technological Devices Used or Could Be Adapted to Warfare

Approximate Dates	Chinese Technology
403–206 B.C.	Chemical warfare, crossbow
206–220 B.C.	Parachute
317–589 A.D.	Propeller
618–907 A.D.	Gunpowder[a]
960–1279 A.D.	Flame-thrower
960–1279 A.D.	Bomb
960–1279 A.D.	Rocket
960–1279 A.D.	Grenade

[a] Some say gunpowder was invented around the Third Century B.C., but the true gunpowder formula was first published around 1040 by Tseng Kung-Liang (http://inventors.about.com/science/inventors).

One can argue that many inventions have been modified to directly support warfare. Some technologies have also indirectly supported warfare, such as through logistics, because many non-combatant technologies had a dual use similar to the way in which computerized weapons of today have a dual use (e.g., commercial off-the-shelf software [COTS]) (see Exhibit 2).

Some of the technological devices invented by the Chinese that could be used indirectly to support the offensive, defensive, and exploitative aspects of warfare include the following:

- Medical needles and medical observations (2205–1766 B.C.)
- Piston bellows; gas and petroleum as fuel; compass[3] (722–481 B.C.)
- Manned kites (403–206 B.C.)
- Steel from cast iron (206–220 B.C.)
- Sliding calipers; belt drive (206–9 B.C.)
- Water, chain pumps; rudder; seismograph (5–220 A.D.)
- Multiple masts; battened sails; watertight ship compartments (220–280 A.D.)
- Use of algebra in geometry (265–316 A.D.)
- Enhanced steel making process (317–420 A.D.)
- Block printing; mechanical clock (589–618 A.D.)

- First use of the symbol for "zero" (in India 876 A.D.)
- Chain drives (907–960 A.D.)
- Mercator map projection; flares (960–1279 A.D.)

Western Technology[4]

Having noted the early inventions of the Chinese, one can easily relate how they have impacted societies and their ability to wage wars on each other. If the technological inventions of the Chinese impacted how wars were prosecuted, the Europeans brought it to the stage of a "fine art." Then, during the last half of the Twentieth Century and into the Twenty-first Century, the United States took the lead in technological development for warfare, and especially for information warfare purposes.

Sixteenth-Century Technology

Using the Gregorian calendar to mark time, it was noted that the Sixteenth Century (1500–1599) showed little in the way of major technological inventions. It also showed little new as to how wars were prosecuted. The main technological invention directly related to warfare during this period was the "hand gonne," actually more of a handheld cannon. It was not very reliable, as is usually the case with the first few attempts at new warfare technology. Later, "miniaturization" took place and improvements led to the matchlock design that included a cover plate and flash pan. This weapon, known as the "matchlock," was the weapon used by early explorers of the empires and nation-states when they set out to colonize the New World, as well as to continue the subjugation of the Old World.

Seventeenth-Century Technology

In approximately 1642 A.D., Pascal designed a mechanical calculator. Thus, from the abacus in 3000 B.C., to the use of the symbol for "zero" in 876 A.D., to Pascal's calculator in 1642 A.D., we see increasing thought about the use of mathematics for developing technology. There also seemed to be more of an inclination to use technology to assist the human race, as well as control some of us. That was about to significantly increase due to the technological developments of the Seventeenth Century and beyond.

During the Seventeenth Century (1600–1699), there were some significant technological inventions. Although some may have had indirect uses in warfare, the majority of them were used to enhance society, including the following:

- Telescope
- Reflecting telescope
- Human-powered submarine

- Steam turbine
- Blood transfusion
- Calculating machine
- Adding machine
- Micrometer
- Barometer
- Air pump
- Pocket watch
- Steam pump

Some of the other technological inventions of this period could be used in warfare, but probably were used more by the non-combatants of the period. The major technological weapon created during this time appeared to be the flintlock rifles and pistols that led to experiments with different weapon designs such as rotating cylinders. This was a key warfare discovery and just the beginning of the revolution of rifles and handguns. The human-powered submarine was just a primitive invention and was of no real value as a warship during this period.

Eighteenth-Century Technology

The Eighteenth Century (1700–1799) saw little change in most of the inventions vis-à-vis warfare technology. However, as in past centuries, the Eighteenth Century did witness technological inventions that assisted in making life a little better for people (excluding the guillotine, but that of course depended on what end of the blade you were on). The inventions of the period included:

- Steam engine
- Fire extinguisher
- Electric telegraph
- Steamship
- Submarine (non-human powered)
- First demonstration of the European parachute
- Hot-air balloon
- Guillotine
- Bicycle
- Ambulance
- Smallpox vaccination
- Battery

Although invented in the 1700s, the non-human powered submarine did not truly come into its own as a great technological weapon until the Twentieth Century. Such things as the electric telegraph and hot-air balloon could be considered indirect weapons of war because they could be used for military communication and surveillance, respectively. While the parachute was interesting, no reliable type of airplane had yet been invented.

Exhibit 3.   Major Technologically Driven Inventions of the Ninteenth Century

Printing press	Blueprint
Amphibious vehicle	Facsimile
Armored warship	Anesthesia for tooth removal
Revolver	Sewing machine with a motor
Ether anesthesia	Demonstration of the principles of fiber optics
Telegraph	Barbed wire
Repeating rifle	Toilet paper
Tin can	Metal detector
Barbed wire	Roll film for cameras
Telephone	Radar
Machine gun	Matchbook
Typewriter	Internal combustion engine
Paper-strip photographic film	Incandescent light bulb
Photography	Improved submarine
Improved propeller	Steam locomotive
Telegraph	Portland cement
Morse code	Dynamite
Bicycle	

Ninteenth-Century Technology

The Ninteenth Century (1800–1899) began the period in which technological inventions of the era that contributed to warfare really began to take off. Although starting slowly, the inventions developed in sophistication and numbers at an ever-increasing rate — a rate that still has not shown any signs of slowing down. In fact, the opposite is true.

However, this century, as with the previous one, also saw continued refinements and "better-invented" technological devices and improvements on previously invented devices. This period also saw an increase in the ability to communicate and conduct surveillance. And as we know, communication is vital to battlefield coordination. Also, surveillance — knowing where the enemy is located, their troop strength, and other battle issues — is a key ingredient to the successful prosecution of warfare.

Even in today's information warfare environment, communication, miscommunication, communication interception, and impeding the adversary's communications are key to defeating that adversary. Surveillance of the adversary remains a key ingredient in warfare today. Some of the major technological inventions of the 1800s are noted in Exhibit 3.

The following is a short summary of the history of communication and the technology used to store, process, and transmit information.[5]

- *Telegraph.* Samuel Morse developed the Morse code and the first really successful electric telegraph in 1837. He is credited with digitizing the alphabet.

- *Pony Express.* In 1860, the Pony Express mail service began but only lasted about 18 months due to telegraph lines through to California. The Pony Express' objective was the fast delivery of communications, through the form of mail. It was one, if not the first, system that was to be eliminated due to a new technology — the telegraph system.
- *Trans-Atlantic cable.* The cable was installed in approximately 1866; and by 1900, there were 15 cables across the Atlantic Ocean. The first successful trans-Atlantic cable brought fast communication across the ocean. Thus, communication by ship that had previously taken weeks and sometimes months now was accomplished in seconds.
- *Typewriter.* In 1867, the keyboard layout of the first practical typewriter — the same, basic layout as the one used today — was created. This provided a faster means to write the communications that were to be transmitted by telegraph, and later as keyboards for inputting to computers.
- *Telephone.* Alexander Graham Bell is credited with inventing the telephone in 1876, and we all know the results of that invention. However, no one could have dreamed of the power of today's telephone, when coupled with the computer. Although the basic telephone was a great technological breakthrough, when coupled with the computer, it exponentially changed the way we share information.

This period saw the evolution and revolution in firearms from rifles and pistols, including such weapons as the Enfield rifle and Colt breech-loading revolver. However, the weapon that truly had a devastating effect on prosecuting wars, especially beginning in the Twentieth Century with World War I in Europe was the machine gun.

Also during the Nineteenth Century, we begin to see the first serious signs of human use of machines to do their calculations for them. It is true that earlier scholars and inventors had used mathematics for centuries and did invent some primitive ways of looking at the use of mathematics to help them better understand their universe. However, when in 1823 Babbage began work on "algebraic solutions mechanisms," we see the first baseline and model of things to come. This was followed in 1888 by Marquand's designing what was called an "electronic logic machine," followed in 1890 by Hollerith's designs of a "tabulating machine." Now, looking back at those three designs, we begin to see signs of the true beginning of modern-day technology, which led to high technology.

Notes

1. See History of Chinese Invention and Discovery, The West's Debt to China, at http://www.computersmith.com/chineseinvention/.
2. See http://qi-journal.com/action.lasso?-Token.SearchID=Abacus&-Response=culture.asp.
3. Some historians believe that the compass which had a magnetic pointer was developed around 1190. Others believe that in China, Shen Kua, in his *Meng Chi Pi Than* (*Dream Pool Essays*), written around 1080, provided an accurate description of a compass (http://inventors.about.com/science/inventors).
4. See the books by Alvin and Heidi Toffler for more information on the transitions of human societies (Appendix A).
5. Based on a communications history (from the *Orange County Register* newspaper, September 21, 1997).

Appendix D

Revolution in Information Affairs: Tactical and Strategic Implications of Information Warfare and Information Operations

Manuel W. Wik*

> *A new era is born — the Information Age. Information technology is our means of reinforcing human knowledge and communications and opens up a revolutionary new world to mankind.*

A revolution in information affairs is envisioned. It is a slow-motion revolution, thus not always clearly visible. Information age and information technology creates dependencies, capabilities, and vulnerabilities that have to be understood and managed. A revivalism for this revolutionary new era is needed.

The purposes of this paper are to cast some light on new rising threats and opportunities and to discuss tactical and strategic implications of information warfare and information operations. It is important to increase awareness within private, government, and military sectors. A picture of what the future might look like is given. The aspects presented include: the meaning of information

---

* Reprinted with permission from *Revolution in Information Affairs: Tactical and Strategic Implications of Information Warfare Operations,* Global Communications, Hanson Cooke (Division of HHC Plc), London. www.globalcomms.co.uk.

operations and information warfare; what is really new; essential drawbacks, driving forces in the development; impact on information infrastructure; and countermeasures. The importance and the necessity to establish a strategy, to introduce processes, and to organize resources for national security are explained.

Part of this presentation is based on an address given at the Danish Supreme Commander's security policy course in 1999. In addition, background material from two major American reports have been used (Reference 1 and 2) as well as a report from the European Parliament (Reference 3). Permission has been obtained to use material from these references.

The New Future

Powershift

The futurist Alvin Toffler has predicted a revolutionary power struggle to come:

> *For we stand at the edge of the deepest powershift in human history.*

He talks about power involving the use of violence or force (which is considered to be the monopoly of military forces), the use of wealth, and the use of knowledge. He talks about a shift from the inflexible and low quality power of force and the versatile medium quality of wealth, towards the application of knowledge being the most versatile and high quality power. Actor Sean Connery, in a film set in Cuba during the reign of the dictator Batista, plays a British mercenary. In a memorable scene, the tyrant's military chief says: "Major, tell me what your favorite weapon is, and I'll get it for you." To which Connery replies: "Brains." From iron power to brainpower, from hardware to wetware. Is this really something new?

Almost 2500 years ago Sun Tzu from China talked about it. Napoleon has said: "The sword is beaten by the mind." Winston Churchill has said: "Empires of the future are the empires of the mind." Are we really standing on the edge of the deepest powershift in human history? We know that knowledge is a force multiplier. By using new means and new technological ideas used in new ways, one does not just make everything a little bit better or more efficient but rather creates new ways of doing things. This multiplier is clearly a tremendous amplifier. *We are not redesigning the past; we are inventing the future.* How is that done?

From time immemorial man has used tools. Tools have been able to strengthen the power of violence and the power of wealth (see Exhibit 1). However, in the past not many tools were invented to strengthen human senses, thoughts, or communications; at least not until the art of printing was invented. Today we experience the beginning of revolutionary changes in this respect — the strengthening of power of information and knowledge — and a historical shift in human development.

Human senses have begun to expand through the technology of a global network of technical sensors with increasing resolution and capability to detect new conditions. *Human thoughts* expand with a global network of computers. The capability to search, navigate, find, examine, understand, assimilate, and refine just the correct information has become a strategic competence. *Human*

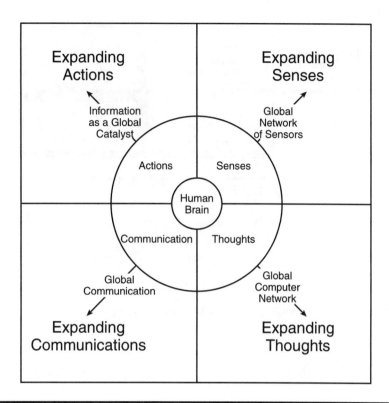

Exhibit 1.   Information Tools Revolutionize Human Capabilities

communications expand with a global communications network carrying speech, text, video, and data. Fiber optic cables, satellites, mobile phones, the Internet, and television media disseminate information globally by the speed of light, far exceeding the speed of our thoughts. Shimon Peres once said:

> *The greatest change in our time has not been effected by armies or states or international organizations; it has been driven by the spread of information.*

Information is a catalyst for our *human actions* and now we can see many means of expanding action far beyond the reach of our hands and to a global level. It changes our life forever, it changes our work and our leisure time, and it changes society, both civil and military. It changes and expands our opportunities. However, it also changes and expands the nature of threats in all conceivable dimensions; and that is also where information warfare occurs. We are only standing at the very beginning of this evolution and are not able to foresee the consequences strategically.

The Knowledge Arena

The traditional map we typically use is geographical. We must now supplement it with the logic map, the world map of ideas, and the human biographical

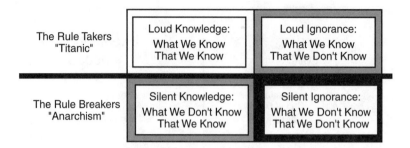

Exhibit 2.   The Knowledge Battlefield

map, which deals with how impressions and perceptions influence the human brain. We must enter the knowledge arena, the focal plane of the Twenty-first Century. The knowledge arena consists of the human consciousness, the subconscious, and the unconscious. The consciousness can be divided into *what you know that you know* and *what you know that you don't know*. They represent the *loud knowledge* and the *loud ignorance*. Together, they form the perfect foundation for the planners, the rule takers, those who want to define everything and stick to it. Mark Twain once said,

> *For those who only have a hammer as a tool, all problems look like nails.*

Personally, I would name this approach "Titanic" because in the end it will always fail (see Exhibit 2).

Sooner or later something unknown comes up; change is the only sure thing in life.

> *Those who do not expect the unexpected are not able to find it.*

> —Herakleitos, Greek philosopher

The more a system is adapted to its purpose, the less it will be able to manage changes. If a competitor in peacetime or an adversary in wartime knows all our rules and we keep to them, we will be lost. It will be as if a chess player has the knowledge of everything the opponent thinks. What you have to do is to do the unexpected and take the opponent by surprise. That is also the essence of war gaming.

We must activate the subconscious, *what you don't know that you know*. This is the *silent knowledge*. Nobody can see the holes where there is no pattern of action. It is hard to animate elusive patterns where many surprises spring up. In order to know how to build a system, you must know how it shall be used; but not until you use it, will you know how it should have been built. You learn by mistakes. Not until you are faced with a new situation and you do something, will you find out that you are actually able to handle it.

How well have we actually been able to predict war? General J. Enoch Powell once said:

> *The history is full of wars that everyone knew would not come.*

Look at all the negative imaginations of new inventions. In 1943, the chairman of the board of IBM, Thomas Watson, said that there would probably only be a need for five computers in the world in the future. In 1876 after the telephone was invented and introduced, the Western Union Telegraph Company said that the telephone could not seriously be used for communication and was of no value to the company. In 1899, Charles H. Duell, who was a civil servant in a leading position at the American Patent Bureau, motivated a proposal to disband the bureau saying that everything that could be invented had now been invented. When the talking picture was invented, the silent film directors thought it would be ridiculous to listen to the actors. And so on. These examples of delusions and lack of imagination are going to multiply in the future.

To find out more about silent knowledge, you need to do mental stretching. Think about those words "mental stretching"; when did you last do so? You were probably jogging and did some body stretching, but when did you do the mental one? When you did, you were the rule breaker, the anarchist according to the vocabulary of the rule taker. This is an important kind of knowledge that you should try to experience more in the future. It is needed in the information and knowledge era and is especially useful in conflicts.

Stretching can be accomplished by turning the way of thinking in new directions. George Bernard Shaw gave a good example of mental stretching:

> *Most people look at the world as it is and ask them why. I am dreaming of things that never existed and ask myself why not?*

The unexplainable that triggers our thoughts and requires mobilization of our intellectual capability, which also combines fragmentary knowledge from different parts of our brain to form new knowledge, leads the way to mental stretching. Bertrand Russell has said:

> *The brain is a remarkable machine that can combine expressions in the most astonishing ways.*

Albert Einstein has expressed it in this way:

> *The most beautiful we may experience is the unexplainable. It is the source of all true science.*

Today and even more tomorrow, intangible assets become more important. A trademark or a special design can prove to be far more valuable than a traditional product.

Creativity, fantasy, innovation, and flexibility are going to be the most wanted characteristics in the future. The new attitude is to mentally stretch so that one may predict changes and catch opportunities, rather than to react on events and solve problems (see Exhibit 3). In an open architecture, rigid constructions become historic buildings. Knowledge wins over strict hierarchical organizations. Organizations must be able to switch in a flexible way

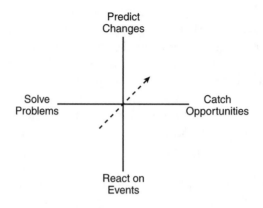

Exhibit 3.   The New Attitudes

between hierarchy and flat networks depending on the tasks. Opaque properties fence in knowledge and counteract the precision of decisions. Hierarchy
restrains creativity. Nor can hierarchies react with sufficient speed due to
changes as the rate of change increases all the time. *Pace wins over hierarchy.*
As human beings we will experience an increasing difficulty to manage this
increasing pace, which will change much of our lives. Furthermore, in conflicts
and war, it will be a matter of life or death.

Last, but not least, the knowledge arena consists of unconsciousness, *what
you don't know that you don't know.* I would call it *silent ignorance.* This is
a particularly interesting region.

We are talking about areas where there is no pattern of action. It is like
dropping your keys at night and only searching for them where there is light
enabling you to see. Now, if you get in contact with people busy in areas
not previously known to you, you will experience new things. This means
that you will be able to make cross-cultural excursions into this type of silent
ignorance and be able to capture new knowledge. Think about it! Information
technology means increased opportunities to transform unconsciousness into
consciousness. Thus we learn more.

So, what should you do? Should you be the rule taker or the rule breaker,
the passenger on the Titanic, or the anarchist? I guess none of them is good
forever, and you will have to try to balance between the bright and the dark
sides according to the situation and the wisdom that you have.

The Rising New Threat

The Changing World

Information and knowledge have always been — and are now more than
ever — part of the economy of a nation, of a corporation, and of a family.
In the information society, "knowing" becomes more important than "owning."
As [the] Information Age evolves, society will be threatened in new ways.
Many people take information systems for granted and do not realize the great

dependence on them, the significance of their vulnerability, and what would happen in case of malfunction or disruption. Previously, all information was hidden that was not open.

Today all information is open that is not hidden. Immensely sophisticated information systems have so far been erected on insecure foundations. Networking capability has outpaced the capability of protecting information.

Data traffic is tripling every year and will overtake voice as the dominant type of traffic over the world's telecommunications networks by [the] year 2005. Seventy-five million new customers signed up for cellular phone service in 1998, bringing the worldwide untethered population to roughly 285 million. In 1998 an average of five million e-mails were sent every minute. Users can listen to e-mail messages over the phone and then reply to them with voice messages. Or they can have e-mail messages and attachments printed as faxes at a fax machine. Intelligent software agents will sort and filter incoming messages and allow callers to retrieve and manage voice mail using spoken commands rather than a telephone keypad.

Today there are more than 100 million users of the Internet and there is a new Web site added every four seconds. Internet traffic is doubling every 100 days. By the year 2005 there may be ten billion people using the Internet with an enormous number of Web sites and information to be found and also to be at risk for intrusions. The real assets will be symbols and bytes, not cash money like notes and coins. Time is money, information is money, and filtering high-speed information for special purposes involves a great deal more money. The Internet is changing the way the world economy functions. By the year 2005 sales over the Internet are expected to reach five trillion U.S. dollars in the United States and Europe.

It is clear that the information technological revolution and the new global economy of information and knowledge will create a number of new threats that will make the criminals of the past look very pale in comparison. We do not yet know the outcome of the changing circumstances but we will need to rapidly create and mobilize forces of information defense to encounter expected cyber-crimes.

Our vulnerability, particularly to cyber attacks, is real and growing.

—U.S. President Bill Clinton

Billions in proprietary secrets have been stolen from high-tech corporations. Most corporations have been penetrated electronically by cyber-criminals. In the United States the FBI estimates that electronic crimes cost victims about ten billion dollars a year. The importance of consumer confidence and shareholder value prevents the reporting of more than a fraction of the actual intrusions to law enforcement agencies. It is also estimated that only a small fraction of all intrusions are known to the owners of the systems under attack.

We have already experienced an arsenal of information warfare weapons: computer viruses, worms, Trojan horses, logic bombs, and software for denial-of-service. Compromising high-powered scanners and sniffers proliferate and are being used to intercept mobile phone calls, faxes, and satellite and landline communications. A number of new methods are being used and further developed to steal information and to camouflage where attacks originate.

There is also a large arsenal of tools for destruction of information, and the information infrastructure. For example, telephone lines can be overloaded by special software, and traffic control of air, sea and land vehicles can be disrupted or given false information. Financial institutions, emergency services, and other government services software can be scrambled. Control of electric power, pipelines, and industrial processes can be altered by remote control and sabotage can be directed against stock exchanges.

"Peace really does not exist in the information age" (Kenneth A. Minihan) and the threat spectrum is constantly changing. Malicious tools are constantly improving and changing.

Password-cracking programs are widely available. Programs to detect weak points in the security of a system can now additionally be used to automate attacks against the identified vulnerabilities. Computer chips with malicious code (i.e., trapdoors, backdoors, logic bombs) can be obtained commercially at low cost. Programs to edit home pages on the World Wide Web can be used to attack network servers. Powerful high capacity malicious servers can attack information systems connected to the Internet.

Computer Threats and Security: Some Threats

There are a number of external and internal threats that must be considered. Most operating systems are shipped with inherent vulnerabilities. Further vulnerabilities are introduced during configuration and usage. Loading files from the Internet can ruin a computer that does not have a very good and recent antivirus program. Exploited information is widely and quickly disseminated.

When it comes to computers connected to neighboring computers, it must be understood that you depend on the neighboring computer's security. When you don't know how that computer is connected, you are not protected from running into a cascade of security problems.

To many organized crime groups, the Internet is a tool and not a target. There are also many powerful software tools available for hackers on the Internet, and the trend is towards even more powerful tools. Very little skill is needed to make use of hacker tools, as hacking techniques have become automated. Software that anybody can run can crash a computer and shut it down.

As an example, Back Orifice (BO) is a backdoor for Microsoft Windows 95 and 98. It allows a remote unauthorized user to take over your system. The software has higher quality than any commercial product seen so far. People playing on the computer and being beaten by other players developed BO. With the software at hand, they were able to come back and shut down the other player's computer. Capabilities include the execution of commands, listing of files, sharing of directories, uploading and downloading of files, monitoring of keystrokes, opening and closing of the CD-ROM door to make people crazy, and the killing of processes.

Programs such as SATAN were originally designed to help computer and network administrators detect weak points in the security of their systems. However, computer intruders who want to rapidly assess the vulnerability of a target can also use it. Coupled with other software programs, it allows a hacker to access an automated reconnaissance and intrusion package without leaving an audit trail. New versions of SATAN also automate attacks against the identified vulnerabilities, and other even more sophisticated software tools can be found.

Security Management

There are several key elements to handle intrusion detection:

- Assess implementations of key architectural elements and advise clients on their application
- Make use of analysis and support tools to understand network topologies and information flow between systems
- Execute vulnerability analysis to identify weaknesses
- Mitigate identified vulnerabilities as far as possible
- Employ monitoring and analysis tools to detect exploitation of residual risks
- Repeat the whole process when new capabilities are proposed — i.e., before a new capability is added

Modern tools automate vulnerability detection. Remote servers can be scanned for vulnerability analysis. The Web site for Internet Security Systems is http://www.iss.net.

Security and Vulnerability Tools

There are a number of tools available to study security and vulnerability:

- Network visualization, Monitors, and Sniffers. Examples are Visio, NetViz, NetPartitioner, NeoTrace, TraceRoute, Ethload, Net X-ray, Etherpeak, TCPDump, Snoop, IP-Watcher, T-sight, Scott/Tkined
- Vulnerability analysis. Examples are ISS Internet Scanner, Kane Security Analyst, Trident IP Toolbox/L3 Expert, Security Profile Inspector (SPI), SNI Ballista, SATAN
- Intrusion detection. Examples are RealSecure, NetRanger, Stalker/CyberCop, Intruder Alert, Network Flight Recorder, SHADOW, NIDS
- Exploitation. Examples are NTSecurity, RootShell, Offline NT Password Utility, Lopht Heavy Industries, AntiOnline, Insecure/Fyodor
- Other useful tools are TCPwrappers, Tripwire, COPS, crack, LophtCrack, ScanNT

A network-sniffing device on a communication line acts as a kind of hub. Traffic is usually carried in cleartext and the sniffer can pick up usernames, passwords, and e-mail. It allows traffic to be filtered, watched and/or recorded. A sniffer like Net X-ray provides a traffic map and identifies important points in the network. A tool like IP-Watcher is capable of session watching, takeover, and close-down.

Defense measures to handle computer security include a number of measures to be implemented concurrently:

- Policy including defense guidance and offensive rules of engagement
- Training and professionalization
- Assignment of resources, responsibility, and authority, including defensive Red Team
- Vulnerability identification and familiarization
- Countermeasure assessment

It is necessary to apply staff, time, and capital resources to vulnerability analysis and to tool assessment. Effective practical and tolerable procedures must be developed. Open-source discussions of vulnerabilities, exploitation mechanisms, and tools must be followed. One should never buy tools without understanding their capabilities and never rely on a single tool and source. Much of the work requires long days and enthusiasm beyond the norm to succeed. It is a big effort to keep up with changing computer environments and new threats.

The financial sector is thought to have the highest degree of computer security, one reason being that it cannot afford a bad reputation. Many military systems are more vulnerable, and personal systems in civil society are among the most vulnerable.

Vulnerability Assessment and Red Teaming

There is a difference between a vulnerability assessment and Red Teaming. Vulnerability assessment includes a review of security posture as part of risk management. The results are reviewed with the security department, and solutions are implemented. This is a regular component of security procedures and a risk management process. Safeguards are evaluated, threats anticipated, and resource allocations are decided on.

Red Teams act as the enemy to exploit vulnerabilities to the fullest possible extent and to provide perspectives to the threat of an information attack. Vulnerability chains are exploited, and INFOSEC, physical, and personnel vulnerabilities are combined. The true implications of known vulnerabilities are thereby demonstrated to management. This supplements a vulnerability assessment as part of risk management, recognizes intangible benefits of information security, and emphasizes dramatic situations. Both vulnerability assessments and Red Teaming are vital components of information security and critical infrastructure protection.

Red Team personnel are selected based on experience and a strong personality.

Communications Threats and Security: Interception

Communications security and the "Development of Surveillance Technology and Risk of Abuse of Economic Information" has been studied and reported to the Director General for Research of the European Parliament in 1999. The study considers the state of the art in communications intelligence (COMINT) of automated processing for intelligence purposes of intercepted broadband multi-language leased or common carrier systems, and its applicability to COMINT targeting and selection, including speech recognition. It includes interception of international communications, the ECHELON and COMINT production, law enforcement, economic intelligence capabilities after year 2000, and policy issues for the European Parliament. The following is a summary from the report:

> *Communications intelligence (COMINT) involving the covert interception of foreign communications has been practiced by almost every advanced nation since international telecommunications became available. COMINT is a large-scale industrial activity providing consumers with intelligence on diplomatic, economic and scientific developments. The capabilities of and constraints on COMINT activity may usefully be considered in the framework of the "intelligence cycle."*

Globally, about 15–20 billion Euro is expended annually on COMINT and related activities. The largest component in this expenditure is incurred by the major English-speaking nations of the UKUSA alliance. The report describes how COMINT organizations have for more than 80 years made arrangements for the interception of communications from commercial satellites, of ground communications using satellites, of undersea cables using submarines, and of the Internet. In excess of 120 satellite systems are currently in simultaneous operation collecting intelligence.

The highly automated UKUSA system for processing COMINT, often known as ECHELON, has been widely discussed within Europe following a 1997 Scientific and Technological Options Assessment (STOA) report. That report summarized information from the only two primary sources then available on ECHELON. This report provides original new documentary and other evidence about the ECHELON system and its involvement in the interception of communications satellites.

COMINT information derived from the interception of international communications has long been routinely used to obtain sensitive data concerning individuals, governments, trade, and international organizations. This report sets out the organizational and reporting framework within which economically sensitive information is collected and disseminated, summarizing examples where European commercial organizations have been the subject of surveillance.

This report identifies a previously unknown international organization — "ILETS" — which has, without parliamentary or public discussion or awareness, put in place contentious plans to require manufacturers and operators of new communications systems to build in monitoring capacity for use by national security or law enforcement organizations.

COMINT organizations now perceive that the technical difficulties of collecting communications are increasing, and that future production may be costlier and more limited than at present. The perception of such difficulties may provide a useful basis for policy options aimed at protective measures concerning economic information and effective encryption.

Key findings concerning the state of the art in COMINT include:

- Comprehensive systems exist to access, intercept, and process every important modern form of communications, with few exceptions
- Contrary to reports in the press, effective "word spotting" search systems to automatically select telephone calls of intelligence interest are not yet available, despite 30 years of research. However, speaker recognition systems — in effect, "voiceprints" — have been developed and are deployed to recognize the speech of targeted individuals making international calls
- Recent diplomatic initiatives by the United States government are seeking European agreement to the "key escrow" system of cryptography masked intelligence collection requirements, and have formed part of a long-term program which has undermined and continues to undermine the communications privacy of non-U.S. nationals, including European governments, companies, and citizens
- There is wide-ranging evidence indicating that major governments are routinely utilizing communications intelligence to provide commercial advantage to companies and trade

The United States National Security Agency (NSA) — the signal intelligence agency — directly under the President, has about 20,000 employees and can command another 20,000 people from the defense. NSA has the capability to intercept all wireless and cable communications including voice telephony. Softly encrypted information (perhaps up to 56 bit coding) is decoded in real time. One billion messages are filtered every day and about 0.1 percent is analyzed manually. There are links between the NSA and the Department of Commerce. One conclusion is that there is no possibility to avoid the risk of fax, telephone, or e-mail messages being intercepted and analyzed, even if they are softly encrypted. Thus the only countermeasure is to encrypt all messages, thereby saturating the system.

The Swedish Security Service claims that mobile GSM phones can be used for industrial espionage, even when switched off, without putting anything special into the phone, and without the owner knowing that the telephone is being bugged. Other sources deny that this is possible. Until a clear statement is at hand, it is recommended not to bring mobile phones into places where highly secret matters are being discussed.

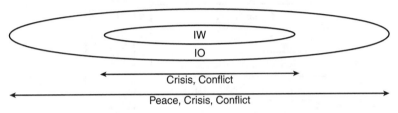

Information Operations (IO)
Actions taken to affect adversary information and information systems
while defending one's own information and information systems.

Information Warfare (IW)
Information operations conducted during times of crises or conflict to achieve
or promote specific objectives over a specific adversary or adversaries.

Ultimate Goal of Information Operations
Human decision-making.

IW

IO

Crisis, Conflict

Peace, Crisis, Conflict

Exhibit 4.   Information Operations Terminology

Information on the location of mobile phones is monitored and there are special computer centers that can handle and record a great number of mobile phone positions. Such information is used by law enforcement agencies.

The New Warfare

What Is Information Operations and Information Warfare?

What is information warfare? The meaning of information warfare changes all the time and brings my thoughts to Greek mythology where Hercules was fighting a hydra with many heads that regrew when cut off. Information warfare has also been compared with the story of blind men touching an elephant to try to describe what they think it really is (see Exhibit 4).

Depending on what they feel, each of them will come up with different ideas. There is no consensus of definitions, and one could even talk about "definition warfare." Definitions establish conditions, which can prevent looking for new patterns of behavior and development in a changing environment. In my view there is no end state; all these terms and definitions are only glimpses of road signs on an endless journey towards the future.

In the United States, *Information Operations* is the superior strategic term integrating various capabilities and activities such as information warfare. Information operations are actions taken to affect adversary information and information systems while defending one's own information and information systems. Information warfare is information operations conducted during times of crises or conflicts to achieve or promote specific objectives against a specific adversary or adversaries. The ultimate goal of information operations is to impact the human decision-making.

To some people information warfare is a generic term for all forms of struggle for control and superiority concerning information (see Exhibit 5).

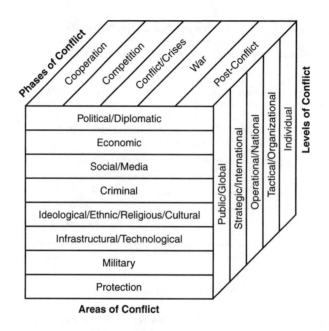

Exhibit 5.   Information Warfare Levels, Areas, and Phases of Conflict

Ultimately it could be the struggle of minds to become the master of a situation. In a very wide view one deals with many levels of interaction, phases of conflict, and arenas. According to one source, there are three classes of conflicts: personal, corporate, and global information warfare. Another example is individual, organizational or tactical, national or operative, international or strategic, and global. In a wide sense, information warfare appears during all phases of conflict from cooperation, competition, crises, war to post-conflict. The arenas can be military, political, diplomatic, economic, social, infrastructure, criminal, ideological and religious, and possibly more. There are several forms or methods of warfare, and a multitude of technologies, goals, and targets.

Human power balances on three fundamental pillars: violence, wealth, and knowledge (see Exhibit 6). In the struggle for power, one of the basic sources is information. The struggles are, generally speaking, information operations and information warfare. They engage the military sector seeking power of violence, the financial sector striving after power of wealth, and humanity aiming at knowledge power. Power can be transformed between violence, wealth, and knowledge; and these transformations also incorporate information operations and information warfare. The model of the Information Ocean from which the three pillars violence, wealth, and knowledge rise to carry power is applicable at all levels: state, corporation, family, and personal.

Information can be used for good or evil purposes. Symmetric information and knowledge (refined information) can maintain peace. Advantages in asymmetric information can make it possible to win in a conflict using knowledge as a weapon and information as bullets and ammunition (see Exhibit 7).

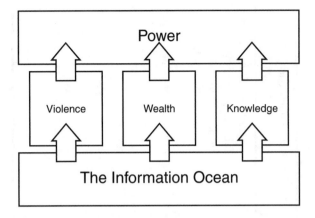

Exhibit 6. Information as a Fundamental Source to Create Power of Violence, Wealth, and Knowledge

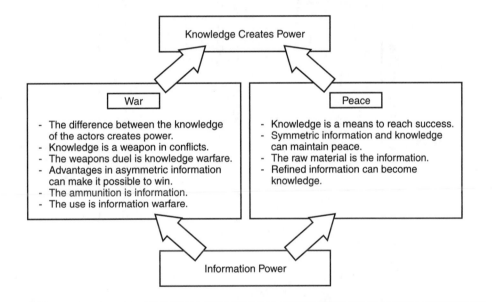

Exhibit 7. Information in Service of Peace and War

Information warfare can begin without a declaration of war and can be fought on a wide front or battlefield — openly or subversively, isolated or widespread. Information operations can stretch over all phases including: peace, crisis, the pursuit of conflict, and actual war. This totality makes the concept complicated and multidimensional and places high demands on the expansion of our field of vision in all directions, to include all the phases of a conflict, at all levels, and in all sectors of society (see Exhibit 8). This forms a palette with greater possibilities and threats, with a greater number of targets and wider battlefields than ever before.

The objectives of information warfare can be masking or unmasking of facts, exploitation, deception (such as disinformation), disruption or denial-

Exhibit 8. Example of IO/IW to Prevent Conflicts: Conflict Prevention Plan

	Peace/Competition IW action initiatives	Conflict/Crisis IW Plan	War IW annex to war plan	Post-conflict/Peace Sequel to IW plan
Political	Engage leadership	Discredit leadership Support democratic leader	Isolate leadership Disrupt military C2	Support elections Stabilize government
Military	Access intelligence C2 and logistics	Deter aggression Disrupt acquisition Deceive leadership	Defeat infosystems Disrupt operations Impede reinforcement	Support demobilization Support arms limitations
Economic	Access financial systems	Disrupt fund transfers Deny international support	Deny support of military operations	Maintain financial systems
Social	Promote democracy Deter migration	Support democratic leader Set expectations	Support democratic leader Set expectations	Support national reconciliation
Infrastructure	Access transportation, communications, POI, and power	Disrupt systems	Deny flow Deny communications	Monitor systems
Protection	Access vulnerabilities Identify adversary IW capabilities	Protect vulnerabilities Manipulate adversary IW	Protect one's own capabilities Deny military adversary IW	Monitor adversary IW

of-service, and destruction of information. The activities can be open or concealed, occur in peacetime or in combination with various conflicts and war. Hackers outside the United States have intruded into computers belonging to the American Department of Defense. Media has in one way or another played a significant role in conflicts like the Gulf War, in Yugoslavia, [and] in Somalia, just to mention a few. The Internet has been used to acquire information for news bulletins to reveal the Indonesian military actions in East Timor.

Intrusions in computer systems may appear as if they come from one part or country, whereas they actually are launched from quite a different place and country. This can become a significant risk in evaluating who has participated in information warfare.

An example of an American plan for information warfare shows escalation and de-escalation of various activities in each arena such as political, military, etc. It is a good illustration of the totality and the requirements for extensive cooperation between representatives of different responsibilities.

There is also talk of a more restricted application of information warfare. The actors can then be separate organizations, groups, or simply individuals instead of being alliances or states. They can achieve their purpose without having to resort to conventional threats or means of pressure. This makes the boundaries unclear between the concepts, competition, rivalry, peace, crisis, and war. It has been said that a few clever hackers or insiders could paralyze a high-tech nation within a very short space of time.

The boundaries are also indistinct in conflicts where old military conditions and weapons are used in old ways, together with old military weapons modified and used in new ways, and with new military conditions and weapons used in new ways. Do not regard a combat aircraft as a combat aircraft any more; look on it as a flying computer center with a weapons cargo.

With less platforms and soldiers and with more information, the development goes from a platform-centric to an information-centric defense. The shift from power of violence to power of knowledge is applicable to military defense as well.

It is not difficult to feel lost in the information warfare environment (see Exhibit 9). The clear grouping national state against national state is gone. The identifiable adversary and uniformed armies are gone. Defined borders and the predictable axis of attack are gone. Mobilization of a considerable part of the population is gone, and the requirements for a large industrial production are gone. Financing over a considerable part of the state budget is gone. Land, sea, and air forces are gone. Declaration of war, Geneva conventions, and diplomatic protocols are gone, as well as a clear adversary. Information warfare can come sneaking in the shape of a single, only a few, or many actors; they come without reliable identification, without uniform, without borders, without [a] defined front, without large industrial production or large financing, without fire power, without rules of engagement and conventions.

It is often said that the truth is the first victim of war. Normally that refers to conventional war, but information war is no exception. The actors do not necessarily even know that they are participating. International media can

Exhibit 9.   How to Get Lost in Unconventional Conflicts

	Conventional War	Information War
Actor	Nation-state versus nation-state	Individual/small group versus nation-state
Adversary	Identifiable uniformed enemy	No certain identification No uniform
Geography	Fixed frontiers	No frontiers
Axis of attack	Predictable	Not definite
Enlistment	Large part of population	Very few individuals
Industrial needs	Large production	No industrial base
Financing	State treasury required	Privately affordable
Military forces	Land, sea, air forces	No firepower
Mobilization	Required	Not required
Deployment	Required	Not required
Rules	Declaration of war Geneva conventions Diplomatic protocols	No declaration of war No conventions No diplomatic protocols
Outcome	Clear victor established	Clear loser established

play a significant role without knowing it. Weapons of information warfare are deployed at low cost and with good accessibility globally. War can start without notice and propagate at [the] speed of light. The theatre is logical and not physical. A domain can be controlled without physical presence. What does the plan of attack and the defense look like when the map is not a geographical map with marked units and lines of attack, but rather the interior of human minds and their knowledge, perceptions, and influences?

What Is New?

During conflict or war, information and knowledge have always been key parameters. This is why information warfare is in fact nothing new as a phenomenon. However, what are new today, are primarily these factors:

- Widespread expansion of fields of activity and global connections
- Interaction and synchronization embracing all important areas
- The effects of amplification and intensification through information technology
- The increasingly high tempo in a course of events
- The emphasis (especially in the U.S.A.) on reduced casualties

The increasing importance of information makes it a clear, primary target for warfare. This is one reason why modern information warfare is a highly topical subject for debate. The opposite of information warfare is not particularly under debate. One major root of conflicts is lack of confidence. Confidence-building measures are becoming extremely important in solving

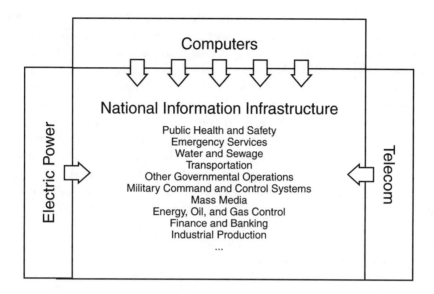

Exhibit 10. The National Information Infrastructure Dependence on Electric Power, Telecom, and Computers

conflicts. Instead of engaging in information warfare, one should develop a number of positive means of using information. For this reason the whole region of producing confidence-building information ought to be studied, developed, and used.

Impact on Infrastructure

Military power and national economy are increasingly reliant on interdependent critical infrastructures. Advances in information technology and competitive pressure to improve efficiency and productivity have created new vulnerabilities to both physical attacks and information attacks as these infrastructures have become increasingly automated and interlinked. Any interruption or manipulation of these critical functions must be brief, infrequent, manageable, isolated, and minimally detrimental to the well-being of a nation.

The national information infrastructure is at the center of the information warfare (see Exhibit 10). Who is the responsible authority for it and who defends it? The objective of the military defense is armed combat. The police force has its traditional responsibility and needs evidence. Commercial service providers might want to hide their deficiencies. Meanwhile, military defense and national assets are increasingly dependent on interdependent critical infrastructures.

The destructive and extensive effect that information warfare can have on an information society is hard to imagine. We need computer support for electrical power supplies, telecommunications, water supplies, sewage, heating, air and rail transport services, daily newspapers, social insurance, taxation and banking procedures, and much more — in fact, for the entire infrastructure.

Electric power networks and communication networks are the blood vessels and the nervous systems of society and are controlled by computers — the brain of the society — without which catastrophe is near. When the networks break down, everyone goes down with them. Only when faced with a complete electric power blackout such as in 1983, did Sweden start to understand the chain reactions of electronic systems blackout and further consequences for the infrastructure. If the telecommunications hub and conduits in the World Trade Center in New York City had been damaged during the 1994 attack on the Trade Center, this would have had a disastrous effect on billions of U.S. dollars propagating throughout the world.

Attacking the command and control systems, telecommunications, and associated infrastructure of a high-tech nation is thought to be an effective way to break down the nation. This can become an even more serious threat in the future. This is possible for a number of reasons:

- The threat increases if an attack of the infrastructure becomes more profitable.
- The threat increases if the probability for intrusion, acquisition, and manipulation increases due to more actors with more IT knowledge on a global arena.
- The vulnerability increases if the reliability decreases. This can be caused by increased technical complexity and interdependent critical hardware or software dependencies. It can also depend on the susceptibility to deviating environmental conditions (e.g., temperature, electromagnetic radiation).
- The vulnerability increases if installation, operation, and supervision becomes more centralized. Economical and practical reasons often lead to centralization.
- The vulnerability increases with additional functions depending on electric power if backup power is missing.
- The effects increase if the dependence on technology increases in the society in more and more areas and the consequences of disruptions thus increase.
- Protective measures are insufficient if knowledge of the need for safety is insufficient or if the economical investments are insufficient. Examples of protective measures are redundancy, alternative procedures, and emergency plans.

Information Warfare Drawbacks

Information warfare has several important drawbacks:

- Trust as one of the most fundamental cornerstones of human civilization is violated. Involvement in information warfare such that the final outcome is loss of trust can be very expensive indeed.
- An underhanded expansion of information warfare in civilian society blurs the boundaries for everyday competition and is a kind of war fought without declaration of war.

- Information warfare legitimizes terrorism. If governments and actors in society, ambitious to be respected by the people, believe it is acceptable for them to engage in information warfare, this will become the final victory for terrorism.
- The effects of information warfare are unpredictable. To achieve the intended effect is virtually impossible. Human behavior is far less predictable than technical circumstances.

Experimenting with a complex information society is certainly easy with the technical means at our disposal today, but to achieve the intended effect is virtually impossible. Here there are clear parallels with nuclear weapons and biological warfare. The extent of dispersion and long-term effects of such operations cannot be predicted nor restricted. Information as a weapon is double-edged and can strike back against the aggressor. It is not unlikely that information warfare activities in the end will have a greater negative effect on the aggressor than on those attacked if the aggressor himself is strongly dependent on information technology and the attacked is a "backward" nation.

Human behavior will always be less predictable than technical circumstances. It would be a grave error in judgment to believe that information warfare is so superior that it will contribute toward solving all military conflicts. This is a real problem but not well recognized by everyone. The will of man depends on human values and cultures and is more difficult to understand than technological systems. The straightforward military aspects of information warfare are not particularly effective in those cases where the cultural differences of the actors are substantial and where the difference in information technology (IT) dependence is great. There are many examples of homogenous groups, ethnic as well as religious with deeply rooted moral codes, which cannot be defeated by information warfare.

On the other hand, their power of mass media is in their favor and thus can reinforce their own opinions.

Driving Forces of Information Warfare and Information Operations

There are three main reasons why information warfare and information operations have attained interest and have become driving forces.

- One thought is that information warfare can lead to less blood-shedding. There is a dislike of casualties, especially if the motivation to participate in a conflict is low. Media is quick and powerful and can intensify the effects of losses and can influence public opinion. Politically, information warfare and information operations may be attractive if trust is not harmed.
- Another thought is that information warfare can lead to less cost than other means of fighting. In some cases there could be a dramatic cost reduction. For example, a treacherous chip in a telephone line can cause more harm than a number of bombs, and the restoration cost after the conflict can be minimal.

- Still another thought is that information technology such as computers and communication systems is a phenomenal tool revolutionizing the way we work and thus both a force multiplier and a source for new capabilities.

Types of Information Warfare

Sources, Forms, and Objectives

The various forms of information warfare can be described by combining sources of attack, forms of attack, and tactical objectives. There are two major sources of attack:

- *Outside:* defense against the outside can be firewalls, physical isolation, and encryption.
- *Inside:* protecting against insiders like disgruntled employees is different and deals with personal security and operating procedures.

There are four major forms of attack:

- *Data attacks* occur when an opponent is inserting data and thus manipulating an information system. Examples are corrupting files, jamming radio transmissions of data, broadcasting deceptive propaganda, and spamming (sending large amounts of input irrelevant data). It must be recognized that information warfare campaigns do not always require sophisticated technology. Disinformation campaigns can be carried out in many unsophisticated ways, and human factors such as decision-making are always of main importance in the operations. The media sector is one major battlefield. Psychological operations can be based on studies of public opinion and analyses as a basis for tailor-made messages. Data-based picture and sound manipulation (morphing) can make the observer believe that the person he sees and hears in reality also does and says what is shown.
- *Software attacks* intend to carry out functions that cause the system to fail, like computer viruses, Trojan horses, logic bombs, and trap doors that allow a hostile party continuing access to a system. Back doors in software can be aimed at attacking built-in safety and security mechanisms. Microchips can have weaknesses that have been programmed beforehand or have hidden added functions that can be used by an adversary in a conflict (chipping).
- *Hacking* (or cracking which is the criminal aspect) is unauthorized entry into an information system to interact with its function to cause deception, theft, fraud, destruction, and other types of harm. One can detect or create tempest radiation — the electromagnetic emanations from computers, monitors, encryption devices, and keyboards. Software can be employed covertly to generate intentional emanations that bypass conventional security mechanisms. This can be used for commercial and national espionage, or even for software compliance.

- *Physical attacks* are attempts to destroy a system physically. Examples are methods for causing fires; using bombs; [and] producing various harmful environments such as electromagnetic pulse, high power microwave, and other directed energy environments, or creating electromagnetic terrorism. High power microwave (HPM) weapons can emanate high power repetitive pulses in a narrow frequency band concentrated at a frequency between a few hundred MHz and several GHz. The duration is short (typically a hundred nanoseconds). Such normally high-tech weapons can cause "front door damage" to equipment. High power impulse ultra wideband (UWB) weapons can emanate very short repetitive pulses (typically a pulse width of hundreds of picoseconds) spreading power over a very broad spectrum. Weapons can be low-tech, cheap, and are likely to cause "back door disruption." Continuous jammers are commercially available that can disrupt Global Positioning Systems (GPS) or mobile phone services; there are other types of radio frequency (RF) and electromagnetic pulse (EMP) weapons as well.

The main objectives of information warfare attacks are:

- *Exploitation:* to make use of the opponent's information for one's own purposes
- *Deception:* to manipulate the opponent's information and to keep him operating
- *Disruption or denial-of-service:* to put the system out of operation for some time or to make it unreliable
- *Destruction:* to harm the system in such a way that it cannot operate any more

Information warfare embraces a wide range of threat magnitudes:

- *Natural hazards and unintended threats.* Natural hazards, errors, and unintended consequences represent the low end of threats. Examples are natural disasters, Year 2000 problems (Y2K), human operator accidents, design and manufacturing errors, and acts of God.
- *Hacker type threats.* Unorganized or loosely coordinated information warfare threats form the next category of threats. Examples are hackers, pranksters, gangs, individual criminals, organized crime, and disgruntled employees.
- *Tactical threats.* Tactical information warfare is the next higher threat magnitude and can be sophisticated in its technology and planning. Examples are operational deception; electronic warfare (EW); electronic countermeasures (ECM); stealth in military operations; arms control camouflage; concealment and deception; industrial espionage; organized crime; limited strikes to demonstrate, resolve, or deter would-be opponents.
- *Strategic threats.* Strategic information warfare is the top of the line. It has not received as much attention as the three previous categories although it is thought to become the most important in years to come. Examples are extended terrorist campaigns, organized crime, industrial espionage, and organized strategically targeted IT operations.

Comparison between Electronic Warfare and Information Warfare

There is a similarity between the structure of electronic warfare and information warfare. In both cases, there is a never-ending duel between measures and countermeasures.

> *In some ways information warfare is like classical electronic warfare. It's a never-ending struggle for advantage between countermeasures and counter-countermeasures. On the other hand, EW has been like an organized street brawl while IW has been more like a mugging.*

> —James Stekert

Electronic warfare support measures intend to facilitate or support warfare missions. From the attacker point of view, information warfare *support measures* refer to tools used to penetrate, misuse, or disrupt system operations. As an example, programs to detect weaknesses in systems (e.g., SATAN programs) can be used. *Countermeasures* are used to prevent the accomplishment of the adversary's mission. Examples can be firewalls, intrusion detection, audit, and strong authentication or antivirus programs. *Counter-countermeasures* are used to defeat the installed countermeasures. Examples can be insider or trusted path access and/or a new and unexpected virus.

Some Scenarios

Natural Hazards and Unintended Threats

Although Y2K problems are among those in the low end of threat magnitudes, they have attracted much attention and unrest concerning what could happen. Many systems have been examined and corrected but it is virtually impossible to guarantee that all of the important fixes have been carried out. Old systems interconnect with new ones and failures can spread and affect new systems.

Hacker Type Threats

A more severe category of threats can evolve from the Y2K problem. Technicians working with [such] problems have had access to all areas of an organization's information system. An organization that lacks total knowledge of its systems could be manipulated and the systems compromised in various ways. Viruses, logic bombs, trapdoors, and backdoors could be installed and triggered, perhaps years later.

New malicious software is born and is introduced on the Internet. Hackers and crackers may use this software to covertly control or disrupt other computers attached to the Internet. This may be far more serious than when a 16-year-old English boy took down some 100 U.S. defense systems in 1994 and other hackers rerouted calls from 911 emergency numbers in Florida to Yellow Pages sex-service numbers in Sweden.

Tactical Threats

According to information warfare specialists at the Pentagon, an electronic Pearl Harbor could result from a properly prepared and well-coordinated attack. It is said that fewer than 30 computer virtuosos located around the world, with a budget of less than 10 million U.S. dollars, could bring the United States to its knees. This scenario, also indicated by acts of hackers around the world, holds true for any high-tech nation and is a very real and increasing danger to national security.

Meanwhile new military capabilities are being developed to launch cyber-attacks or counterattacks making use of the power of information warfare and information operations.

Today, almost a dozen nations have such capabilities — with or without political support — extending over a range of activities, often in conjunction with other measures. Logic bombs can be planted in foreign computer networks to paralyze parts of foreign civil and military information infrastructures. Systems are prepared to take over public radio and television broadcasts.

Strategic Threats

Dependencies on information technology systems, coupled with inadequate defenses, create vulnerabilities. Strategic information warfare (SIW) consists of coordinated, systematic attacks through computers, communication systems, databases, and media. This corresponds to a potential threat unlike any previous threat, just as the Information Age society is unlike the Smokestack Age society. With conventional thinking this new threat is easily underestimated.

Strategic information warfare is the most alarming threat to a whole nation. The enemy could be a hostile nation, a terrorist organization, an organized crime syndicate, or a highly motivated cult. The necessary means and tools to carry out an attack can be found as the proliferation of information technology and expertise is global.

Strategic information warfare to achieve major long-term objectives is thought to start with a careful long-range plan including covert reconnaissance missions to find critical information assets. Other elements would be long-term collection, analysis, and exploitation of open-source intelligence information including but not limited to political, governmental, financial, industrial, societal, and personal activities. Strategic information warfare would then continue to develop as a campaign of covert strikes with a series of coordinated and precisely executed operations for months or even years to gain as much advantage as possible before revealing the hostile operations. Strategic information warfare would finally be combined and integrated with other actions such as terrorism, diplomatic measures, and military operations for maximum synergy. This kind of warfare would more likely resemble Waterloo than an electronic Pearl Harbor. The reason is that there would be an associated strategic long-range plan, ultimately to achieve strategic power in some respect.

Strategic information warfare has special features in comparison with the other threats. Targets are selected for [their] ability to cause systematic collapse of an opponent's capabilities. Many interdependencies among systems could cause failures to spread widely. Goals are strategic in nature and could eventually result in changing the balance of financial markets, the balance of military power, or the stability of an international coalition. Combining information warfare with other military, diplomatic, societal, ethnic, or market tools creates synergy and increases effectiveness. Strategic information warfare represents an interesting and low-cost asymmetric option for adversaries that cannot compete using other types of power.

The very nature of strategic information warfare is such that no hard evidence of activities is revealed until operations have increased to a certain magnitude. This makes it very difficult to convince decision-makers of the importance of investing in alertness and effective countermeasures. In the United States, a number of hearings have pointed at structured attacks from unknown sources and that some of these attacks could be "state sponsored." A present known threat is that hostile organizations might try to exploit the Y2K problem and the connected window of opportunity at the millennium shift. Such operations could be well concealed. In Europe, the combination of Y2K and the start of the Euro present more information warfare opportunities.

Strategic information warfare concealment is expected to be better and detection more difficult the more serious the opponent. Many historical examples point at threats that were undetected. Encryption systems are widely available to permit groups to communicate among themselves and to plan attacks in secret. Nations such as the United States, Russia, China, Britain, France, Australia, and Canada have considerable resources to develop information warfare capabilities. Apart from these states, rogue states such as Iraq, Iran, Libya, and North Korea, would have motives to develop a strategic information warfare capability. Global terrorist networks such as those headed by Osama bin Laden, or by very powerful criminal organizations in countries like Colombia, Russia, and Papua New Guinea, are also significant players. The required technology, knowledge, and tools are readily available. Industrial espionage has long been known and has increased considerably with the new IT tools at hand and with the number of people relieved from Cold War intelligence activities and that now have taken on new assignments. Multinational corporations with loose national bounds are also powerful players [in] the arena — sometimes in cooperation with government intelligence facilities, sometimes not. State-sanctioned industrial espionage exists and may be increasing.

It is easy to agree with the following wording:

...we have created a global village without a police department.

—Frank J. Cilluffo

Countermeasures

Recommendations

Summary

Information warfare is a new type of warfare and makes it vital for new kinds of defense in cooperation among the civil private, civil public, and military parts of a nation, and also between nations on an international scale. Both non-criminal and criminal economies have gone global. New groups of criminal organizations, activists, extremists, and terrorists have emerged worldwide. Rethinking both civil affairs and military operations is a must. On a national scale, recommendations to cope with the new threats could look like this:

- Increase awareness and understanding of the threats, especially among critical infrastructure providers and users.
- Develop and implement a national security policy for the Information Age and re-examine it constantly.
- Make information assurance a national security objective.
- Implement and exercise policies ensuring critical government services.
- Cooperate closely [with] civil and military government, industry and universities on methods, products and management concerning information assurance.
- Create an information age military capability, expected to be very different from traditional military systems and thinking, and re-examine it continuously.
- Create requirements, methods, organizations, and networks for the Information Age intelligence community.

Awareness

Critical infrastructure providers and dependent users are familiar with hacking and computer viruses but largely unaware of the possible results of an information warfare attack. The visibility of the threat must be clear. Awareness must be raised to a point where actions are motivated and taken. Government must provide industry with information and incentives for taking action and make the private sector lead the way toward reasonable security measures.

> *Many leaders and decision-makers are not cyber-fluent. They do not understand the technologies and the realm of possibilities, rendering them less effective in dealing with information warfare issues, policies, and future-focused directions.*
>
> —Brenton Green

> *Our best deterrence to the strategic information warfare threat involves continual emphasis on education, training, and awareness for all users; cyber legislation at all levels of government; and research and development in all areas of information technology.*
>
> —Mark Centra

However, Sun Tzu has stated:

> *The best way to wage war is by attacking the enemy's strategy.*

Policies

It is imperative to respect and incorporate the strategic information warfare threat into overall national defense planning and doctrine. Discussion and policy concerning information warfare are, unfortunately, continuously centered on vulnerabilities, not on a long-term strategy for national security and dominance in national affairs. Such a strategy must cover the total impact on the nation of the Information Revolution. A secure, strong nation is the best foundation for well-being.

> *Just as nuclear dominance was the key to coalition leadership in the old era, information dominance will be the key in the Information Age.*

> —Joseph Nye and William Owens

Information Assurance

Information assurance is only one step in this policy that must cover the broader perspectives of defending against an attack of national interests. This is a new kind of defense that does not fit into existing traditional military and civil defense. It is vital to rethink national defense, the meaning of national security in the Information Age, the methods of achieving it, and to cover all these new perspectives. The United States Defense Science Board has marked the important difference between old and new thinking in this way:

> *Current practices and assumptions are the ingredients in a recipe for a national security disaster.*

Government and the Private Sector

There is a great difference between protection against strategic information warfare and protection against intrusions by a number of hackers. Protection against strategic information warfare requires extensive cooperation between nation organizations. Private companies play the most important role in development of information technologies. Government is not able to influence this to any great extent. However, government is more concerned about the threat than most of the private sector. Many people are not fully aware of the impact of a strategic attack on the telecommunications and information infrastructure. Thus, the private sector must become aware of the problems, be involved in policy planning, and take on a role of producing information assurance systems. To do so, collaborative partnerships between the public and the private sectors are imperative.

There are a number of obstacles to the development of good information security. Government has lost its leadership in information technology development and has difficulty in regulating security. The commercial sector sets standards mainly to gain market dominance and profitability. Market and profit change rapidly, which makes it difficult to maintain a certain standard. It is even a question of whether a company can control standards for its own products.

> *Acceleration is the consequence of reducing everything to its digital form and increasing the power of digital processors. Now, any transaction is instantaneous and markets can turn dramatically in seconds.*

> —David J. Rothkopf

Just like cowboys enter and exit an arena, commercial systems do so on the multinational arena. It appears to be very difficult to ride the market for more than a limited time, and only a few well-known companies remain on the scene as years go by. Microsoft has been criticized for its market dominance, but this dominance has resulted in widespread common software products. On the other hand, unsecure parts of [Microsoft's] software could have widespread negative effects.

Legal obstacles hinder improvement of personnel and operational security. Law has difficulties to keep pace with technological development and appears to be far behind in several security areas. Legal obstacles must be overcome concerning Red-Team exercises.

> *At the bottom, allowing Red-Team testing of computer systems is no different from allowing the cop on the beat to stop and twist the door-knobs on every storefront. No one would tell the police officer to stop doing that for fear of invading the store owner's privacy. We've got to develop a similar set of rules for cyberspace or the only people who will be trying the doorknobs will be the crooks.*

> —Stewart Baker

Military Forces

Today, military forces depend heavily on civil resources and commercial information technology and systems. Examples are systems for mobilization, transportation, logistics support, and communications. Without perfect control of the security of these systems — and there is no such guaranteed control — military operations are at risk. Military services depend to a large extent on commercial off-the-shelf (COTS) hardware and software, and there are no special military requirements on these. The pace of development in the information technology (IT) area is such that items that were not even on the market two years ago represent 80 percent of the profits today. There is no possibility for the military to keep up with this pace in all respects.

In the 1980s, discussions on the revolution in military affairs (RMA) started. Sophisticated information technology showed the way to new forms of warfare with high-capacity computer and high-fidelity sensor networks, and precision guided munitions. However, without guaranteed information assurance, the credibility of this new military power is undermined by information warfare.

The revolution in information affairs (RIA) extends over the whole society and will be even more developed and pronounced in the future. Along with the revolution in information affairs comes the revolution in military affairs. This is expressed in a true shift from platform-centric warfare to information-centric warfare, which requires new doctrines, new organizations, new kinds of training and knowledge, and new technological tools. The primary prerequisites will be to use electrons and photons as bullets, not gunpowder and fire.

The speed of operations will be quite different from what has been known until now. Targets will be the information components in any vital function. War will be permanently going on, mostly concealed and almost always without any declaration of war. Thus, software, hardware, and wetware attacks should be expected — in peace and in declared war — throughout all design, development, manufacturing, deployment, and mission phases.

Life will be quite different from the old days and from conventional thinking. It is likely that the nature of warfare will change, just as most other things are changing in the Information Age. Operations and war will be carried out in diffuse ever-changing networks without any absolute front or identities of actors. In the blurred environment of information warfare, it will be difficult to clearly identify the actor and the real objective behind an intrusion — intentional or accidental — and to respond with appropriate means. There will be diffuse boundaries between civil and military systems, actions, and responsibilities. The most important map of battle-space will be the instantly changing map of information concerning its sources, routes, carriers, receptors, importance, speed, concealment, strategy, and other factors. The most important command leaders in the networks will be those who have the ability to think faster, observe more facts simultaneously, and have the best quality of intuition and innovation. The campaigns will be using the synergy of a combination of hardware military territorial forces, software information forces, and political, societal, ethnical, and similar wetware forces. There will not be a resemblance to today's military and political traditional thinking. There will be enormous conceptual problems, policy questions, and cultural issues. A great challenge will be to abandon deeply rooted traditional thinking and doctrines.

Intelligence

The intelligence community faces competition from new actors. It must form new, continually changing alliances. This implies rethinking organizations and customizing products. There is a certain relationship between information warfare, business intelligence, open sources scanning operations, and civil and military intelligence operations. Boundaries between these are not very clear, and it is likely that one activity can transform into another. The result of long-term intelligence operations can be useful for strategic information warfare although the original objective for the operations was different.

Strategic information warfare requires changes in intelligence organizations and operation, methodologies and information sources, and personnel. A new breed of operators and analysts are needed. Threats throughout the Cold War evolved in incremental steps, and intelligence adapted to this development. Information warfare threats can emerge very fast or instantaneously, which requires different methods of intelligence. New forms of information sharing are needed between the military intelligence community, the law enforcement community, and the private sector. This could be done by stripping off and protecting sensitive details, while sharing generic targets, techniques, and sources of attack, as well as effective assurance methods.

Swedish Activities

The Swedish government has recognized the increasing threat of information warfare. A Cabinet working group has issued recommendations for a strategy and for the assignment of responsibilities for defense and protection against information warfare. Some of the proposals follow:

- A high-level coordination group within the Cabinet Office and the Ministries for problems concerning public administration and for national crisis management should be created. The information infrastructure must be regarded as a national asset concerning safety.
- A national coordination of information warfare under the leadership of the National Board of Civil Emergency Preparedness (ÖCB) should be created.
- A Computer Emergency Response Team (CERT) under the leadership of The National Post and Telecom Agency (PTS) and with support of the National Police Board (RPS) should be created.
- There should be an obligation to report all IT-related incidents within the public administration.
- An active IT control function (Red Team) should be managed by the Armed Forces.
- Media coverage within the IT area at the National Board of Psychological Defense (SPF) should be encouraged.

Internationally, there are challenges for international law, international cooperation, and the use of force. Doctrines concerning the use of information warfare and information operations under the United Nations or other international legal auspices are of interest. What could be discussed are international operations or upholding sanctions. Also of interest are principles of building regimes for defensive actions taken in cyberspace. This could involve actions like tracing and counterhacking. Some of these thoughts have been raised at workshops in Sweden.

According to an American view the basic term is information operations, and information warfare is one part of information operations in crises and

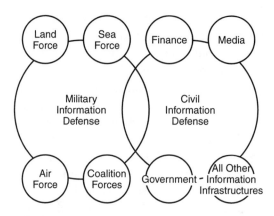

Exhibit 11.   The Information Defense

war. One application of information operations in military operations is command and control warfare. All of these terms comprise both offensive and defensive parts.

The Swedish Armed Forces define command and control warfare as offensive only and has made studies according to this. The Swedish industry is reluctant to use the term "information warfare" with regard to the previously mentioned drawbacks. In my personal view, we ought to agree with international views as far as possible and primarily develop the defensive parts.

Emphasizing the Information Defense

If it is true that we are standing on the edge of the deepest powershift in human history, we need rethinking for both the military and civil business. A new starting point and idea is to emphasize the importance of information forces by creating the Information Defense. The Information Defense consists of new civil and military forces. The military force must be integrated, with distinct and profound ingredients. These ingredients are force multipliers of land, sea, and air forces serving as cement to join together land, sea, air, and information operations.

The Information Defense must also include coalition forces and civil society. The civil information defense must defend finance, media, government, and all other functions associated with the information infrastructure, including those parts which are required by the military defense (see Exhibit 11).

Military information forces have been associated with some kind of "hacker army," but this might not at all be the task, or at least far from the only task. The information forces could be assigned a number of special and qualified tasks requiring new kinds of recruiting and training depending on the nature of the special areas. To begin with, there would be no existing organization to fit the new forces, and existing established organizations might resist the new information defense forces because of the fear that their own budget could be threatened in making room for a new organization.

Concerning the military part of the information defense, there are several reasons to introduce it:

- Command and control systems are primary targets and must be protected and defended. Information warfare is in progress before a declaration of war and is harming the national information infrastructure. Thus, the capability will be reduced to exercise military force in times of war.
- A prerequisite to conquer land, sea and air is to conquer the information and knowledge battle-space. This requires a shift from a platform-centric defense to an information-centric defense with information transparency and interoperability as features for the forces.
- Information forces are needed to fight enemy information forces without having a weak position. If one party is equipped with information weapons and the other is not, the competitive arena is sharply tilted. What should one do when unarmed and in front of such forces? Many special disciplines are needed.
- Emerging new military disciplines and weapons used in new ways require new types of knowledge, organization, and training. Innovations and development of new patterns of behavior are also required.
- There is a gap between the revolution in military affairs and conservative military ideas and organizations. Futurizing and rethinking require a new starting point and new perspectives, not just incremental changes of what is already there. It is a question of inventing the future, not merely improving the past.
- The Information Defense main objective is to achieve information superiority and decision superiority as a foundation for planning and executing cooperative land, sea, air, and information operations. In order to do so, the Information Defense executes information operations to achieve information assurance, produces dominant battle-space knowledge, and acts as a focal point for a true and common battle-space picture.

The following should be noted: information and decision superiority is not an eternal realm, but rather a temporary condition within a limited space. Information operations are performed both as a precondition for joint operations and within joint operations.

Traditional battles involve land, sea, and air forces supported by command and control systems, intelligence, surveillance, and reconnaissance and systems to acquire timely battle-space awareness. With the addition of the third main element — information operations including information warfare — and with information technology as a reinforcement for a vision of battle-space and for command and control systems, the art of war is revolutionized and information superiority becomes a clear objective (see Exhibit 12). Together with the fighting prerequisites, this can in turn provide for command and control and decision superiority. In peacetime only part of the main element information operations and dominant battle-space awareness are being used. In conflict and war, these elements are used to their full extent together with the land, sea, and air forces.

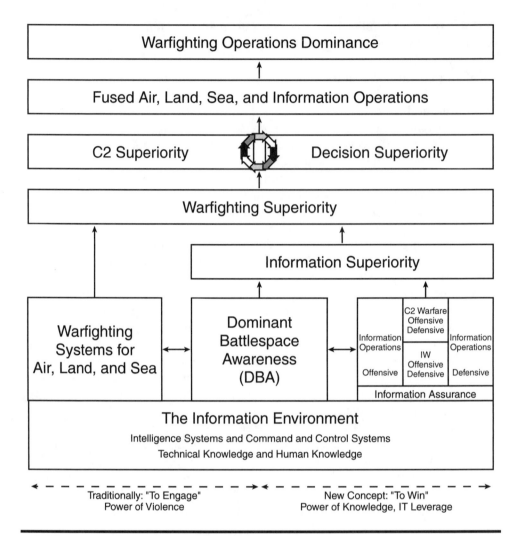

Exhibit 12.   Information Age Warfare

Information operations depend on information assurance and include command and control warfare, network-centric warfare/soft-warfare/cyber-warfare and other signals warfare (see Exhibit 13).

In my view, a conclusion of this is that the geographical map and video images of battle-space must be supplemented by a map of human minds and perceptions, and this requires a new attitude. We must proceed from power of force to power of mind, supplementing power of force with power of knowledge with the aid of information, information assurance, and information defense. We must be able to manage both conventional warfare and information warfare and other more concealed, disguised, or masqueraded forms of conflict.

Conflicts can be complex and hard to grasp. One way to try to analyze different layers or classes within a conflict can be to divide it into hardware, software, and wetware (brainware) (see Exhibit 14).

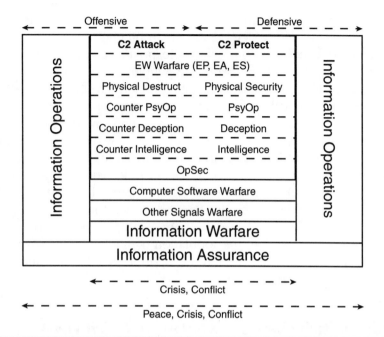

Exhibit 13. Information Operations/Information Warfare/Command and Control Warfare

Exhibit 14. Hardwar, Softwar, and Brainwar

Domain	Target	Battlefield Armed Forces	Power	Examples
Wetware Brainpower Brainwar	Human thoughts Perceptions	Political Religious Ethnic, etc.	Knowledge	Milosevic
Software Softpower Softwar	Information	Cyberspace Media Information forces	Wealth	TV programs
Hardware Hardpower Hardwar	Land	Air, land, sea, and its forces	Violence	Yugoslavia

Within the hardware region, the battle of territory is fought with land, sea, and air forces and power is conquered using violence. In the software region, the battle is about information and information superiority, is fought with information forces — visible or invisible — and power is conquered using the control of media, cyberspace, electromagnetic spectrum, and other carriers of information, together with sources and receptors of information. In the wetware region, the battle is about thoughts and concepts; forces are camouflaged as political, religious, ethnic, and other groups; and power is conquered using control of knowledge. In the worst case, this can mean lack of freedom in an undemocratic environment and, in the very best case, free access to knowledge in a completely democratic environment.

At times of political elections when the politicians are quarrelling, it is a sort of camouflaged wetware war. The whole population is mobilized to participate in the elections. The victorious political group captures the power and occupies the political arena. Certain countries prevent their populations from free access to information and from exercising the free word. That resembles an information blockade in the same way as, for instance, one speaks about an oil blockade. During the blockade, forces are in effect to mentally mobilize and arm according to the objectives of the rulers. When the objective is attained, a mental occupation is prevailing similar to territorial occupation, and the nation is equipped with mental forces not easily engaged using other types of forces. There are many examples of religious and ethnic forces in charge of occupying the population mentally and successfully. It can be very difficult to disengage such an occupation. Similar conditions apply to the political area in several cases.

Points of Similarity between Military and Civil Views in the Information Age

There are some fundamental points of similarity between military and civil activities in the information age, although the wording and terminology are often different. Fundamental is the utilization of information leveraged by information technology. Fundamental also is the agility of business operations, the speed of development, production, and marketing of goods and services. Both sides need reorganization, networking, and concurrent processes. Offensive spirit and action is vital to survive and prosper in the business world.

Both military and civil activities can be said to rest on five fundamental pillars in the Information Age:

> *Information assurance* through:
> - *Authentication* — verification of originator
> - *Non-repudiation* — undeniable proof of participation
> - *Availability* — assured access by authorized users
> - *Confidentiality* — protection from unauthorized disclosure
> - *Integrity* — protection from unauthorized change
>
> *Infrastructure protection* concerns the government, the military, and business as all of them use the same national information infrastructure. Infrastructures are increasingly transnational and global.
> *Information dominance* does not mean total dominance but rather sufficient dominance and superiority at the critical time and place. Information is a catalyst in a business process and the race goes to the swift. Information warfare can be regarded as the struggle for control in a decision space. Information dominance can be achieved through business intelligence, and speed and agility of product development, marketing, and finance.

Perception management is the ultimate goal of information operations and information warfare. In civil business this is done in several ways:

- Broadly informing through effective public affairs, which is a normal corporate function
- Broadly persuading through coordinated public diplomacy, which is lobbying
- Focused persuading through well-targeted psychological operations, which is advertising
- Distorting the opponent's sense of reality through deception

Finally, *operational effectiveness* combines important features:

- In the modern market, any given enterprise can simultaneously be a competitor, customer, supplier, and ally.
- In the marketplace, effectiveness means being adaptable to early changing trends in technology and business practices.
- Being innovative and highly adaptive in product design, production, business practice, and marketing.

An American View
Information Age Forces Information Operations

Telecommunications, automated data processing, sophisticated decision aids, remote sensors, and other types of information technology applications are being rapidly developed today, are globally proliferating and being used, and are creating dependencies within an increasing number of areas for civil and military purposes. There are no firm boundaries in the world of information between civil and military systems, and the development results in a situation where no single authority has full control over the totality. New information systems and new technologies are continuously being introduced and offer almost unlimited possibilities to exploit the value embedded in timely, accurate, and relevant information.

Today, information is a strategic resource of vital importance to national economy and security. This reality extends to civil and military business at all levels. Every system designed and deployed has some inherent weakness and vulnerability. In many cases this is the inevitable result of aspiration for increased user functionality, effectiveness, and convenience. The complexity and vulnerability of the information systems are often disguised by user-friendly software. Technical possibilities, system performance, efficiency, and decreasing cost result in an increasing number of users becoming dependent on them and running the risk of falling victim to latent/concealed vulnerabilities. All development has both good and evil sides, and it is clear that the arsenal of IT tools also finds applications for evil purposes.

The explosive global proliferation of information technology has a considerable effect on our actions in peace, crises, conflict, and war. Our reliance on information technology creates dependencies and vulnerabilities in the whole of our modern infrastructure and generates requirements on information defense capabilities. Information warfare is a central and joint defense responsibility for the whole nation. Our dependence on information and information systems, and the exposure of vulnerabilities to a great number of threats, from computer hackers through criminals, vandals, and terrorists to nation states, makes it compelling and urgent to focus on the emerging disciplines of information warfare. Its unique characteristics demand information defense, and its double-edged nature creates new powerful opportunities to enhance diplomatic, economic, and military efforts and to support conflict solutions in the future. Information warfare can be an important element that contributes to the defusing of conflicts and thus to avoid military confrontations. This requires close cooperation among a number of sectors in the society.

Some Basic Conditions

Information warfare involves actions taken to achieve information superiority by affecting adversary information, information-based processes, information systems, and computer-based networks while defending one's own information, information-based processes, information systems, and computer-based networks.

Information warfare focuses on vulnerability and opportunities offered as a consequence of increasing dependence on information and information systems. Information warfare aims at the information itself, at transmission, collection, and processing of information, and at human decisions based on information. The influence may have its greatest impact in peace and the initial stages of a crisis. If carefully conceived, coordinated, and executed, information warfare can make an important contribution to defusing a crisis. Diplomatic and economic efforts can be enhanced and could forestall or eliminate the need to employ military forces.

Information warfare can be waged both within and beyond the traditional battlefield. Information warfare is applicable in all phases of war, from mobilization to deployment, employment, sustainment, maintenance, and regrouping, in all military operations, and in all levels of a conflict. Defensive information warfare is constantly applicable in both peace and war and is a constant part of all protection and defense. Offensive information warfare may be conducted in a variety of situations in all military operations. When fully developed and integrated, offensive information warfare can offer an enormous potential and force enabler in support of the warfighter.

Defensive and offensive information warfare are two sides of the same coin, engraved by both threats and opportunities. Information and information systems are both targets for attack and for protection. Duels are fought between friendly and adversarial information and information systems. To develop an integrated strategy for information warfare, it is necessary to understand the fundamental parts of offensive and defensive information warfare and their capabilities.

Coordination among military defense, civil government, and industry is imperative. Military defense relies on civil communications and networks, transport systems, and electric power. The technical complexity makes it impossible for a military commander to command and control all information and civil resources.

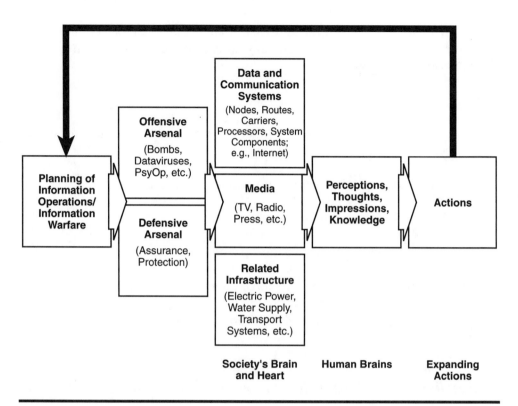

Exhibit 15.   A Comprehensive Process for Information Operations

Information warfare can involve complex legal aspects and policy requiring careful review and national-level coordination and approval.

Command and control warfare is a subset of information warfare and is an application in military operations that specifically attacks and defends the command and control systems.

Apart from this, the capabilities and disciplines employed in command and control warfare, together with other, less traditional ones related to information systems, can be employed to achieve information warfare objectives that are outside the military command and control range of targets (see Exhibit 15). Traditional disciplines in command and control warfare are psychological operations, deception, operations security, and electronic warfare. Examples of other disciplines related to information systems are other signal warfare, computer warfare, and media warfare.

Information operations can have many objectives; for example:

- Protect and defend the information infrastructure
- Deter war
- Affect infrastructure
- Disrupt enemy preparations of attack
- Support peace operations
- Expose enemy deception
- Degrade or destroy command and control systems
- Decapitate enemy national and military leaders from forces

My own thought in relation to this is that the arsenal of offensive and defensive information operations mainly affects three areas, which in turn affects people. These areas are:

- Data and communication systems
- Mass media
- All other infrastructure in society directly or indirectly related to information operations

An organization for information operations could thus comprise:

- Military information forces
- Civil society defense of the national information infrastructure
- The role of media in information operations

In the following, mainly information systems and computer warfare will be discussed. The meaning of computer warfare is computer systems used against other computer systems. The systems can be interconnected through networks.

Information Operations, Information Warfare — Defensive Aspects

The threat against military and civil information systems constitutes a substantial risk for national security and calls for a national security strategy. Defensive information warfare must be organized as a system linking together policy, doctrine, technology, assessment, evaluation, training, simulation, and a mutually supporting national organizational infrastructure. Within military defense, defensive information warfare must be carefully considered, integrated at all levels of conflict, and applied to all phases of military operations.

Along the way, information warfare can be integrated into all of national security and defense with the overall objective of capturing the latent potential of information warfare to enhance warfighting capability.

The objective of defensive information warfare is to achieve information assurance to protect access to timely, accurate, and relevant information wherever and whenever needed (see Exhibit 16). Organizing defensive information warfare as a system begins with a broad vision with collaborative efforts among the military defense, government, and industry. Visions and ideas are moved from abstract concepts to a set of specific questions and answers based on policy and standards. The five critical components that should be included in any attempt to form a defense system have previously been described:

- *Authentication* — verification of originator
- *Non-repudiation* — undeniable proof of participation
- *Availability* — assured access by authorized users
- *Confidentiality* — protection from unauthorized disclosure
- *Integrity* — protection from unauthorized change

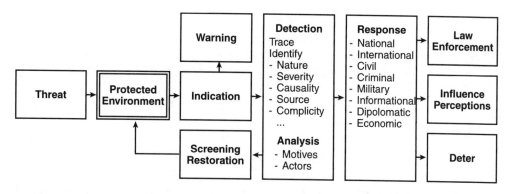

Exhibit 16. A Comprehensive Model for Civil and Military Defense against Information Warfare, Scalable to All Levels of War

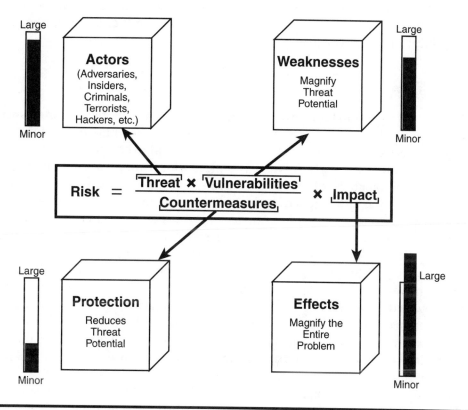

Exhibit 17. Risk Assessments

A model to manage risks and protect the national information infrastructure is described in Exhibit 17 with the aid of a number of processes. The model is scalable and applicable for civil and military use at all levels of conflict and all forms of operations. It begins with a defined and appropriately protected and defended information environment. Those threats that can affect the information environment are continuously surveyed.

Indications of any interaction with the protected environment initiates a process to disseminate warnings and a process to detect, track, identify, and analyze attacks and other degrading conditions. The nature, severity, causality, complicity, and other characteristics are ascertained based on knowledge of the threat built on information from various sources.

Motives and actors are analyzed. This starts a process to select an appropriate response. It can imply law enforcement, the influence of perceptions, and deterrence from further attacks. The detection process also starts a process to screen and restore the original information environment.

Often, security chains are described by protect, detect, and react, including the need for CERT groups (CERT = Computer Emergency Response Team). In several respects, the overall process can also be applied for forms of threat other than from foreign computers and networks; for example, for threats from electromagnetic terrorism. (Electromagnetic terrorism is the unlawful use of electromagnetic energy against property or persons to intimidate or coerce a government, the civilian population, or any segment thereof, in furtherance of political or social objectives.)

Protected Environment

The protected information environment is not an impenetrable fortress that guarantees 100 percent security because that is neither practical, affordable, nor even necessary. The focus is on defining the real needs and dependencies. The environment depends on what is critical with respect to national security and is a combination of physical systems and places, as well as abstract processes such as intelligence analysis. The protected environment shall not only provide protection commensurate with the value of its content but also ensure that there are resources available to respond to a broad range of attacks.

The protected information environment is founded in a valid approach to managing risks. The process to manage risks includes consideration of information needs, the value of information that may be compromised or lost if the protected environment is intruded, the vulnerability of the systems, the threats posed by potential adversaries and natural phenomena, and the resources available for protection and defense. In addition, the value of information changes in each phase of an information process, which must be considered in the risk management process.

Hardly any of these parameters are quantifiable without appropriate methods of measurement. Without measurement values, risk management is a big problem. One solution is to use time as a measure of security. The time during which security and protection functions must work shall be longer than the time it takes to detect an attack and to react in an appropriate way.

Threat Knowledge

The protected environment can be threatened by a number of actors; for example:

- Insiders and authorized users
- Hackers, criminals, and organized crime
- Terrorists
- Foreign countries
- Industrial and economic espionage

Articulation of a threat must neither be overstated nor underestimated and must be as comprehensively and reasonably defined as possible. Intelligence organizations must continuously pay attention to how the threat evolves. It is a dynamic mission that must adjust to changing conditions. To get at the essence of a threat, three basic elements must be understood:

- Identities and intentions of possible attackers
- Possible attack techniques and methods
- Potential targets, extending from the strategic to the tactical levels

Indication and Warning

Threat knowledge is an input to a process indicating deviations, intrusions, and attacks; it is important to disseminate warnings to persons, organizations, and processes that are considered to be at risk or require warning to other decision-making processes. Indication and warning must be executed automatically without any delay, as crippling attacks can occur at speeds exceeding the unaided human capacity to detect, analyze, and disseminate warnings.

A high-quality system for information warfare indication and warning requires a policy establishing authorities, roles, and responsibilities across local and national jurisdictions. This requires close cooperation between law enforcement, intelligence organizations, and private enterprise. Indication and warning cannot be realized without good collaboration between government and industry.

Detection, Tracking, Identification, and Analysis

Automatic detection of attacks and the immediate issuance of an alarm are imperative, considering that the time from the initiation of an attack to its culmination can be extremely short. In addition, automatic protective countermeasures limiting damage and propagation of attacks must be self-initiating. Defense against attacks depends on the quality of threat intelligence; how well associated indication and warning processes function; and the agility of system providers, users, and administrators in implementing protective countermeasures.

Thus, detection, tracking, and identification of foreign activities require close cooperation between government, industry, and other parts of the society. Identification of signs of foreign activity, analysis of the signs, and dissemination of warnings constitute critical elements of the detection process. It requires knowledge based on information from various sources such as law enforcement, intelligence organizations, system providers, and users. The analysis is comprehensive and time-consuming and requires central support.

Restoration

Attack detection initiates reactive processes. The first process is restoration of conditions. Restoration relies on a preestablished understanding of the desired levels and conditions of system performance and functionality. Availability and integrity of information may be prioritized, as well as the detection of when anomalous conditions have degraded system processes below a threshold of acceptable function. An example of an organization of importance with respect to this in the United States is the National Security Telecommunications Advisory Committee (NSTAC).

Response Aimed at Attacker

Attack detection mechanisms initiate the response process. Timely identification of motives and actors is the cornerstone of an effective and properly focused response. Results of the process for identification and analysis are limited to national-level decision makers. Attacks cannot unambiguously point to motives and actors. Apparently, similar events or indications may have quite different causes, complicity, and severity. Different implications for national security call for providing decision makers with the best and most comprehensive information as a basis for the selection of response.

   Knowledge of true motives and actors is very important for the design of response processes. However, there is no clear and unambiguous set of automatic response processes to choose from. The reason is that the seam between civil and military roles is blurred when it comes to national security with respect to defensive information warfare. An attack against a commercial system in service for both civil and military purposes raises legal and policy issues highlighting the need for increased interagency coordination and joint civil–military response operations. Through all of this, the limits of the proper and legitimate role of government to provide for the common defense must be recognized and respected, ensuring no violation of personal freedom and rights of privacy.

   The effectiveness of the response process depends on the efficient integration of attack detection and analysis capabilities. Timely response is essential to influence adversary perceptions, establish user confidence, and maintain public support.

Information Operations, Information Warfare

Offensive Aspects

Military defense relies on information to plan operations, deploy forces, and execute missions. Advances in information technology have significantly changed these processes. Complicated information systems support powerful infrastructures that dramatically enhance defense capabilities. However, joint forces become more vulnerable as a consequence of the increased dependence on these new technologies.

Defensive information warfare incorporates a comprehensive strategy to protect and defend information and information systems. When combined with offensive information warfare, the net result is an opportunity to use information warfare to exploit situations and to win.

As with defensive information warfare, offensive information warfare capabilities can be used at each level of warfare and across the range of military operations. This makes it important to carefully consider how these conditions can be exploited. Employment of information warfare can result in a decisive force enabler for the joint warfighter. Offensive information warfare capability can affect every aspect of an adversary's decision cycle by impacting its information centers of gravity. The focal point of information warfare is the human decision process. Traditional perception methods such as psychological operations and information system attack can produce synergistic effects and affect information systems, information links, and information nodes.

Offensive information warfare can have deterrent effect on a potential adversary during peace and crisis. The same threat applies to one's own nation. To counteract this, both offensive and defensive capabilities are required. A strong information defense limits the adversary's possibilities to attack. The ability to respond quickly, effectively, and decisively will influence the adversary's disposition to use information warfare. Together with economic, political, diplomatic, legal, and military power, information defense constitutes an essential element of total national strength. Information warfare capability has a deterrent effect in the Information Age similar to nuclear weapons deterrence during the Cold War.

Offensive information operations can be used in military operations in peace to deter the development of crises, control crisis escalation, project power, and promote peace. Examples of targets are financial and media systems. Such circumstances may require special authorization and approval with support, coordination, cooperation, and participation of civil agencies. Offensive information warfare can also be employed to disrupt or stop drug cartels and other criminal activities.

There are many targets other than the military for the application of offensive information warfare. Offensive information warfare in wartime is not just a matter of military targets like the adversary's command and control system. There are many other information systems that are of importance to the adversary and that might be easier to attack and might be more vulnerable. Examples are systems for production of critical necessities and systems for command and control of electric power and telecommunications.

Measures for Implementation of Information Operations and Information Warfare

Information warfare focuses on achieving information superiority to achieve a decisive edge in war while information operations form a strategy for peace.

This makes it essential to capture the latent potential of information operations and information warfare. Information superiority requires both offensive and defensive information warfare advantages and builds on amalgamating many traditionally separate disciplines. Five principles must be fulfilled to achieve superiority:

- Establish the necessary relationships, within government and throughout the nation, to secure the information needs of all constituencies. Seal those arrangements in law and policy in order to preserve peace, security, and stability.
- Reduce the opportunities presented to the potential adversaries by educating, training, and increasing the awareness of people to vulnerabilities and protective measures, both military and civil.
- Improve measures to protect against and detect attacks by pursuing emerging technological capabilities in new ways.
- Improve information and information system attack capabilities.
- Increase capabilities of synergy created by integrated defense-in-depth solutions at all levels.

To incorporate information operations as a natural part of defense, they must be integrated in the following six major areas:

- Education, training, and exercises
- Policy
- Doctrine
- Assessments
- Organizational infrastructure
- Technology

It is interesting to note that these six areas are of importance in both military and civil business, and thus are of general interest to the whole society.

Education, training, and exercises offer the greatest return on investment to develop information operations capabilities, and they focus on concepts, policy, doctrine issues, and the role of information operations throughout the range of military operations at all levels of conflict.

Policy issues are developed in cooperation with all parties concerned and deal with roles, instructions, and tasks to achieve capabilities to command and execute information operations. A doctrine for information operations comprises principles for both offensive and defensive information operations and deals with organization, responsibilities, coordination between levels of command, planning considerations, integration and deconfliction of activities, and intelligence support.

Assessment focuses on the judgment of the role of information operations in overall missions. Special processes and their command are the prime analytical tools supporting the articulation of joint requirements.

The organizational infrastructure is based on merging separate disciplines to form a totality. This can be achieved by collaboration between tailored information operation cells. The building of information operation capability implies an amalgamation of traditionally separate disciplines. The intelligence business is one of the areas exposed to new challenges concerning information operations.

Information-based technology is a principal enabler of information operations. Close collaboration with academic and scientific organizations promotes ideas that may influence future warfighting strategy and doctrine.

Information operations and information warfare build by amalgamating traditionally separate areas of knowledge. Such areas are psychological operations, operations security, electronic warfare, network management, counter-psychological operations, counter-intelligence, computer security, deception, intelligence, physical security, counter-deception, public affairs, and information attack.

Conclusions of the American View

Military defense is dependent on information to plan operations, deploy forces, and execute missions. Advanced information technology has decisively altered these processes. Complicated information systems support powerful infrastructures, thereby dramatically enhancing warfighting capabilities. Meanwhile, vulnerabilities increase as a consequence of the dependencies of those rapidly emerging technologies. Conversely, many such deficiencies also concern the adversary. Thus, opportunities are presented to use offensive actions in one's own favor.

The Information Age presents opportunities to nations and military organizations to gain decisive advantages through access to timely, accurate, and relevant information. Information is a strategic resource driving a global competitive environment. This fact permeates every facet of warfighting in the new century.

In many places around the world, information operations are being studied today. Principles for defensive information operations are being developed and introduced to protect and defend information and information systems. Combined with offensive operations, opportunities are at hand to exploit situations and gain advantage.

Information operations and information warfare are a reality today and will be more important in the future. They impact societies, governments, and the whole range of military operations at all levels of conflict. Implementing information operations capabilities is a challenging task.

Man has much to learn to understand the essence of information operations and its relevance to survival and conflict, both now and in the future. When properly developed and applied, information operations can serve as a fundamental strategy for peace and a decisive edge in war. We have not yet arrived at that point.

Epilogue

Concerning the military part of the Information Age vocabulary, what is the result of all these glorious phrases such as "information superiority," "dominant battle-space awareness," and "revolution in military affairs"? What is the answer when considering conflicts such as those in Somalia and Yugoslavia? Maybe it is not such an easy thing to win the information war, to gain information superiority, and to successfully implement new solutions when applied to old types of conflict. Maybe it will take time before development catches up with expectations. Rethinking solutions to conflicts is necessary now that the information age rebalances the proportion between knowledge, wealth, and violence. In that process, international law and human rights must also be taken into new perspectives.

People have not changed appreciably; they have not learned from mistakes made by earlier generations, nor have they learned to talk with one another rather than to or past one another. Those who search for something in the darkness of the night only under the light from the streetlamp are not always lucky to find what they are looking for. Maybe other areas should be illuminated to look for solutions leading to more peaceful relations in the world.

Who is able to point the light in the right direction? The deepest powershift in human history must be met by reinforcing the *knowledge defense*.

Acknowledgments

I wish to thank Dr. William A. Radasky, Metatech Corporation (Goleta, California), for his valuable language review. My sincere thanks also go to my wife Margareta Wik von Bornstedt for her kind support and encouragement.

References

1. U.S. Joint Chiefs of Staff. *Information Warfare. A Strategy for Peace...The Decisive Edge in War.* (Brochure with remark from John M. Shalikashvili, Chairman of the Joint Chiefs of Staff).
2. Center for Strategic and International Studies task force report. *Cybercrime... Cyberterrorism...Cyberwarfare...Averting an Electronic Waterloo,* Washington, D.C., 1998, ISBN: 0-89206-295-9.
3. European Parliament, Scientific and Technological Options Assessment, STOA. Development of Surveillance Technology and Risk of Abuse of Economic Information, Luxembourg, April 1999, PE 168.184/Part ¾.
4. U.S. Joint Chiefs of Staff. *Joint Doctrine for Information Operations,* Joint Pub. 3-13, October 9, 1998.
5. Toffler, Alvin. *Power Shift,* Bantam Books, New York, ISBN 0-553-29215-3.
6. Office of the Under Secretary of Defense for Acquisition & Technology, Washington, D.C. 20301-3140. *Report of the Defense Science Board Task Force on Information Warfare — Defense,* November 1996.
7. The White House. *A National Security Strategy for a New Century,* Washington, D.C., October 1998.
8. Davies, Ian and Parker, Rick. *Information Operations,* NATO C3 Agency, The Hague, The Netherlands, oral presentation, 1999.
9. Devost, Matthew G. *Vulnerability Assessment/Red Team Experience,* Infrastructure Defense Inc., oral presentation, 1999.
10. Wik, Manuel W. *Mobilization for a New Era,* Militaert tidskrift Nummer 1 — 1999, Det Krigsvidenskabelige Selskab, ISSN 0026-3850.

11. Wik, Manuel W. *Global Information Infrastructure: Threats,* Global Communications Interactive 1997, Hanson Cooke limited, ISBN: 0946 393 893, http://www.global-comms.co.uk/interactive/technology/firewall/280.html.
12. Borg, L., Hamrefors, S., and Wik, M. *Information Warfare — A Wolf in Sheep's Clothing!,* Link from <http://www.infowar.com>, also in Swedish: Kungl Krigsvet-enskapsakademiens Handlingar och Tidskrift, 3. häftet, 1998.
13. Wik, Manuel W. Informationsoperationer — en strategi för fred; information-skrigföring — en avgörande spjutspets i krig, Kungl Krigsvetenskapsakademiens Handlingar och Tidskrift, 3 häftet, 1999.

Some Books on Information Warfare:

14. Schwartau, Winn. *Information Warfare: Cyberterrorism: Protecting Your Personal Security in the Electronic Age,* Thunder's Mouth Press, ISBN 1-56025-132-8, 1994.
15. Denning, Dorothy E. *Information Warfare and Security,* Addison-Wesley Longman Inc., ISBN 0-201-43303-6, 1999.
16. Alberts, David S., Garstka, John J., and Stein, Frederick P. *Network Centric Warfare,* CCRP, ISBN 1-57906-019-6, 1999.
17. Campen, Alan D. and Dearth, Douglas H. (Editors). *Cyberwar 2.0: Myths, Mysteries and Reality,* AFCEA International Press, ISBN 0-916159-27-2, June 1998.
18. Greenberg, Lawrence T., Goodman, Seymour E., and Soo Hoo, Kevin J. *Information Warfare and International Law,* CCRP, ISBN 1-57906-001-3, 1997.

About the Author

Manuel W. Wik is Chief Engineer and Strategic Specialist on Future Defence Science and Technology Programs. He is also a member of the Swedish Royal Academy of War Sciences and the Secretary of its Technical Military Sciences Department, and a member of the Swedish National Committee of the International Union of Radio Science (URSI), the Secretary of the International Electromechanical Committee (IEC) Standardization Sub-Committee SC 77C, the Chairman of URSI Committee on Nuclear Electromagnetic Pulse, a senior member of IEEE EMC Society, and has been recognized as an EMP Fellow by the U.S. Summa Foundation. He received the M.Sc. E.E. degree from The Royal Institute of Technology in Stockholm in 1962. He was an active researcher at the Swedish National Defence Research Establishment in the areas of nuclear weapons effects and contributed to books on nuclear electromagnetic pulse effects and electromagnetic compatibility. He later became the head of the Defence Materiel Administration Telecommunications Transmission Network procurement.

Mr. M.W. Wik can be reached at Defence Materiel Administration (FMV), SE-115 88 Stockholm, tel: +46 8 782 67 32, fax: +46 8 782 62 32, e-mail: mawik@fmv.se.

Disclaimer. The opinions expressed in the paper are the author's personal views and do not necessarily reflect those of FMV.

Afterword

"Information warfare" has become an increasingly popular notion over the years, driven — one suspects — by the multimedia age in which we now live, in which war is shown as a computerized, televised, remote-controlled, and somewhat push-button affair.

Information warfare is the translation of combat operations into the realm of computer networks, many of which are now of vital importance to the operation of a nation-state's transport, supply, financial, and utility infrastructure. Attacks can be aimed at disrupting the operation of crucial facets of these networks, denying legitimate access to the systems, or introducing untrustworthy information. And, of course, to address this, governments on both sides of the Atlantic have established organizations to protect the operation of these "critical national infrastructure" elements from so-called cyber-attacks.

This goes further, however; there is a feeling among some proponents that conflict between armed forces can be translated to combat between computer networks, that destruction or denial of government and national computer systems can take the place of the "19-year-old with a rifle" that the infantry traditionally required to control a hill. There is, indeed, a school of thought that argues that future conflicts can be almost bloodless; this, we might remark, is a dangerous perception on the part of the media, the commanders, and of course the politicians.

There is, however, a germ of truth in this notion that warfare can and must change with the arrival of computer networks and computer vulnerabilities: we are faced now with a new way of waging war.

All warfare centers on control and access to crucial information: the disposition of troops; the political will of the government; the morale of the populace; information about plans, feints, and the state of supplies; and information about victories, defeats, and the progress of particular actions. "Traditional" warfare might well be about the application of overwhelming force at a significant point, but it is *information* that is required to tell one where that point is, what constitutes overwhelming force, and whether or not

629

one has been successful. A major element of warfare is therefore concerned with controlling this information, with denying it to an enemy, and with introducing and manipulating false information. Signals, propaganda, and intelligence are all valid aspects of warfare.

Of course, there is nothing new about this aspect of combat. Many credit Genghis Khan with having begun information-based warfare with his orders to intercept any enemy messengers that might be observed. It is probable, however, that the importance of information as a warfare tool was understood long before that: the use of codes, the use of maneuvering instruction signals, the observation of those signals, and the attempts to break those codes... these can be traced back to well before Caesar's time.

Warfare, therefore, has always been about the control of information as much as it has been about the control of maneuvering room. So what, if anything, is new about information warfare that it should merit the attention that it is now being afforded throughout the centers of defense studies on both sides of the Atlantic?

There are several important new developments that make this form of combat particularly apposite. First of all, simply considering the scope of "normal combat" data that is available tells us that there has been a step change in the potential for information-oriented attacks. Satellite monitoring, asset tracking, integrated tactical displays, real-time observation data flows,... the sheer quantity of available data, the vast extent of modern battlefields, and the remoteness of command and control centers — and indeed, the remoteness of political direction — make influence over information a vital element of modern combat.

Second, the nature of combat — or at least of the combatants — has changed over the years: away from the set-piece battlefield and associated strategic units, and toward "low-intensity conflict" and the use of smaller, mobile, tactically oriented units. These terrorist or guerrilla actions are as much about manipulating public and political opinion — that is, as much concerned with *propaganda* — as they are with "true" military or combat-related objectives such as the taking and holding of towns and trade routes. And, of course, such sub-state combatants require flexible force multipliers to operate effectively: terror tactics can sit comfortably alongside non-standard operations such as those involved in the information warfare attacks.

In many ways, however, these two factors are simply the most obvious such elements that drive information warfare. In truth, there is an additional — and many believe, much more important — reason for information warfare to have risen on the military and political agenda. The very nature of war itself now drives one to consider operations to attack or to defend computer networks.

Warfare is first and foremost an *economic* activity, almost regardless of the political ambitions expressed before, or the cultural patina applied after the event, combat is waged to gain control of economic assets. The Germans in World War II fought initially for *lebensraum* — that is, for *land*, the most important of all economic aspects; the Japanese, similarly, fought their wars for control of the Chinese economic assets, and then more broadly for the economic assets of their immediate area, driven by the paucity of their local resources in the face of accelerating population growth. The Gulf War was initiated by the seizure of oil fields, and the Napoleonic Wars by the imperialistic ambitions of a France anxious to rebuild their economic status after the revolution. Perhaps only the most current "War on Terror" is not fuelled on the American allies' part by economic concerns, although the terrorists themselves have clearly attacked primarily economic targets.

Warfare is the process of seizing economic assets that were previously held by one's neighbors. Of course, those assets have traditionally been physical and geographical in nature: land, mining resources, mountain passes, trade goods, city wealth, etc. To seize them, however, has required one physically to go to them, and therefore to have an army, a navy, and air force and the command structures to control them. Equally, it has required defenders to have an army, a navy, and an air force to assist in the holding of those assets. Combat arose from the conflict between the attackers and the defenders... and what we might think of as "traditional" information warfare arose from the conflict over the data flows and intelligence systems used by both parties.

The change in economic activity involved in the so-called "digital economy," and the almost complete dependence on computer networks to control the "traditional economy" of the developed world, means now that those economic assets can be acquired or influenced remotely. For many companies, their economic assets are almost wholly digital in nature; for others, their assets can be wholly controlled and manipulated digitally; and for still others, the coordination of their physical assets can be achieved digitally. Therefore, the process of seizing control of those assets — that is, of warfare — can now be achieved digitally, with little or no exposure of troops to conflict.

This is the promise of the new forms of information warfare: warfare is economic combat. As the economy moves into the digital domain, so too will the conflicts to acquire and retain control of those economic activities.

There is much to consider in this view of future conflicts. Most obviously, the "front line" is no longer a place controlled and inhabited by front-line troops. Instead, it is the domain of network administrators and information security officers — civilians working for primarily commercial concerns. Second, in recent conflicts — the Gulf, Bosnia, Falklands, etc. — homeland economic activities have essentially remained unchanged by distant battle. This is unlikely to be the case in the future, with worms, viruses, hacker intrusions, and denial-of-service attacks striking rapidly and remotely against those very heartland systems, ignoring the distant soldier.

Perhaps most importantly, however, is the third consideration: thus far, other than nuisance attacks against Web sites, we seem not to have had the cyber-Armageddon that has been threatened. This is surprising. As mentioned above, hacking tactics would suit terrorists in particular: effect at a safe distance, with the damage potentially far in excess of the input effort. But yet, while there have been hackers who have daubed slogans on Web sites in support of Irish nationalists, there is no suggestion that they were encouraged or even condoned by the terrorists themselves. The situation appears the same for the other, headline political protests around the world; some hackers who support the political views of the terrorists have mounted cyber-protests that have involved Web site defacement, but the terrorists themselves have not mounted information warfare attacks. And even at a nation-state level, during the recent "spy plane" standoff between America and China, what computer attacks were mounted appear to have been defacements performed by known hackers, known to have been active against those sites before the forced landing of the spy plane.

In information warfare, therefore, we seem to have something quixotic. On the one hand, the perfect future vision of conflict: intangible weapons targeted at equally intangible objectives, operated by remote, invisible opponents safe from detection, discovery and retaliation. But on the other hand, this has not yet apparently proceeded beyond the simplistic defacement of opportunistic targets by coincidental fellow travelers of the true opponents.

The open questions that remain to be addressed include:

- How effective can information warfare techniques be in practice, in support of some economic combat objective?
- How likely is it that such attacks will indeed be mounted, and from which political opposition are they likely to arise?
- What are the skill levels and preparedness of different political opposition — terrorists, protest or nation-state — in the information warfare arena?
- Why — if indeed this is the case — has there been to date little or no use of these tactics in conflicts?

In an age in which vulnerabilities and conflict patterns are unquestionably changing, the most important tasks for military analysts are surely to understand why, how, and in what directions those changes are likely to take us. However, because the vulnerabilities now extend to the *civilian* infrastructure, this is no longer simply a question that should fascinate and excite the military analysts; this is now a question that should equally fascinate and excite IT directors.

Perhaps the best advice is that given to Boy Scouts: *Be prepared*.

Dr. Neil Barrett
November 2001

About the Authors

Andy Jones, MBE is an experienced military intelligence analyst and information technology security specialist. He has considerable experience in the analysis of intelligence material in strategic, tactical, and counter-insurgency operations, and a wide range of information systems management experience. In addition, he has considerable experience in the security of information technology systems, having been responsible for the implementation of information technology security within all areas of the British Army and in some joint service organizations. He has directed both intelligence and security operations, and briefed the results at the highest level. He was awarded the MBE for his work during his service in Northern Ireland and has gained an Open University Bachelor of Science degree in mathematics and technology. After completing 25 years of service with the British Army's Intelligence Corps, he moved into research in information warfare and information security. He has gained considerable experience as a project manager within the U.K. Defence Evaluation and Research Agency (DERA) for the security aspects of the digitization of the battlefield initiative and has gained considerable expertise on the criminal and terrorist aspects of information security. He is currently the business manager for the Secure e-Business department of QinetiQ, the privatized portion of DERA. He holds a lecturership with the U.K. Open University and is a visiting lecturer at the University of Glamorgan, Wales, on a Masters of Science for Network Security and Computer Crime.

Dr. Gerald L. Kovacich graduated from the University of Maryland with a bachelor's degree in history and politics, with emphasis in Asia; the University of Northern Colorado with a master's degree in social science, with emphasis in public administration; Golden Gate University with a master's degree in telecommunications management; the DoD Language Institute (Chinese Mandarin); and August Vollmer University with a doctorate degree in criminology. He is also a retired Certified Fraud Examiner, Certified Protection Professional, and a Certified Information Systems Security Professional.

Dr. Kovacich has over 37 years of industrial security, investigations, information systems security, and information warfare experience in both the U.S. government as a special agent, and in business as a technologist and manager for numerous technology-based, international corporations as an ISSO, security, audit and investigations manager, and consultant to United States and foreign government agencies and corporations. He has also developed and managed several internationally based INFOSEC programs for Fortune 500 corporations, and managed several information systems security organizations, to include providing service and support for their information warfare products and services.

Dr. Kovacich has taught both graduate and undergraduate courses in criminal justice, technology crimes investigations, and security for Los Angeles City College, DeAnza College, Golden Gate University, and August Vollmer University. He has also lectured internationally, presented workshops on these topics for national and international conferences, as well as written numerous published articles on high-tech crime investigations, information systems security, and information warfare, both nationally and internationally.

Dr. Kovacich currently spends his time on Whidbey Island, Washington, where he continues to conduct research, write, and often lectures internationally on such topics as global, nation-state, and corporate aspects of information systems security; fraud; general security; high-tech crime investigations; information assurance; proprietary information protection; economic and industrial espionage; and information warfare. He is also the founder of ShockwaveWriters.com, an informal association of writers, researchers, and lecturers who concentrate on the above noted topics (see http://www.shockwavewriters.com).

Dr. Kovacich has written more than 92 published articles on high-technology crime, information warfare, techno-terrorism, computer crime, and information systems security for various international magazines. He is the author and co-author of five books:

- *Information Systems Security Officer's Guide: Establishing and Managing an Information Protection Program,* May 1998, ISBN 0-7506-9896-9, Butterworth-Heinemann.
- *I-Way Robbery: Crime on the Internet,* May 1999, ISBN 0-7506-7029-0, Butterworth-Heinemann.
- *High-Technology Crime Investigator's Handbook: Working in the Global Information Environment,* September 1999, ISBN 0-7506-7086-X, Butterworth-Heinemann.
- *Netspionage: The Global Threat to Information,* September 2000, ISBN: 0-7506-7257-9, Butterworth-Heinemann.
- *Information Assurance: Surviving in the Information Environment,* September 2001, ISBN: 185233326X, Springer-Verlag Ltd. (London).

Perry Luzwick is Director, Information Assurance Architectures, at Northrop Grumman Information Technology (NGIT), a Northrop-Grumman company. He is a senior consultant throughout NGIT for information warfare (IW), information assurance (IA), critical infrastructure protection (CIP), and knowledge management (KM).

Perry served as a Lieutenant Colonel in the U.S. Air Force and was Military Assistant to the Principal Deputy Assistant Secretary of Defense for Command, Control, Communications, and Intelligence (ASD(C3I)); Deputy Director for Defensive IO, IO Strategy and Integration Directorate, ASD(C3I); Chief, Information Assurance Architecture, Directorate for Engineering and Interoperability, Defense Information Systems Agency (DISA); Deputy Chief, Current Operations and Chief, Operations and Information Warfare Integration, Operations Directorate, DISA; Information Assurance Action Officer, IA Division (J6K), the Joint Staff; and Chief, JCS, CINC, and Defense Agency Communications-Computer Security Support, National Security Agency (NSA).

Perry is a doctoral student in George Washington University's Knowledge Management program. He earned an M.A., and was a Distinguished Graduate, in Computer Resources Management from Webster University; an MBA from the University of North Dakota; and a B.S., psychology, from Loyola University of Chicago. He has taught as an adjunct faculty member at the University of Maryland, the City Colleges of Chicago, and NSA's National Crypotologic School.

Perry has had 20 articles published nationally and internationally; as well, he has been a frequent speaker and lecturer on Coherent Knowledge-based Operations, IW, and KM. He is a 1998 member of the *International Who's Who of Information Technology*.

Index

A

Abacus, 573, 574
Academy of War Sciences, 267
Access control devices, 435
ACCT, see Agency for the French-Speaking Community
ACN, see Andean Community of Nations
ACP Group, see African, Caribbean, and Pacific Group of States
Active unbalancing, examples of, 495
Activist warfare, see Business, government, and activist warfare
Activists, information warfare tactics by, 400
Acts of God, 601
ADSL, see Advanced digital subscriber line
Advanced digital subscriber line (ADSL), 65
Advanced Research Project Agency (ARPA), 42, 50
Advanced Technology Office, projects of, 336
AEA, see American Electronics Association
AfDB, see African Development Bank
AFIWC, see Air Force Information Warfare Center
Africa
    conflicts, 531
    existence of Internet in, 359
    map of, 355
    nation-states of, see Nation-states, African and other
African, Caribbean, and Pacific Group of States (ACP Group), 354
African Development Bank (AfDB), 355
AFRL, see USAF Rome Laboratory
Agency for the French-Speaking Community (ACCT), 356

Agency for the Prohibition of Nuclear Weapons in Latin America and the Caribbean (OPANAL), 296
Age of Technology, 101
AI, see Artificial intelligence
AIDS problem, global, 562
Air Force Information Warfare Center (AFIWC), 329, 330
Airport security authorities, 74
Air traffic control, 23
ajeeb, 64
Alldas.de, extract of information from, 412
ALOHAnet, 63
Al-Qaeda, 388
Alta Vista, 43, 248, 553
AMAN, see Israeli Military Intelligence Services
Amazon.com, 71, 73, 520
American Civil Liberties Union, 186
American Electronics Association (AEA), 38
American Patent Bureau, 583
American Society for Industrial Security (ASIS), 506
America Online, 44, 45
Americas
    nation-states of the, see Nation-states, of the Americas
    potential conflicts of the, 298, 299–300
America's Most Wanted, 110
Anarchists, 541, 582
Andean Community of Nations (ACN), 296
Anonymity, 409
Antarctica
    territorial claim by New Zealand in, 234
    territorial claim by Norway in, 266
Anti-NATO air defense early warning system, 288

D